Ecology of Photosynthesis in Sun and Shade

Ecology of Photosynthesis in Sun and Shade

Edited by J.R Evans S. von Caemmerer W.W. Adams III

National Library of Australia Cataloguing-in-Publication Data

Ecology of photosynthesis in sun and shade.

Includes bibliographies.
ISBN 0 643 04823 5.

1. Photosynthesis. 2. Plants, Effect of solar radiation on. 3. Plants, Effect of shade on. 4. Plants—Photoinhibition. 5. Acclimatization (Plants). 6. Botany—Ecology. I. Evans, J. R. (John R.). II. von Caemmerer, S. (Susanna). III. Adams, W. W. (William W.). IV. Commonwealth Scientific and Industrial Research Organisation (Australia).

581.1′3342

Reprinted from *Australian Journal of Plant Physiology,* Volume 15, Numbers 1 and 2 (1988)

Cover design by John Best CSIRO Editorial Services Unit Melbourne

Contents

Contributors

William W. Adams III	Lehrstuhl für Botanik II, Universität Würzburg, Mittlerer Dallenbergweg 64, 8700 Würzburg, Federal Republic of Germany.
Jan M. Anderson	Division of Plant Industry, CSIRO, G.P.O. Box 1600, Canberra, A.C.T. 2601, Australia.
T. John Andrews	Research School of Biological Sciences, Australian National University, G.P.O. Box 475, Canberra, A.C.T. 2601, Australia.
Marilyn C. Ball	Research School of Pacific Studies, Australian National University, G.P.O. Box 4, Canberra, A.C.T. 2601, Australia.
Olle Björkman	Department of Plant Biology, Carnegie Institution of Washington, 290 Panama Street, Stanford, CA 94305, U.S.A.
W. J. Cai	Rice Research Institute, Guangdong Academy of Agricultural Sciences, Guangzhou, Guangdong, People's Republic of China.
W. S. Chow	Division of Plant Industry, CSIRO, G.P.O. Box 1600, Canberra, A.C.T. 2601, Australia.
Robyn E. Cleland	Research Institute for Photosynthesis, University of Sheffield, Sheffield, S10 2TN, England.
I. R. Cowan	Research School of Biological Sciences, Australian National University, G.P.O. Box 475, Canberra, A.C.T. 2601, Australia.
Christa Critchley	Botany Department, University of Queensland, St Lucia, Qld 4067, Australia.
Barbara Demmig	Lehrstuhl for Botanik II, Universität Würzburg, Mittlerer Dallenbergweg 64, 8700 Würzburg, Federal Republic of Germany.
D. Doley	Botany Department, University of Queensland, St Lucia, Qld 4067, Australia.
John R. Evans	Research School of Biological Sciences, Australian National University, G.P.O. Box 475, Canberra, A.C.T. 2601, Australia.
Graham D. Farquhar	Research School of Biological Sciences, Australian National University, G.P.O. Box 475, Canberra, A.C.T. 2601, Australia.
H. Y. Feng	Rice Research Institute, Guangdong Academy of Agricultural Sciences, Guangzhou, Guangdong, People's Republic of China.
Christopher B. Field	Department of Plant Biology, Carnegie Institution of Washington, 290 Panama Street, Stanford, CA 94305, U.S.A.
Thomas J. Givnish	Department of Botany, University of Wisconsin, Madison, WI 53706, U.S.A.
D. J. Goodchild	Division of Plant Industry, CSIRO, G.P.O. Box 1600, Canberra, A.C.T. 2601, Australia.
Dennis H. Greer	Plant Physiology Division, DSIR, Private Bag, Palmerston North, New Zealand.
Suzan E. Hetherington	Division of Horticulture, CSIRO, P.O. Box 52, North Ryde, Sydney, N.S.W. 2113, Australia.
Jie Hie	Division of Horticulture, CSIRO, P.O. Box 52, North Ryde, Sydney, N.S.W. 2113, Australia.
Q. M. Huang	Rice Research Institute, Guangdong Academy of Agricultural Sciences, Guangzhou, Guangdong, People's Republic of China.
George Koch	Department of Biological Sciences, Stanford University, Stanford, CA 94305, U.S.A.

P. E. Kriedemann	Division of Forest Research, CSIRO, P.O. Box 4008, Canberra, A.C.T. 2601, Australia.
Manfred Küppers	Institut für Botanik, Technische Hochschule, Schnittspahnstrasse 10, D-1600 Darmstadt, Federal Republic of Germany.
A. Laisk	Institute for Astrophysics and Atmospheric Physics, Estonian Academy of Sciences, Tartu, Estonia, U.S.S.R.
X. Z. Lin	Rice Research Institute, Guangdong Academy of Agricultural Sciences, Guangzhou, Guangdong, People's Republic of China.
M. M. Ludlow	Division of Tropical Crops and Pastures, CSIRO, 306 Carmody Road, St Lucia, Qld 4067, Australia.
Luping Qian	Shanghai Institute of Plant Physiology, Academia Sinica, Shanghai, People's Republic of China.
Harold A. Mooney	Department of Biological Sciences, Stanford University, Stanford, CA 94305, U.S.A.
Keith A. Mott	Department of Biology, College of Science, Utah State University, Logan, Utah 84322-5305, U.S.A.
Robyn Nott	Division of Horticulture, CSIRO, P.O. Box 52, North Ryde, Sydney, N.S.W. 2113, Australia.
V. Oja	Institute for Astrophysics and Atmospheric Physics, Estonian Academy of Sciences, Tartu, Estonia, U.S.S.R.
C. B. Osmond	Research School of Biological Sciences, Australian National University, G.P.O Box 475, Canberra, A.C.T. 2601, Australia.
Robert W. Pearcy	Department of Botany, University of California, Davis, CA 95616, U.S.A.
S. B. Powles	Department of Agronomy, Waite Agricultural Research Institute, University of Adelaide, Glen Osmond, S.A. 5064, Australia.
Robert M. Smillie	Division of Horticulture, CSIRO, P.O. Box 52, North Ryde, Sydney, N.S.W. 2113, Australia.
G. C. Stocker	Tropical Forest Research Centre, CSIRO, P.O. Box 780, Atherton, Qld 4887, Australia.
W. A. Thompson	Division of Forest Research, CSIRO, P.O. Box 4008, Canberra, A.C.T. 2600, Australia.
Emmanuel F. Torquebiau	International Council for Research in Agroforestry, P.O. Box 30677, Nairobi, Kenya.
Z. P. Tu	Rice Research Institute, Guangdong Academy of Agricultural Sciences, Guangzhou, Guangdong, People's Republic of China.
G. L. Unwin	Tropical Forest Research Centre, CSIRO, P.O. Box 780, Atherton, Qld 4887, Australia.
Klaus Winter	Lehrstuhl für Botanik II, Universität Würzburg, Mittlerer Dallenbergweg 64, 8700 Würzburg, Federal Republic of Germany.
Ian E. Woodrow	Division of Plant Industry, CSIRO, G.P.O. Box 1600, Canberra, A.C.T. 2601, Australia.
D. J. Yates	Botany Department, University of Queensland, St Lucia, Qld 4067, Australia.
L. Y. Ye	Rice Research Institute, Guangdong Academy of Agricultural Sciences, Guangzhou, Guangdong, People's Republic of China.

Preface

In 1987 the Research School of Biological Sciences at the ANU decided to honour the role of Professor Sir Rutherford Robertson in biological research in Australia and, more generally, in a wide range of matters related to science and technology, by establishing a series of Symposia in particular fields of biology to be named The Robertson Symposia. In the process the School also wished to contribute to one of the national aims of the ANU, of providing a focus for interaction between Australian and foreign scientists from a range of different institutions and locations.

The first symposium took place from 18 to 20 February 1987. Appropriately, given Professor Robertson's distinguished career in plant physiology and biochemistry, and his deep interest in the biological context within which physiological responses take place, it was focused on a topic of central importance in plant science—the ecology of photosynthesis in sun and shade. It is timely that this volume coincides with a wave of enthusiasm for function analysis of the photosynthetic apparatus in response to irradiance, and the attendant problems of acclimation and photoinhibition.

Professor Robertson gave the opening address and participated actively throughout. Some 40 scientists, including visitors from North America, Europe and Asia, participated in a program that enabled a thorough evaluation of the subject in an atmosphere that was informal throughout. The papers which were presented appear in this special volume of *Australian Journal of Plant Physiology*.

I am indebted to Professor Ian Cowan who was the prime organiser of the Symposium, to Dr John Evans and Dr Susanne von Caemmerer who prepared the papers for publication, and their colleagues, together with Dr Laurie Martinelli, Managing Editor of the journal, who ensured that the Symposium would be a success and that this record of its Proceedings would be produced.

R. O. Slatyer

(Left to right): Sir Rutherford and Lady Robertson: Professor Ralph Slatyer.

Participants in the First Robertson Symposium, on 'Ecology of Photosynthesis in Sun and Shade', Canberra, 18–20 February 1987.

Standing (left to right): J. Hie, L. K. Huang, Z. P. Tu, C. B. Osmond, R. Hodgson, C. Critchley, M. M. Ludlow, K. Winter, D. Doley, I. E. Woodrow, M. C. Ball, P. E. Kriedemann, C. Field, W. S. Chow, R. M. Smillie, D. H. Greer, E. F. Torquebiau, G. D. Farquhar.
Seated (left to right): M. Küppers, J. R. Evans, W. W. Adams III, B. Demmig, R. E. Cleland, I. R. Cowan, R. Robertson, R. O. Slatyer, S. E. Hetherington, R. W. Pearcy, O. Björkman, J. M. Anderson, J. Landsberg, T. J. Givnish.

Ecology of Photosynthesis in the Sun and Shade: Summary and Prognostications

C. B. Osmond[A] *and W. S. Chow*[B]

[A] Plant Environmental Biology Group, Research School of Biological Sciences, Australian National University, G.P.O. Box 475, Canberra, A.C.T. 2601, Australia.
[B] Division of Plant Industry, CSIRO, G.P.O. Box 1600, Canberra, A.C.T. 2601, Australia.

This has been a most timely symposium, coinciding with a wave of enthusiasm for functional analysis of the photosynthetic apparatus in response to irradiance, and the attendant problems of acclimation and photoinhibition. Eugene Garfield of 'Current Contents' recently asked why a review on comparative photosynthesis of sun and shade plants (Boardman 1977) should be cited 50% more frequently in 1986 than in 1985 (Garfield 1987). This, and the astonishing number of presentations at the 1986 International Congress on Photosynthesis, dealing with photoinhibition and acclimation (Biggins 1987), testify to the extent of current interest in this field. Another of the questions in Garfield's correspondence was 'Why is it that Australian research in plant sciences should so predominate?'. It is fair to say that studies of the ecology of photosynthesis in the sun and shade, which were so dramatically stimulated by Olle Björkman's collaborations in Australia in 1971, represent a significant component of both the cause and the effect behind these questions.

Perhaps the germ of work described at this meeting has been in the air longer than we realised. Some years before he took up the first Australian Chair in Botany to be filled by a physiologist (University of Melbourne, 1904), Ewart (1896) described some of the earliest investigations of photoinhibition. In his studies in Pfeffer's laboratory in Leipzig, with algae and with leaves of *Pyrus*, *Aesculus* and *Ficus*, Ewart assessed O_2 evolution in cells and tissues under a microscope, using Engelmann's bacterial method. An extract from Ewart's paper shows that our approach has changed little, in spite of better tools:

> 'Leaves exposed, attached to the parent plant, to full sunlight till 4PM show no assimilation, though living, green, and with normal chlorophyll grains. After 2h. in diffuse light still no assimilation, but next day the same preparations show, if living, a weak but distinct power of assimilation, and fresh preparations made from the same leaves now show a fairly active power of assimilation'.

Papers at this symposium show we have come a long way, and plotting the course of these developments provides one of the most instructive examples of the interplay between studies of structure and function, of the states and processes, which underlie ecophysiological research (Osmond *et al.* 1980). These papers provide contemporary insights into a range of states and processes that extends over a scale of 15 or more orders of magnitude (Fig. 1). Our creative instincts have been challenged at every step in this continuum. Within each black box, defined by about three orders of magnitude, we have been encouraged to explore functional relationships that are relevant to the states and processes upscale and downscale. At the small scales and at the large scales we seem to be better equipped to define the states than to measure the processes. This

0310-7841/88/010001$03.00

simply reflects the fact that at small scales separation of signal from noise is often difficult (see Cleland 1988) and, at the large scale, long-term observation of stochastic processes remains a problem (see Torquebiau 1988).

It is in the middle range of the continuum, at the level of the individual leaf, that we seem to have the best integration of states and processes as they relate to the ecology of photosynthesis in the sun and shade. Major tasks ahead are upward integration of leaf processes to organism performance (Pearcy 1988; Field 1988; and others, this volume) and the reductionist probing of the dynamics of chloroplast states, such as nitrogen budgets and membrane organisation (Evans 1988; Anderson *et al.* 1988; and others, this volume), as they limit the capacity for acclimation and susceptibility to photoinhibition.

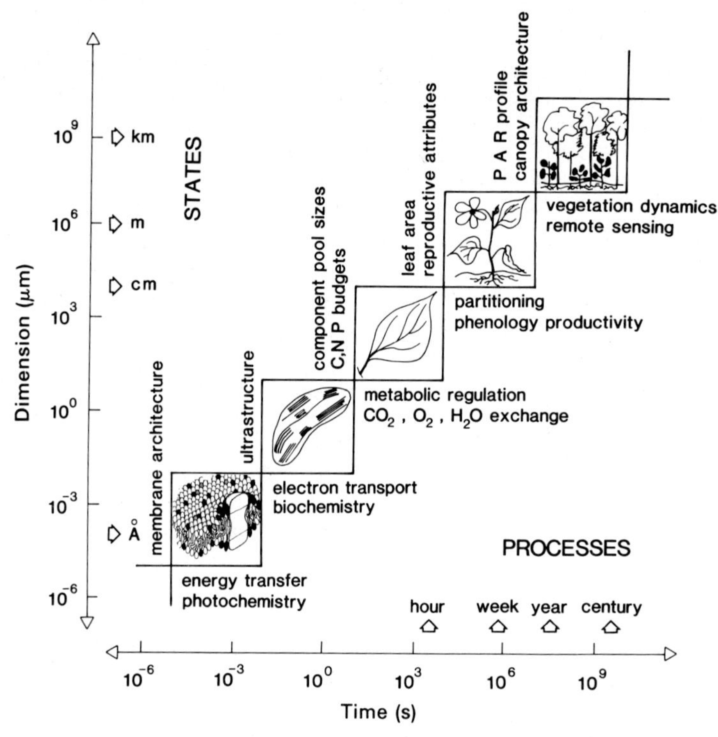

Fig. 1. An arbitrary scaling of relevant states (defined by approximate dimensions in μm) and processes (defined by approximate relaxation times, the $t_{\frac{1}{2}}$ to return to a steady state after perturbation) discussed in the course of this volume. The scheme is derived from that of Osmond *et al.* (1980) following the suggestions of Dr J. A. Berry, and seems to be in conformity with recent treatments in dynamic ecology (Delcourt *et al.* 1983) and landscape ecology (Urban *et al.* 1987).

Leaf Processes to Organism Performance

Individual leaves are the productive units of the organism, as the individual is the productive unit of the community. We have long known that the carbon assimilation processes of leaves of herbaceous plants in the sun have higher light-saturated rates in air than those formed in the shade. Yet we know very little of the capacity for a fully expanded leaf formed in the sun or shade to acclimate to a change in light environment. In shade leaves transferred to sun, this seems to depend on availability of nitrogen (Ferrar and Osmond 1986), almost certainly associated with the nitrogen demands of

RuP_2* carboxylase–oxygenase. Yet the response of different leaves to such a transfer from shade to sun is not well known and the balance between photoinhibitory damage, photosynthetic reconstruction, and accelerated senescence may well differ throughout the leaf population. Leaf responses after sun to shade transitions may be dominated by the need to contain respiratory costs and redirection of nitrogen from underutilised RuP_2 carboxylase–oxygenase. Before expanding research in this direction, it would be well to be sure that such transitions are ecologically meaningful. Studies of sunfleck responses (Pearcy 1988) and gap formation (Torquebiau 1988) represent two extremes of such transitions in the rainforest environment. How the sum of these processes determines the performance and survival of the individual organism cannot be inferred. We need observations and integrations of the sort described by Pearcy (1983).

In non-herbaceous plants, it is not at all clear if the same rules apply. Some observations of non-acclimation to sun and shade (Langenheim *et al.* 1984; Thompson *et al.* 1988) may reflect more on our ability to grow rainforest plants in the laboratory than on suboptimal response. Perhaps, too, these observations reflect transition states in processes with time constants an order of magnitude or more larger than in herbaceous plants. Slowly changing states in long-lived leaves of rainforest plants may tell us a good deal about what is, and what is not, important during acclimation (Chow *et al.* 1988). Although the complexity of these large-scale processes is daunting, papers in this volume give encouragement that integration of leaf processes to organism performance and community dynamics is being achieved (Field 1988). Nevertheless, we need to consider that less than perfect tuning of the photosynthetic apparatus in relation to sun and shade habitats may be quite common. Every distribution of organisms, even if it is normal, has one peak and two tails. It may be easier to identify physiological processes that contribute to performance at the peak, but those perhaps different processes that determine presence at the tail are equally significant in an ecological sense (Osmond 1987).

Although processes at the large scale will be driven by the frequency and amplitude of abiotic changes in light environment, and are limited by what is physically feasible at the molecular level, the patterns of response may be qualified by a host of biological factors. For example, in our studies of acclimation to bright light, the efficacy of nitrogen nutrition in *Solanum dulcamara* may be related to viral infection. In many shaded habitats throughout Europe, this species is heavily infested with several viruses, sometimes without overt symptoms. Clones that appeared to show obligate shade behaviour in the laboratory (shade ecotypes, Gauhl 1976) were found to be virus-infected, and to be capable of full shade–sun acclimation after being rendered virus-free, if provided with adequate nitrogen (Osmond 1983; Ferrar and Osmond 1986). The notion of internal regulation of leaf nitrogen budgets, and impaired ability to reconstruct the photosynthetic apparatus, due to diversion of resources by viral infection, is being tested.

Leaf Processes to Ultrastructural States and Processes

Terashima has elegantly shown that the light gradient through a leaf can lead to a gradient in chloroplast properties that is similar to the range of properties found in leaves of sun- and shade-grown plants (Terashima and Inoue 1985). We have expanded this observation to make analogies between the leaf and rainforest-canopy light gradients, and with chlorophyll composition and thylakoid membrane organisation through the leaf in sun and shade plants (Anderson and Osmond 1987). These wide-ranging

*Abbreviations used: Chl, chlorophyll; HEPES, *N*-2-hydroxyethylpiperazine-*N'*-2-ethanesulfonic acid; RuP_2, ribulose 1,5-bisphosphate.

analogies in state descriptions illustrate the ease with which it is possible to move through the continuum of Fig. 1. However, many of our inferences are poorly supported by evidence as to the dynamic range of processes *in vivo*. Thus, we are confronted with a wide array of *in vitro* evidence as to the mechanistic basis of thylakoid stacking (Barber 1982), and its potential functional significance (Anderson 1982; Evans 1986), but very few observations of the relevant processes *in vivo* (e.g. change in granal stacking with time, Terashima *et al.* 1986) during shade–sun acclimation.

At even smaller scales, at the molecular level, this volume has provided novel insights into processes of excitation utilisation in primary photochemistry, especially in relation to excess light energy. There is no doubt that excessive light energy is damaging to the photosynthetic apparatus of higher plants. Bright light can quickly lead to slowly reversible inhibition of photosynthesis in leaves that have developed in the shade, and of leaves that have developed in the sun if their ability to dispose of excitation in normal photosynthetic processes is impaired. In the laboratory, this latter response can be demonstrated by withholding CO_2 and minimising photorespiration (Cornic 1976; Powles and Osmond 1978). In the field, illumination of sun-grown plants in bright light at low temperature (Greer 1988) or under certain conditions of water or salinity stress (Björkman *et al.* 1988) has a similar effect. It is now evident also that substantial changes in photosynthetic efficiency can occur, apparently without damage to primary photochemistry. One of the high points of the conference was evidence for the role of the xanthophyll cycle in the harmless dissipation of excitation energy (Demmig *et al.* 1987). Estimates of the partitioning of excitation energy between protective (deflection of energy from reaction centres), productive (transduction of energy by reaction centres and electron transport) and photoinhibitory processes (damage to the reaction centres) may soon be possible.

These advances have led us into semantic difficulties with respect to the term 'photoinhibition'. French (1967) reminded us that the way we pose, or answer, questions in science is constrained 'within the limits of contemporary knowledge and vocabulary'. It is important that vocabulary not impede the quest for knowledge, and for this reason it is necessary to reconsider definitions of photoinhibition based on reduced photosynthetic efficiency alone. Thus, it seems unreasonable to consider as photoinhibition those processes that promote the deflection of excess excitation from the reaction centre, and thereby lower quantum yield, but that at the same time protect the reaction centre from damage (Björkman 1987). Notions that one process of photoinhibition protects against other processes leading to photoinhibitory damage are not helpful. That the former processes may be the first line of defence against the latter is entirely tenable. In the same vein, it seems misleading to regard photoinhibitory damage itself as a 'last ditch' mechanism for the disposal of excess excitation (Horton 1987). It is preferable to reserve the term photoinhibition for light-dependent damage to the photosynthetic apparatus, even though we do not yet have a unique set of criteria to describe this damage, or its repair.

Criteria for and Primary Sites of Photoinhibition

As is evident in the preceding contributions, photoinhibition of the photosynthetic apparatus by exposure of leaves of shade-grown plants to excessive light may involve many components. The extent of photoinhibition depends on prior light history, contemporary irradiance and other environmental stresses, time of exposure and capacity for recovery from damage. In this sense, photoinhibition *in vivo* is closely related to the processes that underlie acclimation. In terms of leaf gas exchange, photoinhibition is probably best characterised by a decrease both in the initial slope and the saturated rate of photosynthesis in air, as a function of irradiance (Fig. 2). If treatments lead to a decrease in photosynthetic efficiency (reduced quantum yield) due to deflection of excess excitation without damage, then the light-saturated rate should be unaltered. Although the former response is commonly observed in leaves, evidence for the latter, and its association with deflection of excess excitation, needs to be sought. Time-course studies with protoplasts (Horton *et al.* 1987) suggest that the latter response may be observed in early stages of photoinhibition.

Perhaps because it was more readily measured than most other photosystem II components and because of its rapid turnover, the herbicide-binding protein became a favourite target for research into photoinhibition and now occupies a central role in many accounts of photoinhibitory processes (Kyle and Ohad 1986). It seems increasingly likely that certain redox components in the reaction centre of photosystem II, and the herbicide-binding site are located on the same 32 kDa polypeptide (Arntz and Trebst

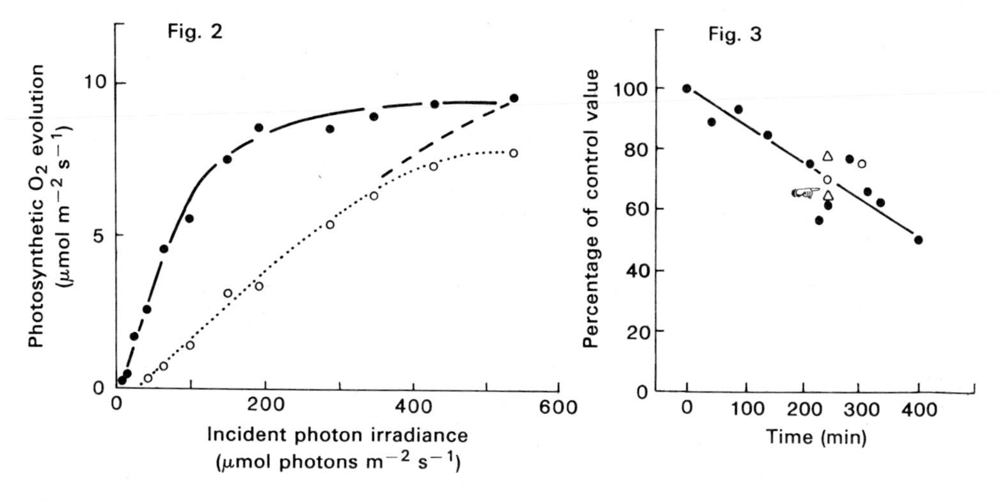

Fig. 2. Photoinhibition of photosynthesis in shade-grown (100–150 μmol photons m^{-2} s^{-1}) spinach leaves, shown by change in light response curves before (●) and after (○) treatment for 5 h at 1800 μmol photons m^{-2} s^{-1} (data of Walker and Osmond 1986). Damage of the photosynthetic apparatus is indicated by decrease of initial slope (apparent quantum yield) and light-saturated rate. The dashed line extrapolates the response expected if the decrease in the initial slope was due to deflection of excitation without damage.

Fig. 3. Photoinhibition of spinach leaves at approximately 25°C. Spinach grown under shade (100–150 μmol photons m^{-2} s^{-1}) in a glasshouse was illuminated at approximately 1600 μmol photons m^{-2} s^{-1} for the durations indicated. Functional photosystem II centres were determined by the O_2 yield per single-turnover flash in leaf discs in the presence of continuous background far-red light (●) and in isolated thylakoids (○), while atrazine-binding sites were determined in isolated thylakoids (△), with the following respective mean values ± s.e.m.: 2·46 ± 0·17(6), 2·26±0·05(5), and 2·25 mmol (mol Chl)$^{-1}$. The finger of authority points to the mean decline in quantum yield of O_2 evolution in seven separate experiments.

1986; Barber *et al.* 1987; Nanba and Satoh 1987). Thus, during photoinhibition, the loss from the thylakoid membrane of the herbicide-binding polypeptide via rapid turnover may be a natural consequence of damage to the photosystem II reaction centre, since a damaged reaction centre would have little requirement for functional quinone-binding proteins. Recovery from photoinhibition *in vivo* presumably reflects the resynthesis and insertion of this polypeptide. Such notions go a long way towards resolving conflicting hypotheses as to photoinhibitory mechanisms (Kyle *et al.* 1984; Cleland and Critchley 1985). If chloroplasts are isolated from leaves of shade-grown spinach after photoinhibition, we find a surprisingly good coincidence between *in vitro* and *in vivo* estimates of photosystem II function using the O_2 flash yield technique. *In vivo* quantum yield changes to about the same extent, and so does the concentration of binding sites for atrazine measured on isolated thylakoids (Fig. 3).

In his opening remarks to the symposium, Sir Rutherford Robertson recounted a tale of heat production in wheat storage and of Q_{10} (the temperature coefficient of complex processes). A brief survey of the literature on photoinhibition suggests that we may have lost sight of this simple technique for analysis of primary photophysical and secondary biochemical processes. Kok *et al.* (1965) demonstrated quite clearly that photoinhibition of chloroplast thylakoids had a Q_{10} of near 1·0. We would expect this of primary photochemical processes that might underlie light-dependent loss of reaction-centre function. On the other hand, depletion of atrazine-binding sites in the thylakoid presumably involves much more complex biochemistry of protein synthesis, insertion and proteolytic removal, and hence would display a higher Q_{10}. Yet many of the exper-

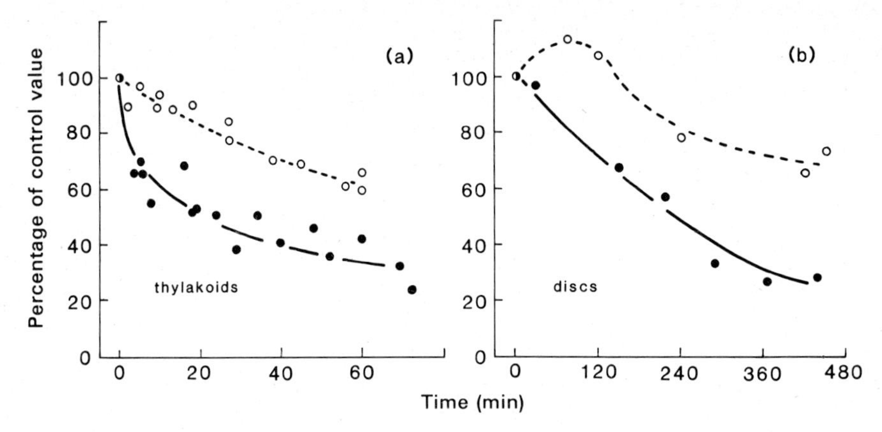

Fig. 4. (*a*) Photoinhibition of isolated thylakoids at 0°C. Thylakoids isolated from spinach grown in full glasshouse light were suspended in a medium containing 330 mM sorbitol, 5 mM $MgCl_2$, 10 mM NaCl and 10 mM HEPES (pH 7·8) at a chlorophyll concentration of 100 μM, and illuminated with white light at approximately 5000 μmol photons $m^{-2} s^{-1}$. At the times indicated, aliquots were taken for the determination of functional photosystem II centres (●) and atrazine-binding sites (○), in assays maintained at 0°C. Control values for functional photosystem II centres and atrazine-binding sites were 3·09 and 2·83 mmol (mol Chl)$^{-1}$, respectively. (*b*) Photoinhibition of spinach leaf discs at 10°C. Spinach was grown in a cabinet with a photoperiod of 12 h (100–120 μmol photons $m^{-2} s^{-1}$, 25°C) and a dark period of 12 h (17°C). Leaf discs were floated on water kept at 10°C and illuminated with white light at approximately 1500 μmol photons $m^{-2} s^{-1}$ for the durations shown. Control values for functional photosystem II centres *in vivo* (●) and atrazine-binding sites in isolated thylakoids (○) were 2·22 ± 0·14(4) and 1·83 ± 0·04(4) mmol (mol Chl)$^{-1}$, respectively.

iments that underlie the above controversy as to the roles of loss of the herbicide-binding protein and loss of reaction-centre function in photoinhibition were done under different temperature conditions. The evidence for the importance of herbicide binding is largely based on *in vivo* studies of intact cells at 25–30°C. The evidence for reaction-centre involvment is largely based on *in vitro* studies of isolated components at 0–5°C.

Although we have not yet determined the Q_{10}s precisely, experiments with thylakoids isolated from glasshouse-grown spinach indeed suggest that temperature can be used to separate primary and secondary events in photoinhibition. At 25°C it is difficult to identify any marked difference in the kinetics of loss of photosystem II function and the concentration of atrazine-binding sites (not shown). At 0°C it is quite clear that photosystem II function is lost ahead of reduction in the concentration of atrazine-binding sites (Fig. 4*a*). Interpretation of data from *in vivo* experiments is complicated by the temperature coefficients of recovery processes, considerations that almost certainly do not apply in the thylakoid experiments. However, it is interesting that much the same relationships obtain when experiments are carried out in leaf discs at 10°C (Fig. 4*b*). These experiments suggest that loss of photosystem II reaction-centre function during photoinhibition of shade-grown spinach is not directly related to loss of herbicide-binding sites from the thylakoid membranes.

Whatever explanation of these relationships is subsequently uncovered, it does seem that new tools may be at hand to probe the mechanisms of 'light rigor' (Ewart 1896), the mechanisms of photoinhibition *in vivo*. Estimation of the concentration of functional photosystem II centres *in vivo* by means of saturating, single-turnover flashes with steady background photosystem I light (Chow *et al.* 1988) provides one approach to analysis of photoinhibition and the efficacy of acclimation. In conjunction with modern methods of fluorescence analysis (Demmig and Winter 1988), we may soon have the tools necessary to refine our vocabulary and to follow changes in photosynthetic function *in vivo*. These tools may also help us understand the mechanisms used to cope with changing light environment, and point to ways we may extend, through genetic means, the dynamic range of response and the tolerance limits in different species. Ultimately, our accounts of the ecology of photosynthesis in the sun and shade depend on application of these tools and insights to large populations of leaves and organisms under changing light environments.

Acknowledgments

Two kinds of acknowledgment are necessary. The first is evident from the content of this paper, which testifies again to the unique opportunities for collaborations in Canberra between the Research School of Biological Sciences (RSBS) and CSIRO. One of us (CBO) recalls his first experiments in Canberra in 1967 were done as a guest in the laboratory of Boardman and Anderson and this hospitality has continued for two decades. One decade ago, it was in the same laboratory that the other of us (WSC) received his first post-doctoral training, and the opportunity for interaction with collaborators at RSBS. Without such collaboration, many of us would have had far less to say during the symposium.

The second kind of acknowlegment is a personal one to Bob Robertson, but one that we feel with confidence can be made on behalf of all the participants. We are indebted, directly or indirectly, to Sir Rutherford Robertson for the opportunities he so wisely and widely made available for the practice of our profession in Australia. At the Universities of Sydney and Adelaide, as founding chairman of the Australian Research Grants Commission, and most especially as Master of University House and then Director of the Research School of Biological Sciences, his generous and tolerant advice has been a major force behind much of the excitement in Australian plant sciences research.

References

Anderson, J. M. (1982). The significance of grana stacking in chlorophyll *b*-containing chloroplasts. *Photobiochem. Photobiophys.* **3**, 225–41.

Anderson, J. M., Chow, W. S., and Goodchild, D. J. (1988). Thylakoid membrane organisation in sun/shade acclimation. *Aust. J. Plant Physiol.* **15**, 11–26.

Anderson, J. M., and Osmond, C. B. (1987). Shade–sun responses: compromises between acclimation and photoinhibition. In 'Photoinhibition'. (Eds D. J. Kyle, C. B. Osmond and C. J. Arntzen.) Topics in Photosynthesis, Vol. 9, pp. 1–38. (Elsevier: Amsterdam.)

Arntz, B., and Trebst, A. (1986). On the role of the Q_B protein of PSII in photoinhibition. *FEBS Lett.* **194**, 43–9.

Barber, J. (1982). Influence of surface charges on thylakoid structure and function. *Annu. Rev. Plant Physiol.* **33**, 261–95.

Barber, J., Chapman, D. J., and Telfer, A. (1987). Characterization of a PSII reaction centre isolated from the chloroplasts of *Pisum sativum*. *FEBS Lett.* **220**, 67–73.

Biggins, J. (Ed.) (1987). 'Progress in Photosynthesis Research.' Vols 1–4. (Martinus Nijhoff: Dordrecht.)

Björkman, O. (1987). High-irradiance stress in higher plants and interaction with other stress factors. In 'Progress in Photosynthesis Research'. (Ed. J. Biggins.) Vol. 4, pp. 11–18. (Martinus Nijhoff: Dordrecht.)

Björkman, O., Demmig, B., and Andrews, T. J. (1988). Mangrove photosynthesis: response to high-irradiance stress. *Aust. J. Plant Physiol.* **15**, 43–61.

Boardman, N. K. (1977). Comparative photosynthesis of sun and shade plants. *Annu. Rev. Plant Physiol.* **28**, 355–77.

Chow, W. S., Luping Qian, Goodchild, D. J., and Anderson, J. M. (1988). Photosynthetic acclimation of *Alocasia macrorrhiza* (L.) G. Don to growth irradiance: structure, function and composition of chloroplasts. *Aust. J. Plant Physiol.* **15**, 107–22.

Cleland, R. E. (1988). Molecular events of photoinhibitory inactivation in the reaction centre of photosystem II. *Aust. J. Plant Physiol.* **15**, 135–50.

Cleland, R. E., and Critchley, C. (1985). Studies on the mechanism of photoinhibition in higher plants. II. Inactivation by high light of photosystem II reaction centre function in isolated spinach thylakoids and O_2 evolving particles. *Photobiochem. Photobiophys.* **10**, 83–92.

Cornic, G. (1976). Effet exercé sur l'activité photosynthétique du *Sinapis alba* L. par une inhibition temporaire de la photorespiration se déroulant dans un air sans CO_2. *C. R. Acad. Sci. Paris (Sér. D)* **282**, 1955–8.

Delcourt, H. R., Delcourt, P. A., and Webb III, T. (1983). Dynamic plant ecology: the spectrum of vegetation change in space and time. *Q. Sci. Rev.* **1**, 153–75.

Demmig, B., and Winter, K. (1988). Characterisation of three components of non-photochemical fluorescence quenching and their response to photoinhibition. *Aust. J. Plant Physiol.* **15**, 163–78.

Demmig, B., Winter, K., Krüger, A., and Czygan, F.-C. (1987). Photoinhibition and zeaxanthin formation in intact leaves. A possible role of the xanthophyll cycle in the dissipation of excess light energy. *Plant Physiol.* **84**, 218–24.

Evans, J. R. (1986). A quantitative analysis of light distribution between the two photosystems, considering variation in both the relative amounts of the chlorophyll–protein complexes and the spectral quality of light. *Photobiochem. Photobiophys.* **10**, 135–47.

Evans, J. R. (1988). Acclimation by the thylakoid membranes to growth irradiance and the partitioning of nitrogen between soluble and thylakoid proteins. *Aust. J. Plant Physiol.* **15**, 93–106.

Ewart, A. J. (1896). On assimilatory inhibition in plants. *J. Linn. Soc.* **31**, 364–461.

Ferrar, P. J., and Osmond, C. B. (1986). Nitrogen supply as a factor influencing photoinhibition and photosynthetic acclimation after transfer of shade-grown *Solanum dulcamara* to bright light. *Planta (Berl.)* **168**, 563–70.

Field, C. B. (1988). On the role of photosynthetic responses in constraining the habitat distribution of rainforest plants. *Aust. J. Plant Physiol.* **15**, 343–58.

French, C. S. (1967). Photosynthesis. In 'The Search for Understanding'. (Ed. C. P. Haskins.) pp. 155–78. (Carnegie Institution: Washington D.C.)

Garfield, E. (1987). Citation classics in plant sciences and their impact on current research. *Curr. Contents* **27**(40), 3–13.

Gauhl, E. (1976). Photosynthetic response to varying light intensity in ecotypes of *Solanum dulcamara* L. from shaded and exposed habitats. *Oecologia (Berl.)* **22**, 274–86.

Greer, D. H. (1988). Effect of temperature on photoinhibition and recovery in *Actinidia deliciosa*. *Aust. J. Plant Physiol.* **15**, 195–205.

Horton, P. (1987). Interplay between environmental and metabolic factors in the regulation of electron transport in higher plants. In 'Progress in Photosynthesis Research'. (Ed. J. Biggins.) Vol. 2, pp. 681–8. (Martinus Nijhoff: Dordrecht.)

Horton, P., Lee, P., and Hague, A. (1987). Photoinhibition of isolated chloroplasts and protoplasts. In 'Progress in Photosynthesis Research'. (Ed. J. Biggins.) Vol. 4, pp. 59–63. (Martinus Nijhoff: Dordrecht.)

Kok, B., Gassner, E. B., and Rurainski, H. J. (1965). Photoinhibition of chloroplast reactions. *Photochem. Photobiol.* **4**, 215–27.

Kyle, D. J., and Ohad, I. (1986). Photoinhibition. In 'Photosynthesis III'. (Eds L. A. Staehelin and C. J. Arntzen.) Encycl. Plant Physiol. (New Ser.) Vol. 19, pp. 468–75. (Springer-Verlag: Berlin.)

Kyle, D. J., Ohad, I., and Arntzen, C. J. (1984). Membrane protein damage and repair: selective loss of a quionone-protein function in chloroplast membranes. *Proc. Natl Acad. Sci. U.S.A.* **81**, 4070-4.

Langenheim, J. H., Osmond, C. B., Brooks, A., and Ferrar, P. (1984). Photosynthetic responses to light in seedlings of selected Amazonian and Australian rainforest tree species. *Oecologia (Berl.)* **63**, 215–24.

Nanba, O., and Satoh, K. (1987). Isolation of a photosystem II reaction center consisting of D-1 and D-2 polypeptides and cytochrome *b*-559. *Proc. Natl Acad. Sci. U.S.A.* **84**, 109–12.

Osmond, C. B. (1983). Interactions between irradiance, nitrogen nutrition and water stress in the sun–shade response of *Solanum dulcamara*. *Oecologia (Berl.)* **57**, 316–21.

Osmond, C. B. (1987). Photosynthesis and carbon economy. *New Phytol.* **106**, 161–75.

Osmond, C. B., Björkman, O., and Anderson, D. J. (1980). 'Physiological Processes in Plant Ecology. Ecological Studies'. Vol. 36. (Springer-Verlag: Heidelberg.)

Pearcy, R. W. (1983). The light environment and growth of C_3 and C_4 tree species in the understory of a Hawaiian forest. *Oecologia (Berl.)* **58**, 19–25.

Pearcy, R. W. (1988). Photosynthetic utilisation of lightflecks by understory plants. *Aust. J. Plant Physiol.* **15**, 223–38.

Powles, S. B., and Osmond, C. B. (1978). Inhibition of the capacity and efficiency of photosynthesis in bean leaflets illuminated in a CO_2-free atmosphere at low oxygen: a possible role for photorespiration. *Aust. J. Plant. Physiol.* **5**, 619–29.

Terashima, I., and Inoue, Y. (1985). Vertical gradient in photosynthetic properties of spinach chloroplasts dependent on intra-leaf light environment. *Plant Cell Physiol.* **26**, 781-5.

Terashima, I., Sakaguchi, S., and Hara, N. (1986). Intra-leaf and intracellular gradients in chloroplast ultrastructure of dorsiventral leaves illuminated from the adaxial or abaxial side during their development. *Plant Cell Physiol.* **27**, 1023-31.

Thompson, W. A., Stocker, G. C., and Kriedemann, P. E. (1988). Growth and photosynthetic response to light and nutrients of *Flindersia brayleyana* F. Muell., a rainforest tree with broad tolerance to sun and shade. *Aust. J. Plant Physiol.* **15**, 299–315.

Torquebiau, E. F. (1988). Photosynthetically active radiation environment, patch dynamics and architecture in a tropical rainforest in Sumatra. *Aust. J. Plant Physiol.* **15**, 327–42.

Urban, D. L., O'Neill, R. V., and Shugart, H. H. (1987). Landscape ecology. A hierarchial perspective can help scientists understand spatial patterns. *BioScience* **37**, 119–27.

Walker, D. A., and Osmond, C. B. (1986). Measurement of photosynthesis *in vivo* with a leaf disc electrode: correlations between light dependence of steady-state photosynthetic O_2 evolution and chlorophyll *a* fluorescence transients. *Proc. R. Soc. Lond. B* **227**, 267–80.

Thylakoid Membrane Organisation in Sun/Shade Acclimation

Jan M. Anderson, W. S. Chow and D. J. Goodchild

Division of Plant Industry, CSIRO, G.P.O. Box 1600, Canberra, A.C.T. 2601, Australia.

Abstract

The photosynthetic apparatus of plants responds to changing light quantity and quality with coordinated changes in both the light-harvesting antennae of the photosystems and the amounts of electron transport components and ATP synthase. These compositional modulations are accompanied by changes in thylakoid membrane organisation and photosynthetic capacity. It is now clear that there is a dynamic continuum of organisation and function of the photosynthetic apparatus from the appressed granal and non-appressed stroma thylakoids within a chloroplast, to different chloroplasts within a leaf, to leaves within and between species. While it is very unlikely that there is a unique solution to photosynthesis in the sun or shade, substantial changes in composition, and hence thylakoid membrane organisation and function, are elicited as part of sun/shade responses.

Prologue

'Evidently, photosynthetic adaptation to high light involves a balanced increase in the capacity of a number of component processes — biophysical, biochemical, and photochemical — so that no single leaf factor exclusively limits the overall rate of photosynthesis.

Adaptation to low light appears to be primarily a question of economics. The capacity of the light-harvesting system (together with the reaction centers) is increased or at least kept constant, whereas there is a decrease in the level of constituents of the electron transport chain and those involved in CO_2 fixation. High levels would be of little use to the plant in low-light environments. They would not be disadvantageous per se, *but the synthesis and maintenance of such levels would require an increased expenditure of energy and would result in a greater carbon loss through respiration and hence a higher light compensation point.'*

O. Björkman, Carnegie Institution of Washington Year Book 71, p. 134 (1972).

Introduction

The superb ability of plants to adapt to changes in their light environment allows them to succeed in habitats ranging from the arid sun-soaked deserts to the deeply shaded floors of dense tropical forests. Plants must cope with contrasting light environments that may vary by three orders of magnitude in irradiance. The irradiance available varies seasonally, diurnally and spatially, not only at the level of the plant canopy itself, but even at the level of the leaf (e.g. across leaves of fixed orientation). Shade plants on the forest floor receive less than 0·5% of the irradiance experienced above the

0310-7841/88/010011$03.00

plant canopy in a Queensland tropical rainforest (Björkman and Ludlow 1972). This irradiance on the forest floor comprises two very different radiations: the continuous, diffuse, low flux radiation (40%; 400–700 nm) is markedly enriched in far-red light, and sunflecks, short intermittent bursts of direct radiation (60%, 400–700 nm). This complex spectral distribution of radiation is not readily achieved in the laboratory and most studies of the sun/shade responses use plants that have been grown under artificial fluorescent and tungsten lights at different irradiances, or natural light with varying degrees of shading with shade cloth in a glasshouse.

It is well established that plants growing in open sunny habitats (sun plants) have much greater photosynthetic capacities, which saturate at higher irradiance, than those of plants growing in shaded habitats (shade plants) (Boardman 1977; Wild 1979; Björkman 1981). However, shade plants use low irradiance more effectively for net photosynthetic CO_2 uptake than do sun plants, due to the lower dark respiration rates of shade plants. Perceiving the need to define the sun/shade responses of leaves at the molecular level, Olle Björkman, during his 1971 visit to Australia, persuaded nine scientists from three laboratories to embark on a massive collaborative project which was published in the Carnegie Institution of Washington Year Book (1972) Vol. 71, pp. 82–141. It is remarkable that the findings, and often conclusions, of these classical studies need little subsequent alteration, despite the dramatic increase in our knowledge of the molecular organisation of the components of the photosynthetic apparatus.

In this overview, we discuss the changing concepts relating to the molecular organisation and spatial distribution of PS I and PS II.* An appreciation of modern concepts of lateral heterogeneity in the distribution of the supramolecular protein complexes along thylakoid membranes provides a basis for understanding not only the modulations in the structure, composition and function of photosynthetic membranes in response to sun or shade, but also the regulatory mechanisms which optimise light interception and electron flow in response to light. Together with brief descriptions of the structure, function and composition of the thylakoid membranes of sun and shade plants and plants adapted to high and low irradiance, we address some of the outstanding questions relating to the functional consequences of the differences in their membrane composition and organisation.

Molecular Organisation of Plant Thylakoid Membranes

The thylakoid membrane network of most higher plants and green algae have characteristic interconnected regions of stacked and unstacked membranes. The appressed membranes of the granal thylakoid stacks are in close contact and not directly exposed to the stroma. In contrast, the non-appressed membranes, whose outer surfaces are in direct contact with the chloroplast stroma, include the stroma lamellae and the end membranes and margins of the grana stacks. The inner surfaces of both appressed and non-appressed membranes enclose a narrow, continuous, aqueous space, the lumen.

Freeze–fracture electron microscopy has shown distinct differences in both the size and number of particles in the four fracture faces of appressed and non-appressed membranes (cf. Staehelin and Arntzen 1983). The variation in particle distribution implies differences in the location of the multiprotein, membrane-spanning complexes of thylakoid membranes. This differentiated structure is maintained *in vitro* only in the presence of divalent cations (>3 mM) or monovalent cations (>150 mM) (Izawa and Good 1966). In low-salt media, the thylakoid membrane network destacks to form single sheets of only non-appressed membranes; concomitantly, there is a lateral

* Abbreviations used: Chl, chlorophyll; cyt, cytochrome; PS I, II, photosystems I, II; Rubisco, ribulosebisphosphate carboxylase/oxygenase.

migration and randomisation of the freeze–fracture particles of different sizes. Hence lateral segregation of the thylakoid multiprotein complexes is a basic feature of membrane stacking. These structural differences between appressed and non-appressed membrane regions are paralleled by compositional and functional differences.

The concept of photosynthesis consisting of two light reactions catalysed by PS I and PS II has been immortalised in the Z scheme of Hill and Bendall (1960). Following its publication, biochemical fractionation studies of thylakoids fragmented with detergents or mechanical shearing showed that the photosystems were indeed structural entities as they could be physically separated from one another. Stacked granal membrane fractions were enriched in PS II compared to the non-appressed stromal thylakoid fraction, which contained mainly PS I complex. Thus it was proposed by Sane *et al.* (1970), and generally accepted through the 1970s, that both PS II and some PS I were located close together in the appressed grana regions to participate in linear electron transport, while other PS I, involved in cyclic photophosphorylation, was distributed in the non-appressed stroma thylakoids. However, the implied heterogeneity of PS I was not substantiated by biochemical, structural or biophysical studies.

The application of the aqueous polymer two-phase partition technique, pioneered by Albertsson (1971), to the mechanically fragmented granal thylakoid membrane fraction produced inside-out and right-side-out thylakoid vesicles that were shown to be derived from appressed and non-appressed membranes, respectively. Moreover, the appressed membrane fractions were enriched in PS II complex and PS II activity (Ackerlund *et al.* 1976). Following analyses of the chlorophyll–proteins in appressed and non-appressed membrane fractions, Andersson and Anderson (1980) presented a new model for thylakoid membrane organisation (Fig. 1) which proposed an extreme lateral heterogeneity in the distribution of the photosystems. Consequently, most PS II complexes and their associated chlorophyll *a/b*-proteins (LHCII) are located in the appressed membrane regions and are laterally segregated from the PS I complex; ATP synthase is found exclusively in the non-appressed regions (Miller and Staehelin 1976). This was initially a startling idea as it had been accepted that most of the light-harvesting pigment arrays of PS II and PS I were in contact with each other, with the regulation of excitation energy distribution between the photosystems being controlled by spillover. Further, the idea of mobile electron shuttlers had not been seriously considered since the original suggestion of Sane *et al.* (1970). Nevertheless, the new model was quickly accepted because the functional consequences of a lateral separation of the photosystems made sense (Anderson 1981, 1982).

The new model also has an heterogeneity of PS II (Fig. 1). Indeed Melis and coworkers (cf. Melis 1984) had independently shown by biophysical measurements that PS II was heterogeneous. The major PS II_α pool, with a larger light-harvesting antenna, is located in the appressed membrane fraction, and the minor PS II_β pool, with a smaller light-harvesting antenna, is present in stroma thylakoid vesicles.

Comparative structural and biochemical analyses in many laboratories supported the new model, although there was a lack of agreement on the location of the cyt *b/f* complex which functions as an electron carrier between the two photosystems (cf. Haehnel 1984). It was important to resolve this question because of the implications for long-range electron transfer, and also to determine the degree of separation of PS I complex from PS II complex. Recently, immunogold electron microscopy has allowed the direct detection of antigens in appressed and non-appressed membranes (e.g. Goodchild *et al.* 1985; Vallon *et al.* 1986; Anderson and Goodchild 1987). This has enabled the model with extreme lateral heterogeneity of PS I to be verified, as there is a total exclusion of PS I complex from the appressed granal thylakoids. It is also clear that the cyt *b/f* complex is present in both membrane regions.

This molecular organisation of thylakoid membranes should not be regarded as fixed and static. Certain environmental stimuli can evoke controlled, reversible lateral movements of some protein complexes between the appressed and non-appressed regions, often in a matter of minutes. These movements include the reversible phosphorylation of part of LHCII and its subsequent migration from the edges of the grana stacks to stroma thylakoids (Kyle *et al.* 1983, 1984), and the heat-induced movement of core PS II complex from appressed to non-appressed regions (Andersson *et al.* 1987). Additional long-term modulations in the biosynthesis and degradation of thylakoid components can result in changes of membrane composition, which in turn will influence thylakoid stacking patterns and function. These short-term and long-term mechanisms for regulation of the photosynthetic apparatus at the membrane level are important to optimise both light-harvesting and ATP and NADPH production.

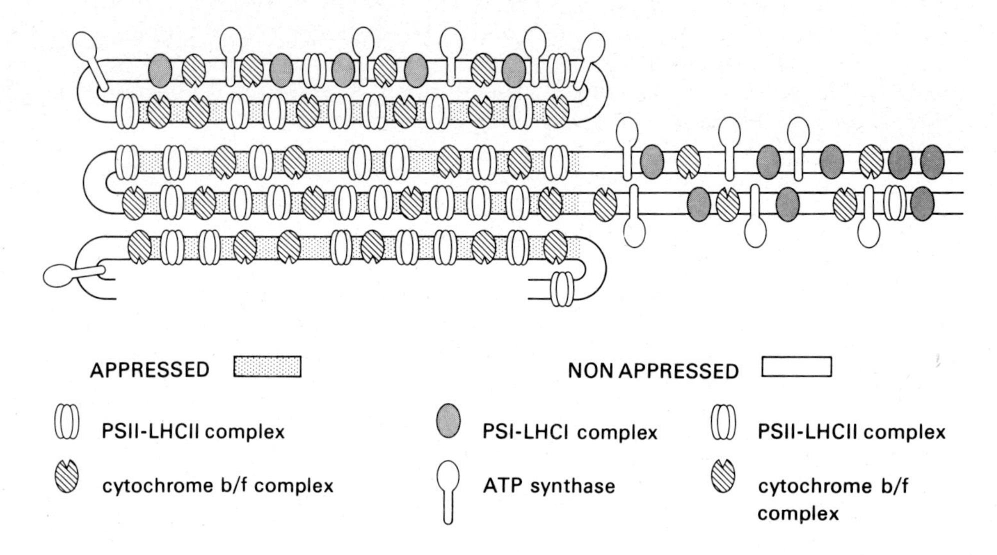

Fig. 1. Scheme of the lateral distribution of the supramolecular protein complexes in plant thylakoid membranes. The thylakoid membrane network is made up of appressed and non-appressed regions.

Adaptation of Thylakoid Structure in Sun and Shade Plants

There are striking morphological differences between the chloroplasts of plants adapted to sunlight or high irradiance and shade or low irradiance (Table 1). The larger shade plant chloroplasts have not only a much greater area of thylakoid membranes, but also have many more thylakoids per granum relative to sun-plant chloroplasts (Goodchild *et al.* 1972; Lichtenthaler 1981; Melis and Harvey 1981). The very large, irregularly orientated grana of shade plants may have as many as 100 thylakoids per grana stack and extend across the entire chloroplast width (see figure 4, Chow *et al.* 1988). Similarly, there is an increase in membrane content and the size of chloroplasts in leaves from the upper, middle to lower part of trees (Goryshina 1980). Typically, the ratio of the length of appressed to non-appressed membrane is greater for shade (4–5) than for sun chloroplasts (1–1·5). Also, as the leaves of most plants have a transmittance for photosynthetically active radiation of only 1–5%, the chloroplasts on the shaded side of leaves with fixed orientations are exposed to light qualities and quantities very similar to those of deeply shaded habitats. With spinach leaves grown in a fixed orientation, the ratio of the length of appressed to non-appressed membranes and the total membrane length per unit chloroplast area increase linearly as a function of the distance from the upper leaf surface without an apparent increase in chloroplast size (Terashima and Inoue 1985).

Table 1. Overview of adaptation of the structure, composition and function of thylakoid membranes of sun and shade chloroplasts

	Sun	Shade
Size of chloroplast	Normal	Large
Stromal / thylakoid volume	High	Low
Thylakoid membranes per chloroplast	Low level	High level
Number of thylakoids per grana	Few	Many
Appressed to non-appressed membrane ratio	Low	High
Chlorophyll content per chloroplast	Low	High
Chlorophyll *a* / *b* ratio	High	Low
Xanthophylls / β-carotene	Low	High
P680 / Chl	High	Low
Cyt *f* / Chl	High	Low
P700 / Chl	Little change	Little change
CF_1 / Chl	High	Low
PS II and PS I activities	High	Low
Intensity required for saturation of electron transport	High	Low
Quantum yields for whole-chain electron transport	Same	Same
ATP synthase activity / Chl	High	Low

Interestingly, differences exist also in the density of the freeze–fracture particles of the appressed thylakoid membranes as a response to sun / shade (Table 2). In appressed membranes, the large EF_s particles are thought to represent core PS II complexes each containing P680, while 80% of the small PF_s particles may represent LHCII (Simpson 1981) and the remainder, the cyt *b/f* complex. A significant feature of the sun / shade response is shown in Table 2. The ratio of PF_s/EF_s particles is higher in the shade plants, indicating an adjustment in the amounts of LHCII relative to core PS II complexes in shade and low irradiance. The preferred interparticle distance (centre to centre) of the core PS II complex particles on the EF_s fracture face was obtained from the radial distribution function plot of the data for each species as described in Goodchild *et al.* (1983). The greater separation of the core PS II complexes in shade species and the low-light *Atriplex* (Table 2) may reflect a decreased amount of core PS II complexes per unit area of appressed membranes.

Table 2. Distribution of freeze–fracture particles/μm^2 in appressed membranes, and the separation distance between EF_s particles in response to growth in the sun or shade

PF_s (small) particles mainly represent LHCII, and EF_s (large) particles represent core PS II complex (Simpson 1981)

	No. of particles / μm^2		PF_s/EF_s	Preferred interparticle distance on EF_s fracture face (nm)
	PF_s	EF_s		
Alocasia	5100[A]	1290[A]	4.0	24
Lomandra	5000[A]	1380[A]	3.6	26
Atriplex				
Low light	3800[B]	1650[B]	2.3	24
Intermediate light	3900[B]	1850[B]	2.1	—
High light	4100[B]	2100[B]	1.95	22
Spinach (high light)	3400	1571	2.2	22

[A] Goodchild *et al.* (1972). [B] Björkman *et al.* (1972).

Sun/Shade Acclimation of the Light-Harvesting Antennae of Thylakoid Membranes

Pigment Composition

Given the striking differences in the extent of thylakoid membrane appression as part of the sun / shade response, the morphological differences produced imply differences in the actual stoichiometries of the PS II–LHCII complexes located in appressed membranes relative to the PS I complexes and ATP synthase of non-appressed regions (Fig. 1). This indeed is the case as significant changes in pigment composition, chlorophyll-proteins, and other proteins occur during acclimation (Table 1). Canopy shade-acclimated plants contain more chlorophyll per chloroplast (Anderson *et al.* 1973; Goryshina 1980), as might be expected from their enhanced membrane content. It is well known that the pigment composition is also affected by sun / shade acclimation. Typically, shade plant thylakoids have Chl *a/b* ratios of 2·0–2·2 compared to 2·6–3·6 for sun plant thylakoids, and also have higher xanthophyll / β-carotene ratios. Indeed, an inverse relationship exists between the extent of membrane appression and the Chl *a/b* ratio (Boardman *et al.* 1974; Aro *et al.* 1986). These changes in pigment content also occur across spinach leaves grown in a fixed orientation, with chloroplasts at the upper surface having a Chl *a/b* ratio of 3·5 compared to 2·5 at the lower shaded surface (Terashima and Inoue 1985).

Table 3. Sun/shade acclimation of chlorophyll-protein complexes

The percentage distribution of chlorophyll between the chlorophyll-proteins resolved by non-denaturing gel electrophoresis. For peas, irradiances during growth (μmol quanta m^{-2} s^{-1}) were: high, 840; intermediate, 165; low, 42; data from Leong and Anderson (1984*a*). *Alocasia* data are from Chu and Anderson (1984)

Pigment–protein complex	Spinach	Peas			*Alocasia*
		High	Inter-mediate	Low	
PS I–LHCI	31	26	25	23	23
LHCII	43	42	49	56	57
PS II core	12	13	11	10	8
Chl *a* / *b*	2·2	3·2	2·8	2·7	2·3

Chlorophyll-Protein Content

The variations in pigment content as a response to sun / shade acclimation reflect differences in the complement of specific chlorophyll-proteins. These include the β-carotene–Chl *a* proteins of PS II core complex (47 and 43 kDa apoproteins) and PS I core complex (68 kDa apoproteins) that are universally conserved. The light-harvesting pigment-protein antennae complexes of plants and green algae include several xanthophyll-Chl *a/b*-proteins serving PS II (LHCII, 30–25 kDa apoproteins) and the minor xanthophyll-Chl *a/b*-proteins serving PS I (LHCI, 25–20 kDa proteins) (Thornber 1986). There is a qualitative increase in the amount of total chlorophyll associated with LHCII in shade and low irradiance-adapted plants, and a decrease in the PS I–LHCI complexes (Table 3). Accompanying the increases in total LHCII, there are relatively more appressed to non-appressed membranes (Table 1), which might be expected because LHCII is thought to mediate thylakoid appression (cf. Staehelin and Arntzen 1983). Thus, the inverse relationship between the Chl *a/b* ratios and the extent of thylakoid stacking is consistent with the enhanced LHCII content and reduced ATP synthase content.

The variations in pigment–proteins elicited in the sun/shade response may be associated with modulations in both the stoichiometries of P680 and P700, and the number of antennae pigment molecules associated with each reaction centre; that is, the photosynthetic unit size of PS II and PS I. As the strategies employed to permit maximal light-harvesting within thylakoid membranes are part of the sun/shade response, it is important to consider both parameters.

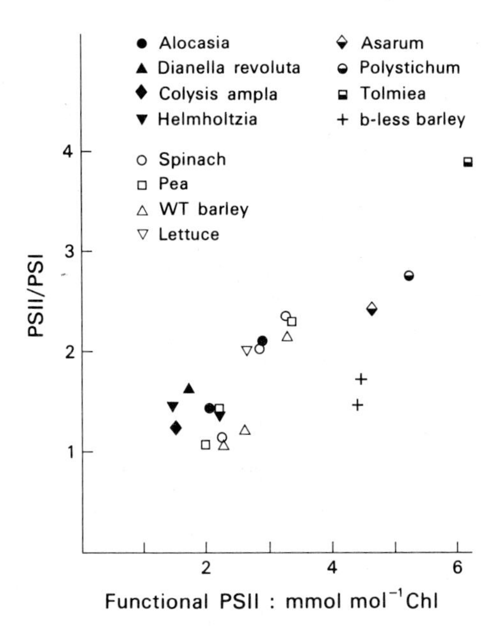

Fig. 2. The concentration of functional PS II in thylakoid membranes from sun and shade-adapted plants and the Chl *b*-less barley mutant. Some plant species have also been grown at both high and low irradiances. PS II was determined from the yield of O_2 from flash-induced oxidation of water from leaf discs, and PS I was measured as the light-induced oxidation of P700 in isolated thylakoids (Chow and Hope 1987). Functional PS II centres were estimated in the Chl *b*-less barley mutant by dividing the DCMU-binding sites by the factor 1·14. Data on the right for three shade plants (half-open symbols) are from Melis and Harvey (1981).

Stoichiometries of Reaction Centres

It is agreed that variations in growth irradiance do not induce substantial changes in the Chl/P700 ratios of many sun species (cf. Anderson 1986) or *Alocasia* (Chow *et al.* 1988). However, the amount of P680 can increase by 70% with increasing irradiance (Leong and Anderson 1984*b*; Wild *et al.* 1986; Chow and Anderson 1987*b*; Chow and Hope 1987; Evans 1987*b*). The situation in shade plants is conflicting. Compared to sun plants, Melis and Harvey (1981) found increased concentrations of P680 in shade plants. They measured P680 as the absorbance change at 320 nm due to the semiquinone radical of the primary electron acceptor, Q_A, of PS II. On the other hand, we found the reverse situation with four other shade plants which have lower P680 concentrations compared to those of sun plants. Since the concentration of P700 remains relatively constant, while PS II increases in sun plants relative to shade plants, the ratio of PS II to PS I reaction centres varies as shown in Fig. 2. The PS II/PS I ratios fall into three groups; a lower one for shade plants, a middle one for sun plants, and an upper one for the three values determined by Melis and Harvey (1981) for shade plants (Fig. 2). As the Chl *b*-less barley mutant without LHCII has a much reduced light-harvesting capacity in PS II, Ghirardi *et al.* (1986) suggest that the barley mutant needs more PS II units relative to PS I units to balance excitation between the photosystems. Our results (Fig. 2) are in agreement with this suggestion. However, the high PS II / PS I centre ratios obtained by Melis and Harvey (1981) for their shade plants are puzzling, as shade plants with more LHCII and less core PS II would scarcely be employing the same strategy as the Chl *b*-less barley mutant without LHCII. Instead we suggest that shade plants increase the light-harvesting capacity in PS II by decreasing P680 concentration, and simultaneously increasing their LCHII content; that is, by making larger though fewer PS II units.

Photosynthetic Unit Size

Determinations of the photosynthetic unit size of the individual photosystems are not easy; they are further complicated by the heterogeneity of PS II. A useful index of the 'apparent' size of PS II is given by the LHCII / core PS II ratio which can be calculated from the proportion of chlorophyll in each pigment–protein complex resolved by non-denaturing gel electrophoresis (Table 3; Fig. 3) (Anderson 1986). Both shade and low irradiance-acclimated plants have more LHCII and less core PS II compared to sun plants, indicating larger 'apparent' PS II photosynthetic units (Chu and Anderson 1984; Lichtenthaler and Meier 1984; Anderson 1986; Wild *et al.* 1986). However, in pea plants grown in far-red-enriched and -deficient illumination, Melis (1984) found that the number of chlorophyll molecules transferring excitation energy to PS II_α, PS II_β and PS I were unchanged, despite marked changes in their PS II / PS I reaction centre ratios.

The marked variations seen in the amounts of Chl–proteins (Table 3; Fig. 3) support the concept of dynamic long-term adjustments of Chl–proteins in response to light. While Melis (1984) suggests that the sun / shade acclimation involves alterations in the relative number of PS II_α, PS II_β and PS I units (each of unchanged antenna size), others favour the notion of a more variable PS II antenna size in the shade (Chu and Anderson 1984; Lichtenthaler and Meier 1984; Anderson 1986; Wild *et al.* 1986). Thus, it is unresolved whether acclimation to low irradiance or shade occurs by increasing the relative number of PS II_α to PS II_β units, or alternatively, the PS II_α in the appressed thylakoids of shade plants may have more peripheral LHCII associated with each PS II core complex.

Sun/Shade Acclimation of Electron Transport Components

Our earlier research clearly demonstrated that significant changes in the amounts of electron transport components occurred between sun and shade-grown plants (Boardman *et al.* 1972). Shade plant chloroplasts have less cyt *b*-559, cyt *b*-563, cyt *f* and plastoquinone per unit Chl, while the P700 content is unaltered. The reduction of cyt *f* on a Chl basis is striking; typically, the cyt *f* / P700 ratios are 0·33–0·5 for shade plants compared to ratios near 1 for sun plant thylakoids.

Comparisons of the amounts of thylakoid multiprotein electron transport complexes of mustard, spinach, pea and *Alocasia* grown at high and low irradiances are shown in Table 4. Extensive adjustments in the relative amounts of electron transport complexes and ATP synthase occur, with the cyt *b/f* complexes and ATP synthase being the most changed, followed by PS II complex and little or no change in PS I complex on a chlorophyll basis.

These results contradict the ideas of an invariant stoichiometry of electron transport complexes and ATP synthase, and constant P680 / P700 ratios advanced by Whitmarsh and Ort (1984) and Graan and Ort (1984). Their results seem to be inconsistent with current concepts of an extreme lateral heterogeneity in the distribution of thylakoid complexes between appressed and non-appressed regions.

Sun/Shade Acclimation of Function of Thylakoid Membranes

Given the compositional differences of thylakoids evoked in the sun / shade response, it is scarcely surprising that marked variations occur in their photosynthetic capacities.

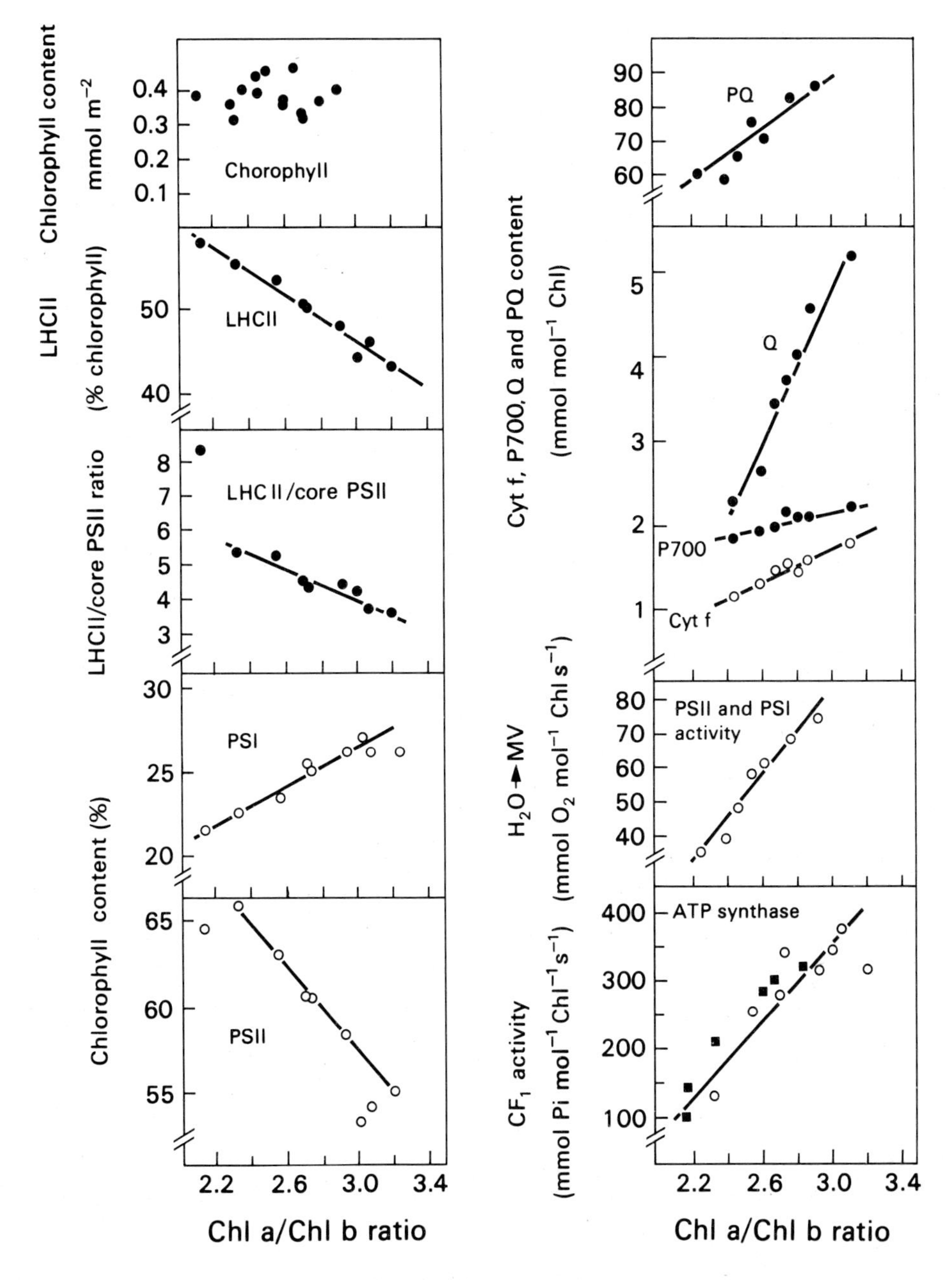

Fig. 3. Correlations between the Chl *a* / *b* ratios of pea thylakoid membranes adapted to varying growth irradiances and the chlorophyll content, concentrations and ratios of components, and photosynthetic activities (redrawn from Leong and Anderson 1984*a*, 1984*b*, 1986).

Shade plants have much lower light-saturated rates of PS II and PS I activities as well as photophosphorylation, which saturate at very low irradiance (Boardman *et al.* 1972). These changes parallel those seen in the capacity of their leaves for CO_2 assimilation at light saturation. Striking differences in overall electron transport capacity and ATP synthase activity (Table 4) are also observed with several sun genotypes, and the shade plant *Alocasia* (Chow *et al.* 1988) grown at different irradiances. Similar results are also mirrored by the chloroplasts from the upper surface of leaves of fixed orientation that have higher electron transport capacities than those at the lower surface (Terashima and Inoue 1985). Significantly, the quantum yields of oxygen evolution of sun and shade leaves are comparable (Björkman and Demmig 1987; Evans 1987*a*).

Comparisons with peas grown over a wide range of irradiances show that linear relationships exist between the Chl *a*/*b* ratios of isolated pea thylakoids and their properties (Fig. 3). Although the relationship between growth irradiance and Chl *a*/*b* ratios is not linear (Leong and Anderson 1984*a*), the correlations shown in Fig. 3 are remarkable. They demonstrate that the many changes occurring in the distribution of chlorophyll between the Chl *a*/*b*-proteins and Chl *a*-proteins, electron transport components and overall electron transport capacity are extremely well coordinated. Clearly

Table 4. Stoichiometries of electron transport complexes and ATP synthase activity in response to sun/shade acclimation

ATP synthase activity is expressed as mmol P_i mol^{-1} Chl s^{-1}

Plant	Irradiance	Chl *a* / Chl *b*	PS II / PS I	Cyt *b*/*f* / PS I	ATP synthase activity
Mustard[A]	High	—	1·0	0·84	—
	Low	—	0·7	0·36	—
Spinach[B]	High	2·7	1·62	1·02	245
	Low	2·2	1·2	0·7	145
Pea[C]	High	3·0	1·69	1·25	330
	Low	2·5	1·25	0·7	170
Alocasia[D]	High	3·3	1·9	0·7	—
	Low	2·4	1·3	0·5	—

[A]Wild *et al.* (1986). [B]Chow and Hope (1987). [C]Chow and Anderson (1987*b*). [D]Chow *et al.* (1988).

irradiance is very important in regulating photosynthetic capacity. However, the extent to which light quality, rather than light quantity, is involved in the sun–shade response is not clear (Melis 1984; Anderson and Osmond 1987).

The basis for the flexibility in electron transport capacity and photophosphorylation observed in the sun/shade response lies in the dynamic relationship between the thylakoid supramolecular complexes and mobile electron carriers, together with adjustment in the relative amounts of appressed and non-appressed membranes. Changing patterns in the synthesis and degradation of thylakoid components take several hours to days to occur. For example, the transfer of pea plants (previously acclimated to low irradiance) to high irradiance doubles the photosynthetic capacity (Fig. 4). Once again, this increased capacity is accompanied by changes in many components of the photosynthetic apparatus, including the Chl *a*/*b* ratios, P680 (measured by atrazine binding) and cyt *f* contents, ATP synthase and Rubisco activities (Fig. 4). There is, however, no change in either total chlorophyll or P700 content (Chow and Anderson 1987*a*, 1987*b*). These dynamic experiments, where plants previously acclimated to low irradiance are transferred to high irradiance, support the notion of highly plastic functional relationships in sun/shade responses.

Acclimation in Relation to the Utilisation of Resources

The light acclimation of thylakoid membranes involves the coordinated allocation of resources, proteins, pigments and lipids, to achieve and maintain optimal rates of photosynthesis. In low irradiance or shade adaptation, where light is the limiting factor, a substantial amount of photosynthetic resources must be invested in the synthesis and

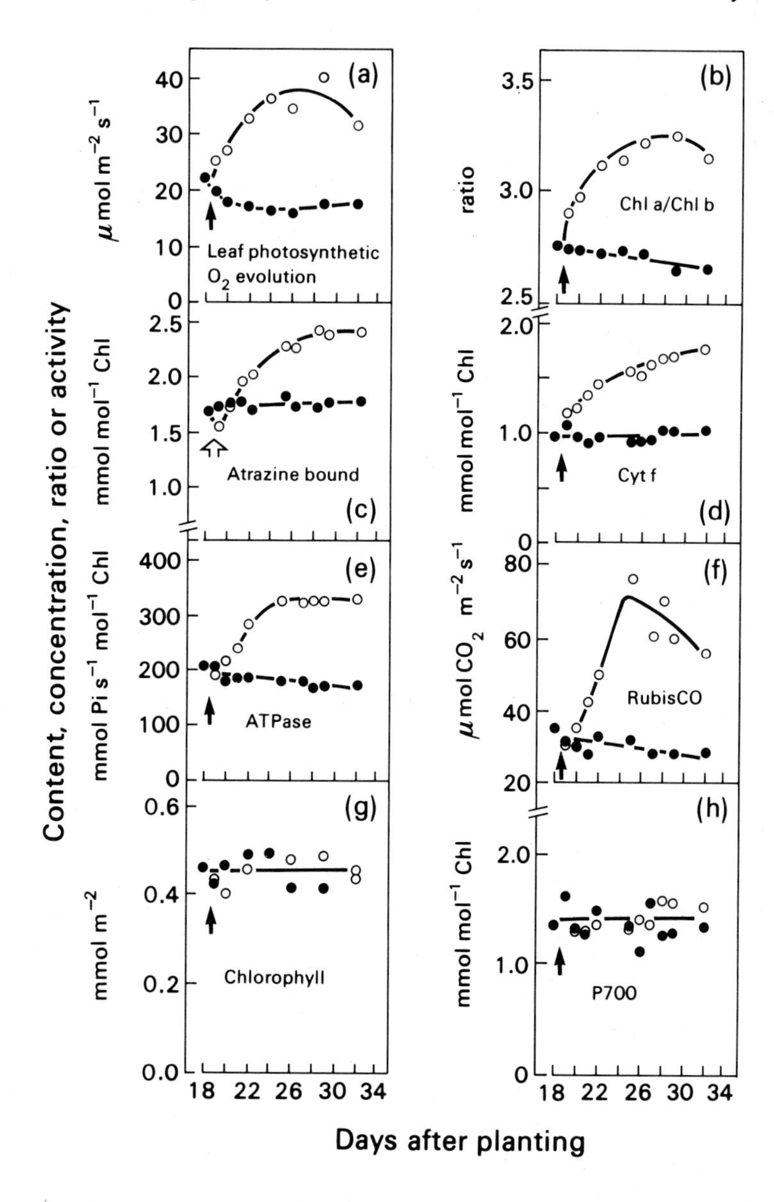

Fig. 4 Acclimation of the photosynthetic apparatus of peas grown in low irradiance (60 μmol photons m^{-2} s^{-1}) for 19 days, and then transferred to high irradiance (390 μmol photons m^{-2} s^{-1}). ● Control low light peas. ○ Peas acclimated to high light. Solid arrows indicate the time of transfer. The open arrow in (*c*) points to a temporary decrease in atrazine binding which was accompanied by a temporary decrease of quantum yield of O_2 evolution (redrawn from Chow and Anderson 1987*a*, 1987*b*).

maintenance of the light-harvesting assemblies of both photosystems. On the other hand, large amounts of electron transport components, ATP synthase or the stromal CO_2 fixation enzymes are not required. At low irradiance, relatively more chlorophyll is mainly associated with the apoproteins of LHCII and LHCI. These Chl *a*/*b*-proteins surround the core PS II and PS I core complexes of fixed subunit composition, leading to larger PS II and PS I units. Another consequence is that, as relatively more Chl *a*/*b*-proteins are synthesised, more thylakoids become appressed, and more membranes can be accommodated in the chloroplast. Evans (1987*b*) has suggested that investment in the Chl *a*/*b*-proteins, rather than the Chl *a*-proteins, for enhanced light-harvesting capacity, is also consistent with the nitrogen costs of each complex. The increased number of pigment molecules associated with each P680 offsets the lower number of PS II reaction centres so that the total light-harvesting capacity of PS II balances that of PS I, despite the lower PS II/PS I reaction centre ratios (Fig. 2).

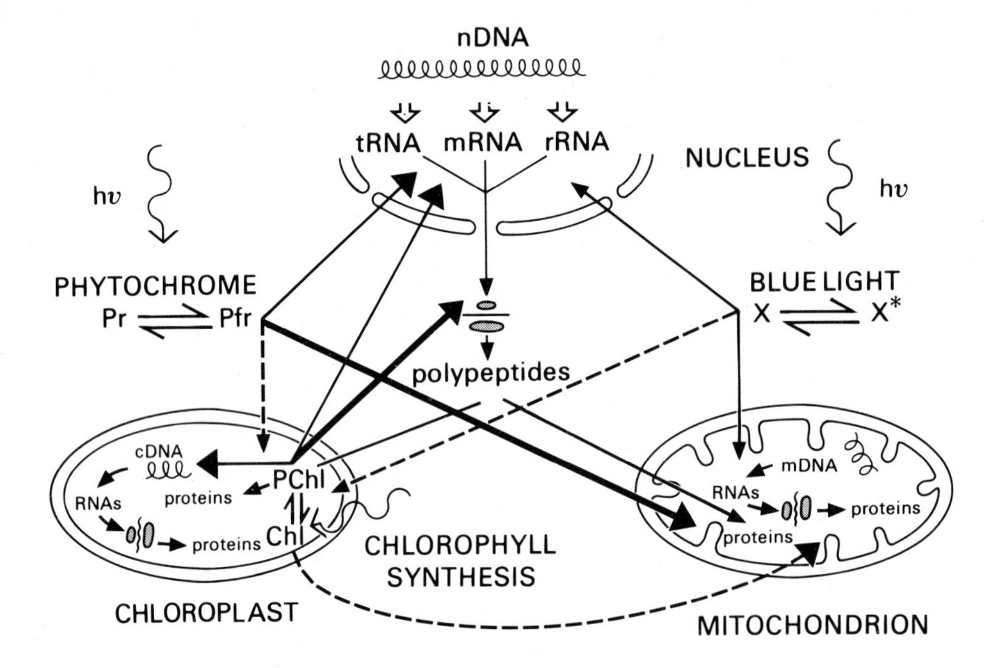

Fig. 5. A scheme showing the complex interactions between the plant genomes and the photoreceptors.

Conversely, under sunlight or high irradiance conditions, the rate of light absorption by either photosystem is not limiting. Rather the potentially limiting factors in bright light are the electron transport steps. Hence acclimation to sun or high irradiance leads to a much higher content of the cyt *b*/*f* complex, plastoquinone, plastocyanin and ferredoxin, as well as ATP synthase which support faster rates of electron transport and photophosphorylation.

The photosynthetic apparatus is faced with two compromises: those required to bring about the most effective use of light, and those required to withstand short- and long-term changes in light quantity and quality (Anderson and Osmond 1987). This is accomplished by significant variations in the light-harvesting components of both photosystems. At low irradiance, thylakoids have larger, though fewer, PS II photosynthetic units with more Chl *a*/*b*-proteins, fewer Chl *a*-proteins and fewer P680 to maximise light-harvesting. On the other hand, at high irradiance thylakoids have smaller, though more, PS II photosynthetic units with more PS II reaction centres and fewer Chl *a*/*b*-proteins, and the smaller PS II antennae will help to minimise the deleterious effects of photoinhibition.

Future Challenges

The synthesis and assembly of thylakoid supramolecular complexes is a cooperative venture shared by the nuclear and chloroplast genomes, as some proteins of each complex are encoded by the nuclear genome and others by the chloroplast genome (Herrmann *et al.* 1985). Indeed in the cell there is a three-ringed circus, with interactions between the nucleus, mitochondria and chloroplasts being dictated by three ring-masters, the light receptors (Fig. 5). Plants have at least two photoreceptors, phytochrome and the blue-light receptors, and much of the chlorophyll synthesis is light-dependent. The complex molecular mechanism of interactions between the photoreceptors and gene expression is a fascinating area that is now being explored. It is evident already that complicated networks of regulation involving transcriptional, translational and post-translational events may occur for each thylakoid multiprotein complex, as well as diverse mechanisms to coordinate interactions between the nuclear and chloroplast genomes. In order to understand how the photosynthetic apparatus acclimates to its light environment, we must define the molecular mechanisms of photoregulation.

Epilogue

'I have just begun on this fascinating theatre of the membranes. I feel like a drama critic who begins to write about a huge theatrical company with an almost endless repertoire. He can see only some of their plays, can place only his interpretations on the character parts as they unfold before his eyes but their art will continue even after he is dead. On the membrane stage, the company approaches astronomical numbers, the individual molecular actors are playing many roles, and we have not yet seen a fraction of their plays. The gradual unfolding of their many-sided art will continue to intrigue scientific theatre writers for generations.'

R. N. Robertson (1983). In 'The Lively Membranes'. (Cambridge University Press: Cambridge.)

References

Ackerlund, H. E., Andersson, B., and Albertsson, P.-A. (1976). Isolation of photosystem II enriched membrane vesicles from spinach chloroplasts by phase partition. *Biochim. Biophys. Acta* **449**, 525–33.

Albertsson, P.-A. (1971). 'Partition of Cell Particles and Macromolecules.' 2nd ed. (Wiley-Interscience: New York.)

Anderson, J. M. (1981). Consequences of spatial separation of photosystem 1 and 2 in thylakoid membranes of higher plant chloroplasts. *FEBS Lett.* **124**, 1–10.

Anderson, J. M. (1982). The role of chlorophyll–protein complexes in the function and structure of chloroplast thylakoids. *Mol. Cell. Biochem.* **46**, 161–72.

Anderson, J. M. (1986). Photoregulation of the composition, function, and structure of thylakoid membranes. *Annu. Rev. Plant Physiol.* **37**, 93–136

Anderson, J. M., and Goodchild, D. J. (1987). Lateral distribution of the photosystem I complex between the appressed and non-appressed regions of spinach thylakoid membranes: an immunocytochemical study. In 'Progress in Photosynthesis Research'. (Ed. J. Biggins.) Vol. 2, pp. 301–4. (Martinus Nijhoff: Dordrecht.)

Anderson, J. M., Goodchild, D. J., and Boardman, N. K. (1973). Composition of the photosystems and chloroplast structure in extreme shade plants. *Biochim. Biophys. Acta* **325**, 573–85.

Anderson, J. M., and Osmond, C. B. (1987). Shade–sun responses: compromises between acclimation and photoinhibition. In 'Photoinhibition'. (Eds D. J. Kyle, C. B. Osmond and C. J. Arntzen.) Topics in Photosynthesis, Vol. 9, pp. 1–38. (Elsevier: Amsterdam.)

Andersson, B., and Anderson, J. M. (1980). Lateral heterogeneity in the distribution of chlorophyll–protein complexes of the thylakoid membranes of spinach chloroplasts. *Biochim. Biophys. Acta* **593**, 427–40.

Andersson, B., Sundby, C., Larsson, U. K., Mäenpää, P., and Melis, A. (1987). Dynamic aspects of the organization of the thylakoid membrane. In 'Progress in Photosynthesis Research'. (Ed. J. Biggins.) Vol. 2, pp. 669–76. (Martinus Nijhoff: Dordrecht.)

Aro, E.-M., Rintamäki, E., Korhonen, P., and Mäenpää, P. (1986). Relationship between chloroplast structure and O_2 evolution rate of leaf discs in plants from different biotypes in south Finland. *Plant Cell Environ.* **9**, 87–94.

Björkman, O. (1981). Responses to different quantum flux densities. In 'Physiological Plant Ecology I. Responses to the Physical Environment'. (Eds O. L. Lange, P. S. Nobel, C. B. Osmond and H. Ziegler). Encycl. Plant Physiol. New Ser. Vol. 12A, pp. 57–107. (Springer-Verlag: Berlin.)

Björkman, O., Boardman, N. K., Anderson, J. M., Thorne, S. W., Goodchild, D. J., and Pyliotis, N. A. (1972). Effect of light intensity during growth of *Atriplex patula* on the capacity of photosynthetic reactions, chloroplast components and structure. *Carnegie Inst. Wash. Year Book* **71**, 115–35.

Björkman, O., and Demmig, B. (1987). Photon yield of O_2 evolution and chlorophyll fluorescence characteristics at 77K among vasular plants of diverse origin. *Planta* **170**, 489–504.

Björkman, O., and Ludlow, M. M. (1972). Characterization of the light climate on the floor of a Queensland rainforest. *Carnegie Inst. Wash. Year Book* **71**, 85–94.

Boardman, N. K. (1977). Comparative photosynthesis of sun and shade plants. *Annu. Rev. Plant Physiol.* **28**, 355–77.

Boardman, N. K., Anderson, J. M., Thorne, S. W., and Björkman, O. (1972). Photochemical reactions of chloroplasts and components of the photosynthetic electron transport chain in two rainforest species. *Carnegie Inst. Wash. Year Book* **71**, 107–14.

Boardman, N. K., Björkman, O., Anderson, J. M., Goodchild, D. J., and Thorne, S. W. (1974). Photosynthetic adaptation of higher plants to light intensity: relationship between chloroplast structure, composition of the photosystems and photosynthetic rates. In 'Proc. 3rd Int. Congr. Photosynthesis'. (Ed. M. Avron.) pp. 1809–27. (Elsevier: Amsterdam.).

Chow, W. S., and Anderson, J. M. (1987*a*). Photosynthetic responses of *Pisum sativum* to an increase in irradiance during growth. I. Photosynthetic activities. *Aust. J. Plant Physiol.* **14**, 1–8.

Chow, W. S., and Anderson, J. M. (1987*b*). Photosynthetic responses of *Pisum sativum* to an increase in irradiance during growth. II. Thylakoid membrane components. *Aust. J. Plant Physiol.* **14**, 9–19.

Chow, W. S., and Hope, A. B. (1987). The stoichiometries of supramolecular complexes in thylakoid membranes from spinach chloroplasts. *Aust. J. Plant Physiol.* **14**, 21–8.

Chow, W. S., Qian, L., Goodchild, D. J., and Anderson, J. M. (1988). Photosynthetic acclimation of *Alocasia macrorrhiza* (L.) G. Don to growth irradiance: structure, function and composition of chloroplasts. *Aust. J. Plant Physiol.* **15**, 107–22.

Chu, Z. X., and Anderson, J. M. (1984). Modulations of the light-harvesting assemblies of a shade plant, *Alocasia macrorrhiza*. *Photobiochem. Photobiophys.* **8**, 1–10.

Evans, J. R. (1987*a*). The dependence of quantum yield on wavelength and growth irradiance. *Aust. J. Plant Physiol.* **14**, 69–79.

Evans, J. R. (1987*b*). The relationship between electron transport components and photosynthetic capacity in pea leaves grown at different irradiances. *Aust. J. Plant Physiol.* **14**, 157–70.

Ghirardi, M. L., McCauley, S. W., and Melis, A. (1986). Photochemical apparatus organization in the thylakoid membrane of *Hordeum vulgare* wild type and chlorophyll *b*-less chlorina f_2 mutant. *Biochim. Biophys. Acta* **851**, 331–9.

Goodchild, D. J., Anderson, J. M., and Andersson, B. (1985). Immunocytochemical localization of the cytochrome *b*/*f* complex of chloroplast thylakoid membranes. *Cell Biol. Int. Rep.* **9**, 715–21.

Goodchild, D. J., Björkman, O., and Pyliotis, N. A. (1972). Chloroplast ultrastructure, leaf anatomy and soluble protein in rainforest species. *Carnegie Inst. Wash. Year Book* **71**, 102–7.

Goodchild, D. J., Duniec, J. T., and Anderson, J. M. (1983). The lateral displacement of intramembraneous particles in chloroplast membranes as a function of light intensity. *FEBS Lett.* **154**, 243–6.

Goryshina, T. K. (1980). Structural and functional features of the leaf assimilatory apparatus in plants of a forest–steppe oakwood. *Acta Ecol./Ecol. Plant.* **1**, 47–54.

Graan, T., and Ort, D. R. (1984). Quantitation of the rapid electron donors to P700, the functional plastoquinone pool, and the ratios of the photosystems in spinach chloroplasts. *J. Biol. Chem.* **259**, 14003–10.

Haehnel, W. (1984). Photosynthetic electron transport in higher plants. *Annu. Rev. Plant Physiol.* **35**, 659–93.

Herrmann, R. G., Westhoff, P., Alt, J., Tottgen, J., and Nelson, N. (1985). Thylakoid membrane proteins and their genes. In 'Molecular Form and Function of the Plant Genome'. (Eds L. van Vloten Doting, G. S. P. Groot and T. C. Hall.) pp. 233–56. (Plenum: New York.)

Hill, R., and Bendall, F. (1960). Function of the two cytochrome components in chloroplasts: a working hypothesis. *Nature* (*London*) **86**, 136–7.

Izawa, S., and Good, N. E. (1966). Effect of salts and electron transport on the conformation of isolated chloroplasts. II. Electron microscopy. *Plant Physiol.* **41**, 544–52.

Kyle, D. J., Kuang, T.-Y., Watson, J. L., and Arntzen, C. J. (1984). Movement of a sub-population of the light harvesting complex (LHCII) from grana to stroma lamellae as a consequence of its phosphorylation. *Biochim. Biophys. Acta* **765**, 89–96.

Kyle, D. J., Staehelin, L. A., and Arntzen, C. J. (1983). Lateral mobility of the light-harvesting complex in chloroplast membranes controls excitation energy distribution in higher plants. *Arch. Biochem. Biophys.* **222**, 527–41.

Leong, T. Y., and Anderson, J. M. (1984*a*). Adaptation of the thylakoid membrances of pea chloroplasts to light intensities. I. Study on the distribution of chlorophyll-protein complexes. *Photosynth. Res.* **5**, 105–15.

Leong, T. Y., and Anderson, J. M. (1984*b*). Adaptation of the thylakoid membranes of pea chloroplasts to light intensities. II. Regulation of electron transport capacities, electron carriers, coupling factor (CF_1) activity and rates of photosynthesis. *Photosynth. Res.* **5**, 117–28.

Leong, T. Y., and Anderson, J. M. (1986). Light-quality and irradiance adaptation of the composition and function of pea thylakoid membranes. *Biochim. Biophys. Acta* **850**, 57–63.

Lichtenthaler, H. K. (1981). Adaptation of leaves and chloroplasts to high quanta fluence rates. In 'Photosynthesis'. (Ed. G. Akoyunoglou.) Vol. 6, pp. 278–88. (Balaban Int. Sci. Serv.: Philadelphia.)

Lichtentaler, H. K., and Meier, D. (1984). Regulation of chloroplast photomorphogenesis by light intensity and light quality. In 'Chloroplast Biogenesis'. (Ed. R. J. Ellis.) pp. 261–81. (Cambridge University Press: Cambridge.)

Melis, A. (1984). Light regulation of photosynthetic membrane structure, organization and function. *J. Cell. Biochem.* **24**, 271–85.

Melis, A., and Harvey, G. W. (1981). Regulation of photosystem stoichiometry, chlorophyll *a* and chlorophyll *b* content and relation to chloroplast ultrastructure. *Biochim. Biophys. Acta* **637**, 138–45.

Miller, K. R., and Staehelin, L. A. (1976). Analysis of the thylakoid outer surface. Coupling factor is limited to unstacked membrane regions. *J. Cell Biol.* **68**, 30–47.

Sane, P. V., Goodchild, D. J., and Park, R. B. (1970). Characterization of chloroplast photosystems 1 and 2 separated by a non-detergent method. *Biochim. Biophys. Acta* **216**, 162–78.

Simpson, D. J. (1981). The ultrastructure of barley thylakoid membranes. In 'Photosynthesis'. (Ed. G. Akoyunoglou.) Vol. 3, pp. 15–22. (Balaban Int. Sci. Serv.: Philadelphia.)

Staehelin, L. A., and Arntzen, C. J. (1983). Regulation of chloroplast membrane function: protein phosphorylation changes the spatial organization of membrane components. *J. Cell Biol.* **97**, 1327–37.

Terashima, I., and Inoue, Y. (1985). Vertical gradient in photosynthetic properties of spinach chloroplasts dependent on intra-leaf light environment. *Plant Cell Physiol.* **26**, 781–5.

Thornber, J. P. (1986). Biochemical characterization and structure of pigment–proteins of photosynthetic organisms. In 'Photosynthesis III. Photosynthetic Membranes and Light-harvesting Systems'. (Eds L. A. Staehelin and C. J. Arntzen.) Encycl. Plant Physiol. New Ser., Vol. 19, pp. 98–142. (Springer-Verlag: Berlin.)

Vallon, O., Wollman, F. A., and Olive, J. (1986). Lateral distribution of the main protein complexes of the photosynthetic apparatus in *Chlamydomonas reinhardtii* and in spinach: an immunocytochemical study using intact thylakoid membranes and a PS II enriched membrane preparation. *Photobiochem. Photobiophys.* **12**, 203–20.

Whitmarsh, J., and Ort, D. R. (1984). Stoichiometries of electron transport complexes in spinach chloroplasts. *Arch. Biochem. Biophys.* **231**, 378–89.

Wild, A. (1979). Physiology of photosynthesis in higher plants. The adaptation of photosynthesis to light intensity and light quality. *Ber. Dtsch. Bot. Ges.* **92**, 341–64.

Wild, A., Höpfner, M. Rühle, W., and Richter, M. (1986). Changes in the stoichiometry of photosystem II components as an adaptive response to high-light and low-light conditions during growth. *Z. Naturforsch.* **41c**, 597–603.

The Molecular Mechanism of Photoinhibition — Facts and Fiction

Christa Critchley

Botany Department, Australian National University, G.P.O. Box 4, Canberra, A.C.T. 2601, Australia; present address: Botany Department, University of Queensland, St Lucia, Qld 4067, Australia.

> 'In one aspect chloroplasts proved to be a simpler and more convenient material to study than whole algae. The latter, especially if cultured in strong light, are equipped with a mechanism which restores the damage of photoinhibition and thus complicates matters. No such mechanism was noticed in chloroplasts.'
>
> Kok *et al.* (1965)

Abstract

In this paper, the evidence supporting two different models for the molecular mechanism of photoinhibition is discussed. One hypothesis centres around the suggestion that photoinhibition is due to the loss of the herbicide-binding D1 polypeptide of photosystem II. The other model suggests that damage to a functional group in the reaction centre is the primary cause of photoinhibition. In order to put the apparent controversy into context, recent developments in our understanding of the structure and function of the photosystem II reaction centre are described. Interpretation and judgement of all available evidence suggest primary photoinhibitory damage to be incurred by the reaction-centre chlorophyll P680 destabilising the apoprotein(s) and eventually resulting in their proteolytic degradation and removal from the photosystem II complex and the thylakoid membrane.

Introduction

Some aspects of the phenomenology of photoinhibition at the level of the whole organism have been described in other papers in this volume. The underlying mechanism is not well understood and the subject of much research activity and controversy. The paper by Cleland (1988) in this volume goes a long way towards providing evidence in favour of a molecular mechanism, which suggests the primary site of damage to be the reaction-centre chlorophyll P680. In this contribution, that and other recent concepts concerning light-induced damage (photoinhibition) to chloroplast membranes and the attempts to discern the initial site(s) and biochemical mechanism(s) of the damage are discussed.

In the pioneering work of Myers and Burr (1940) using the unicellular green alga *Chlorella*, and of Trebst (1962) and Jones and Kok (1966*a*) with isolated chloroplasts, high-light-induced damage was shown to occur in two stages: the first involved an inhibition of photosynthetic activity with no loss of chlorophyll, and the second was characterised by a severe photo-oxidation or photobleaching of chlorophyll and general cell disruption. Photoinhibition is the term now used for the first stage of photodamage (Powles 1984) and I shall deal only with this phenomenon here.

Photosystem II as the Primary Target

It was found very early that the action spectrum of light-induced damage to photosynthesis was coincident with the chlorophyll absorption spectrum (Jones and Kok 1966*b*). This indicated that the thylakoid membrane system and the 'light reactions' were involved. Indeed, many *in vivo* measurements as well as experiments using isolated chloroplasts *in vitro* demonstrated that this was the case.

0310–7841/88/010027$03.00

Chlorophyll Fluorescence

Chlorophyll fluorescence is a sensitive albeit ambiguous indicator of any perturbation in the chloroplast membrane and its functions. Changes in both the fluorescence yield and kinetics of fluorescence transients in intact leaves (Critchley and Smillie 1981; Powles and Björkman 1982), algal cells (Kyle *et al.* 1984; Whitelam and Codd 1984), isolated intact chloroplasts (Barényi and Krause 1985) and thylakoids (Critchley 1981*a*,

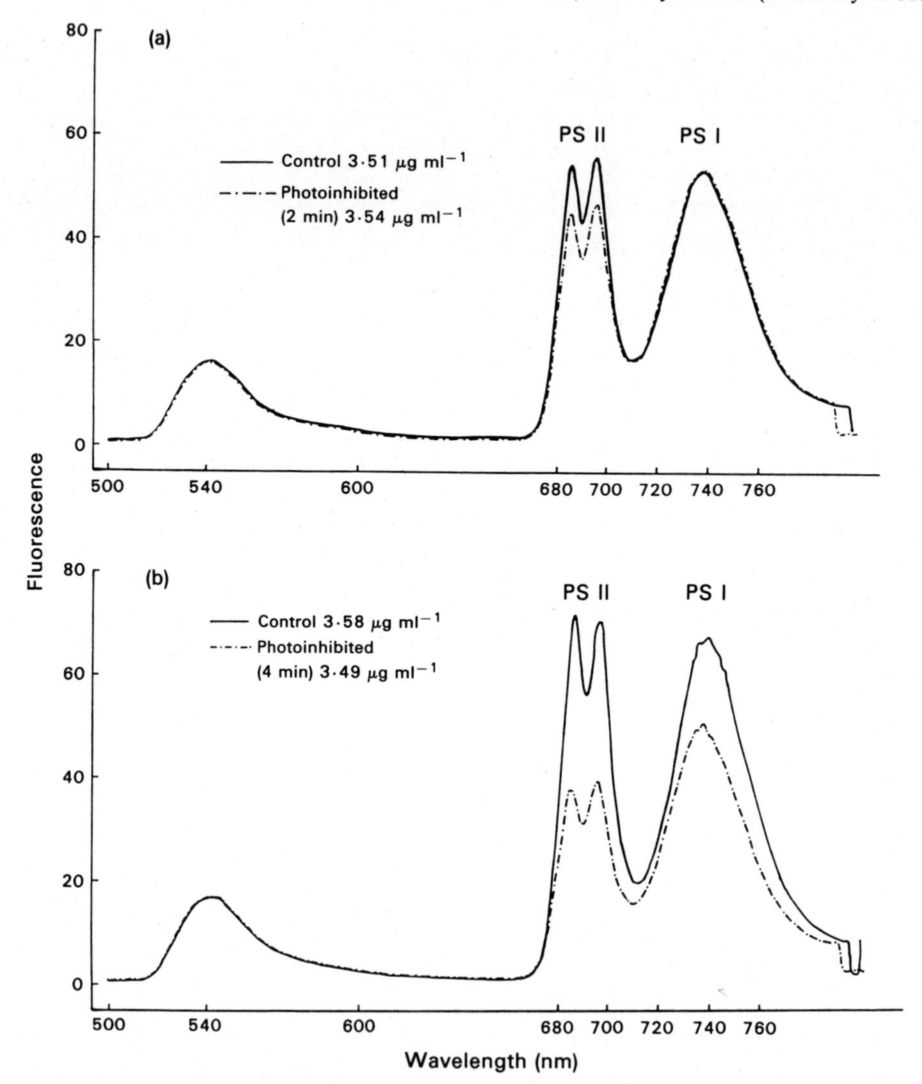

Fig. 1. 77K emission spectra ± 2 min photoinhibition (*a*) and 77K emission spectra ± 4 min photoinhibition (*b*) of isolated spinach thylakoids. Normalised to Na-fluorescein at 540 nm.

1981*b*; Cleland and Critchley 1985; Krause *et al.* 1985; Nedbal *et al.* 1986) exposed to excess photon irradiance suggest that the observed decrease in quantum yields for a given photosynthetic reaction may be the result of a disruption of the photosynthetic membrane. Algal cells and intact leaves show similar symptoms following excess light treatment, indicating that CO_2 deprivation from stomatal closure is not the cause of the reduced CO_2 fixation rates. Furthermore, the similarity of these responses in isolated chloroplasts shows that the loss of carbon fixation is not due primarily to any metabolic disruption localised in the cytoplasm or the mitochondria. Thus, the primary target of photoinhibition damage is the chloroplast thylakoid membrane.

Fluorescence emission measurements at liquid nitrogen temperatures (77K) can be used to characterise the perturbations in the pigments. Alterations associated with photosystem I can be distinguished from those of PS* II since their emission wavelengths are somewhat different (Fig. 1). In broken spinach chloroplasts, fluorescence emission from PS II occurs at wavelengths of 685 and 695 nm and from PS I at 735 nm. As shown in Fig. 1*a*, thylakoids that had been exposed to excess light of short duration exhibited a substantial decrease in the fluorescence yield from PS II with little effect on that from PS I. Barényi and Krause (1985), however, observed a marked decrease in PS I fluorescence also, consistent with results obtained by Cleland and Critchley (1985 and Fig. 1*b*) following longer exposure to high light. Powles *et al.* (1979) and Powles and Björkman (1982) observed a major effect on the 77K F_{692} induction transient with only a slight perturbation occurring in the PS I F_{735} transient following exposure of *Nerium oleander* leaves to high light levels.

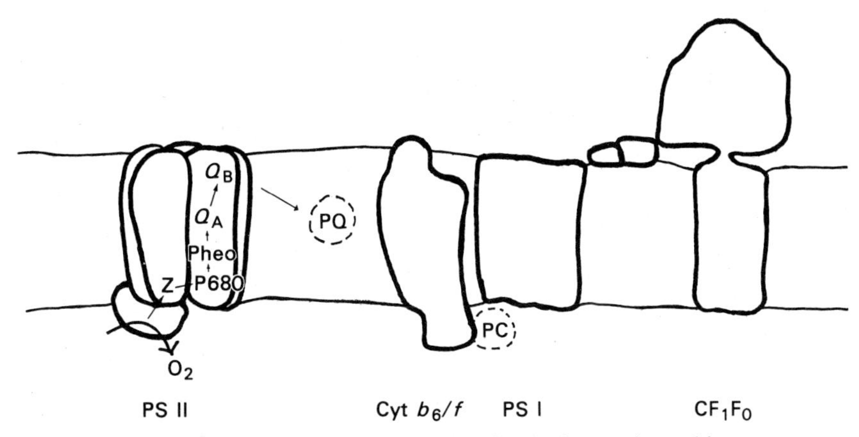

Fig. 2. Current model of the thylakoid membrane with the four major multicomponent complexes designated PS II, PS I, cytochrome b_6/f complex and ATP synthase. PS II is represented in more detail, including the flow of electrons through this complex.

Electron Transport

An alternative method to distinguish between effects on PS I and PS II involves isolating chloroplast membranes from treated leaves and characterising the partial reactions of photosynthetic electron transport by using assays specific to either PS I or PS II. In such experiments using thylakoids isolated from photoinhibited cucumber leaves, Critchley (1981*a*) showed that PS II activity was significantly reduced compared to control samples, whereas PS I activity did not change. Barényi and Krause (1985) on the other hand observed substantial inhibition of PS I electron transport reactions in experiments with isolated intact chloroplasts or thylakoids. These discrepancies may be due to the photoinhibition treatments being administered on intact plants or isolated systems, respectively. There is, however, general consensus that PS II is more sensitive and the primary site of damage.

PS II is one of four multicomponent complexes in the thylakoid membrane. It catalyses the oxidation of water, producing molecular oxygen, and reduces plastoquinone (plastoquinol) in several steps (Fig. 2). The most important of those is

*Abbreviations used: CF_1, coupling factor 1; DCMU, 3′-(3,4-dichlorophenyl)-1, 1′-dimethylurea; P680, reaction centre in photosystem II; Pheo, pheophytin; PS, photosystem; SDS–PAGE, sodium dodecyl sulfate–polyacrylamide gel electrophoresis.

the charge separation event mediated by quanta transferred to the reaction-centre chlorophyll, generating the oxidising power to hydrolyse water and the reducing power to transfer an electron to plastoquinone. This step is an obvious target for photoinhibition. However, in order to determine the biochemical mechanism and the primary site of damage within this complex, a reasonably detailed knowledge of the

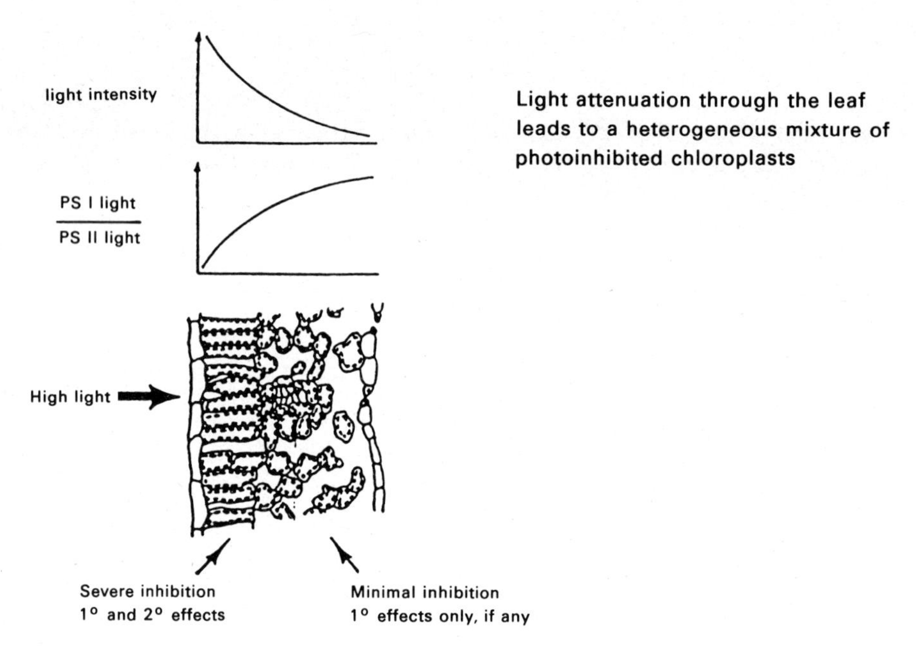

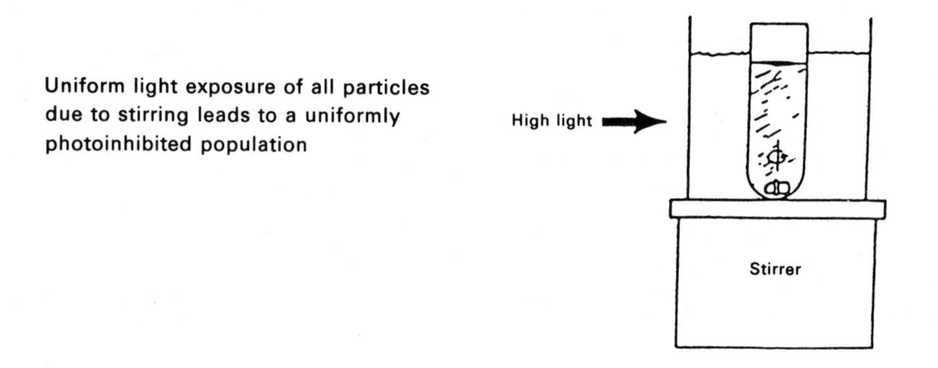

Fig. 3. Geometry of photoinhibition set-ups *in vivo* (*a*) and *in vitro* (*b*) and possible consequences. 1°, Primary effect; 2°, secondary effect.

molecular architecture of PS II as well as the physical principles of excitation energy transfer and electron transport are required. Attempts to characterise photoinhibition in thylakoid membranes isolated from leaves photoinhibited *in vivo* are problematic because of light attenuation across the many-cell-layer thickness of most leaves (Fig. 3*a*). This effect was demonstrated by Powles and Björkman (1982) when the characteristic loss of the F_{692} PS II fluorescence signal was observed only on the exposed surface of a *Nerium oleander* leaf. The self-shaded surface exhibited little or no effect. Preparations of chloroplast membranes from such a treated leaf presumably contain a heterogeneous mixture of severely damaged and almost unaffected membranes (Critchley 1981*a*, 1981*b*). In such samples, it may be difficult to distinguish primary from secondary and tertiary effects of damage at the level of the PS II complex. Alternative, somewhat less complex experimental systems are unicellular algae in a stirred culture or suspensions of protoplasts, isolated chloroplasts or thylakoid membranes (Fig. 3*b*). Such a set-up will ensure uniform exposure of all the photosynthetic membranes to the light, allowing for less ambiguous interpretation of the results, although the responses may not be the same, kinetically or mechanistically.

Structure of the Photosystem II Reaction-centre Complex

Before discussing the various lines of evidence for the primary target of light damage within PS II, it is necessary to review the present state of understanding of the complex's substructure itself. The molecular architecture of PS II has not been examined at a high degree of resolution since a complex this large has not yet been crystallised. Our present model of the PS II substructure can therefore only be inferred from pertinent biochemical structure–function analyses, specific mutant work and, perhaps less convincingly, from homologies with similar structures for which X-ray crystallographic data are available, namely the purified reaction-centre complex of the purple sulfur bacterium *Rhodopseudomonas viridis* (see Satoh 1985, for review).

Polypeptide Composition

Biochemically purified functional PS II reaction-centre complexes of photosynthetic eukaryotes and cyanobacteria consist of approximately eight polypeptides. Three of these polypeptides are extrinsic membrane proteins mainly involved in the oxygen-evolving function of PS II (Critchley 1985). The other five polypeptides are hydrophobic, intrinsic membrane proteins including two chlorophyll-binding proteins with approximate molecular weights of 47 and 43 kDa, two colourless, diffusely staining (by Coomassie blue) proteins at approximately 34 kDa (D2) and 32 kDa (D1), and a low-molecular-weight *b*-type cytochrome with a spectral absorption maximum at 559 nm (cyt *b*559). The roles of all of these polypeptides within PS II are ill-defined and at present we can only speculate as to their function.

The 47/43 kDa Polypeptide Model

Some early biochemical evidence suggested that the 47 kDa protein may bind the reaction-centre chlorophylls P680 and pheophytin (Pheo). Nakatani (1983) and Green and Camm (1984) have shown that purified 47 kDa protein exhibits a low-temperature fluorescence (77K) at 695 nm, whereas the 43 kDa protein fluorescence peak occurs at 685 nm. Breton (1982) has used linear and circular dichroism techniques to show that the F_{695}, F_{685} and F_{735} signals arise from chromophores oriented parallel to the plane of the membrane. In addition, he showed that the bulk of the antenna chlorophylls was also so oriented, whereas Pheo was oriented perpendicular to the plane of the membrane. The conclusion from these data was that the F_{695} signal arises from a recombination event involving Pheo and, therefore, from a reaction-centre chlorophyll. This interpretation is not conclusive, however, because a small population of chlorophylls not associated with the reaction centre may also be oriented perpendicular to the plane of the membrane, their signal may be masked by the bulk chlorophylls, and the correlation of the Pheo orientation with the F_{695} signal may therefore be misleading. If we assume, however, that the F_{695} signal does emanate from the reaction centre, then the observation that chromophores associated with the 47 kDa polypeptide emit the F_{695} signal (Nakatani 1983) would indicate that this protein represents the reaction centre. Further support for this view was provided when a light-induced optical change at 685 nm (characteristic of Pheo) was identified in the purified 47 kDa protein preparation (Nakatani *et al.* 1984).

If the 47 kDa protein represents the apoprotein of the PS II reaction centre, what then is the role of the 43 kDa protein? Breton (1983) suggests that this protein is an antennae-containing polypeptide in very close proximity to the reaction centre such that the energy generated in the recombination of the charged radical pair may be transferred to its antennae chlorophylls, and that it is the de-excitation of these antennae chlorophylls that results in the fluorescence emission at 685 nm.

In neither of these studies was any attempt made to detect 'contamination' by D1 or D2, because these proteins only later became regarded as possible candidates for the apoprotein(s) of the PS II reaction centre. However, more recent data obtained by Satoh (1986) suggest that oxygen-evolving PS II particles from a thermophilic cyanobacterium retain their capacity for primary charge separation and Q_A reduction after depletion of two hydrophobic polypeptides of 31 and 28 kDa. Vermaas *et al.* (1986) reported the isolation of functionally inactive PS II particles from a genetically engineered mutant of *Synechocystis* 6803 in which a 47 kDa polypeptide is absent.

The D1/D2 Model

The 32 kDa protein (D1) has been comparatively well studied because of its herbicide-binding capacity (Kyle 1985*a*) and its high turnover rates (Mattoo *et al.* 1984). The role of the protein involves plastoquinone reduction using electrons supplied by the first stable electron-acceptor quinone Q_A and originating from the reaction-centre chlorophyll P680 (see Fig. 2).

The role of the 34 kDa protein (D2), however, is not at all clear. Its amino acid sequence exhibits a remarkable degree of homology with that of D1, and since D1 is probably a quinone-binding protein, it has been suggested that the D2 protein may also be one (Hearst and Sauer 1984). Babcock *et al.* (1984) have suggested that the ESR signal associated with the primary donor Z on the oxidising side of PS II (signal II) is typical of that expected from a quinone cation. Consequently, a quinone may be involved also in electron donation to P680 and the D2 protein may bind or stabilise this quinone cation.

Consistent with some data and interpretations discussed above, the PS II reaction centre may be represented as shown in Fig. 4*a*. An alternative (Fig. 4*b*) may be more appropriate if sequence homology arguments, currently much in vogue, can be substantiated by experimental evidence. The model in Fig. 4*b* is based on a comparison of the putative PS II complex of photosynthetic eukaryotes and cyanobacteria with that of the purple sulfur bacterium *R. viridis*, for which the crystal structure has been reported (Deisenhofer *et al.* 1984). The reaction centre of these bacteria bears some resemblance to PS II in terms of functionality since a similar radical pair results in a charge separation followed by electron transfer to primary and secondary quinone acceptors. The attractive functional analogy of the reducing side of the bacterial reaction centre with PS II, however, does not hold for the oxidising components: in PS II, water is oxidised whereas in the bacterial reaction centre a *c*-type cytochrome is oxidised. Also unlike PS II, the bacterial reaction centre consists of only three integral membrane proteins, the H, M and L subunits. The crystal structure of the complex has localised the radical-pair chromophores as well as the Q_A- and Q_B-binding sites in domains defined by both the L and M subunits. This contention is supported by biochemical measurements of Brown *et al.* (1984) and de Vitry and Diner (1984) who have shown that the L subunit becomes labelled when using the Q_B-binding site-specific [^{14}C]azidoatrazine. It also explains the earlier observation by Marinetti *et al.* (1979) that the quinone analogue azidoanthraquinone specifically labels the M subunit.

Primary sequence homologies are not necessarily the determinant of structural (and functional) homology (Hudson *et al.* 1987). Some authors, however, have suggested that there is a significant amino acid sequence homology (28%) between the D1 protein of PS II and the L subunit, and between the D2 protein and the M subunit, i.e. 32% (Hearst and Sauer 1984). This homology is somewhat more striking when comparing polypeptide domains by using hydropathy plots of these sequences. Homology arguments put forward originally by Hearst and Sauer (1984) and now espoused by many

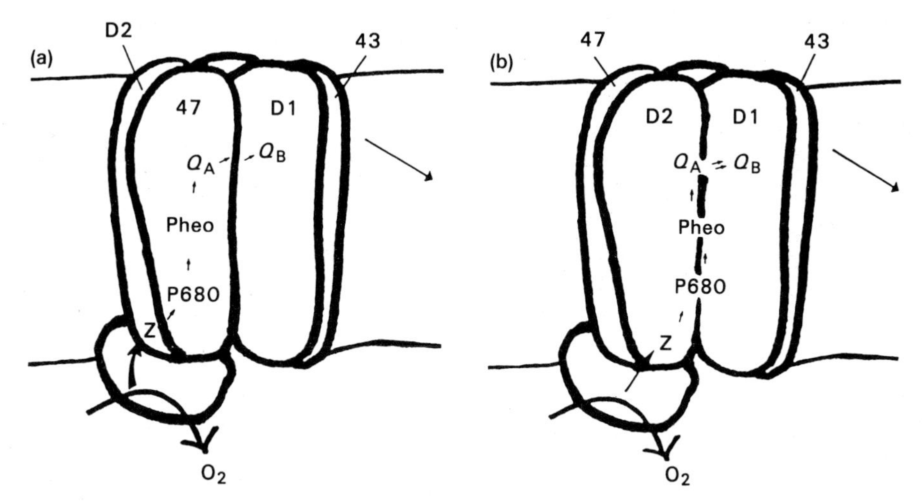

Fig. 4. PS II model with 47 / 43 kDa polypeptides as reaction-centre apoproteins (*a*) and PS II model with D1 / D2 (32 / 34 kDa) polypeptides as reaction-centre apoproteins (*b*). The numbers indicating molecular weights of individual polypeptides are those most commonly obtained from SDS–PAGE, which differ somewhat in most instances from those obtained by deduction from gene sequences (Critchley and Bottomley 1986).

others suggest that the D2 and D1 proteins may, in fact, be the functional equivalent of the M and L subunits, respectively, of the bacterial reaction centre. Trebst and Depka (1985) have speculated that the D2 and D1 proteins would represent the reaction centre of PS II and that the 47 and 43 kDa chlorophyll-binding proteins may only be peripherally but tightly associated proteins binding antennae but not reaction-centre chlorophylls (as shown in Fig. 4*b*, but see a recent review, Evans 1987). It should be pointed out, however, that similar primary sequence homologies and hydropathy-plot overlaps can be found when comparing M and H subunits with the 47 kDa polypeptide.

Very recent results have provided biophysical and biochemical evidence that the D1 / D2 pair may constitute the protein matrix for the PS II reaction centre. Reaction-centre complexes containing only D1, D2 and cyt *b*559 have been isolated and shown to undergo flash-induced absorbance changes indicative of the radical pair, $P680^+$ $Pheo^-$ (Danielius *et al.* 1987; Nanba and Satoh 1987), as well as attain the EPR-detectable light-induced triplet state (Okamura *et al.* 1987).

The crystal structure of the bacterial reaction centre indicates that the transmembrane helices of the L and M subunits are tightly interwoven. If the analogy holds for D1 and D2, then damage or destruction of D1 would undoubtedly render the reaction centre non-functional. According to Cleland and Critchley (1985) and Cleland *et al.* (1986), such inactivation of the reaction centre is the case during photoinhibition. Using trypsin treatments, which selectively abolish the quinone- and herbicide-binding functions of D1 (Renger 1976), Arntz and Trebst (1986) showed that most of these functions can be removed with little loss of sensitivity to photoinhibition.

Photoinhibition Damage within the Photosystem II Complex

Site

In the previous section, I discussed two current models of PS II structure in order to put into context data pertaining to the site(s) of photoinhibitory damage within the complex itself. The multiple functions of this complex can be assayed separately by the use of specific biochemical manipulations, biophysical measurements, and/or electron donors and acceptors. For example, the water-oxidising function of PS II can be bypassed using artificial electron donors such as diphenyl carbazide, hydroxylamine or catechol. Until recently it was thought that electrons can leave PS II from either the Q_B site using quinone-based acceptors (anthraquinone or benzoquinone) or from the Q_A site using the soluble polyanion, silicomolybdate (Girault and Galmiche 1974). Silicomolybdate was suggested to disrupt the complex physically in a time-dependent fashion, displacing the Q_B protein and exposing an otherwise hidden Q_A site to be oxidised by the polyanion. New results (Graan 1986) indicate, however, that silicomolybdate accepts electrons from Q_B also and that it displaces DCMU from the herbicide-binding site (Boger 1982), thereby rendering the assay apparently DCMU-insensitive. It appears that only mild trypsinisation abolishes the Q_B-binding site and exposes the Q_A site to oxidation by other soluble acceptors such as ferricyanide (Renger 1976).

Overall PS II activity with water as the endogenous donor and diaminodurene as the acceptor was lost following photoinhibition of *Chlamydomonas reinhardtii* cells (Kyle *et al.* 1984). However, very little photoinhibition-induced loss of activity was seen with silicomolybdate as acceptor. Kyle *et al.* (1984) have used this observation to argue that the primary site of damage to PS II in photoinhibited algal cells is the Q_B site. Furthermore, they reported a loss in the number of atrazine-binding sites in the membranes of photoinhibited cells, confirming that the Q_B protein is affected by the high-light treatment. In contrast to these results, Tytler *et al.* (1984) reported that photoinhibition of the unicellular cyanobacterium *Microcystis aeruginosa* resulted in damage to the reaction centre itself, since a light-induced loss of both diaminodurene- and silicomolybdate-stimulated PS II activity was observed. The latter results are consistent with those obtained with isolated thylakoids (Cleland and Critchley 1985; Ohad *et al.* 1985). It is difficult to reconcile these contrasting data and although kinetic studies of the damage to these two activities in intact algal cells (Kyle *et al.* 1985) and in isolated thylakoids (Cleland and Critchley 1985; Ohad *et al.* 1985) attempted to clarify the issue of the primary site of damage, they failed to do so because none of the correlations could be shown to be linear. Furthermore, in the light of the results by Boger (1982) and Graan (1986) concerning the acceptor site for silicomolybdate, this experimental approach is no longer useful.

The kinetics of the chlorophyll *a* fluorescence (which at room temperature arises almost exclusively from PS II) may be an indicator of the site of a perturbation. In order to study the kinetics involving the PS II complex alone in a photosynthetic membrane, the exit of electrons from the complex has to be blocked using an inhibitor (typically DCMU), or the temperature has to be very low (typically 77K). Under these circumstances, PS II fluorescence exhibits an instantaneous rise to a low level (F_0) followed by a relatively rapid rise to a maximum (F_m). The difference between F_0 and F_m represents the variable component of fluorescence (F_v), which arises from the energy (emitted as light) generated from the charge recombination of the radical pair (Klimov *et al.* 1980). If the reduced Pheo can donate its electron to Q_A, no or very little charge recombination occurs and fluorescence is low. When Q_A is already reduced, Pheo can no longer donate its electron to Q_A^-, recombination with $P680^+$ occurs, and a lower-energy photon is released (recombination luminescence or fluorescence). Consequently, the variable fluorescence rise reflects Q_A reduction in the DCMU-inhibited or low-temperature complex.

Perturbations resulting in a lowered variable fluorescence indicate that either no charge separation is taking place or that the radical pair is de-excited by a mechanism that does not involve recombination luminescence. Perturbations leading to an increased variable fluorescence have almost always been shown to be due to a blockage beyond the Q_A site (similar to adding DCMU). Detailed examination of the variable fluorescence (Butler 1977) demonstrates that the overall fluorescence yield (ϕF) is affected by the rate constant for fluorescence (k_F), and the competing processes of non-radiative thermal deactivation (k_D) and energy transfer to an adjacent non-radiative chlorophyll (k_T). Any increase in the rate constants for either thermal deactivation or energy transfer would effectively decrease k_F, and thereby fluorescence yield. Although there is no evidence to either support or refute these possibilities, it is conceivable that light-induced perturbations might alter chromophore orientations in a way that would influence these rate constants.

Two alternative explanations of the reduced ϕF have been suggested, both of which are speculative and without any experimental support. Renger *et al.* (1984) have noted that ADRY reagents added to isolated thylakoids induce a rapid stable fluorescence quenching due to the formation of the reaction-centre state (P680.Pheo$^-$). In such a situation, Pheo$^-$ cannot recombine with P680 to give the fluorescence observed at 695 nm. Kyle *et al.* (1984) have suggested that, by analogy with Renger's model, the photoinhibited PS II centre still functional in the water oxidation component can reduce $P680^+$ while stabilising Pheo$^-$, thereby forming a reaction-centre state (P680.Pheo$^-$) similar to that produced with the ADRY reagents. An alternative by Cleland and Critchley (1985) suggested that an efficient fluorescence quencher is generated during photoinhibition, which may be equivalent to the 'T' component described by Den Haan *et al.* (1973). The 'T' component may be equivalent to Pheo$^-$ and the quenching may, therefore, be attributed to the formation and stabilisation of the same reaction-centre state described above (P680.Pheo$^-$). Unfortunately, the interpretation of the loss of variable fluorescence is quite complex, and inferences as to the site of damage based on fluorescence parameters are open to different interpretations.

The extinction coefficient of a reduced quinone is somewhat lower than that of its oxidised form and therefore the reduction of Q_A can be measured directly by an absorption change at 320 nm (van Gorkom 1974; Melis and Brown 1980). Thus, perturbations before the Q_A site can be identified by the absence of, or alteration in, the kinetics of the light-induced A_{320} change. Cleland *et al.* (1986, 1987) have provided direct evidence for such an inhibition of photoreduction of Q_A.

Using the types of measurements described above, there is still no consensus about the primary site of damage within the complex. There is some evidence to suggest that photoinhibitory damage may manifest itself as accelerated turnover of the D1 protein. On the other hand, the decrease in the variable fluorescence associated with photoinhibitory damage in all systems studied so far suggests that the damage is located in or very close to the reaction centre. The latter suggestion accounts for the fluorescence data, which are particularly important in view of the recent measurements by Björkman and Demmig (1987) showing the linear correlation between fluorescence (F_v/F_m) and quantum yield in a wide range of higher plants measured under various photoinhibitory conditions. In photoinhibited membrane preparations in which both PS II activity and the variable fluorescence are reduced by 80–90%, the activity cannot be recovered by the addition of artificial electron donors (Critchley 1981*a*, 1981*b*; Arntzen *et al.* 1984; Tytler *et al.* 1984). Inactivation of the water-oxidising function also results in low activities and quenched fluorescence. In this case, however, activity and fluorescence can be reconstituted using artificial donors such as DPC. Donor experiments with negative results in photoinhibited samples are not conclusive and cannot exclude damage to the water-oxidising component. Theg *et al.* (1986) have shown that photoinhibition of chloride-deficient or Tris-washed chloroplasts, i.e. chloroplasts devoid of oxygen-evolving activity, resulted in a lesion between Z and Q_A.

Oxidation of the primary electron donor Z to Z^+ by the radical pair results in an observable ESR signal at $g = 2{\cdot}0$ (signal II). The rate at which signal II disappears is proportional to the rate at which it can be reduced by the water oxidation components of PS II. Thus, a slow signal II decay indicates inhibition of oxygen evolution (and hence reduction of Z^+), whereas a fast signal II decay indicates normal reaction-centre activity. Disappearance of signal II would be indicative of an inability of the radical pair to oxidise Z or, in other words, loss of reaction-centre (charge separation) activity. ESR signal II measurements on chloroplast membranes isolated from photoinhibited algal cells indicate that although some of the signal II is lost, its kinetics are unaffected (Kyle *et al.* 1984). The decrease in signal amplitude in the photoinhibited samples probably indicates a decrease in the number of active reaction centres. This decrease, however, was only about 20% compared with an 80% loss in overall photochemical activity.

Most of the data discussed above suggest that the initial photodamage occurs in the reaction centre itself, presumably involving the chromophores P680 or Pheo, followed by destabilisation of the chromophore–protein association, leading to the disintegration of the apoprotein.

Mechanism

Whether or not oxygen plays a role in the mechanism underlying photoinhibition is an unresolved question, although the burden of the evidence at present rests with the view that it does not — or not in any mechanistic sense. Photoinhibition under anaerobic conditions in thylakoids appears to proceed in a very similar manner to that observed in aerobic conditions (Cleland and Critchley 1985; Arntz and Trebst 1986). Evidence for oxygen involvement in the damaging process has been presented by Cornic *et al.* (1982) and Krause *et al.* (1978) who have reported that the extent of photodamage is decreased as media oxygen concentration is lowered in the isolated chloroplast system.

In vivo experiments with intact leaves exposed to light in the absence of CO_2 (Powles *et al.* 1979) have shown that photoinhibition damage increases as the concentration of oxygen around the leaf is lowered. Krause and Laasch (1987) presented similar evidence in experiments with intact chloroplasts, and Osmond (1981) has suggested that photorespiration, a process that consumes oxygen, may be a metabolic mechanism for reducing the energy load on the photosynthetic light reactions when the reductive dark reactions are limiting. If this is the case, an increase in susceptibility to photoinhibition when lowering oxygen concentrations would be predicted and is indeed observed, and this process might outweigh any effects of changing oxygen concentrations on the local environment of PS II. Quite clearly the question is difficult to access experimentally because local O_2 concentrations are impossible to determine.

If oxygen is mechanistically involved, the active species may be the superoxide radical, O_2^-, the hydroxyl radical, .OH, or singlet oxygen, O_2^1. According to Halliwell (1984), the superoxide radical would mainly and primarily be expected to oxidise membrane lipids. One of the most damaging reactions of the hydroxyl radical is the removal of a hydrogen atom from a hydrocarbon side chain of a membrane lipid, starting off the chain reaction of lipid peroxidation. The polyunsaturated fatty acids of the chloroplast lipids are perfect targets and especially sensitive (Halliwell 1984). The same is true for singlet oxygen, which is generated by the transfer of energy from excited pigment molecules, for example chlorophyll *a* or *b*, onto O_2, and which can directly oxidise unsaturated fatty acids. Illumination of thylakoids *in vitro* causes photoinhibition and marked lipid peroxidation (Takahama and Nishimura 1975) and since the trapping of light by chlorophylls takes place in a membrane with a high internal O_2 concentration, it seems likely that singlet oxygen is formed *in vivo*. Unfortunately, there are no more recent reports on chloroplast membrane lipid peroxidation and photoinhibition — an area in urgent need of further experimental exploitation.

Repair

Careful analysis of the components that are repaired during the recovery process may shed some light on the question of the requirement for protein synthesis. The finding that protein synthesis is required for recovery (Greer *et al.* 1986) merely indicates that a functional unit has to be newly made and assembled. It says nothing at all about the nature of the damage in the first place. The damage may only involve the reactive prosthetic group, i.e. the chlorophyll or quinone, and not necessarily the protein itself, which is probably only the matrix for the active cofactor. Indeed, the synthesis of the polypeptide may not be the limiting factor in the recovery process at all. Ohad *et al.* (1984) were able to demonstrate that the physiological recovery in *Chlamydomonas* required chloroplast- but not nucleus-directed protein synthesis. This was also shown to be the case in several other species of microalgae (Setlik *et al.* 1984) and *Phaseolus vulgaris* (Greer *et al.* 1986). These observations support the biochemical data indicating that PS II is the primary target of photoinhibition, since all the intrinsic PS II polypeptides are chloroplast-encoded (Satoh 1985). Furthermore, the requirement for protein synthesis indicates that structural damage occurs during photoinhibition, which cannot be repaired *in situ* so that the damaged elements need to be replaced by the products of *de novo* protein synthesis. None of these observations has any immediate bearing on the molecular mechanism of the primary damage, and the various labelling experiments with the D1 protein (Ohad *et al.* 1985; Reisman and Ohad 1986) are not particularly revealing in this respect.

Indeed, the turnover of the Q_B protein under non-photoinhibitory conditions has been shown by several laboratories (Edelman *et al.* 1984; Kyle 1985*b*; Nedbal *et al.* 1986) to be light-intensity dependent although it saturates at comparatively low light. The capacity for synthesis of this protein is appropriately high since its mRNA is one of the most abundant and stable mRNA species in the chloroplast. The suggestion, arising from the results obtained with *Chlamydomonas* cells, but not consistent with a number of other data, that D1 (Q_B)-dependent function is lost before reaction-centre activity (Kyle *et al.* 1984), and that D1 is removed before the loss of reaction-centre activity (Ohad *et al.* 1985), is difficult to reconcile with the reaction-centre model of PS II depicted in Fig. 4*b*. If D1 was part of the reaction centre itself, then its removal as a consequence of photoinhibition, or indeed its fast turnover under non-photoinhibitory conditions, would certainly render the reaction centre highly unstable under any conditions.

Greer *et al.* (1986) have recently suggested that photoinhibition in the physiological sense and manifest as loss of photosynthetic activity and reduction in quantum yield, only occurs when the rate of damage exceeds the rate of repair. Although damage may increase with light intensity, under most conditions cellular repair mechanisms can match this rate of damage and there is no net loss of photosynthetic capacity, i.e. no photoinhibition. Many environmental factors can affect the light intensity required for photoinhibition by altering either the rate of damage or the rate of repair. For example, some higher plants (van Hasselt and van Berlo 1980) and some algae (Kyle 1985*b*) are significantly more photosensitive at low temperatures. This may possibly be due to the Q_{10} effect on protein synthesis and the recovery mechanism, but it may also depend on the kinetics of development of secondary damage.

The protein synthesis-requiring repair mechanisms are absent in isolated thylakoids, and the rate of light-induced PS II damage is accordingly rapid. Under these conditions, any damage and loss of prelabelled D1 protein from the membrane can be studied without the complicating effects of newly synthesised D1 protein insertion. No net loss of the 32 kDa polypeptide was observed during photoinhibition of spinach thylakoids derived from intact prelabelled chloroplasts (Cleland and Critchley 1985). The simple but revealing conclusion from these observations is that photoinhibition, fulfilling all other criteria in terms of its manifestations in activity, quantum yield and fluorescence decline, does not depend on the proteolytic breakdown of the D1, and probably any other, protein but is primarily due to modification of the functionality of the radical pair, P680 and Pheo.

Acknowledgments

Support from the National Research Fellowship Scheme and a CSIRO Extra-mural Grant is gratefully acknowledged. I wish to thank in particular Ms Joanne Perks for expert and patient handling of the manuscript.

References

Arntz, B., and Trebst, A. (1986). On the role of the Q_B protein of PS II in photoinhibition. *FEBS Lett.* **194**, 43–9.

Arntzen, C. J., Kyle, D. J., Wettern, M., and Ohad, I. (1984). Photoinhibition: a consequence of the accelerated breakdown of the apoprotein of the secondary electron acceptor of photosystem II. In 'Biosynthesis of the Photosynthetic Apparatus: Molecular Biology, Development and Regulation'. (Eds R. Hallick, L. A. Staehelin and J. P. Thornber.) UCLA Symposium Series No. 14, pp. 313–24. (A. R. Liss Inc.: New York.)

Babcock, G. T., Buttner, W. J., Ghanotakis, D. F., O'Malley, P. J., Yerkes, C. T., and Yocum, C. F. (1984). Structures and electron transfer reactions on the oxidizing side of photosystem II. In 'Advances in Photosynthesis Research. Vol. 1'. (Ed. C. Sybesma.) pp. 243–52. (Nijhoff/Junk: The Hague.)

Barényi, B., and Krause, G. H. (1985). Inhibition of photosynthetic reactions by light. A study with isolated spinach chloroplasts. *Planta* (*Berl.*) **163**, 218–26.

Björkman, O., and Demmig, B. (1987). Photon yield of O_2 evolution and chlorophyll fluorescence characteristics at 77K among vascular plants of diverse origins. *Planta* (*Berl.*) **170**, 489–504.

Boger, P. (1982). Replacement of photosynthetic electron transport inhibitors by silicomolybdate. *Physiol. Plant.* **54**, 221–4.

Breton, J. (1982). The 695 nm fluorescence (F_{695}) of chloroplasts at low temperature is emitted from the primary acceptor of photosystem II. *FEBS Lett.* **147**, 16–20.

Breton, J. (1983). The emission of chlorophyll in vivo: antenna fluorescence or ultrafast luminescence from the reaction center pigments. *FEBS Lett.* **159**, 1–5.

Brown, A. E., Gilbert, C. W., Guy, R., and Arntzen, C. J. (1984). Triazine herbicide resistance in the photosynthetic bacterium *Rhodopseudomonas sphaeroides*. *Proc. Natl Acad. Sci. U.S.A.* **81**, 6310–14.

Butler, W. (1977). Chlorophyll fluorescence: a probe for electron transfer and energy transfer. In 'Photosynthesis I. Photosynthetic Electron Transport and Photophosphorylation'. (Eds A. Trebst and M. Avron.) Encycl. Plant Physiol. New Ser., Vol. 5, pp. 149–67. (Springer-Verlag: Berlin.)

Cleland, R. E. (1988). Molecular events of photoinhibitory inactivation in the reaction centre of photosystem II. *Aust. J. Plant Physiol.* **15**, 135–50.

Cleland, R. E., and Critchley, C. (1985). Studies on the mechanism of photoinhibition in higher plants. II. Inactivation by high light of photosystem II reaction center function in isolated spinach thylakoids and O_2 evolving particles. *Photobiochem. Photobiophys.* **10**, 83–92.

Cleland, R. E., Critchley, C., and Melis, A. (1987). Alteration of electron flow around P_{680}: the effect of photoinhibition. In 'Progress in Photosynthesis Research'. (Ed. J. Biggins.) Vol. 4, pp. 27-30. (Martinus Nijhoff: Dordrecht.)

Cleland, R. E., Melis, A., and Neale, P. J. (1986). Mechanisms of photoinhibition: photochemical reaction center inactivation in system II of chloroplasts. *Photosynth. Res.* **9**, 79–88.

Cornic, G., Woo, K. C., and Osmond, C. B. (1982). Photoinhibition of CO_2 dependent O_2 evolution by intact chloroplasts isolated from spinach leaves. *Plant Physiol. (Bethesda)* **70**, 1310–15.

Critchley, C. (1981*a*). Studies on the mechanism of photoinhibition in higher plants. I. Effects of high light intensity on chloroplast activities in cucumber adapted to low light. *Plant Physiol. (Bethesda)* **67**, 1161–5.

Critchley, C. (1981*b*). The mechanism of photoinhibition in higher plants. In 'Photosynthesis IV'. (Ed. G. Akoyunoglou.) pp. 297–305. (Balaban Int. Sci. Services: Philadelphia.)

Critchley, C. (1985). The role of chloride in photosystem II. *Biochim. Biophys. Acta* **811**, 33–46.

Critchley, C., and Bottomley, W. (1986). Mutants as tools for the elucidation of photosynthetic processes. In 'Plant Gene Research. A Genetic Approach to Plant Biochemistry'. (Eds A. D. Blonstein and P. J. King.) pp. 55–72. (Springer-Verlag: Wien.)

Critchley, C., and Smillie, R. M. (1981). Leaf chlorophyll fluorescence as an indicator of high light stress (photoinhibition) in *Cucumis sativus* L. *Aust. J. Plant Physiol.* **8**, 133–41.

Danielius, R. V., Satoh, K., van Kan, P. J. M., Plijter, J. J., Nuijs, A. M., and van Gorkom, H. J. (1987). The primary reaction of photosystem II in the D1–D2–cytochrome *b*-559 complex. *FEBS Lett.* **213**, 241–4.

Deisenhofer, J., Epp, O., Miki, K., Huber, R., and Michel, H. (1984). X-ray structure analysis of a membrane protein complex. Electron density map at 3A resolution and a model of the chromophores of the photosynthetic reaction centre from *Rhodopseudomonas viridis*. *J. Mol. Biol.* **180**, 382–98.

Den Haan, G. A., Warden, T. T., and Duysens, L. N. M. (1973). Kinetics of the fluorescence yield of chlorophyll a_2 in spinach chloroplasts at liquid nitrogen temperature during and following a 16 μs flash. *Biochim. Biophys. Acta* **325**, 120–5.

Edelman, M., Mattoo, A. K., and Marder, J. B. (1984). Three hats of the rapidly-metabolized 32 kilodalton protein of the thylakoids. In 'Chloroplast Biogenesis'. (Ed. R. J. Ellis.) pp. 283–302. (Cambridge University Press.)

Evans, M. C. W. (1987). Plant reaction centre defined. *Nature (Lond.)* **327**, 284–5.

Girault, A., and Galmiche, J. M. (1974). Restoration by silicotungstic acid of DCMU-inhibited photoreactions in spinach chloroplasts. *Biochim. Biophys. Acta* **333**, 314–19.

Gorkom, H. J. van (1974). Identification of the reduced primary electron acceptor of photosystem II as a bound semiquinone anion. *Biochim. Biophys. Acta* **374**, 439–42.

Graan, T. (1986). The interaction of silicomolybdate with the photosystem II herbicide-binding site. *FEBS Lett.* **206**, 9–14.

Green, B. R., and Camm, E. L. (1984). Evidence that CP 47 (CPa-1) is the reaction centre of photosystem II. In 'Advances in Photosynthesis Research. Vol. 2'. (Ed. C. Sybesma.) pp. 95–8. (Nijhoff/Junk.: The Hague.)

Greer, D. H., Berry, J. A., and Björkman, O. (1986). Photoinhibition of photosynthesis in intact bean leaves: role of light and temperature, and requirement for chloroplast-protein synthesis during recovery. *Planta (Berl.)* **168**, 253–60.

Halliwell, B. (1984). Oxygen-derived species and herbicide action. *What's New Plant Physiol.* **15**, 21–4.

Hasselt, P. R. van, and Berlo, H. A. C. van (1980). Photooxidative damage to the photosynthetic apparatus during chilling. *Physiol. Plant.* **50**, 52–6.

Hearst, J. E., and Sauer, K. (1984). Protein sequence homologies between portions of the L and M subunits of reaction centres of *Rhodopseudomonas capsulata* and the Q_B-protein of chloroplast thylakoid membranes: a proposed relation to quinone-binding sites. *Z. Naturforsch. Sect. C. Biosci.* **39**, 421–4.

Hudson, G. S., Mason, J. G., Holton, T. A., Koller, B., Cox, G. B., Whitfeld, P. R., and Bottomley, W. (1987). A gene cluster in the spinach and pea chloroplast genome encoding one CF_1 and three CF_0 subunits of the H^+-ATP synthase complex, and the ribosomal protein S2. *J. Mol. Biol.* (In press.)

Jones, L. W., and Kok, B. (1966*a*). Photoinhibition of chloroplast reactions. I. Kinetics and action spectra. *Plant Physiol.* **41**, 1037–43.

Jones, L. W., and Kok, B. (1966*b*). Photoinhibition of chloroplast reactions. II. Multiple effects. *Plant Physiol.* **41**, 1044–9.

Klimov, V. V., Dolan, E., and Ke, B. (1980). EPR properties of an intermediary electron acceptor (pheophytin) in photosystem-II reaction centers at cryogenic temperatures. *FEBS Lett.* **112**, 97–100.
Kok, B., Gassner, E. B., and Rurainski, H. J. (1965). Photoinhibition of chloroplast reactions. *Photochem. Photobiol.* **4**, 215–27.
Krause, G. H., Kirk, M., Heber, U., and Osmond, C. B. (1978). O_2-dependent inhibition of photosynthetic capacity in intact isolated chloroplasts and isolated cells from spinach leaves illuminated in the absence of CO_2. *Planta (Berl.)* **142**, 229–33.
Krause, G. H., Köster, S., and Wong, S. C. (1985). Photoinhibition of photosynthesis under anaerobic conditions studied in leaves and chloroplasts of *Spinacia oleracea* L. *Planta (Berl.)* **165**, 430–8.
Krause, G. H., and Laasch, H. (1987). Photoinhibition of photosynthesis. Studies of mechanisms of damage and protection in chloroplasts. In 'Progress in Photosynthesis Research'. (Ed. J. Biggins.) Vol. 4, pp. 19–20. (Martinus Nijhoff: Dordrecht.)
Kyle, D. J. (1985*a*). The 32000 dalton Q_B protein of photosystem II. *Photochem. Photobiol.* **41**, 107–16.
Kyle, D. J. (1985*b*). Light-induced turnover of the 32000 dalton Q_B protein. In 'Molecular Biology of the Photosynthetic Apparatus'. (Eds K. E. Steinback, S. Bonitz, C. J. Arntzen and L. Bogorad.) pp. 33–8. (Cold Spring Harbor Laboratory.: Cold Spring Harbor, N.Y.)
Kyle, D. J., Ohad, I., and Arntzen, C. J. (1984). Membrane protein damage and repair: selective loss of a quinone-protein function in chloroplast membranes. *Proc. Natl Acad. Sci. U.S.A.* **81**, 4070–4.
Kyle, D. J., Ohad, I., and Arntzen, C. J. (1985). Molecular mechanism of compensation to light stress in chloroplast membranes. In 'Cellular and Molecular Biology of Plant Stress'. (Eds J. L. Key and T. Kosuge.) UCLA Symposium Series No. 22, pp. 51–70. (A. R. Liss, Inc.: New York.)
Marinetti, T. D., Okamura, M. Y., and Feher, G. (1979). Localization of the primary quinone binding site in reaction centers from *Rhodopseudomonas sphaeroides* R-26 by photoaffinity labelling. *Biochemistry* **18**, 3126–33.
Mattoo, A. K., Hoffman-Falk, H., Marder, J. B., and Edelman, M. (1984). Regulation of protein metabolism: coupling of photosynthetic electron transport to *in vivo* degradation of the rapidly metabolised 32-kilodalton protein of the chloroplast membranes. *Proc. Natl Acad. Sci. U.S.A.* **81**, 1380–4.
Melis, A., and Brown, J. S. (1980). Stoichiometry of system I and system II reaction centres and of plastoquinone in different photosynthetic membranes. *Proc. Natl Acad. Sci. U.S.A.* **77**, 4712–16.
Myers, J., and Burr, G. (1940). Studies on photosynthesis. Some effects of light of high intensity on *Chlorella*. *J. Gen. Physiol.* **24**, 45–67.
Nakatani, H. Y. (1983). Correlation of the low temperature 695 nm fluorescence emission with the reaction center of PS II (CP47). In 'The Oxygen Evolving System of Photosynthesis'. (Eds Y. Inoue *et al.*) pp. 49–54. (Academic Press: Tokyo.)
Nakatani, H. Y., Ke, B., Dolan, E., and Arntzen, C. J. (1984). Identity of the photosystem II reaction center polypeptide. *Biochim. Biophys. Acta* **765**, 347–52.
Nanba, O., and Satoh, K. (1987). Isolation of a photosystem II reaction center consisting of D-1 and D-2 polypeptides and cytochrome *b*-559. *Proc. Natl Acad. Sci. U.S.A.* **84**, 109–12.
Nedbal, L., Setlikova, E., Masojidek, J., and Setlik, I. (1986). The nature of photoinhibition in isolated thylakoids. *Biochim. Biophys. Acta* **848**, 108–19.
Ohad, I., Kyle, D. J., and Hirschberg, J. (1985). Light-dependent degradation of the Q_B-protein in isolated pea thylakoids. *EMBO J.* **4**, 1655–9.
Okamura, M. Y., Satoh, K., Isaacson, R. A., and Feher, G. (1987). Evidence of the primary charge separation in the D_1D_2 complex of photosystem II from spinach: EPR of the triplet state. In 'Progress in Photosynthesis Research'. (Ed. J. Biggins.) Vol. 1, pp. 379–81. (Martinus Nijhoff: Dordrecht.)
Osmond, C. B. (1981). Photorespiration and photoinhibition: some implications for the energetics of photosynthesis. *Biochim. Biophys. Acta* **639**, 77–98.
Powles, S. B. (1984). Photoinhibition of photosynthesis induced by visible light. *Annu. Rev. Plant Physiol.* **35**, 15–44.
Powles, S. B., and Björkman, O. (1982). Photoinhibition of photosynthesis: effect of chlorophyll fluorescence at 77K in intact leaves and in chloroplast membranes of *Nerium oleander*. *Planta (Berl.)* **156**, 97–107.
Powles, S. B., Osmond, C. B., and Thorne, S. W. (1979). Photoinhibition of intact attached leaves of C_3 plants illuminated in the absence of both carbon dioxide and of photorespiration. *Plant Physiol.* **64**, 982–8.

Reisman, S., and Ohad, I. (1986). Light-dependent degradation of the thylakoid 32 kDa Q_B protein in isolated chloroplast membranes of *Chlamydomonas reinhardtii*. *Biochim. Biophys. Acta* **849**, 51–61.
Renger, G. (1976). Studies on the structural and functional organization of system II of photosynthesis. The use of trypsin as a structurally selective inhibitor at the outer surface of the thylakoid membrane. *Biochim. Biophys. Acta* **440**, 287–300.
Renger, G., Koike, H., Yuasa, M., and Inoue, Y. (1984). Studies on the mechanism of the fluorescence decline induced by strong actinic light in PS II particles under different redox conditions. *FEBS Lett.* **163**, 89–93.
Satoh, K. (1985). Protein-pigments and photosystem II reaction center. *Photochem. Photobiol.* **42**, 845–53.
Satoh, K. (1986). Photosystem II particles largely depleted in the two intrinsic polypeptides in the 30 kDa region from *Synechococcus* sp. Identification of a subunit which carries the photosystem II reaction center. *FEBS Lett.* **204**, 357–62.
Setlik, I., Nedbal, L., Masojidek, J., and Setlikova, E. (1984). Irradiance dependent changes in photosystem 2 caused by chloramphenicol and uncouplers in photosynthesizing cells. In 'Advances in Photosynthesis Research. Vol. 3'. (Ed. C. Sybesma.) pp. 259–62. (Nijhoff/Junk: The Hague.)
Takahama, U., and Nishimura, M. (1975). Formation of singlet molecular oxygen in illuminated chloroplasts. Effects on photoinactivation and lipid peroxidation. *Plant Cell Physiol.* **16**, 737–48.
Theg, S. M., Filar, L. J., and Dilley, R. A. (1986). Photoinactivation of chloroplasts already inhibited on the oxidizing side of photosystem II. *Biochim. Biophys. Acta* **849**, 104–11.
Trebst, A. (1962). Lichtinaktivierung der O_2-Entwicklung in der Photosynthese. *Z. Naturforsch. Sect. B* **17**, 660–3.
Trebst, A., and Depka, B. (1985). The architecture of photosystem II in plant photosynthesis: which peptide subunits carry the reaction center of PS II? In 'Antennas and Reaction Centers of Photosynthetic Bacteria — Structure, Interactions and Dynamics'. (Ed. M. E. Michel-Beyerle.) Springer Series in Chemical Physics 42, pp. 216–24. (Springer-Verlag: Berlin.)
Tytler, E. M., Whitelam, G. C., Hipkins, M. F., and Codd, G. A. (1984). Photoinactivation of photosystem II during photoinhibition in the cyanobacterium *Microcystis aeruginosa*. *Planta (Berl.)* **160**, 229–34.
Vermaas, W. F. J., Williams, J. G. K., Rutherford, A. W., Mathis, P., and Arntzen, C. J. (1986). Geneticallly engineered mutant of the cyanobacterium *Synechocystis* 6803 lacks the photosystem II chlorophyll-binding protein CP-47. *Proc. Natl Acad. Sci. U.S.A.* **83**, 9474–7.
Vitry, C. de, and Diner, B. A. (1984). Photoaffinity labelling of the azidoatrazine receptor site in reaction centers of *Rhodopseudomonas sphaeroides*. *FEBS Lett.* **167**, 327–31.
Whitelam, G. C., and Codd, G. A. (1984). Photoinhibition of photosynthesis and in vivo chlorophyll fluorescence in the green alga *Ankistrodesmus braunii*. *Plant Cell Physiol.* **25**, 465–71.

Mangrove Photosynthesis: Response to High-irradiance Stress*

Olle Björkman[AB], *Barbara Demmig*[AC] *and T. John Andrews*[D]

[A]Carnegie Institution of Washington, Department of Plant Biology,
290 Panama Street, Stanford, California 94305, U.S.A.
[B]To whom all correspondence should be addressed.
[C]Present address: Lehrstuhl fur Botanik II der Universität, Mittlerer Dallenbergweg 64, D-8700, Wurzburg, Federal Republic of Germany.
[D]Australian Institute of Marine Science, Cape Ferguson, P.M.B. No. 3,
Townsville M.S.O., Qld 4810, Australia; present address: Research School of Biological Sciences, Australian National University, G.P.O. Box 475,
Canberra, A.C.T. 2601, Australia.

Abstract

Efficiencies of photosynthetic energy conversion were determined in sun and shade leaves of several mangrove species, growing in an open intertidal habitat in North Queensland, by measuring the maximum photon yield of O_2 evolution and 77K chlorophyll fluorescence characteristics. Preliminary measurements confirmed that mangrove leaves have low water potentials, low stomatal conductances and low light-saturated CO_2 exchange rates. Mangrove sun leaves therefore received a very large excess of excitation energy.

Mangrove shade leaves had as high a photon yield of O_2 evolution as non-mangrove leaves and their fluorescence characteristics were normal, showing that the energy conversion efficiency was unaffected by the high salinity. Mangrove sun leaves had markedly depressed photon yields and fluorescence was severely quenched showing that the efficiency of the photochemistry of photosystem II was reduced. The efficiency of energy conversion decreased with an increased radiation receipt. No such depression was detected in sun leaves of non-mangrove species growing in adjacent non-saline sites. Shading of mangrove sun leaves resulted in an increase in the efficiency of energy conversion but, in most species, more than 1 week was required for these leaves to reach the efficiency of shade leaves. Leaves exposed to direct sunlight had somewhat higher efficiencies in mangrove plants cultivated in 10% seawater as compared with full-strength seawater but the salinity of the culture solution had little effect on the increase in the efficiency upon shading.

Field and laboratory fluorescence measurements indicated that the reduced efficiency of energy conversion in mangrove sun leaves resulted from a large increase in the rate constant for radiationless energy dissipation in the antenna chlorophyll rather than from damage to the photosystem II reaction centres. We propose that this increase in radiationless energy dissipation serves to protect the reaction centres against damage by excessive excitation.

Introduction

The high salinity of intertidal mangrove habitats imposes at least two potential restrictions on the photosynthetic rate of mangrove leaves: high leaf water deficits and low stomatal conductances. The high leaf water deficit (low leaf water potential) is obviously a direct consequence of the high salinity of the soil solution. The reason for low stomatal conductance is less obvious. Briefly, low stomatal conductance is a requisite for a low ratio of transpiration to carbon fixation (= high water use efficiency), which is thought to be needed for maintenance of a physiologically acceptable salt/carbon balance within the leaves (Andrews and Muller 1985; Ball 1986).

*C.I.W.-D.P.B. Publication No. 989.

0310-7841/88/010043$03.00

Irrespective of the underlying function experiments show that, compared to non-halophytic C_3 plants, mangrove species do indeed have high water use efficiency, accompanied by low leaf intercellular CO_2 concentration, low stomatal conductance, and low light-saturated photosynthetic rate (Farquhar *et al.* 1982; Andrews *et al.* 1984; Ball and Farquhar 1984*a*, 1984*b*; Andrews and Muller 1985). Earlier measurements of stomatal conductances and CO_2 exchange rates in mangrove leaves in the field also yielded low values (Moore *et al.* 1972, 1973; Lugo *et al.* 1975; Attiwill and Clough 1980).

Low photosynthetic rates at light saturation would inevitably result in a large excess of excitation energy whenever a leaf becomes exposed to direct sunlight. Excessive excitation energy is known to be potentially injurious to the photosynthetic system, especially photosystem II (PS II) (for a review see Powles 1984). The main objective of the present study was to investigate how leaves of mangroves respond to excessive light in an open tropical habitat characterised by abundant sunshine for most of the year. Two principal methods were used to assess the efficiency of photosynthetic energy conversion by the leaves: (1) measurements of the photon yield of O_2 evolution under saturating CO_2 and strictly rate-limiting photon irradiances, and (2) determinations of the relative yield of low temperature (77K) chlorophyll fluorescence at 692 and 734 nm in the state where all PS II reaction centres are open and in the state where they are all closed.

Materials and Methods

Plant Materials

Cape Ferguson, Queensland

Studies in the natural habitat were made on mangrove species growing in the intertidal zone at Cape Ferguson, North Queensland (19°16′S., 148°4′E.) during the period May through October 1983. This time of the year is usually characterised by predominantly clear skies, low humidity, and moderate temperatures and such conditions also prevailed during the period of this study. The photon irradiance incident on a plane ± normal to solar beam often reached 2300 μmol $m^{-2} s^{-1}$. Daytime leaf temperatures were usually in the range 26–31°C. Relative humidity during midday was typically in the range 35–55%.

The mangrove species used in the field studies were *Aegialitis annulata* R. Br. (Plumbaginaceae), *Aegiceras corniculatum* (L.) Blanco (Myrsinaceae), *Avicennia marina* (Forsk.) Vierh. (syn. *A. eucalyptifolia* Zipp. ex Miq.) (Avicenniaceae), *Rhizophora stylosa* Griff. (Rhizophoraceae) and *Sonneratia alba* J. Sm. (Sonneratiaceae). Some field studies were also made on plants of *Lysiana subfalcata* (Hook.) Barlow subsp. *maritima* Barlow (Loranthaceae), a mistletoe growing on *S. alba* at Cape Ferguson.

Mangrove seedlings for experimental purposes were also raised from propagules of *R. stylosa* and *A. marina* (both from Cape Ferguson) and *R. apiculata* Blume (from Hinchinbrook Island, N. Qld.) in a naturally lit, screened shadehouse at the Australian Institute of Marine Science as described by Clough (1984). These seedlings were approx. 1 year old. The screen reduced the photon irradiance to about one-third of that outside the shadehouse. However, because of their position and orientation, the leaves used in the present experiments rarely received more than 150 μmol photons $m^{-2} s^{-1}$ during their development.

For comparative purposes certain measurements were also made on some non-halophytic species growing in fully exposed locations above the high tide mark where salinity was much lower than in the intertidal zone occupied by mangroves. These species were: *Colubrina asiatica* Brogn. (Rhamnaceae), *Ipomoea pes-caprae* (L.) R. Br. (Convolvulaceae), *Macroptilium atropurpureum* (DC.) Urban cv. Siratro (Leguminosae), *Passiflora foetida* L. (Passifloraceae), *Premna obtusifolia* R. Br. (Verbenaceae), *Vigna marina* (Burm.) Merrill (Leguminosae), and *Xanthium* sp. (Compositae).

Stanford, California

Propagules of the mangroves *Avicennia marina* and *Rhizophora stylosa* were kindly collected in the native habitat at Cape Ferguson and shipped by air to California by Dr Barry Clough. Upon arrival at Stanford (26 March 1984), the *Avicennia* propagules were planted out into plastic tubs containing a continuously aerated culture solution with 25% natural seawater added. After 3 weeks, when roots had formed and the first leaf pair had developed, the salinity was decreased to 10% of full-strength seawater for one-half of the plants and gradually increased to 100% seawater for the other half over a 2-week period. The plants were then transferred to 1·5-litre cylinders through which a flowing culture solution (either 10% or 100% seawater salinity) from common 200-litre tanks was passed at a rate of approx. 1·5 litres min^{-1}. The nutrient culture solution used to supplement the 10 and 100% seawater solutions contained 3·0 mM K, 1·3 mM Ca, 1 mM Mg, 2·5 mM N (as nitrate), 1 mM P, 1 mM S, 0·1 mM Fe and standard micronutrients. The protocol for establishment of *Rhizophora* seedlings was identical to that for *Avicennia*, except that the *Rhizophora* propagules (arrival date 23 February 1984) were kept in 25% seawater for 5 weeks before the change in salinity was begun.

The establishment as well as the continued growth of most of the plants took place in naturally lit, temperature-controlled glasshouses ('phytocells'). The daily photon irradiance received by the plants was approx. 83% of that in the open. During the period when the effects of light and salinity on chlorophyll fluorescence of these plants were studied (2 June–24 July 1984) the mean daily total photon receipt inside the glasshouses was approx. 50 mol photons m^{-2}; noon photon irradiance (horizontal plane) was approx. 1600–1700 $\mu mol\ m^{-2}\ s^{-1}$. Day and night air temperatures were maintained at 27 ± 1°C and 20 ± 1°C, respectively. The temperature of the culture solution varied from a minimum of 19°C in the morning to a maximum of 31°C in the afternoon. The conductivity and nitrate concentration of the culture solutions were monitored weekly and adjusted as needed. The solutions were replaced with freshly prepared ones every 2–3 weeks.

Rhizophora seedlings, used for determinations of the effect of excessive light on the relationship between the photon yield of oxygen evolution and chlorophyll fluorescence characteristics, were also grown in 10% and 100% seawater under artificial light, in a controlled-growth room. Light was provided for 12 h day^{-1} from banks of multivapour lamps (Metalarc 1000 W C/V, phosphor-coated; GTE Sylvania, Winchester, Ky, USA). The photon irradiance incident on different leaves ranged from 250 to 1200 $\mu mol\ m^{-2}\ s^{-1}$. Leaf temperature ranged from 28 to 31°C during the light period and from 19 to 21°C during the dark period. Relative humidity during the light period was 27–34%. The flowing culture solution system was similar to that used in the glasshouses. Some of these leaves were subsequently exposed to photon irradiances up to 2000 $\mu mol\ m^{-2}\ s^{-1}$ provided from a multivapour lamp (Metalarc 1500 W, GTE Sylvania). In each case a single attached leaf was sealed into a gas-exchange cuvette and exposed to the desired photon irradiance in a stream of humidified air at a leaf temperature of 25°C. Exposure times varied from 20 min to 17 h.

Methods

Measurements of photon yields of photosynthetic O_2 evolution and 77K fluorescence emission were made as described in detail by Björkman and Demmig (1987). Leaves collected in the native habitat were used for photon yield determination. The leaves were brought to the laboratory in Petri dishes on moist filter paper. Usually the time between collection and the start of the measurements was approx. 30 min. The time required to complete a photon yield measurement was about 1 h. All photon yield measurements were made at 25·0°C. 77K fluorescence measurements were conducted both at the field site and in the laboratory. Measurements of stomatal conductance were made in the field site with a steady-state porometer (model LI 1600, Li-Cor, Lincoln, Neb., USA). CO_2 exchange measurements were made with a field-portable system as described by Pearcy and Calkin (1983). Light measurements were made with quantum sensors (model LI 190 SB, Li-Cor). A data logger (model CR7, Campbell, Logan, Ut., USA) together with appropriate sensors was used for recording of light, temperature and relative humidity both in the field and in the controlled-growth facilities. Xylem water potentials were determined on twigs before sunrise and in the afternoon with a laboratory-built pressure bomb. Leaf water potentials were determined with leaf disc psychrometer chambers (model C52, Wescor, Logan, Ut., USA). Solute potentials were determined with the same apparatus on leaf discs frozen in liquid N_2 and then thawed.

Results

Plant Water Relations, Stomatal Conductance and CO_2 Exchange in the Native Habitat

A summary of the xylem water potentials of mangrove plants growing in the intertidal zone is given in Fig. 1. The mistletoe *Lysiana subfalcata* attached to *Sonneratia alba* is included with the mangroves. For a comparison water potentials of some non-mangrove species, growing on the shore above the intertidal zone, are also given. Predawn water potentials of the mangroves ranged from $-2\cdot8$ MPa in *Sonneratia* (seaward) to $-3\cdot5$ MPa in *Aegialitis* (landward side). Water potentials measured in the early afternoon on clear days ranged from $-3\cdot4$ MPa in *Sonneratia* to $-3\cdot9$ MPa in *Aegialitis*. For the five mangrove species the mean value of the difference between predawn and afternoon water potentials was $-0\cdot44$ MPa. *Rhizophora* showed the smallest (0·33 MPa) and *Sonneratia* the largest (0·63 MPa) diurnal variation among the mangroves. As expected, the water potential of *Lysiana* was consistently somewhat lower than that of its host with the difference being smallest at dawn (0·08 MPa) and greatest in the afternoon (0·36 MPa). The non-mangrove species had much higher water potentials than the mangroves. Since the insolation and potential evaporation were similar for the two groups, undoubtedly the large difference in water potential between them was primarily caused by differences in the salinity of the soil solutions.

In Fig. 2 are summarised the results of stomatal conductance measurements made in the field site with a Li-Cor steady-state porometer during June and July 1983. All values

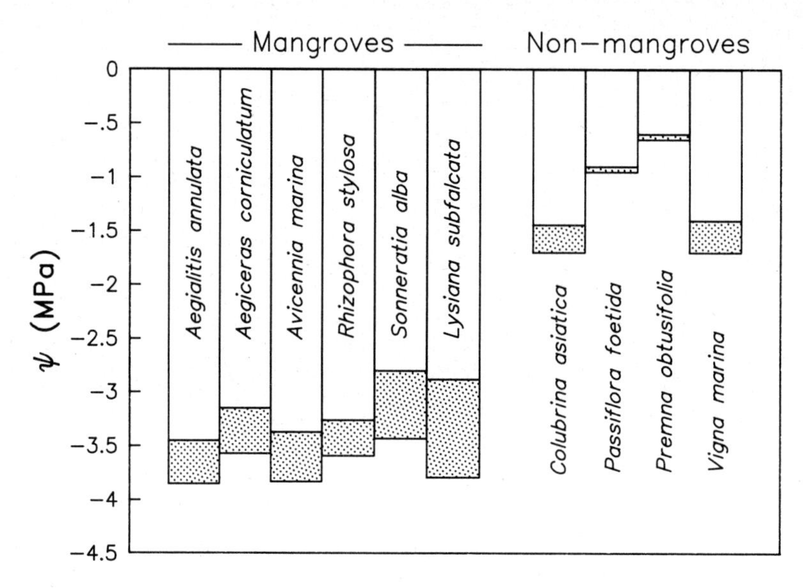

Fig. 1. Xylem water potentials of mangrove species growing in the intertidal zone and of non-mangrove species growing on the shore above this zone. The total heights of the bars depict Ψ values measured in the early afternoon, the blank parts of the bars depict Ψ values measured just before sunrise. Cape Ferguson, N. Qld, 4 June–1 July and 11 September–28 October 1983.

are for clear days when the photon irradiance incident on the leaves was between 1600 and 2000 μmol $m^{-2} s^{-1}$. Leaf temperatures ranged from 24 to 31°C and relative humidity from 44 to 60%. *Sonneratia* and its mistletoe, *Lysiana*, have amphistomatous leaves and the conductance values shown here represent the sum of the two leaf surfaces. The remaining four mangrove species have hypostomatous leaves. The mean stomatal conductance ± s.e. for the five mangrove species was 163 ± 15 mmol $m^{-2} s^{-1}$. Concurrent measurements on *Lysiana* and its host invariably showed higher conductance values for the mistletoe. This is consistent with the observation that the mistletoe had slightly lower potentials than the host and in agreement with previous comparative studies of the relative transpiration rates of mistletoes and their hosts.

The stomatal conductances presented here were measured under conditions of light, temperature and humidity that should promote stomatal opening. Yet these values are low in comparison with those of sun leaves of non-mangrove species under favourable water relations. The conductances clearly are much lower than those here determined for three non-mangrove species growing above the high-tide mark. One would therefore expect that the light-saturated rate of CO_2 uptake of these mangrove leaves would be relatively low.

The results of CO_2 exchange measurements on intact attached leaves of four mangrove species in the field site were in agreement with these expectations (Table 1). Light-saturated rates of net CO_2 uptake were low, ranging from 5·4 to 11·6 μmol CO_2 $m^{-2} s^{-1}$. Especially noteworthy is the observation that leaves, fully exposed to the sun during their development, did not have higher light-saturated rates than leaves which had developed in the shade. On the contrary, these 'sun' leaves had markedly lower rates than the shade leaves. The sun leaves chosen for these measurements had an almost horizontal orientation. The CO_2 exchange rates of the sun leaves shown in Table 1 therefore may be somewhat lower than those of most leaves in exposed situations in

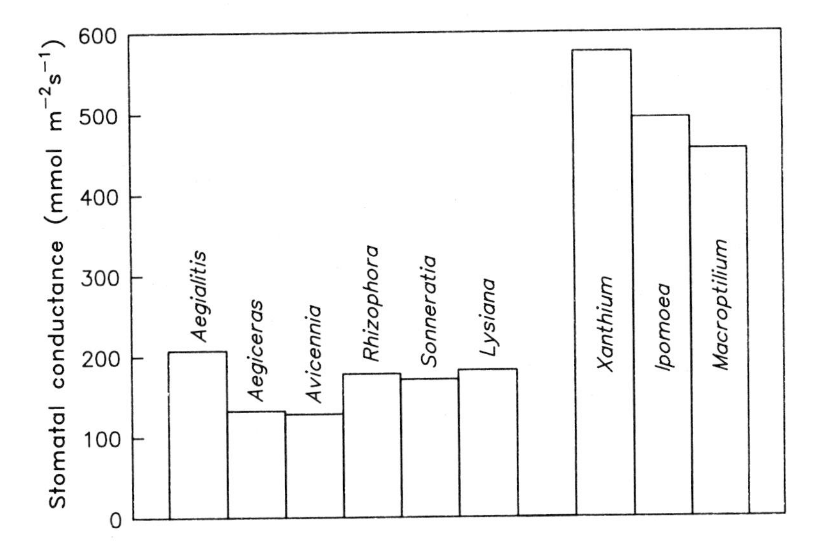

Fig. 2. Stomatal conductance of intertidal mangroves (left) and some non-mangrove species (right) growing above the intertidal zone. All measurements were on clear days. Cape Ferguson, 3 June–4 July and 26 August–13 September 1983.

the canopy. Nevertheless, these results do indicate that the high level of excitation energy encountered by the mangrove leaves may greatly exceed the capacity with which energy can be used for photosynthesis. Such conditions are known to create a potential for photoinhibition.

Photon Yield of O_2 Evolution

The photon yield (quantum yield) of photosynthetic O_2 evolution, determined under conditions where the photosynthetic rate is fully CO_2 saturated and strictly light limited, is a direct measure of the intrinsic and maximum efficiency of light energy conversion by a leaf. The photon yield on the basis absorbed light, ϕ_{O_2}, is remarkably constant among leaves of C_3 species of vascular plants as long as they have not been exposed to excessive light or other severe environmental stress (Björkman and Demmig 1987). Leaves of mangroves growing under high salinity at the present study site are no exception. Shade leaves of *Aegialitis, Aegiceras, Avicennia, Rhizophora,* and *Sonneratia* exhibited as high ϕ_{O_2} values as leaves of non-halophytes grown in non-saline environments (Björkman and Demmig 1987). Hence, high salinity alone causes no impairment of the photon yield in mangrove leaves. However, leaves of these mangrove species developed in the sun had markedly reduced photon yields. This is illustrated in Fig. 3, which shows the relationship between the rate of O_2 evolution and incident photon irradiance for a leaf of *Rhizophora stylosa* taken from a position in the canopy to which little or no direct sunlight penetrated (shade) and for a leaf from the same plant taken

Table 1. Light-saturated rate of CO_2 uptake for shaded and exposed leaves of mangrove species in their native sites

Measurements were made on intact, attached leaves in normal air (0·034% CO_2) at 25–26° leaf temperature. Cape Ferguson 26 June–1 July 1983

Species	CO_2 uptake (μmol m^{-2} s^{-1}) Shade leaf	Sun leaf
Aegialitis annulata	10·8	6·8
Avicennia marina	11·6	6·4
Rhizophora stylosa	7·8	5·4
Sonneratia alba	8·9	6·3

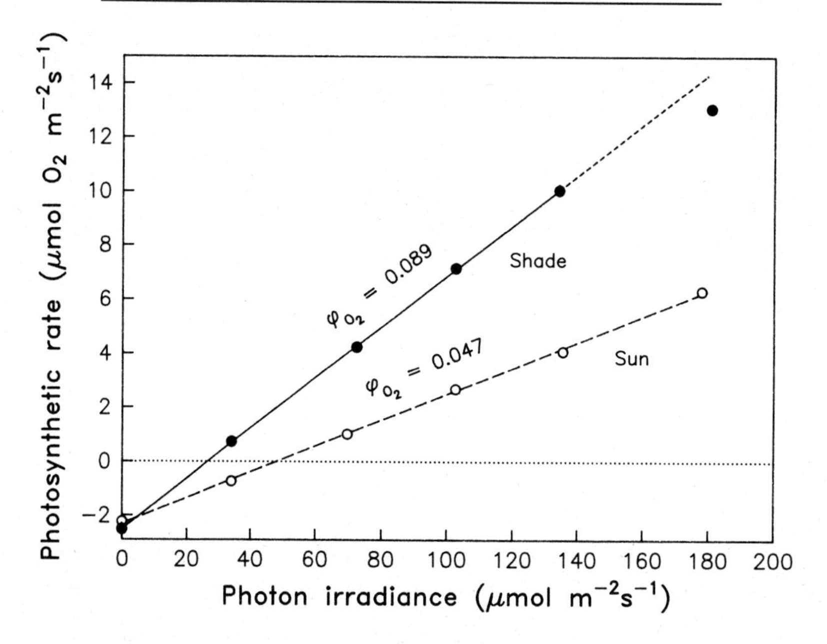

Fig. 3. Initial part of the curve for the relationship between the rate of O_2 evolution and incident photon irradiance in a sun and a shade leaf of *Rhizophora stylosa* growing in the intertidal zone. The slopes of the lines give the photon yields (ϕ_{O_2}) of O_2 evolution on an incident light basis. Measurements were made at 3–5% CO_2 and 25°C. Cape Ferguson, 2–3 August 1983.

from a fully exposed position in the canopy (sun). The ϕ_{O_2} value of this sun leaf was only about one-half of that of the shade leaf. Similar results were obtained with the other mangrove species (Table 2). These results further indicate that the depression of the photon yield increased with increased radiation receipt. Of the leaves that had developed in exposed positions in the canopy, those with steep leaf angles and/or azimuths which decrease the intensity and duration of exposure to direct sunlight exhibited smaller depression of ϕ_{O_2} than those whose orientation permitted a high interception of direct sunlight. No active diurnal leaf movement was detected in any of these mangrove species.

To determine the effect on the photon yield of exposing mangrove shade leaves to direct sunlight, shadehouse-grown plants of *Rhizophora apiculata*, grown and maintained in full-strength seawater, were moved outdoors and single leaves were oriented to give maximum light interception. These leaves were held at the desired angle with aluminium wires. The results, summarised in Table 3, show that even brief exposures to direct sunlight caused a marked decline in ϕ_{O_2} and an exposure time of 1·5 h was sufficient to reduce ϕ_{O_2} by 50%. Maximum inhibition apparently occurred within 1 day as no further reduction was obtained after 2 days of exposure. Recovery in the dark

Table 2. Effect of exposure and leaf orientation on ϕ_{O_2} (incident light basis) in five mangrove species in their native habitat at Cape Ferguson, North Queensland

Values are listed in the order of increased radiation receipt. Measurements were made during the period 2 August–9 October 1983

Species	Exposure	Leaf angle (°)	Leaf azimuth	ϕ_{O_2} (mol O_2 mol^{-1} photon)
Aegialitis annulata	Partial shade	—	—	0·078
	Exposed	25	S	0·060
	Exposed	35	N	0·058
	Exposed	35	N	0·054
Aegiceras corniculatum	Shade	—	—	0·084
	Exposed	35	N	0·044
Avicennia marina	Shadehouse	—	—	0·095
	Partial shade	—	—	0·078
	Exposed	55	SE	0·071
	Exposed	33	W	0·049
	Exposed	2	N	0·043
Rhizophora stylosa	Shadehouse	—	—	0·094
	Partial shade	—	—	0·075
	Exposed	2	N	0·052
	Exposed	35	N	0·047
Sonneratia alba	Open	80	E	0·078
	Exposed	70	E	0·058
	Exposed	45	NW	0·051
	Exposed	47	N	0·036

from short-term exposure to direct sunlight was very slow. It is noteworthy that leaves which were kept fully exposed to direct natural sunlight for several weeks exhibited higher ϕ_{O_2} values than those which had been exposed for 1–2 days, indicating that the leaves are capable of a partial adjustment to the high light environment over time. The ϕ_{O_2} value of the *R. apiculata* leaves that had been maintained in direct sunlight for 42 days was approx. the same as those of similarly exposed leaves of the other mangrove species in their native habitat.

As illustrated in Fig. 4, high-light-treatment of *R. apiculata* shade leaves caused a reduction in the rate of O_2 evolution at all measurement irradiances. However, the reduction in the light-saturated capacity of O_2 evolution was considerably smaller than the reduction in ϕ_{O_2}. Similar results were obtained in many other experiments with mangrove leaves (data not shown). However, while ϕ_{O_2} proved to be remarkably constant among individual untreated shade leaves, the variation in light-saturated capacity of O_2 evolution was often considerable (up to some ±30%). This makes it difficult to obtain an accurate determination of the effect of excessive light on light-saturated capacity relative to its effect on ϕ_{O_2}.

Chlorophyll Fluorescence Measurements (77K)

Determinations of chlorophyll fluorescence yield can provide valuable information on the efficiency of light conversion by PS II. In the present study the relative fluorescence yield from sun and shade leaves was measured at 77K after keeping the leaf samples in darkness for a sufficiently long time to cause complete reoxidation of the PS II reaction centres and to permit relaxation of any fluorescence quenching associated with the presence of ΔpH across the thylakoid membrane in the light (e.g. Krause *et al.* 1983). Values for the fluorescence listed in subsequent figures and tables refer to two extreme states: F_o, representing the state where all reaction centres are open (oxidised) and F_M, the state where all centres are closed (reduced). F_V is the difference between F_M and F_o. Fluorescence emission was simultaneously measured at 692 nm and 734 nm. F_{692} emanates from PS II. F_{734} is composed of at least two components. Most of the $F_{o,734}$ emission is thought to emanate from PS I, whereas $F_{V,734}$ ($=F_{M,734}$ *minus* $F_{o,734}$) is thought to originate from excitation of PS II. According to Kitajima and Butler (1975), the ratio $F_V/F_{M,692}$ may be used as a direct measure of the photochemical efficiency of PS II. For a recent review of the application of 77K chlorophyll fluorescence to studies of effects of excessive light on leaves, see Björkman (1987*b*).

Table 3. Effect of exposing shadehouse-grown *Rhizophora apiculata* leaves to bright sunlight for different times on the photon yield of O_2 evolution

Attached leaves were oriented to give a maximum light interception at noon. Photon irradiance during a 5-h period centred around noon was 1800–2100 μmol m^{-2} s^{-1} and leaf temperature during the same period was 28–32°C. Treatments were made outdoors and the 2-day and 42-day treatments refer to the total duration of exposure to the natural diurnal light cycle. Recovery time includes a 0·5-h period at a low photon irradiance required to measure the photon yield. The balance of the recovery time was in darkness. The mean ϕ_{O_2} value of untreated control leaves was 0·093 O_2 photon^{-1} (incident light basis)

Exposure time	Recovery time	ϕ_{O_2} (% of control)
0·75 h	0·75 h	82
1·5 h	1 h	50
5 h	1 h	23
5 h	20 h	35
2 days	0·75 h	37
42 days	0·75 h	45

In Table 4 are listed the fluorescence emission characteristics for mangrove leaves that had either developed in direct sunlight (Ex) or in shaded positions (Sh) in the canopy. The values are means for a number of leaves oriented at various angles and azimuths but exposed leaves with a nearly vertical orientation were not used. In addition to the five mangrove species, Table 4 also includes values for the mistletoe *Lysiana*.

There is a striking difference in fluorescence emission between exposed and shaded leaves in all of these species, exposed leaves showing a much lower overall fluorescence.

The similarity in the response of the different species is also striking. The percentage reduction in the F_o level of PS II fluorescence ($F_{o,692}$) of exposed compared with shaded leaves ranged from some 33% in *Sonneratia* to 60% in *Rhizophora*. The F_M level of PS II fluorescence ($F_{M,692}$) of exposed leaves was reduced by as much as 77 to 83%, and the reduction in $F_{V,692}$ ranged from 83% in *Sonneratia* to 90% in *Aegialitis*. As a result of the decline in $F_{o,692}$, the reduction in the F_V/F_o ratio (mean reduction 74%) was somewhat smaller than the reduction in F_V (Table 5). Although there was also a marked reduction in the $F_V/F_{M,692}$ ratio in all of the species, the mean reduction in this ratio was only about 32% (Table 5).

Table 4 shows that exposed leaves also had a much lower fluorescence emission at 734 nm than shade leaves. The reduction of $F_{o,734}$ which primarily emanates from PS I was roughly similar to that of $F_{o,692}$. This could indicate that exposure of the leaves to direct sunlight caused a simultaneous decline in the yield of both PS II and PS I fluorescence. The reduction in the measured fluorescence emission could also be caused

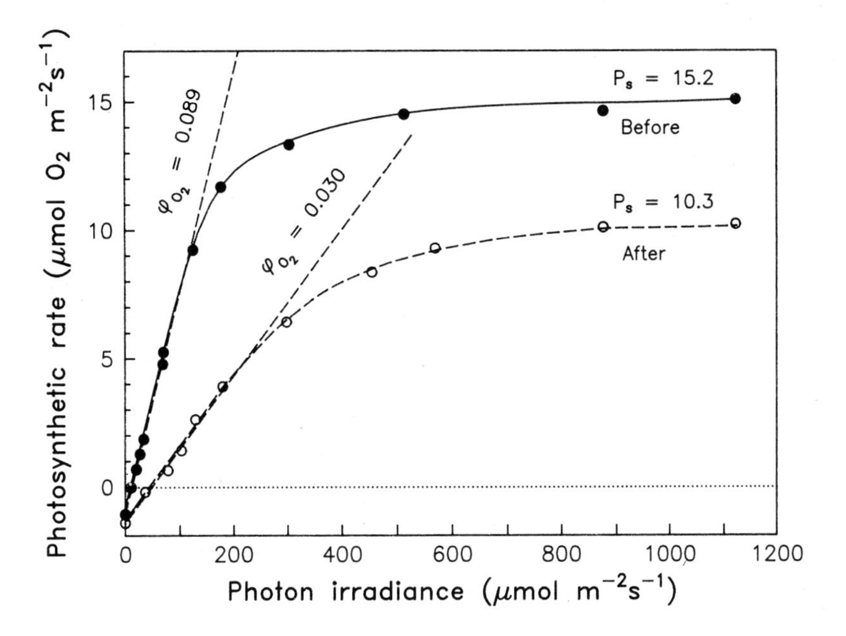

Fig. 4. Rate of O_2 evolution *versus* incident photon irradiance of a shade leaf of *Rhizophora apiculata* before and after exposure to direct sunlight for 3 h. Measurement conditions were as given in Fig. 3. Cape Ferguson, 30 August 1983.

by a decrease in the optical transmittance of the leaves. Such changes in leaf optical properties would attenuate the intensity of the excitation beam reaching the chloroplasts as well as the intensity of fluorescence as it passes from the sites of fluorescence to the leaf surface. Results from short-term exposures of leaves, presented below (Table 6), indicate that most of the observed reduction in $F_{o,692}$ and $F_{o,734}$ was not attributable to changes in leaf optical properties but reflects a reduction in fluorescence emission by the chloroplasts. (The ratios F_V/F_o and F_V/F_M of course would be totally independent of any changes in optical properties.)

The difference between exposed and shaded leaves in $F_{V,692}$ in each species was accompanied by a similar difference in $F_{V,734}$ (Table 4). This is consistent with the view that $F_{V,734}$ directly or indirectly results from excitation of PS II.

Comparison of Exposed and Shaded Surfaces of Mangrove and Non-mangrove Leaves

In Fig. 5 the $F_V/F_{M,692}$ ratios of exposed and shaded leaf surfaces of mangrove leaves ($\Psi = -3\cdot0$ to $-3\cdot8$ MPa) are compared with those of non-mangroves growing above the intertidal zone ($\Psi = -0\cdot7$ to $-1\cdot4$ MPa). Since shade leaves were not available for some of the non-mangrove species, fluorescence emission was measured from the self-shaded abaxial surfaces. $F_{M,692}$ and $F_{M,734}$ emission of lower surfaces of sun leaves of non-mangroves was 1·7–2·0 times that of the upper surfaces. By comparison, $F_{M,692}$ of the lower surfaces of *Rhizophora* sun leaves was as much as 8–12 times higher than the upper leaf surfaces, and $F_{M,734}$ was 3·5–5 times higher. In the other mangrove species the corresponding values were 3·5–5 for $F_{M,692}$ and 3·1–3·8 for $F_{M,734}$. The $F_V/F_{M,692}$ ratio of the lower surfaces of mangrove sun leaves was nearly identical to that of the upper surfaces of shade leaves. As shown in Fig. 5, the $F_V/F_{M,692}$ ratio of the lower leaf surfaces showed little variation among species and there was no significant difference between mangroves and non-mangroves. Similarly, the $F_V/F_{M,692}$ ratio of the upper leaf surfaces of non-mangrove leaves was at most 3% lower than that of the lower surfaces of the same leaves. By contrast, in sun leaves of the mangrove species, $F_V/F_{M,692}$ of the upper surfaces was 33–40% lower than the values for the lower surfaces of the same leaves.

Table 4. 77K fluorescence emission at 692 and 734 nm from exposed and shaded mangrove leaves

Measurements were made in the native habitat at Cape Ferguson during 6 May–29 August 1983. Each value represents the mean for 3–14 leaves of each species. Fluorescence determinations at different times of the day on leaves that were naturally exposed to direct sunlight (Ex) or in shaded positions in the canopy (Sh). Leaf samples were kept in darkness for 20–60 min before measurement

Species		692 nm			734 nm		
		Ex	Sh	Ex/Sh	Ex	Sh	Ex/Sh
Aegialitis annulata	F_o	5·2	9·0	0·58	46·5	160·0	0·29
	F_M	9·9	54·4	0·18	55·6	241·1	0·23
	F_V	4·7	45·4	0·10	9·1	81·0	0·11
Aegiceras corniculatum	F_o	4·1	6·2	0·66	35·9	109·0	0·33
	F_M	8·9	38·5	0·23	42·7	165·7	0·26
	F_V	4·8	32·2	0·15	6·8	56·8	0·12
Avicennia marina	F_o	4·0	9·0	0·44	19·5	71·5	0·27
	F_M	10·9	50·8	0·21	26·8	115·3	0·23
	F_V	6·9	41·8	0·17	8·4	43·8	0·19
Rhizophora stylosa	F_o	3·3	8·2	0·40	33·4	56·9	0·59
	F_M	7·7	46·3	0·17	42·7	88·0	0·49
	F_V	4·4	38·1	0·12	9·1	31·7	0·29
Sonneratia alba	F_o	5·2	7·7	0·67	63·6	131·7	0·48
	F_M	13·6	60·2	0·23	122·6	193·0	0·64
	F_V	8·4	52·5	0·16	17·1	61·3	0·28
Lysiana subfalcata	F_o	7·0	13·5	0·51	43·1	89·2	0·48
	F_M	13·5	59·9	0·23	48·9	136·7	0·36
	F_V	6·6	44·9	0·14	5·8	47·4	0·12

Changes in Fluorescence Characteristics in Response to Exposing Shade Leaves to Direct Sunlight

In the preceding paragraphs fluorescence characteristics were compared between mangrove leaves that had developed either in direct sunlight or in the shade. This section deals with the changes that take place when leaves of *Rhizophora apiculata*, developed in a shadehouse, are suddenly transferred to direct sunlight for periods ranging from 45 min to 9 days. The results, summarised in Table 6, show that marked quenching of $F_{M,692}$ and $F_{V,692}$ occurred within 1 h of exposure. After a 5-h exposure, $F_{M,692}$ was only 14% of its initial value, and $F_{V,692}$ was barely detectable. The quenching of $F_{M,692}$ was accompanied by a smaller, yet substantial, quenching of $F_{o,692}$. The emission at 734 nm was quenched as well. $F_{o,734}$ declined in proportion to the decline in $F_{o,692}$ and $F_{V,734}$ declined in proportion to the decline in $F_{V,692}$. After 5 h in direct sunlight (up to 2100 μmol m^{-2} s^{-1}), the $F_V/F_{M,692}$ ratio had fallen to 0·135. This value is much lower than was found in mangrove leaves that had *developed* in direct sunlight (Table 5). Visual comparisons between exposed and covered parts of the same leaves gave no indication of any apparent changes in leaf optical properties in exposures lasting up to 5 h.

Table 5. Mean values of 77K fluorescence emission characteristics at 692 nm for exposed and shaded mangrove leaves

Values are means ± s.d. for all species listed in Table 4, except *Lysiana subfalcata*

	F_o	F_M	F_V	F_V/F_o	F_V/F_M
Exposed	4·4 ± 0·8	10·2 ± 2·2	5·8 ± 1·7	1·35 ± 0·34	0·567 ± 0·064
Shaded	8·0 ± 1·2	50·0 ± 8·2	41·9 ± 7·5	5·27 ± 0·90	0·838 ± 0·020
Ex/Sh	0·55	0·22	0·14	0·26	0·68

Table 6. Changes in 77K fluorescence characteristics caused by exposing shade leaves of *Rhizophora apiculata* to direct sunlight

Single attached leaves which had developed in a shadehouse were exposed to direct sunlight outdoors for periods ranging from 45 min to 9 days while the other leaves were shaded. Fluorescence measurements were preceded by a 40–60 min period in darkness. To facilitate comparison, the $F_{M,692}$ and $F_{M,734}$ emission of untreated control leaves were set equal to 100. (Cape Ferguson, September 1983)

Exposure time	692 nm				734 nm			
	F_o	F_M	F_V	F_V/F_M	F_o	F_M	F_V	F_V/F_M
0	22·6	*100*	77·4	0·774	67·4	*100*	32·6	0·326
0·75 h	27·6	70·4	42·8	0·608	68·3	89·2	20·9	0·234
1·5 h	17·5	32·0	14·5	0·456	49·7	57·2	7·5	0·131
5·0 h	12·2	14·1	1·9	0·135	34·1	34·8	0·7	0·020
2 days	10·0	12·2	2·2	0·180	42·6	45·5	2·9	0·064
4 days	11·5	14·9	3·4	0·228	46·7	50·2	3·5	0·070
9 days	7·4	18·1	10·7	0·591	30·4	37·8	7·4	0·196

Longer exposures (2–9 days) did not result in any further reduction in $F_V/F_{M,692}$ (or $F_V/F_{M,734}$). Instead a gradual rise took place and, after 9 days in direct sunlight, $F_V/F_{M,692}$ and $F_V/F_{M,734}$ had risen to approx. the same values as found in sun leaves of *R. stylosa* and other mangrove species in the native habitat.

A dark period of 40–60 min preceded the fluorescence measurements presented in Table 6. When the dark period following exposure for 5 h to direct sunlight was extended from 1 h to 20 h, $F_{M,692}$ rose from 14·1 to 19·7 and $F_V/F_{M,692}$ from 0·135 to 0·299. Little further change occurred when the dark period was extended to 42 h. Thus, the extent of recovery was small.

Relationship between Changes in Fluorescence Characteristics and in the Photon Yield of O_2 Evolution

The results obtained in the field experiments show that there exists a relationship between the photon yields of O_2 evolution in sun and shade leaves of mangroves and fluorescence emission. A depression in ϕ_{O_2} was always accompanied by a reduction in fluorescence, especially $F_{V,692}$ and $F_{V,734}$. A positive correlation was obtained when ϕ_{O_2} of individual leaves were plotted *versus* F_V, F_V/F_o or F_V/F_M at both 692 and 734 nm, but it was difficult to ascertain which of these measures of variable fluorescence would provide the best quantitative estimate of the efficiency of energy conversion. To further elucidate these relationships, concurrent measurements of ϕ_{O_2} and 77K fluorescence characteristics were subsequently made in the laboratory at Stanford. Single leaves of *Rhizophora stylosa* grown at a photon irradiance of 250 μmol m^{-2} s^{-1} were exposed to 2000 μmol m^{-2} s^{-1} at 25°C for periods up to 17·5 h. In other cases the plants were

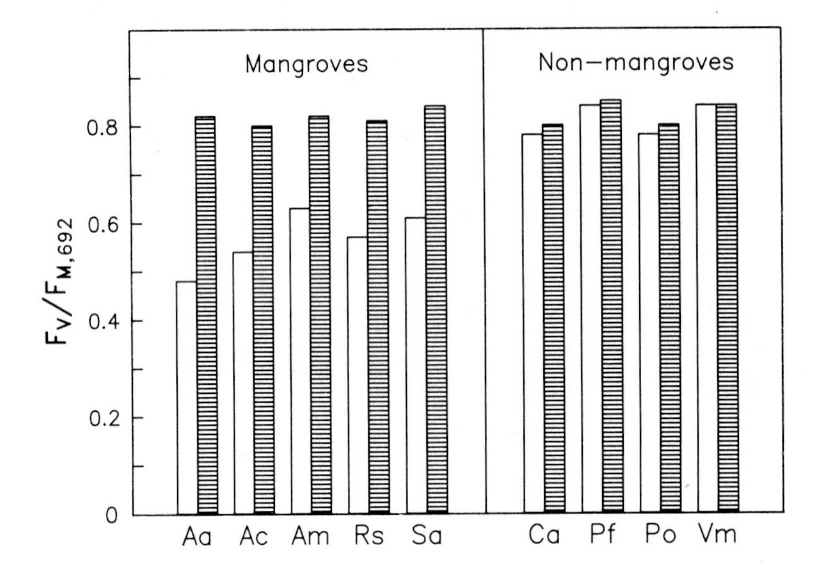

Fig. 5. $F_V/F_{M,692}$ ratio of the exposed adaxial (open bars) and shaded abaxial (hatched bars) surfaces of sun leaves of mangrove and non-mangrove species. Water potentials and species names were as in Fig. 1.

moved from a growth photon irradiance of 250 to one of 1200 μmol m^{-2} s^{-1} (12-h photoperiod) for up to 9 days. Fluorescence and photon yield measurements were determined for the same leaves after a 30–60 min recovery period at approx. 20 μmol m^{-2} s^{-1}. The results of these experiments are shown in Figs 6*a* and 6*b*.

The relationship between $F_{V,692}$ and ϕ_{O_2} was clearly curvilinear so that a small initial reduction in ϕ_{O_2} was accompanied by a relatively large reduction in F_V. A similar curvilinear relationship was obtained between $F_{M,692}$ and ϕ_{O_2} (not shown). $F_{o,692}$ showed a less steep decline than $F_{V,692}$. The scatter of the data points precludes any definite conclusion regarding the exact shape of the relationship but the decline in $F_{o,692}$ appears to be roughly linearly related to the decline in ϕ_{O_2} in this species. Like F_V, the $F_V/F_{o,692}$ ratio was a curvilinear function of ϕ_{O_2}. The data points for the $F_V/F_{M,692}$ ratio fitted a linear regression with ϕ_{O_2} (R^2 = 0·923), although the regression line does not quite extrapolate to the origin. $F_V/F_{M,734}$ likewise showed a linear correlation (R^2 = 0·919) with ϕ_{O_2} (data not shown).

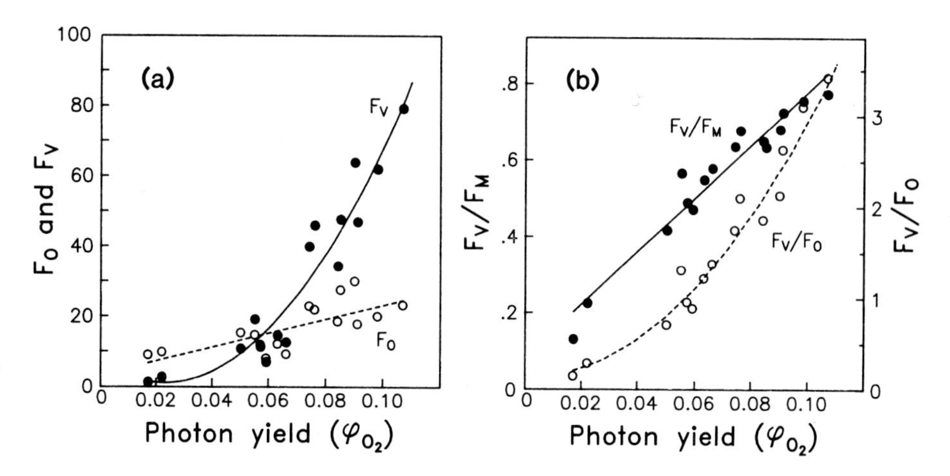

Fig. 6. Relationship between fluorescence characteristics at 692 nm and photon yield of O_2 evolution (on an absorbed light basis) for *Rhizophora stylosa* leaves, exposed to a wide range of light treatments. See text for details. Stanford, California, 15 November–19 December 1984.

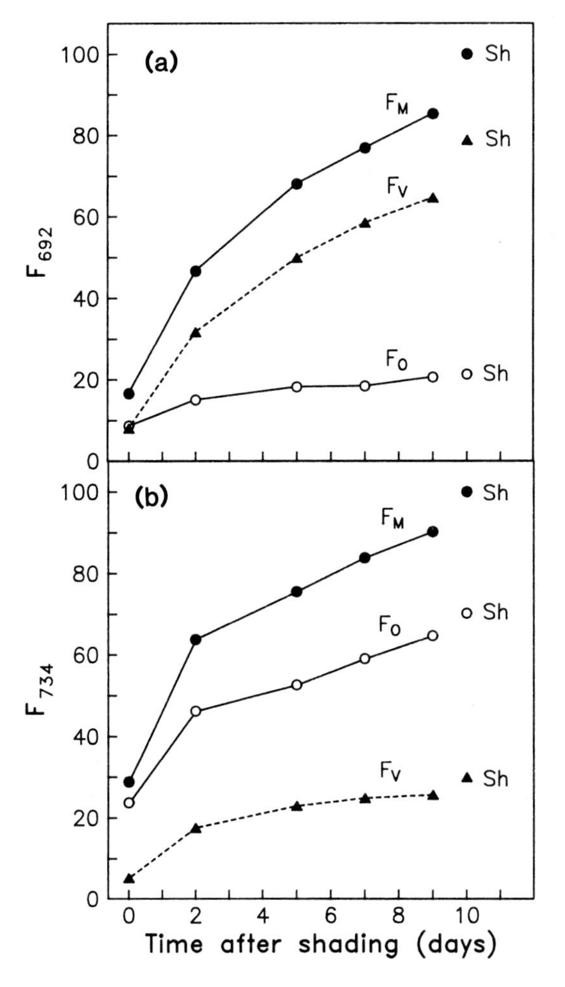

Fig. 7. Time course of recovery of 77K fluorescence emission at 692 nm (*a*) and 734 nm (*b*) of the upper leaf surface of *Rhizophora stylosa* sun leaves after shade leaves in the field. 'Sh' denotes values for leaves developed in the shade. To facilitate comparison the $F_{M,692}$ and $F_{M,734}$ of leaves developed in the shade were set equal to 100. Cape Ferguson, 15 June–24 June 1983.

Recovery of Fluorescence Emission after Shading of Mangrove Sun Leaves

Fully exposed leaves of *Rhizophora stylosa, Avicennia marina, Aegiceras corniculatum* and *Sonneratia alba*, growing in their native habitat, were shaded so that they received diffuse light only. The fluorescence characteristics were then followed for up to 9 days. As shown in Figs 7*a* and 7*b*, shading resulted in a gradual increase in fluorescence emission of the upper leaf surfaces of *R. stylosa* at both the F_o and the F_M levels, both at 692 and 734 nm; eventually, the fluorescence characteristics of these leaves closely approached those of leaves that had developed in the shade. However, these changes were quite slow. The time required for fluorescence emission to reach the values of shade leaves would exceed 9 days. Time courses for the recovery of F_o, F_M and F_V at 692 and 734 nm upon shading of *Avicennia* and *Aegiceras* sun leaves (not shown) were similar to those shown for *Rhizophora* (Fig. 7).

Time courses for recovery of the $F_V/F_{M,692}$ ratio after shading in three mangrove species are shown in Fig. 8. Curves for recovery in *Aegialitis* and *Aegiceras* (not shown) fell between those for *Rhizophora* and *Avicennia*. Recovery of the $F_V/F_{M,692}$ was slow

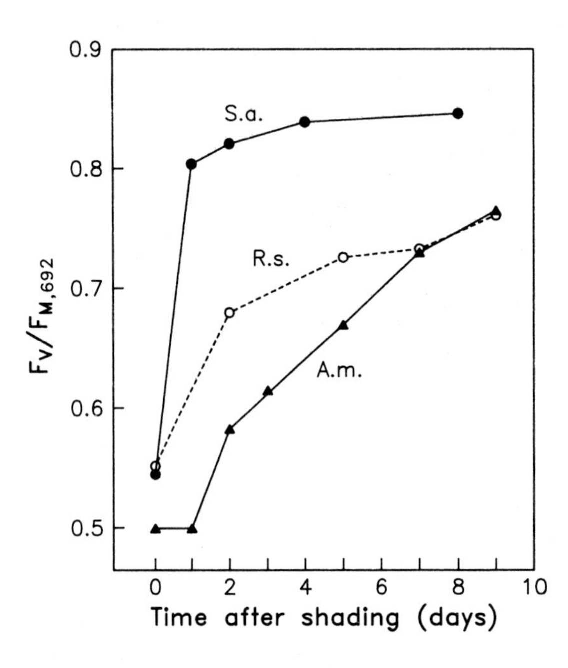

Fig. 8. Time courses of recovery of $F_V/F_{M,692}$ of the upper surface of sun leaves of *Avicennia marina, Rhizophora stylosa* and *Sonneratia alba* after shading in the field. Cape Ferguson, 15 June–2 July 1983.

in all of these mangrove species with the apparent exception of *Sonneratia*. Although the initial $F_V/F_{M,692}$ ratios were nearly identical for *Sonneratia* and *Rhizophora*, in *Sonneratia* this ratio increased to 95% of that of leaves that had developed in the shade within 1 day after shading (Fig. 8). It is estimated that about 1 week would be required for the same degree of recovery to occur in *Rhizophora* and the other mangrove species. It should be noted that the data for recovery in *Sonneratia* are from a single experiment. Unfortunately, the opportunity to repeat this shading experiment did not arise.

Influence of Salinity on Fluorescence Characteristics in Leaves Exposed to Direct Sunlight and on Recovery after Shading

The following experiments were undertaken to investigate the extent to which the high leaf water deficits imposed by the high salinity of the intertidal habitat may be responsible for the severe fluorescence quenching and lowered F_V/F_M ratio in exposed mangrove leaves as well as the slow recovery upon shading observed in most of these species. For this purpose, *Rhizophora* and *Avicennia* plants were grown at 10% and 100% full-strength seawater under high insolation. The water relations and the fluorescence characteristics of *Rhizophora* leaves developed under the two treatments are summarised in Table 7.

Table 7. Effect of growth salinity on leaf water relations and 77K fluorescence characteristics in *Rhizophora stylosa*

Leaves were restrained to a horizontal position 3–4 weeks before measurement. Daily photon irradiance was approx. 50 mol photon m^{-2} and noon irradiance 1600–1700 μmol m^{-2} s^{-1}. Ion concentrations are as NaCl equivalents. Ψ_w, Ψ_s and Ψ_p denote in this order: water, solute and turgor potential. Fluorescence values are mean ± s.d. for nine leaves from each treatment. Stanford, California 9–25 June 1984

	Low salt	High salt
Ion concn (mM)	60	565
Ψ_w medium (MPa)	−0·27	−2·54
Ψ_w leaf (MPa)	−1·74	−3·23
Ψ_s leaf (MPa)	−2·82	−4·06
Ψ_p leaf (MPa)	1·08	0·83
$F_{o,692}$	3·74 ± 0·23	3·49 ± 0·45
$F_{M,692}$	7·87 ± 0·56	5·09 ± 0·82
$F_V/F_{M,692}$	0·523 ± 0·027	0·344 ± 0·064
$F_{o,734}$	19·9 ± 1·6	15·5 ± 2·4
$F_{M,734}$	24·4 ± 1·8	18·3 ± 2·7
$F_V/F_{M,734}$	0·188 ± 0·012	0·137 ± 0·011

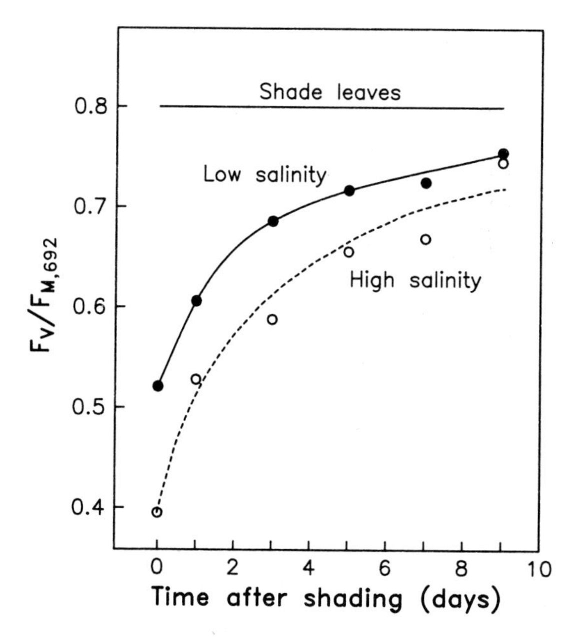

Fig. 9. Time courses of recovery of $F_V/F_{M,692}$ of the upper surfaces of horizontally restrained sun leaves of *Rhizophora stylosa* grown under either low (10% seawater) or high (100% seawater) salinity in 'phytocells'. See Materials and Methods for details. Stanford, California, 26 June–5 July 1984.

It is noteworthy that, while there was a nearly 10-fold ratio in water potential between the two flowing culture solutions, the corresponding ratio in leaf water potential was only about 2. The disproportionally low leaf water potential of the plants grown at low salinity was largely caused by an accumulation of high salt concentration in the leaves (approx. 70% of leaves developed under full-strength seawater; data not shown). As a result the estimated difference in leaf turgor potential (Ψ_p) between the two treatments was only 0·25 MPa. Nevertheless, the fluorescence quenching was significantly less severe and the $F_V/F_{M,692}$ (and $F_V/F_{M,734}$) ratio significantly higher in 10% than in 100% seawater salinity (Table 7). The mean $F_V/F_{M,692}$ value of exposed leaves of *Rhizophora* plants grown at 100% seawater (Table 7) was somewhat lower than in the field at Cape Ferguson (Fig. 5). This difference may have been caused by the fact that all of the leaves used at Stanford were held in a horizontal position and the daily photon receipt by these leaves may, on the average, have been higher than at Cape Ferguson. The somewhat lower leaf water potentials at Stanford might have been a contributing factor.

Similarly exposed leaves of *Avicennia marina* grown in 100% seawater salinity also showed a greater depression of fluorescence emission and lower $F_V/F_{M,692}$ ratios in comparison with leaves of plants grown at 10% salinity. For example, mean values of $F_{M,692}$ were 11·1 and 6·6 for the 10% and 100% salinity treatments, respectively. The corresponding mean $F_V/F_{M,692}$ ratios were 0·605 and 0·544. This difference in the $F_V/F_{M,692}$ ratio between the two salinity treatments was smaller in *Avicennia* than in *Rhizophora*. Nevertheless repeated experiments with *Avicennia* showed that the F_V/F_M ratio was consistently 10–14% higher under low than under the high salinity.

Time courses of recovery of fluorescence after shading of exposed *Rhizophora* and *Avicennia* leaves closely resembled those obtained in the field. As shown in Fig. 9, recovery of the $F_V/F_{M,692}$ ratio in *Rhizophora* leaves was slow and incomplete even after 9 days in shade. While the initial $F_V/F_{M,692}$ was less depressed in the 10% than in the 100% salinity treatment, there was no indication that the rate or extent of *recovery* was promoted by a decreased salinity in *Rhizophora* (Fig. 9). The time courses of recovery after shading of *Avicennia* leaves are not shown as they closely followed those for *Rhizophora*, both in the 10% and the 100% salinity treatments.

Discussion

The water potentials of five mangrove species and their diurnal variation were similar to those previously reported for mangroves in the intertidal zone (Scholander *et al.* 1966; Attiwill and Clough 1980). The light-saturated rates of CO_2 uptake and stomatal conductances were also substantially similar to those previously reported for Australian mangroves (Ball and Critchley 1982; Andrews *et al.* 1984; Ball and Farquhar 1984a; Andrews and Muller 1985) but higher than reported for three Florida mangrove species (Moore *et al.* 1972, 1973; Lugo *et al.* 1975). The light-saturated CO_2 uptake rates and stomatal conductance of mangrove leaves, developed in positions where they were protected from direct sunlight, had at least as high light-saturated rates as non-mangrove leaves developed in moderate shade. However, the light-saturated rates of mangrove leaves exposed to direct sunlight were much lower than previously determined values of sun leaves of non-mangroves growing in warm sunny habitats with adequate water supply (Björkman 1981). Although this is the first report that fully exposed mangrove sun leaves can have *lower* light-saturated rates of CO_2 uptake than leaves of the same plants developed in moderate shade, Ball and Critchley (1982) noted that light-saturated rates of understory leaves of *Avicennia marina* were as high as high as those of exposed leaves. As was mentioned in the Introduction it seems likely that the low stomatal conductances and associated low light-saturated CO_2 uptake rates in mangrove leaves are related to the need to maintain a high water use efficiency (appropriate carbon/salt balance).

In any event, there is no doubt that, because of their low light-saturated photosynthetic rates, mangroves are able to use only a small fraction of the solar radiation incident on fully exposed leaves in this tropical habitat during the season of predominantly clear skies. These leaves therefore received an abundance of excess excitation energy. The marked decline in the photon yield of O_2 evolution with increased radiation receipt and the severe quenching of PS II fluorescence emission (observed after transfer of the leaves to a low photon irradiance or to darkness) obviously were responses to the excess excitation energy. The high salinity and the resulting leaf water deficits had no negative effects on the efficiency of photosynthetic energy conversion on mangrove leaves protected from direct sunlight. Photon yields of oxygen evolution in these leaves were as high as in other species of vascular plants growing under non-saline conditions (Björkman and Demmig 1987). Similarly, 77K fluorescence characteristics of mangrove shade leaves give no indication of any reduction in the photochemical efficiency of PS II.

The various mangrove species showed very similar responses to direct sunlight irrespective of whether their leaves are capable of salt secretion (*Aegialitis, Aegiceras, Avicennia*) or incapable of salt secretion (*Rhizophora, Sonneratia*). Exposed leaves of the mangrove mistletoe, *Lysiana*, which had water potentials similar to those of the mangroves, also exhibited similar fluorescence characteristics. By comparison, sun leaves of non-mangrove species growing under favourable water relations showed no significant reduction in the photon yield of O_2 evolution (cf. Björkman and Demmig 1987; Demmig and Björkman 1987). Although the upper surfaces of exposed leaves of these species had generally lower fluorescence emission than the lower surfaces, the difference between the two surfaces was much smaller than for the mangrove sun leaves. Moreover, exposed surfaces of non-mangrove sun leaves showed little or no reduction in the $F_V/F_{M,692}$ (or $F_V/F_{M,734}$) ratio, whereas exposed surfaces of the mangrove sun leaves all showed a marked reduction in this ratio.

Sudden exposure of *Rhizophora stylosa* shade leaves to direct sunlight caused a rapid decline in the photon yield of O_2 evolution and fluorescence emission. A 5-h exposure to direct sunlight caused very severe inhibition of both, greatly exceeding that found in mangrove sun leaves. The partial recovery that took place after the leaves were maintained in direct sunlight for several days indicates that even mangrove leaves are capable of significant adaptive changes in their ability to withstand excessive light in the longer term. The mechanistic basis of these changes is not known.

Shading of mangrove leaves which had developed in direct sunlight resulted in recovery of fluorescence emission which gradually increased to levels approaching those of leaves that had developed in the shade. With the apparent exception of *Sonneratia* this recovery was quite slow. The time required for full recovery in *R. stylosa* (in the order of 1 week) was similar to that required for the partial recovery observed upon prolonged exposure of *R. apiculata* shade leaves to direct sunlight, discussed above. These time courses also resemble that obtained after *Nerium oleander* plants, grown in direct sunlight under water stress, were rewatered (Björkman and Powles 1984). Such slow recovery could indicate that repair of high-light damaged chloroplasts is involved (however, see below). If so, high salinity of the root solution does not appear to have any adverse effect on the rate of repair. Although low salinity significantly reduced the effect of direct sunlight on fluorescence quenching and the $F_V/F_{M,692}$ ratio in both *Rhizophora* and *Avicennia*, there was no evidence that the rate of recovery after shading was increased by low salinity.

It is now well known that the PS II fluorescence yield at ambient temperature in the state of closed reaction centres (F_M) is rapidly quenched in leaves during exposure to bright light. This quenching, often termed 'energy quenching', is thought to be associated with a build-up of a proton gradient across the thylakoid membrane. It relaxes very fast upon darkening (e.g. Krause *et al.* 1983) and it is improbable that such a ΔpH-dependent quenching could be responsible for the reduction in fluorescence emission observed in the present study.

Until a few years ago any sustained reduction in $F_{M,692}$ and $F_{V,692}$ of the kind observed here was widely considered to reflect a high-light induced *damage* to the photochemistry of PS II (see review by Powles 1984). While this possibility cannot be excluded, we presently prefer an alternative interpretation, namely that the kind of fluorescence quenching observed here is primarily caused by a massive increase in the rate constant for radiationless energy dissipation in the antenna (see Demmig and Björkman 1987; Björkman 1987*a*, 1987*b*). The degree of quenching of $F_{o,692}$ relative to the quenching of $F_{M,692}$, observed in both short-term and long-term exposures to excessive light, as well as during recovery, is in substantial agreement with this concept. Such an increase in the rate constant of radiationless energy dissipation would also give a linear relationship between the $F_V/F_{M,692}$ ratio and the photon yield of the photochemistry of PS II. This would result in a reduced photon yield of O_2 evolution which was indeed observed here. An increase in the fraction of excitation energy that is dissipated in the antenna would reduce the energy reaching the reaction centres, thereby protecting them from excessive excitation. Therefore, the observed responses of mangrove leaves to excessive light levels may reflect a regulatory and protective response rather than reflecting a damage to the reaction centre complex of PS II. The inevitable cost is a reduced efficiency of energy conversion during periods when the light temporarily falls to low levels.

It seems likely that this kind of control of the energy reaching the reaction centres would be common to all green plants, although plants may differ in the extent of maximum energy dissipation and in the rapidity of the induction and relaxation of the response. Such a mechanism of energy dissipation would have profound implications for our interpretations of commonly determined relationships between photosynthetic rate and light, temperature, CO_2 and stomatal conductance. It could help to explain why the photosynthetic rate often appears to be colimited by different component processes such as 'electron transport' and 'carboxylation' or why different external factors such as light and CO_2 often appear to be colimiting.

Acknowledgments

One of us (O.B.) is indebted to Dr John Bunt (Australian Institute of Marine Sciences) for providing field and laboratory facilities at Cape Ferguson and to Dr Robert W. Pearcy (Department of Botany, University of California, Davis) for providing equipment and help in conducting the CO_2 exchange measurements at Cape Ferguson. We also thank Dr Barry F. Clough (Australian Institute of Marine Science) for his expert advice and help at Cape Ferguson, and for collecting and sending mangrove propagules to Stanford. Post-doctoral fellowship support from the Deutsche Forschungsgemeinschaft, and the Carnegie Corporation of New York to B.D. is gratefully acknowledged.

References

Andrews, T. J., Clough, B. F., and Muller, G. J. (1984). Photosynthetic gas exchange properties and carbon isotope ratios of some mangroves in North Queensland. In 'Physiology and Management of Mangroves'. (Ed. H. J. Teas.) pp. 15–23. (Dr. W. Junk Publishers: The Hague.)

Andrews, T. J., and Muller, G. J. (1985). Photosynthetic gas exchange by the mangrove, *Rhizophora stylosa* Griff. in its natural environment. *Oecologia* **65**, 449–55.

Attiwill, P. M., and Clough, B. F. (1980). Carbon dioxide and water vapour exchange in the white mangrove. *Photosynthetica* **14**, 40–7.

Ball, M. C. (1986). Photosynthesis in mangroves. *Wetlands (Australia)* **6**, 12–22.

Ball, M. C., and Critchley, C. (1982). Photosynthetic responses to irradiance by the grey mangrove, *Avicennia marina*, grown under different light regimes. *Plant Physiol.* **70**, 1101–6.

Ball, M. C., and Farquhar, G. D. (1984*a*). Photosynthetic and stomatal responses of the grey mangrove, *Avicennia marina*, to transient salinity conditions. *Plant Physiol.* **74**, 7–11.

Ball, M. C., and Farquhar, G. D. (1984*b*). Photosynthetic and stomatal responses of two mangrove species, *Aegiceras corniculatum* and *Avicennia marina*, to long term salinity and humidity conditions. *Plant Physiol.* **74**, 1–6.

Björkman, O. (1981). Responses to different quantum flux densities. In 'Physiological Plant Ecology I. Responses to the Physical Environment'. (Eds O. L. Lange, P. S. Nobel, C. B. Osmond and H. Ziegler.) Encycl. Plant Physiol. New Ser., Vol. 12A, pp. 57–107. (Springer-Verlag: Heidelberg.)

Björkman, O. (1987*a*). High-irradiance stress in higher plants and interaction with other stress factors. In 'Progress in Photosynthesis Research'. (Ed. J. Biggins.) Vol. 4, pp. 1.11–1.18. (Martinus Nijhoff: Dordrecht.)

Björkman, O. (1987*b*). Low-temperature chlorophyll fluorescence in leaves and its relationship to photon yield of photosynthesis in photoinhibition. In 'Photoinhibition'. (Eds D. J. Kyle, C. B. Osmond and C. J. Arntzen.) Topics in Photosynthesis, Vol. 9, pp. 123–44. (Elsevier: Amsterdam.)

Björkman, O., and Demmig, B. (1987). Photon yield of O_2 evolution and chlorophyll fluorescence characteristics at 77K among vascular plants of diverse origins. *Planta* **170**, 489–504.

Björkman, O., and Powles, S. B. (1984). Inhibition of photosynthetic reactions under water stress: interaction with light level. *Planta* **161**, 490–504.

Clough, B. F. (1984). Growth and salt balance of the mangroves *Avicennia marina* (Forsk.) Vierh. and *Rhizophora stylosa* Griff. in relation to salinity. *Aust. J. Plant Physiol.* **11**, 419–30.

Demmig, B., and Björkman, O. (1987). Comparison of the effect of excessive light on chlorophyll fluorescence (77K) and photon yield of O_2 evolution in leaves of higher plants. *Planta* **171**, 171–84.

Farquhar, G. D., Ball, M. C., Caemmerer, S. von, and Roksandic, Z. (1982). Effect of salinity and humidity on $\delta^{13}C$ values of halophytes — evidence for diffusional isotope fractionation determined by the ratio of intercellular/atmospheric partial pressure of CO_2 under different environmental conditions. *Oecologia* **52**, 121–37.

Kitajima, M., and Butler, W. L. (1975). Quenching of chlorophyll fluorescence and primary photochemistry in chloroplasts by dibromothymoquinone. *Biochim. Biophys. Acta* **376**, 105–15.

Krause, G. H., Briantais, J.-M., and Vernotte, C. (1983). Characterization of chlorophyll fluorescence quenching in chloroplasts by fluorescence spectroscopy at 77K. I. ΔpH-dependent quenching. *Biochim. Biophys. Acta* **723**, 169–75.

Lugo, A. E., Evink, G., Brinsan, M. M., Broce, A., and Snedaker, S. C. (1975). Diurnal rates of photosynthesis, respiration and transpiration in mangrove forest of South Florida. In 'Ecological Studies II. Tropical Ecological Systems'. (Eds F. B. Golley and E. Medina.) pp. 335–50. (Springer-Verlag: New York.)

Moore, R. T., Miller, P. C., Albright, D., and Tieszen, L. L. (1972). Comparative gas exchange characteristics of three mangrove species in winter. *Photosynthetica* **6**, 387–93.

Moore, R. T., Miller, P. C., Ehleringer, J., and Lawrence, W. (1973). Seasonal trends in gas exchange characteristics of three mangrove species. *Photosynthetica* **7**, 387–94.

Pearcy, R. W., and Calkin, H. W. (1983). Carbon dioxide exchange of C_3 and C_4 tree species in the understory of a Hawaiian forest. *Oecologia* **58**, 26–32.

Powles, S. B. (1984). Photoinhibition of photosynthesis induced by visible light. *Annu. Rev. Plant Physiol.* **35**, 15–44.

Scholander, P. F., Bradstreet, E. D., Hammel, H. T., and Hemmingsen, E. A. (1966). Sap concentrations in halophytes and some other plants. *Plant Physiol.* **41**, 529–32.

Adaptation to Sun and Shade: A Whole-plant Perspective

Thomas J. Givnish

Department of Botany, University of Wisconsin,
Madison, WI 53706, U.S.A.

Abstract

Whole-plant energy capture depends not only on the photosynthetic response of individual leaves, but also on their integration into an effective canopy, and on the costs of producing and maintaining their photosynthetic capacity. This paper explores adaptation to irradiance level in this context, focusing on traits whose significance would be elusive if considered in terms of their impact at the leaf level alone. I review traditional approaches used to demonstrate or suggest adaptation to irradiance level, and outline three energetic tradeoffs likely to shape such adaptation, involving the economics of gas exchange, support, and biotic interactions. Recent models using these tradeoffs to account for trends in leaf nitrogen content, stomatal conductance, phyllotaxis, and defensive allocations in sun *v.* shade are evaluated.

A re-evaluation of the classic study of acclimation of the photosynthetic light response in *Atriplex*, crucial to interpreting adaptation to irradiance in many traits, shows that it does not completely support the central dogma of adaptation to sun *v.* shade unless the results are analysed in terms of whole-plant energy capture. Calculations for *Liriodendron* show that the traditional light compensation point has little meaning for net carbon gain, and that the effective compensation point is profoundly influenced by the costs of night leaf respiration, leaf construction, and the construction of associated support and root tissue. The costs of support tissue are especially important, raising the effective compensation point by 140 μmol $m^{-2} s^{-1}$ in trees 1 m tall, and by nearly 1350 μmol $m^{-2} s^{-1}$ in trees 30 m tall. Effective compensation points give maximum tree heights as a function of irradiance, and shade tolerance as a function of tree height; calculations of maximum permissible height in *Liriodendron* correspond roughly with the height of the tallest known individual. Finally, new models for the evolution of canopy width/height ratio in response to irradiance and coverage within a tree stratum, and for the evolution of mottled leaves as a defensive measure in understory herbs, are outlined.

Introduction

A central objective of plant ecology is to understand the causes of patterns in the distribution and abundance of species. Physiological ecologists advance this goal by studying how various morphological and physiological properties permit a plant to survive and compete successfully in certain environments but not in others. Physiological ecology thus provides an important window on the proximal mechanisms that underlie species differences in distribution and habitat-specific competitive ability.

Photosynthetic energy capture provides green plants with almost all of their chemical energy, and is central to their ability to compete and reproduce. Photosynthesis, in turn, is directly and dramatically influenced by the amount of light striking a plant's leaves. Many investigators have therefore studied how different levels of irradiance by photosynthetically active radiation affect photosynthesis, and how the leaf traits that develop under different levels of irradiance influence a plant's photosynthetic response to light level. Comparative studies of the photosynthetic response and leaf characteristics of plants grown under high and low levels of irradiance have provided crucial

0310-7841/88/010063$03.00

insights into the significance of several leaf-level traits seen in plants adapted to sunny *v.* shady conditions (see reviews by Boardman 1977 and Björkman 1981). The key to most of these insights, in turn, has been the study of the photosynthetic light response of individual leaves and the impact that various leaf traits have on it. Species, ecotypes, or acclimated forms with higher rates of leaf photosynthesis under specific light levels have, by and large, been inferred to have an edge in energy capture and competitive ability under those conditions.

Yet, whole-plant growth and competitive ability depend not only on the photosynthetic rate of individual leaves, but also on the geometry and dynamics of a plant's canopy, and the pattern of energy allocation among all organs. Insofar as many traits characteristic of sun- *v.* shade-adapted plants entail energetic costs involving non-photosynthetic organs, or influence the physical environment experienced by other leaves, the significance of such traits may prove difficult to understand if energy capture is considered at the leaf level alone. In this paper, I outline three basic energetic tradeoffs at the whole-plant level likely to shape the evolution of adaptations for energy capture in sun *v.* shade, and illustrate each with leaf- or canopy-level traits whose significance would be elusive, or simply not evident, if considered at the leaf level alone. I begin by briefly examining the central question of how differences between sun- and shade-grown plants can be inferred to be adaptations to irradiance level. Throughout I assume that natural selection favours plants whose form and physiology tend to maximise their net rate of energy capture, because such plants often have the greatest resources with which to reproduce and compete for additional space (Horn 1971; Orians and Solbrig 1977; Givnish 1982, 1986*a*; Cowan 1986).

Sun/Shade Differences as Adaptations

Several features of plant form, physiology, and resource allocation vary with the level of irradiance to which plants are acclimated and/or ecologically restricted (Table 1). Traditionally, three approaches have been used to identify variations in such features as *adaptations* to a specific level of irradiance, based on (1) convergence, (2) correlation with photosynthetic impact, and (3) detailed cost-benefit analysis. Convergence among species from different families or orders in the expression of a given trait in plants restricted to, or grown under, a specific level of irradiance is usually taken as *prima facie* evidence that such behaviour is a result of natural selection. This approach can be applied to either fixed or developmentally plastic traits; in the latter case, convergence in the pattern of response to different irradiance levels would identify the pattern of acclimation—presumably itself genetically determined—as adaptive. Arguments based solely on convergence are, however, limited because they cannot identify how or why variation in a given trait contributes to competitive ability.

A second, more mechanistic approach to identifying traits as adaptations to irradiance level is based on a detailed study of the photosynthetic light response of leaves acclimated to different light levels, together with an analysis of how various features of their morphology and physiology contribute to photosynthetic performance under those levels (e.g. Björkman 1968*a*, 1981; Björkman *et al.* 1972*a*, 1972*b*; Boardman *et al.* 1972; Nobel 1976; Björkman and Powles 1984; Ludlow and Björkman 1984). This approach generally involves two tacit assumptions: (i) that the photosynthetic rates of leaves acclimated (or ecologically restricted) to a specific irradiance level are greater *at that level* than the photosynthetic rates of leaves grown under other irradiance levels; and (ii) that if variation in a given trait enhances leaf photosynthesis—expressed almost invariably per unit area—at a specific irradiance level, then it is an adaptation to that level.

The first of these assumptions seems valid, at least in extreme cases: leaves of plants grown under, or ecologically restricted to, high irradiance levels generally have higher photosynthetic rates per unit area at those levels than do leaves of plants restricted or acclimated to low irradiance levels, and vice versa (Björkman *et al.* 1972*a*, 1972*b*; Jurik *et al.* 1979; Björkman 1981; but see analysis to contrary below). The second assumption seems more questionable. For example, Björkman (1981) argues that because leaves with higher concentrations of Rubisco (RuP_2 carboxylase–oxygenase) have higher photosynthetic rates at high irradiance levels—where carboxylation is likely to limit photosynthesis—the high levels of Rubisco in sun-adapted leaves are adaptive. Conversely, he argues that low Rubisco levels are adaptive under shady conditions because Rubisco content is correlated with rates of dark respiration, and hence inversely correlated with net photosynthesis at low irradiance levels. However, the problem with such

Table 1. Characteristic differences between plants adapted or acclimated to sunny *v.* shady extremes in irradiance level

Derived from Boardman (1977), Björkman (1981), Bazzaz *et al.* (1987) and Givnish (1987)

Trait	Sun	Shade
Leaf-level		
Photosynthetic light response		
Light-saturated rate	High	Low
Compensation irradiance	High	Low
Saturation irradiance	High	Low
Biochemistry		
N, Rubisco, and soluble protein content / mass	High	Slightly lower
Chlorophyll *a* / chlorophyll *b* ratio	High	Low
Chlorophyll / soluble protein ratio	Low	High
Anatomy and ultrastructure		
Chloroplast size	Small	Large
Thylakoid / grana ratio	Low	High
Morphology		
Leaf mass / area	High	Low
Leaf thickness	High	Low
Stomatal size	Small	Large
Stomatal density	High	Low
Palisade / spongy mesophyll ratio	High	Low
Mesophyll cell surface / leaf area ratio	High	Low
Leaf orientation	Erect	Horizontal
Iridescence, lens-shaped epidermal cells	None	Rare
Reddish leaf undersides	Very rare	Infrequent
Canopy-level		
Leaf area index	High to low	Low
Phyllotaxis	Spiral	Distichous
Twig orientation	Erect	± Horizontal
Asymmetric leaf bases	Very rare	Infrequent
Plant-level		
Fractional allocation to leaves	Low	High
Fractional allocation to roots	High	Low
Reproductive effort	High	Low

analyses is that, although they are based firmly on the functional impact of particular traits on leaf photosynthesis, they do not explain why a specific expression of any given trait (e.g. Rubisco content) is adaptive. What factors favour a specific, finite level of Rubisco in leaves? What sets the upper limit on the Rubisco content that would enhance photosynthesis at high irradiance levels, or the lower limit on what would be adaptive at low irradiance levels?

One means of addressing such questions is cost–benefit analysis, the third common approach to identifying adaptations to irradiance level. Cost–benefit analysis involves the assessment of the net effect of a variant of a trait on energy capture, balancing the variant's impact on energy gain against the energetic costs incurred by the plant in producing it, and then analysing which variant would maximise the net rate of energy capture. For example, higher levels of Rubisco may enhance photosynthesis under sunny conditions, but entail increased energetic costs of nutrient capture and enzyme synthesis (Mooney and Gulmon 1979); the resulting increase in leaf N content may also increase exposure to damage by herbivores, resulting in increased rates of leaf loss and/or increased allocation to anti-herbivore defences (Mooney and Gulmon 1982; Givnish 1986*b*). The 'optimal' level of Rubisco under those conditions would maximise the differences between these benefits and costs; other things being equal, plants with this Rubisco level should have an edge in competition under those conditions.

The validity of the assumption that underlies cost–benefit analysis—namely, that competition in a given environment favours plants whose form and physiology maximise their net rate of carbon gain there—has been discussed at length by Horn (1971, 1979) and Givnish (1982, 1986*a*). However, the specific application of cost–benefit analysis to adaptations to irradiance level requires three comments. First, such analyses simply carry mechanistic studies of the impact of various traits on photosynthesis to their logical conclusion, explicitly detailing and (at least potentially) quantifying costs and benefits that would otherwise be discussed in vague or qualitative terms. The specific costs and benefits included in any particular analysis may be disputed on conceptual or empirical grounds, but their stark exposure in a model affords the clearest test of the assumptions that underlie them.

Second, many cost–benefit analyses focus on a trait's impact on leaf photosynthesis, not whole-plant energy capture. Yet, this approach ignores any effect a trait has on whole-plant carbon gain through its impact on canopy geometry and dynamics, on the microenvironments experienced by other leaves, or on energy allocation between photosynthetic and non-photosynthetic tissue. However, analysis at the leaf level alone can be justified if it asks how a fixed amount of energy should be allocated between two or more leaf compounds (e.g. Rubisco *v.* chlorophyll or carbonic anhydrase, chlorophyll *a v.* chlorophyll *b*), tissues (spongy mesophyll *v.* palisade), or organelles (plastid thylakoids *v.* grana) (see Björkman 1968*a*, 1981; Cowan 1986; Parkhurst 1986). Focusing on the *ratio* of investment in different leaf subunits is a valid means of circumscribing the problem of optimal allocation, and should yield sensible results within the limits specified. This approach, however, cannot solve the broader problem of what determines the optimal *total* investment in different leaf subunits, or the *absolute* energy invested in each. Absolute levels of investment in leaf traits, by virtue of their association with costs involving non-photosynthetic organs, almost inevitably involve energetic tradeoffs at the whole-plant level.

Finally, many traits that vary in response to irradiance level are also known to respond to other environmental factors, many of which are themselves correlated with irradiance level. Clough *et al.* (1979) observe that this raises a profound problem for all approaches to the study of adaptation to irradiance level, namely: Are the traits seen in sun (or shade) plants adaptations specifically to irradiance level, to factors correlated with irradiance level, or to the synergistic effects of both? Cost–benefit models can contribute to resolving this difficult question by analysing how plants should respond if only irradiance level, or relative humidity, or some other environmental parameter were to vary, and then comparing the magnitudes of the resultant responses with each other, and with the responses expected if such parameters were to vary in concert. Analyses incorporating the interactive effects of irradiance and other environmental factors (e.g. Osmond 1983; Ludlow and Björkman 1984) are in their infancy.

Energetic Tradeoffs at the Whole-plant Level

In analysing the net contribution of a trait to the whole-plant rate of net carbon gain, three basic kinds of energetic tradeoffs are likely to arise and influence the evolution of that trait and the distribution of species bearing it. These involve the economics of gas exchange, the economics of support, and the economics of biotic interactions:

(1) The economics of gas exchange (Givnish 1986*c*) arise from the unavoidable link between carbon gain and water loss: any passive structure that permits the passage of large, slow-moving CO_2 molecules will also allow the diffusion of smaller, faster molecules of water vapour. As a result, the photosynthetic benefit of any trait that increases the rate at which CO_2 can diffuse into a leaf must be weighed against the energetic costs associated with increased water loss. Such transpirational costs might include a reduction in mesophyll photosynthetic capacity caused by a decrease in leaf water potential, an increased allocation of energy to unproductive roots or xylem, and/or a shortened period of photosynthetic activity (Givnish and Vermeij 1976; Orians and Solbrig 1977; Givnish 1979, 1984). A complementary tradeoff results from the inevitable conflict between leaf photosynthetic capacity and the energetic costs of constructing and maintaining tissue capable of high photosynthetic rates (Mooney and Gulmon 1979; Gulmon and Chu 1981). Highly productive leaves require large inputs of nitrogen, phosphorus, and other mineral nutrients to create the pools of enzymes and pigments needed to sustain high rates of CO_2 uptake (Field and Mooney 1986).

Tradeoffs involving the economics of gas exchange have been implicated in the evolution of several traits that differ between sun- and shade-adapted plants and that influence both photosynthesis and transpiration. Such traits include effective leaf size (Givnish and Vermeij 1976; Givnish 1987), stomatal conductance (Cowan 1977, 1986; Cowan and Farquhar 1977; Givnish 1986*d*), leaf absorptance (Ehleringer and Mooney 1978), leaf orientation (Ehleringer and Forseth 1980; Ehleringer and Werk 1986; Nobel 1986), leaf nitrogen content and mesophyll photosynthetic capacity (Mooney and Gulmon 1979; Gulmon and Chu 1981; Field 1983), chlorophyll/protein and chlorophyll/Rubisco ratios (Björkman *et al.* 1972*b*; Björkman 1981; Cowan 1986), chlorophyll *a*/chlorophyll *b* ratio (Björkman *et al.* 1972*b*; Björkman 1981), internal leaf architecture (Parkhurst 1986) and leaf area index (Horn 1971).

(2) The economics of support arise because, among the leaf and crown forms that have equivalent effects on photosynthesis and transpiration, many differ in the efficiency with which the leaves can be mechanically supported (Givnish 1986*e*). Such differences imply tradeoffs between photosynthetic benefits and mechanical costs. Such tradeoffs, in turn, have been implicated in the evolution of several aspects of leaf shape and arrangement that are likely to affect whole-plant energy capture in sun *v.* shade. These include leaf shape (Givnish 1979, 1984), stem branching angles (Honda and Fisher 1978, 1979; Borchert and Tomlinson 1984; Givnish 1986*f*), compound *v.* simple leaves (Givnish 1979, 1984), and asymmetric leaf bases, anisophylly, and opposite *v.* alternate leaf arrangements (Givnish 1984). Both the economics of support and of gas exchange appear to be involved in the evolution of spiral *v.* distichous phyllotaxes (Givnish 1984).

(3) The economics of biotic interactions arise because many traits that enhance a plant's potential rate of growth—such as high leaf nitrogen content, heavy allocation to foliage, low allocation to defensive compounds or mutualists, or an erect growth habit—may also increase its potential attractiveness to herbivores, implying a tradeoff

between photosynthetic benefits and biotic costs (Janzen 1974; Chew and Rodman 1979; Mooney and Gulmon 1982; Givnish 1986*b*; Gulmon and Mooney 1986). Such tradeoffs, together with those associated with gas exchange and support, may underlie visual mimicry or divergence in leaf form (Gilbert 1975; Barlow and Wiens 1977; Rausher 1978, 1980; Givnish 1984; Ehleringer *et al.* 1986), allocation to defensive compounds (Janzen 1974; McKey *et al.* 1978; Chew and Rodman 1979; Mooney and Gulmon 1982; Bazzaz *et al.* 1987), and leaf flushing (Coley 1983), and may also influence the evolution of such leaf traits as toughness, pubescence, and nitrogen content (Lincoln *et al.* 1982; Coley 1983). Similar tradeoffs may also be involved in the evolution of carnivory and mutualisms with ants or nitrogen-fixing symbionts (Givnish *et al.* 1984; Givnish 1988*a*).

Several leaf and canopy traits that vary with irradiance level can be understood only in terms of their impact on whole-plant carbon gain, not leaf-level photosynthesis. The following three sections address the significance of a few key such traits in terms of their impact on the economics of gas exchange, the economics of support, and the economics of biotic interactions.

Economics of Gas Exchange

Leaf nitrogen content, stomatal conductance and photosynthetic light response show characteristic responses to irradiance level, and are thought to exert a profound influence on plant carbon gain under different light regimes (Björkman 1981; Gulmon and Chu 1981; Schulze and Hall 1982). Models to account for trends in the first two traits are briefly reviewed below.

Variation in photosynthetic light response with irradiance level has been used to support the most tightly held dogma of adaptation to sun *v.* shade: namely, that leaves acclimated to a given light level photosynthesise more rapidly at that level than do those acclimated to other levels (e.g. Björkman and Holmgren 1963; Boardman 1977; Björkman 1981). This dogma has been crucial to interpreting the responses of several traits to irradiance as adaptive. Yet, a critical analysis of the classic study of acclimation to irradiance level (Björkman *et al.* 1972*b*) reveals that the data, as originally analysed, do not completely support the dogma that they had almost universally been seen to support. This analysis, presented in the final portion of this section, shows that support for the dogma re-emerges when the costs of leaf construction are incorporated.

Optimal Leaf Nitrogen Content

Leaves of plants acclimated or adapted to high irradiance levels generally have higher maximum rates of photosynthesis, expressed per unit area, than leaves acclimated or adapted to lower irradiance levels (Björkman *et al.* 1972*b*; Boardman 1977; Björkman 1981). To help identify the mechanistic bases for this fundamental difference between sun and shade leaves, Björkman (1981) analysed the correlations between peak photosynthetic rate and several aspects of leaf biochemistry, physiology and morphology. He found that, across species, light-saturated rates of photosynthesis showed little relation to factors that determine the efficiency of light absorption (such as chlorophyll content) but a strong relation to factors likely to limit the rate of dark reactions (such as Rubisco or total soluble protein content). Wong *et al.* (1979, 1985*a*, 1985*b*) and Schulze and Hall (1982) also found that, within species, stomatal conductance varies in direct proportion to photosynthesis as either short-term irradiance or conditioning irradiance during leaf ontogeny varies. Analyses by Björkman *et al.* (1972*b*), Björkman (1981), and Farquhar and Sharkey (1982) indicated that the higher photosynthetic rates of sun leaves are a result of both higher stomatal conductance, and higher intrinsic photosynthetic capacity of the mesophyll at a given conductance. Presumably, the higher mesophyll capacity of sun leaves reflects, at least in part, their higher concentration of Rubisco and other photosynthetic enzymes. Indeed, maximum photosynthetic rate tends to increase linearly with leaf nitrogen concentration across species adapted to different levels of irradiance or soil fertility (Mooney *et al.* 1978; Field and Mooney 1986).

If mesophyll photosynthetic capacity and maximum photosynthetic rate increase with total leaf nitrogen or soluble protein, why don't shade leaves also have high levels of nitrogen or soluble protein? Mooney and Gulmon (1979) presented a conceptual model to answer this and related questions, based on the impact of leaf nitrogen content on whole-plant growth. They argued that as leaf nitrogen content increases, so does leaf photosynthetic rate, but so also do the root costs the plant must pay in order to obtain that nitrogen. The extent to which additional nitrogen—in the form of various dark reaction enzymes, including Rubisco—can enhance photosynthesis is likely to be greater in sunny than in shady environments, because carboxylation is more likely to limit photosynthesis at high irradiance. Specifically, Mooney and Gulmon (1979) argued that the leaf nitrogen level at which photosynthesis begins to plateau should be higher in sunny than in shady environments. Thus, the optimal leaf nitrogen content—at which the difference between photosynthesis and the energetic investment in roots needed to obtain a given amount of nitrogen is maximised—should be higher in sun than in shade. Based on the higher returns expected from a given investment in nitrogen if water availability does not limit photosynthesis, and on the lower costs of obtaining a given amount of nitrogen on more fertile soils, Mooney and Gulmon (1979) concluded that optimal leaf nitrogen content should also be higher on moister or more fertile sites.

Gulmon and Chu (1981) present data supporting one assumption of this model as it applies to sun *v.* shade adaptation: photosynthesis does increase more rapidly with leaf nitrogen ($g\ N\ m^{-2}$), and saturate at higher levels of leaf nitrogen, at higher levels of irradiance in *Diplacus aurantiacus*. Provided that root costs increase with nitrogen uptake, optimal leaf nitrogen content should thus increase with irradiance. However, although the Mooney–Gulmon model is consistent with this finding, and accords qualitatively with trends seen in leaf nitrogen content in plants exposed to different levels of light, moisture, and soil fertility, the model has five important shortcomings:

(1) Quantifying the root costs associated with nutrient uptake has remained difficult, presumably because roots have other functions (e.g. water uptake) and because nutrient uptake involves active transport, necessitating measurements of both root construction and maintenance costs. This has prevented any quantitative test of the model to date.

(2) The model as originally advanced does not incorporate the fact that, at least across leaves acclimated or adapted to different irradiance levels, maximum rates of photosynthesis are strongly correlated with rates of dark respiration (Fig. 1). Dark respiration averages about 7% of peak photosynthesis, so that every increase in the latter of 1 $\mu mol\ m^{-2}\ s^{-1}$ decreases photosynthesis at low irradiance levels by 0·07 $\mu mol\ m^{-2}\ s^{-1}$, and increases the instantaneous leaf compensation point by 1·4 $\mu mol\ m^{-2}\ s^{-1}$, given the average quantum yield of 0·05 mol CO_2 mol^{-1} absorbed quanta in C_3 plants (Ehleringer and Björkman 1977). This raises the possibility that low irradiance may favour low leaf nitrogen contents and peak photosynthetic rates mainly because they maximise net leaf-level photosynthesis under shady conditions, irrespective of the costs of obtaining a given amount of nitrogen. The impact of leaf nitrogen content on photosynthesis at low and intermediate irradiance levels via its effect on dark respiration should be incorporated in any updated analysis. Note, however, that dark respiration is not tightly coupled to leaf nitrogen content in plants exposed to the same light environment, but differing in soil nitrogen supply (Gulmon and Chu 1981) or intrinsic leaf nitrogen content (Armond and Mooney 1978).

(3) Variation in leaf nitrogen content reflects variation in both leaf mass per unit area and nitrogen concentration per unit mass; at least in certain cases (e.g. Gulmon and Chu 1981), most of the difference between sun and shade leaves in nitrogen content per

unit area reflects a change in leaf density, not nitrogen concentration. Applying the Mooney–Gulmon model to such cases is inappropriate, insofar as the observed variation in leaf nitrogen content per unit area involves no difference in the total nitrogen costs associated with a given investment in leaf tissue. Analysis should instead focus on the significance of packaging a given amount of leaf tissue in an extensive photosynthetic surface of relatively thin leaves with a low mass/area ratio, or in a less extensive surface

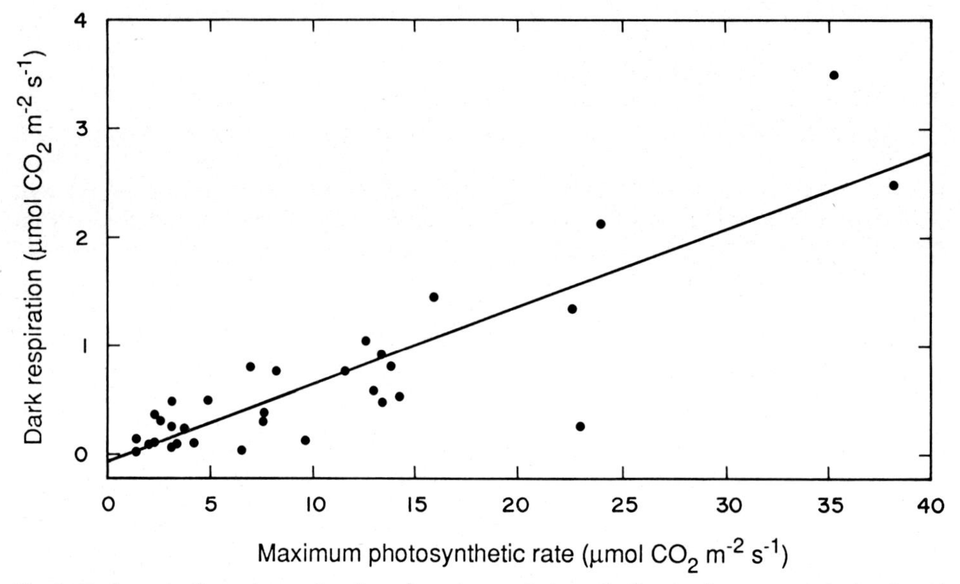

Fig. 1. Dark respiration rate as a function of maximum photosynthetic rate, for sun- and shade-adapted species and individual species acclimated to different irradiance levels. The line is $y = 0{\cdot}071x - 0{\cdot}065$, $r^2 = 0{\cdot}755$, $P<0{\cdot}001$ for 31 d.f. The outlier is *Mimulus cardinalis*; excluding it yields $y = 0{\cdot}078x - 0{\cdot}088$ ($r^2 = 0{\cdot}855$, $P<0{\cdot}001$ for 30 d.f.). Sources of data: Björkman 1968*b* (*Adenocaulon bicolor, Aralia californica, Echinodorus berteroi, Mimulus cardinalis, Plantago lanceolata, Trillium ovatum*); Björkman *et al.* 1972*b* (*Atriplex triangularis*); Boardman *et al.* 1972 (*Alocasia macrorrhiza, Cordyline rubra*); Bunce *et al.* 1977 (*Glycine max*); Clough *et al.* 1979 (*Solanum dulcamara*); Ehleringer and Björkman 1978 (*Encelia californica*); Ludlow and Wilson 1971 (*Phaseolus atropurpureus*); Patterson *et al.* 1978 (*Abutilon theophrasti, Gossypium hirsutum*); Wallace and Dunn 1980 (*Acer rubrum, Cornus florida, Liriodendron tulipifera*); Wilmot and Moore 1973 (*Silene alba, S. dioica*).

of thicker leaves with a higher mass/area ratio. Such an analysis, incorporating the effects of leaf thickness on internal self-shading, competition among chloroplasts for CO_2, total evaporative surface, and root costs associated with transpiration, has been provided in qualitative form by Givnish (1979). Any comprehensive model must include these effects, those modelled by Mooney and Gulmon (1979), and the impact of leaf mass per unit area on dark respiration and net photosynthesis at low and intermediate irradiance levels.

(4) The Mooney–Gulmon model does not incorporate the effects of stomatal conductance explicitly. Stomatal conductance also affects transpiration and associated root costs, and helps determine the photosynthetic benefit associated with a given investment in leaf nitrogen (Givnish 1986*d*). Thus, a comprehensive model for optimal leaf nitrogen content must also include selection on stomatal conductance and incorporate the costs associated with transpiration.

(5) Finally, the Mooney–Gulmon model analyses costs and benefits only at the level of individual leaves, and does not analyse their integration into an effective canopy. Field (1983) predicts that leaf nitrogen content at different points within a canopy should be adjusted so that the marginal rate of return per unit investment in nitrogen (i.e. the

increment in photosynthesis for a given increment in nitrogen content) is equal for all microsites within the canopy. Canopies satisfying this criterion would have the maximum photosynthetic return for a given total investment in nitrogen. Field (1983) presents data for the chaparral shrub *Lepechinia calyculata*, showing that the reduction in nitrogen content and photosynthetic capacity of leaves as they age is consistent with maximising canopy carbon gain, given the increased shading experienced by older leaves. Hirose and Werger (1987) present a similar analysis for *Solidago altissima*, and conclude that nitrogen retranslocation is particularly important in plants with dense canopies.

The predictions produced by these canopy-level models should diverge from those of leaf-level models, because the presence of less shaded microsites within a canopy creates an opportunity cost of maintaining nitrogen in shaded leaves. Yet, an assessment of the precise extent to which the predictions of leaf- and canopy-level models diverge must await a quantification of the root costs associated with obtaining a given amount of nitrogen. Measurement of these costs is also needed to demonstrate that the *absolute* levels of nitrogen observed by Field (1983) and Hirose and Werger (1987) actually maximise whole-plant carbon gain.

Optimal Stomatal Conductance

Stomata are the principal conduits through which CO_2 diffuses into the leaf and water vapour diffuses out. Consequently, potential increases in photosynthesis resulting from greater stomatal conductance must be weighed against the costs associated with increased transpiration, such as increased root allocation, decreased mesophyll photosynthetic capacity, and/or shortened season of photosynthetic activity (Givnish and Vermeij 1976; Givnish 1986*d*). In recent years, two different groups of models have been advanced to address this fundamental tradeoff and account for variation in stomatal conductance. Cowan and Farquhar (1977) and Cowan (1977, 1986) analysed how stomatal conductance $g(t)$ should vary diurnally in response to changing environmental conditions, based on maximising total daily photosynthesis $\int A(g,t).dt$ for a given daily total amount of transpiration $\int E(g,t).dt$. They derived the necessary criterion

$$\partial E(g,t)/\partial g(t) = \lambda \cdot \partial A(g,t)/\partial g(t) \tag{1}$$

or

$$\lambda = \frac{\partial E}{\partial A} = \frac{\partial A(g,t)/\partial g(t)}{\partial E(g,t)/\partial g(t)}, \tag{2}$$

where λ is an unspecified constant Lagrangian multiplier. Several studies have now shown that, within a given species, $\partial E/\partial A$ remains roughly constant as stomatal conductance varies in response to shifts in irradiance, relative humidity, water stress, soil fertility, and ambient CO_2 concentration (Farquhar 1979; Wong *et al.* 1979, 1985*a*, 1985*b*, 1985*c*; Farquhar *et al.* 1980*a*; Hall and Schulze 1980; Field *et al.* 1982; Meinzer 1982; Schulze and Hall 1982; Mooney *et al.* 1983; Ball and Farquhar 1984*a*, 1984*b*). These results are consistent with the Cowan–Farquhar criterion, and hence maximise total photosynthesis for a given total amount of transpiration; the question is whether the total amounts of transpiration corresponding to the observed values of λ—and hence, $g(\lambda,t)$—maximise whole-plant growth. Cowan (1986) addresses this crucial question with a model for λ based on optimal allocation of energy to roots. This model analyses how root allocation affects whole-plant growth via its effects on (i) a plant's proportion of productive tissue, (ii) its chance of exhausting its water supply between rainfalls in a stochastic climate, and thus (iii) the probable length of its period of photosynthetic activity. At present, however, there are no independently derived data to enable this model to predict a specific value of λ—and hence, a specific time course of stomatal conductance $g(\lambda,t)$—for any set of environmental conditions. Furthermore, Cowan's model clearly cannot predict a finite stomatal conductance for plants with essentially constant access to soil moisture, such as waterlilies or mangroves.

Givnish (1986*d*) analysed the question of optimal stomatal conductance and allocation to roots from a different perspective. He asked how variations in stomatal conductance and root *v.* leaf allocation would affect whole-plant growth in a *constant* environment, through their effects on (i) a plant's proportion of productive tissue, (ii) its leaf water potential, and thus (iii) its photosynthetic rate per unit leaf mass. This formulation leads to an independent derivation of a constant value of $\partial E / \partial A$, and a direct means of calculating the value of $\partial E / \partial A$ that maximises whole-plant carbon gain. Observed values of root hydraulic conductivity per unit mass, and of the sensitivity of mesophyll photosynthetic capacity to leaf water potential, permitted the quantitative prediction of both stomatal conductance and root *v.* leaf allocation in *Phaseolus vulgaris* as a function of several environmental and physiological parameters. For the single set of conditions for which data on actual stomatal conductance and root allocation were available, predicted values accord with those observed.

Both the Cowan–Farquhar and Givnish models predict that stomatal conductance should increase with irradiance unless water availability is limited, reflecting the greater potential for carbon gain at a given conductance (i.e. mesophyll photosynthetic capacity) and associated level of transpirational costs. Both predict that other factors that increase mesophyll photosynthetic capacity (e.g. nutrient supply), or decrease transpiration (e.g. relative humidity) or the costs associated with a given amount of transpiration (e.g. root hydraulic conductivity), should favour higher conductances. Both predict a linear relationship between photosynthesis and conductance as mesophyll photosynthetic capacity varies, and a curvilinear relationship as factors affecting the costs of transpiration vary. However, the benefits they assign to increased root allocation differ, involving increased length of the period of photosynthetic activity in the Cowan–Farquhar model, and increased leaf water potential and photosynthetic capacity in the Givnish model. These benefits are likely to apply in pure form only at either end of a xeric–mesic gradient of environmental conditions, with the Cowan–Farquhar model capturing the most important tradeoffs for plants growing in xeric sites whose photosynthesis can be cut short by excessive transpiration or inadequate roots, and the Givnish model capturing the most important tradeoffs for plants growing in mesic sites with essentially continuous access to soil water (Givnish 1986*d*). A comprehensive model for stomatal conductance must incorporate both sets of tradeoffs. Further work on optimal conductance must also resolve the fundamental problem of separating the root costs of water uptake from those of nutrient absorption, and take the benefits of each into account. This suggests the need for an integrated approach to the study of optimal stomatal conductance, leaf nitrogen content, leaf thickness, and root allocation (see previous section).

Photosynthetic Light Response

The photosynthetic response of individual leaves to irradiance level has been studied extensively and is fundamental to our understanding of adaptation to sun and shade (Björkman 1981). The paradigmatic study of Björkman *et al.* (1972*b*) on acclimation of the photosynthetic light response to irradiance in *Atriplex triangularis* (then known as *A. patula*) provides perhaps the best illustration of the characteristic differences in response seen in leaves grown under sunny *v.* shady conditions, and has been widely used to support the view that such differences (and related morphological and physiological traits) are adaptive. Yet, an unnoticed aspect of this classic study seems to undercut its support for adaptive variation in photosynthetic response; as shown below, this support re-emerges if the data are reanalysed in terms of energetic tradeoffs at the whole-plant level.

Björkman *et al.* (1972*b*) grew seedlings of *Atriplex triangularis* in growth chambers under a 16 h photoperiod at one of three different irradiance levels: high (920 μmol

$m^{-2} s^{-1}$), intermediate (290 μmol $m^{-2} s^{-1}$), and low 92 μmol $m^{-2} s^{-1}$). They then measured the net photosynthetic rate per unit area of leaves acclimated to these conditions as a function of irradiance ranging from 0 to 2500 μmol $m^{-2} s^{-1}$.

Björkman *et al.* (1972*b*) regarded the photosynthetic responses of leaves acclimated to different irradiance levels to be adaptive to those levels for three reasons. First, leaves grown at high irradiance had a higher maximum photosynthetic rate, and higher rate at high irradiances (>*c.* 450 μmol $m^{-2} s^{-1}$), than leaves grown at intermediate or low irradiance. Second, leaves grown at low irradiance had lower respiration rates than those acclimated to other irradiance levels. Given that all leaves showed the same initial slope (quantum yield) relating photosynthesis to irradiance, the lower respiration rate of leaves acclimated to low irradiance resulted in their having a lower light compensation point (i.e. the irradiance at which the instantaneous leaf rates of gross photosynthesis and respiration just balance). Consequently, at the lowest irradiances leaves grown at low irradiance had the highest net photosynthetic rate. Finally, leaves grown at intermediate irradiance had a higher respiration rate and light compensation point, and their photosynthesis saturated at higher irradiance, than did those grown at low irradiance; leaves grown at high irradiance had the highest respiration rate and light compensation point, and their photosynthesis saturated at the highest irradiance. These findings suggested that leaves acclimated to each irradiance level are those best adapted to that level, insofar as they appear to have the highest rate of leaf photosynthesis under those conditions.

Close examination of the photosynthetic curves published by Björkman *et al.* (1972*b*) reveals that at the low irradiance of 92 μmol $m^{-2} s^{-1}$, leaves acclimated to that irradiance appear to have a net photosynthetic rate similar to those acclimated to intermediate irradiance. At the intermediate irradiance of 290 μmol $m^{-2} s^{-1}$, leaves acclimated to that irradiance also appear to have a rate similar to those acclimated to high irradiance. Thus, in two of three instances, there is no clear indication that leaves acclimated to a given irradiance level have the highest photosynthetic rate under those conditions. This paradox cuts to the very heart of previous interpretations of adaptation to irradiance level.

Part of the problem is that comparisons based on leaf photosynthesis per unit area fail to include the energetic costs associated with night leaf respiration and leaf construction. Such costs must be considered in calculating the net benefit to the plant of different kinds of leaves, even in the absence of other differences (e.g. leaf orientation, canopy architecture, root allocation) that could override sun/shade differences in the photosynthetic rate of individual leaves.

For plants grown in growth chambers with an on/off light regime (like those studied by Björkman *et al.* 1972*b*), night leaf respiration can be incorporated by calculating leaf carbon balance based on a 24 h cycle. Leaf construction costs are more difficult to quantify because they entail not only the cost, in terms of fixed carbon, of synthesising various leaf compounds, but also the cost of constructing and maintaining roots and stems needed to obtain the nutrients required to synthesise such leaf compounds (Mooney and Gulmon 1979). These construction costs (C) must be set against 24 h leaf carbon balance (P) and expected leaf lifetime (T) to yield the net energetic return ($PT-C$) expected from a leaf, exclusive of other associated root and stem costs and the effects of intracanopy shading. The net *rate* of return per unit investment in leaf tissue is an important determinant of the overall rate of plant growth, and is given by $(PT-C)/(TC) = P/C-1/T$. Leaf lifetime T is not independent of the ratio P/C, because more rapidly growing plants with higher ratios of leaf photosynthetic rate to construction costs will more rapidly shade their lower leaves, favouring their senescence (Field 1983) at a more rapid rate than in less rapidly growing plants. The ratio P/C may therefore be a useful index of the rate of energetic return per unit investment in leaves.

Thus, from an economic point of view, it might prove more illuminating to express photosynthesis not per unit leaf area, but per unit leaf mass (or, perhaps, per unit leaf nitrogen content). Leaf construction costs per unit area should scale like leaf biomass per unit area, provided the leaves in question do not vary much in composition (Osmond *et al.* 1980); if the latter were true, P/C would be directly proportional to photosynthesis per unit leaf mass. However, leaf composition does vary with irradiance, notably in the fraction devoted to soluble protein (Björkman 1981) and nitrogen (Field and Mooney 1986). In the limiting case in which the costs of acquiring nitrogen for protein are very large compared with the cost of synthesising other leaf compounds, photosynthesis per unit leaf nitrogen (or protein) might be a better measure of P/C.

Based on these considerations, I re-analysed the data of Björkman *et al.* (1972*b*) to determine whether the paradox that emerges if photosynthesis is measured per unit area disappears if it is instead measured per unit leaf mass or per unit leaf protein, as more appropriate measures of net contribution to whole-plant carbon gain. In the absence of additional data that would permit the use of more sophisticated photosynthetic models (e.g. Farquhar *et al.* 1980*b*), daytime photosynthesis P_d was approximated using the standard Michaelis–Menten model:

$$P_d = P_{max}\ I/(I+k) - R, \qquad (3)$$

where I is irradiance, k is the Michaelis–Menten constant, R is the dark respiration rate, and $P_{max} - R$ is the maximum photosynthetic rate. Published data were fitted to the model by substituting the reported respiration rate for R, and replacing k with the reported irradiance required to achieve half the maximum photosynthetic rate. The value of P_{max} was then adjusted to yield the reported maximum photosynthetic rate at 2500 μmol m^{-2} s^{-1}, which is the irradiance at which the maximum rate was measured for leaves acclimated to high and intermediate irradiance; the light-saturated level of leaves acclimated to low irradiance was measured at a somewhat lower irradiance in order to avoid the effects of photoinhibition. The values of P_{max}, k, and R used in the calculations presented below are given in Table 2. Leaf carbon balance over a 24 h period was calculated using the formula $P = (16P_d - 8R)/24$, based on the 16 h on/off photoperiod. The average values reported by Björkman *et al.* (1972*b*) for total dry mass and soluble protein content per unit area for leaves acclimated to each irradiance level were used to convert carbon balance per unit area to carbon balance per unit mass and per unit soluble protein, respectively.

Table 2. Parameters used to fit data of Björkman *et al.* (1972*b*) on photosynthetic light response of *Atriplex triangularis* grown at high (920 μmol m^{-2} s^{-1}), intermediate (290 μmol m^{-2} s^{-1}), or low (92 μmol m^{-2} s^{-1}) irradiance

Conditioning irradiance	P_{max} (μmol m^{-2} s^{-1})	R (μmol m^{-2} s^{-1})	k (μmol m^{-2} s^{-1})
High	43·72	3·48	429
Intermediate	28·32	2·12	264
Low	7·73	0·83	77

Calculated leaf carbon balance per unit leaf area, mass, and soluble protein are plotted as a function of incident and conditioning irradiance in Fig. 2. The daily carbon balance per unit area of leaves acclimated to low irradiance is 29% lower at low irradiance than that of leaves acclimated to intermediate irradiance (Fig. 2*a*). Similarly, carbon balance per unit area of leaves acclimated to intermediate irradiance is 6% lower at intermediate irradiance than that of leaves acclimated to high irradiance. Thus, when carbon balance is expressed per unit area, leaves acclimated to a specific irradiance do *less* well than those acclimated to another irradiance in two of three cases.

However, if carbon balance is expressed per unit leaf mass, or per unit soluble leaf protein, a different picture appears (Figs 2*b* and 2*c*). In both cases, leaves acclimated to low irradiance do better at low irradiance than those acclimated to other irradiances; leaves acclimated to intermediate irradiance do better than others at intermediate irradiance; and leaves acclimated to high irradiance do better than others at high irradiance.

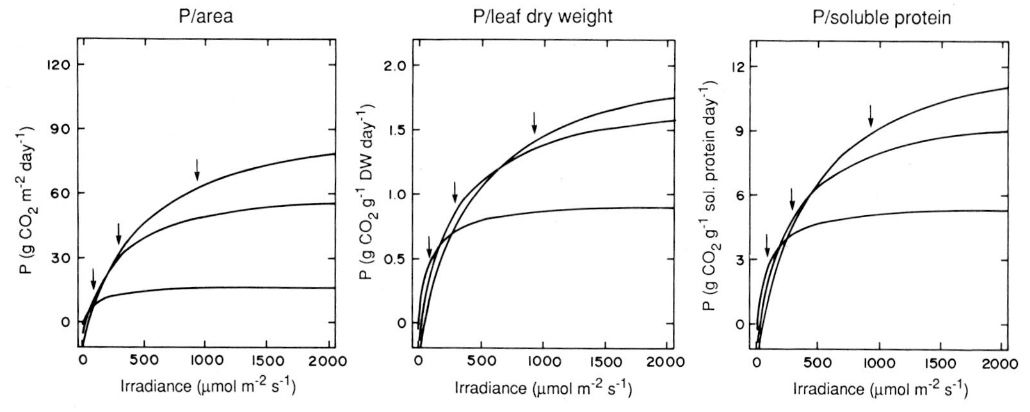

Fig. 2. Photosynthesis as a function of irradiance in *Atriplex triangularis*, expressed as daily carbon balance per unit leaf area (*a*), mass (*b*), and soluble protein content (*c*) (see text). Within each graph, arrows indicate the high, intermediate, and low irradiance levels to which leaves were acclimated. In each graph, leaves acclimated to high irradiance have the greatest photosynthetic rates at high irradiance levels; those acclimated to intermediate irradiance have intermediate rates; and those acclimated to low irradiance have the lowest rates. However, leaves acclimated to a given irradiance have the greatest photosynthetic rate at that irradiance in all cases only if photosynthesis is expressed per unit investment in mass or soluble protein.

This suggests that expressing photosynthesis and respiration as a function of leaf mass or protein content, and thereby indirectly incorporating leaf construction costs, may be more useful in assessing adaptation to light level than expressing them as a function of leaf area. Reporting photosynthesis per unit area affords ease of analysis for such traits as quantum yield, but may bias one's perspective by ignoring the effect of leaf construction costs on whole-plant energy capture. The reason for differences in the results based on leaf area *v.* leaf mass or soluble protein is clear: leaves acclimated to low irradiance levels have low photosynthetic rates per unit area at those levels, but also have a much lower biomass and soluble protein content per unit area than leaves acclimated to higher irradiance levels. When photosynthesis is expressed per unit investment in either leaf mass or soluble protein, the higher rates of return at low irradiances of leaves acclimated to those conditions become apparent.

Four points should be made in connection with the preceding analysis. First, it does not incorporate differences between plants acclimated to different irradiances in root or stem costs, leaf orientation, or total leaf area. Although data do not exist to quantify root costs (*sensu* Givnish 1986*d*), they seem unlikely to affect this particular analysis, given that calculated transpiration rates per unit leaf mass at different irradiances differ little between leaves acclimated to those irradiances.

Second, the Michaelis–Menten equations used in the preceding calculations (equation 3, Table 2) provide a reasonable but inexact fit to the data of Björkman *et al.* (1972*b*). Consequently, the conclusions drawn should be viewed as tentative until confirmed using more sophisticated modelling techniques. Third, although a comparative analysis of photosynthesis of sun- and shade-acclimated leaves under *constant* low irradiance may be warranted for plants grown in growth chambers, such as *Atriplex triangularis* in the case considered above, such an approach would greatly oversimplify the factors affecting relative gas exchange under the dynamic, highly variable light regimes prevailing in natural shaded habitats. Research by Pearcy and his colleagues (Pearcy *et al.* 1985; Chazdon and Pearcy 1986*a*, 1986*b*; Chazdon 1986) on trees growing in tropical forest understories has elegantly demonstrated the importance of photosynthetic induction and non-steady-state behaviour in sunflecks for gas exchange under 'shady' conditions.

Finally, Osmond *et al.* (1980) presented an analysis similar to that given here, but reached different conclusions based on two inappropriate assumptions. They found that leaves acclimated to a given irradiance have a greater 24 h carbon balance at that irradiance than leaves acclimated to another irradiance, regardless of whether photosynthesis is expressed per unit area or per unit mass. However, they assumed a sinusoidal variation in irradiance during the photoperiod, rather than the on/off switch to which the leaves were actually acclimated. A sinusoidal variation introduces a daytime period during which leaves acclimated to high irradiance would be below their compensation point in chambers at 'low' irradiance, and artificially inflates the advantage of leaves acclimated to low irradiance under such conditions. More importantly, comparisons were made only between leaves acclimated to low and high irradiances, at only those two irradiance levels. Thus, Osmond *et al.* (1980) did not recognise the crucial complications introduced by leaves acclimated to intermediate irradiance which—when carbon balance is expressed per unit leaf area—do better than leaves acclimated to low irradiance at that irradiance, and worse than leaves acclimated to high irradiance at intermediate irradiance (Fig. 2*a*).

Economics of Support

This section addresses three aspects of leaf arrangement likely to be shaped by tradeoffs involving the economics of support: leaf area index, relative crown width, and spiral *v.* distichous phyllotaxis. All three traits involve the integration of individual leaves into an effective photosynthetic canopy. Analysis of optimal variation in the first trait requires a whole-plant perspective on how to measure light compensation point and photosynthetic light response, and bears on the issue of maximum plant height at a given irradiance. The second trait entails allocation to leaves *v.* support tissue in dynamic canopies. The third involves a key trait at the twig, rather than the canopy, level. Additional constraints on the evolution of phyllotaxis are reviewed by Givnish (1979, 1984).

Leaf Area Index, Effective Compensation Point and Maximum Plant Height

Leaf area index (LAI) is the ratio of a plant's total leaf area to the horizontal area covered by its canopy. In other words, LAI is the average number of leaf layers over each point occupied by a canopy. Horn (1971) presented an early, highly simplified model for evaluating the relative photosynthetic outputs at a given irradiance from canopies that differ in their number of leaf layers (see also Blackman and Black 1959). This model does not incorporate many traits and processes that help determine whole-canopy photosynthesis and whole-plant carbon gain, including: within-crown variation in leaf orientation, distribution, temperature, and photosynthetic capacity; movement of the sun across the sky; shading by adjacent competitors; shading by a plant's own trunk and branches; costs of building and maintaining support tissue; and eddy transfer of CO_2 through canopy. Nevertheless, Horn's model has the notable advantage of exposing and analysing in the clearest possible terms one tradeoff central to adaptive variation of LAI, involving the balance between number of leaf layers and photosynthesis per layer. Canopies with more leaf layers have a greater potential for carbon gain because they contain more photosynthetic tissue, but they also entail more self-shading, reducing the rate of photosynthesis in some of the lower layers and raising the possibility that some leaves may not yield a net energetic return.

Horn (1971) concluded that optimal LAI depends on irradiance and the photosynthetic light response of individual leaves. To maximise the net rate of instantaneous canopy photosynthesis, a plant should add leaf layers until the net return from the bottom, most heavily shaded layer is zero. That is, leaf area index should be adjusted until the irradiance impinging on the most heavily shaded leaves equals their compensation point, at which their instantaneous rates of photosynthesis and respiration just balance. If LAI were any smaller than this optimal value, enough light would penetrate the plant's canopy that adding more leaves would be profitable; if LAI were any larger, too little light would penetrate to the lowest leaves to permit positive net photosynthesis, and shedding some leaves would be profitable. Higher levels of ambient irradiance increase the irradiance that penetrates through a given number of leaf layers, so that optimal LAI should increase with ambient irradiance. Sunlit plants should thus arrange their leaves diffusely in several layers, shaded plants should arrange their leaves in fewer layers, and in extreme shade plants should pack their leaves tightly into a single layer with no overlap.

This model helps explain why tree species with multilayered canopies grow rapidly in sunlit conditions and thus compete successfully in open habitats and early in succession, and why trees with more nearly monolayered canopies grow rapidly relative to competitors in denser shade later in succession and in forest understories. Horn (1971) also showed how this model can help account for successional trends in species composition and diversity. Although the model could be made more precise by incorporating some of the traits and processes listed earlier, it already serves to explain, at least in qualitative terms, many of the ecologically salient features regarding the canopy form, successional status, and diversity of temperate tree species adapted to different irradiance levels.

Yet, there are two assumptions inherent to this simplest model, involving leaf compensation point and photosynthetic light response, that must be modified if the model is to apply—even on its own terms—to whole plants. The first assumption—that all leaves in a given canopy share the same photosynthetic light response—can be easily relaxed to take into account acclimation of leaves to different irradiance regimes within the canopy. All that is needed is to replace the photosynthetic light response of a leaf acclimated and adapted to a single irradiance regime, with photosynthesis as a function of the irradiance to which a leaf is acclimated (i.e. photosynthesis at the irradiance to which a leaf is acclimated, measured over a series of leaves acclimated to different irradiances).

The second assumption—that the irradiance required for an energetic 'break-even' for a leaf is given by its light compensation point—is clearly inaccurate. Such a view does not take into account several costs—such as night leaf respiration and leaf construction—that must be balanced against leaf photosynthesis to determine a leaf's net return to the plant (Givnish 1984). Insofar as a leaf's energetic break-even point determines not only how many layers of leaves a plant should hold, but also whether it can survive in a given irradiance regime (cf. Björkman *et al.* 1972*a*, 1972*b*; Björkman 1981), it is natural to ask if these additional costs would significantly affect the break-even irradiance, and if these costs would swamp the effect of observed variation in instantaneous leaf light compensation points.

To answer these questions, let us consider some model calculations using photosynthetic data gathered by Wallace and Dunn (1980) on the relatively shade-intolerant *Liriodendron tulipifera* (Magnoliaceae), a multilayered tree native to forests in the eastern United States (Table 2). Instantaneous net leaf photosynthesis P_{net} (g CO_2 g^{-1} leaf s^{-1}) can be roughly modelled using Michaelis-Menten kinetics as:

$$P_{net} = P_{max}\ I/(I+k) - R_{inst}, \quad (4)$$

where R_{inst} is the instantaneous rate of leaf respiration (g CO_2 g^{-1} leaf s^{-1}), $P_{max} - R$ is the maximum net rate of photosynthesis (g CO_2 g^{-1} leaf s^{-1}), P_{max}/k is the initial slope (g CO_2 g^{-1} leaf s^{-1} μmol^{-1} photons m^2 s) of the photosynthetic light response curve, and I is irradiance ($\mu mol\ m^{-2}\ s^{-1}$). For *Liriodendron* leaves grown at 10% of full solar irradiance in a lathhouse in Georgia, U.S.A., $R_{inst} = 8{\cdot}27 \times 10^{-8}$ g g^{-1}, $P_{max} = 1{\cdot}63 \times 10^{-6}$ g g^{-1} s^{-1}, and $k = 236\ \mu mol\ m^{-2}\ s^{-1}$ (Wallace and Dunn 1980). Leaves grown in full sunlight show no significant differences from shade-grown leaves in P_{max}, P_{max}/k or R_{inst}.

The instantaneous leaf light compensation point I_{cp}, usually viewed as *the* light compensation point, is defined by $P_{net} = 0$. Equation (4) implies that $I_{cp} = R_{inst}\ k/ (P_{max} - R_{inst})$. For *Liriodendron* leaves grown at 10% sunlight, I_{cp} is thus roughly 13 $\mu mol\ m^{-2}\ s^{-1}$ (Fig. 3). To calculate the effective energetic break-even irradiance for a leaf—its *ecological compensation point*—at which total leaf benefits and costs just balance, R_{inst} must be augmented by six additional costs (Givnish 1984). These include (i) night leaf respiration; (ii) effective daily cost of leaf construction, amortised over the

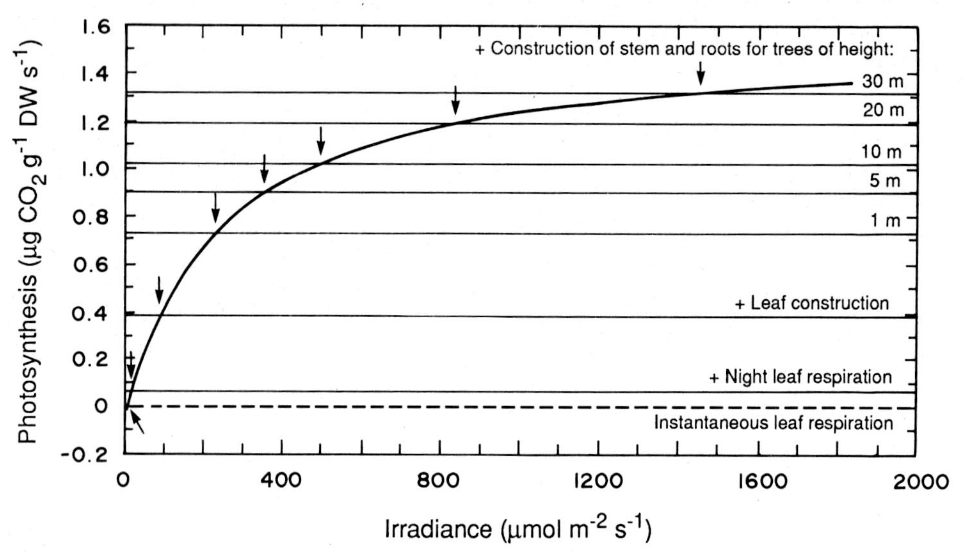

Fig. 3. Effective leaf compensation point in *Liriodendron tulipifera* as a function of the inclusion of various respiratory costs. The curve is the instantaneous rate of net leaf photosynthesis, plotted as a function of irradiance; the dashed line is the instantaneous rate of leaf respiration; the arrow at their intersection marks the traditional compensation point. The solid lines represent the cumulative respiration rates associated with night leaf respiration, leaf construction, and construction of support and root tissue (see text); the corresponding arrows mark the effective compensation points associated with including each additional source of respiration.

life of a leaf; (iii) marginal costs of roots, xylem, and phloem needed to supply an additional leaf; (iv) marginal mechanical cost of supporting an additional leaf in a given position; and (v) expected loss of productivity due to herbivory or disease. The ecological compensation point must be further increased to account for the fact that, although a leaf may be operating above its compensation point as determined by the preceding five costs, its net photosynthesis may be so low that it would pay the plant to extract nutrients from the leaf and place them in a new, well-lit leaf (Field 1983). This sixth energetic decrement in leaf productivity might best be considered an opportunity cost; Field (1988) provides an innovative approach to incorporating this cost in the calculation of optimal LAI.

The five remaining costs are incorporated in the following, highly simplified calculation of an ecological compensation point for *Liriodendron* leaves, assuming constant irradiance during the photoperiod and constant temperature. Data for shade-grown leaves are used in order to estimate compensation points as conservatively as possible, and to produce estimates applicable to the lowermost, shaded leaves of a multilayered canopy and, thus, to Horn's (1971) model. Compensation points calculated for sun-grown leaves are slightly higher than those reported for shade-grown leaves, but the underlying differences between them in photosynthetic parameters are not significant (see above).

Night leaf respiration

If the dark period is a fraction α of the photoperiod, the instantaneous rate of leaf respiration should be multiplied by a factor $1+\alpha$ to correct for nocturnal respiration. This results in an augmented instantaneous respiration rate R_{inst} which can be balanced against the instantaneous daytime rate of gross photosynthesis to yield P_{net}. Our model yields $I'_{cp} = R(1+\alpha)k/(P_{max} - R(1+\alpha))$. For a dark period which fluctuates from about 66 to 100% of the length of the photoperiod during the growing season in Georgia, an average of $\alpha = 0{\cdot}83$ may be substituted, yielding $I'_{cp} = 24\ \mu mol\ m^{-2}\ s^{-1}$ (Fig. 3).

Leaf construction

The instantaneous daytime respiration rate equivalent to the amortised cost of leaf construction may be approximated as the initial cost of leaf construction β (g CO_2 g^{-1} leaf), divided by the lifetime T (s) of the leaf during the daylight hours. This approximation is nearly exact when leaf production is continuous during the growing season, and there is little growth in total leaf mass. Estimates of β differ somewhat: Mooney (1972) estimates that the uptake of 1 g CO_2 yields 0·68 g glucose, which in turn yields 0·28 g leaf; the conversion efficiencies cited by Raven (1986) suggest instead that 1 g CO_2 yields 0·48 g leaf. Averaging these values yields $\beta = 2{\cdot}63$ g CO_2 g^{-1} leaf; estimating T as roughly $8{\cdot}5 \times 10^6$ s (= 6 months $\times$ 30 photoperiods per month $\times$ 13·1 h per photoperiod $\times$ 3600 s h^{-1}), we thus obtain $R_{const} = 3{\cdot}10 \times 10^{-7}\ g\ g^{-1}\ s^{-1}$. This additional respiration adds 69 $\mu mol\ m^{-2}\ s^{-1}$ to the compensation point, yielding $I''_{cp} = 93\ \mu mol\ m^{-2}\ s^{-1}$ (Fig. 3).

Allocation to stem and root tissue

The fraction of production allocated to stem tissue increases with plant height, reflecting the disproportionate increase in support tissue required to maintain mechanical stability (Givnish 1982). Whittaker and Woodwell (1968) present allometric equations relating plant height and annual production of trunk, branch, bark, twig and leaf tissue to basal diameter in woody species spanning a height range from short shrubs to trees roughly 30 m tall, including *Liriodendron*. These equations can be used to calculate the fraction of biomass production allocated to leaves as a function of height in woody plants (Fig. 4). As can be seen, the fractional allocation to leaves declines roughly linearly with the logarithm of plant height, from 46·9% in plants 1 m tall to 24·8% in plants 30 m tall. For purposes of comparison, data on fractional allocation of shoot biomass to leaves in forest herbs from Givnish (1982) are also plotted in Fig. 4; although these data are not completely comparable to those for woody plants because they do not include net annual root production, that omission should not have a large effect. Note that herbs seem mechanically more efficient than woody plants at canopy heights less than about 0·5 m, allocating less of their annual production to support tissue and more to leaves, but are less efficient at greater heights (Fig. 4). This is because herbs have less durable, less costly support structures than woody plants at any given canopy height, but must produce their entire support structure each year, whereas woody plants merely add small increments to their existing structure. The greater mechanical efficiency of woody plants taller than 0·5 m is probably an important reason why there are so few species of forest herbs much greater than that height, together with the inability of taller herbs to exceed their ecological compensation points and achieve positive carbon balance in deep shade (see Givnish 1984; Raven 1986; and analysis below).

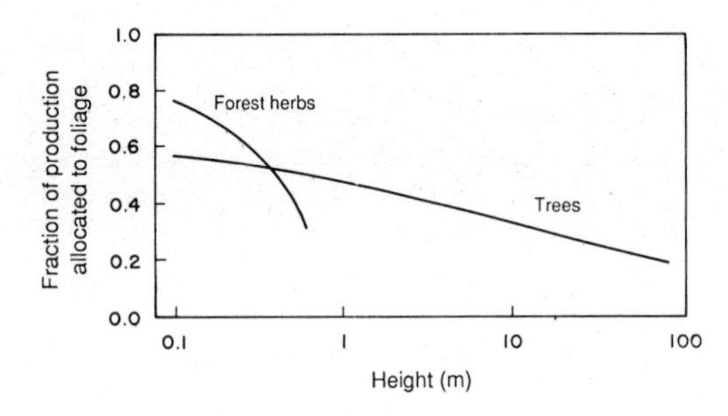

Fig. 4. Proportion of annual biomass production allocated to foliage as a function of plant height. The curve for trees is based on allometric regressions given by Whittaker and Woodwell (1968), root production estimated as 20% of stem production, and a 71% allocation to leaves in the leaf–twig fraction of *Liriodendron* (Whittaker *et al.* 1963). The curve for forest herbs is based on data of Givnish (1982), but excludes root production.

If we assume energy allocation parallels biomass allocation, then if γ is the fraction of biomass allocated to leaves, the energy allocated to non-leaf tissue should be $(1-\gamma)/\gamma \cdot R_{const}$. Given the biomass allocation patterns shown in Fig. 4, the additional respiratory cost associated with stem and root production range from $1 \cdot 13 R_{const}$ in woody plants 1 m tall, to $3 \cdot 03 R_{const}$ in trees 30 m tall. Adding this cost to those associated with night leaf respiration and leaf construction, we find that the ecological compensation point (I_{ecp}) of a *Liriodendron* leaf increases from roughly 233 μmol m^{-2} s^{-1} in plants 1 m tall, to 355 μmol m^{-2} s^{-1} in plants 5 m tall, to 1438 μmol m^{-2} s^{-1} in plants 30 m tall (Fig. 3). In other words, even though *Liriodendron* has an instantaneous leaf compensation point of 13 μmol m^{-2} s^{-1}, its saplings must receive at least 233 μmol m^{-2} s^{-1} to survive to 1 m, and at least 355 μmol m^{-2} s^{-1} to survive to 5 m; even very short seedlings must receive at least 24 μmol m^{-2} s^{-1} (see calculation of I'_{cp} above).

The steeply non-linear rise in I_{ecp} with tree height results from the precise nature of the plateauing of the photosynthetic light response given by Michaelis–Menten kinetics; if the actual response were to saturate more sharply, I_{ecp} would not rise as rapidly. Even so, given *Liriodendron*'s maximum photosynthetic rate, costs of leaf respiration and construction, and scaling of support and root costs with height, *Liriodendron* exposed to light-saturating irradiance should grow no taller than 50 m. Reduced allocation to root production on unusually moist or fertile sites could raise this to an absolute maximum height of 77 m. (Although these conclusions are based on the values of P_{max} and R_{inst} for individual shade-grown leaves, in *Liriodendron* they should apply to entire crowns, given that P_{max} and R_{inst} do not vary significantly between sun- and shade-grown leaves. *Liriodendron* was chosen for analysis based on this latter fact, which obviates many potential complications arising from differences in irradiance and acclimation at different levels within a crown.) These estimates compare with a height of 60 m for the tallest *Liriodendron* individual known (Preston 1976), found growing in Great Smoky Mountains National Park, tending to confirm Whittaker and Woodwell's (1968) hypothesis that the height-dependent scaling of support costs may determine maximum plant height. Note that both estimates underestimate stem and root costs by ignoring respiration by these organs.

Herbivory

Insects consume roughly 7% of annual leaf production in broad-leaved temperate forests in North America (Whittaker and Woodwell 1968). If we assume that such consumption occurs, on average, midway through the growing season, the average instantaneous rate of net photosynthesis must be discounted by roughly 3·5%. Except when other costs raise the ecological compensation point to irradiances near saturation, herbivory should thus have a relatively small effect on I_{ecp}.

Conclusions

For *Liriodendron* the inclusion of the costs of night leaf respiration, leaf construction, and associated stem and root tissue results in a staggering 7- to 110-fold increase in the compensation point over that expected based on instantaneous leaf respiration alone. The increase from 12 to 1438 μmol m^{-2} s^{-1} in compensation point due to such factors is huge compared to the range of instantaneous leaf compensation points seen among vascular plants, and suggests that certain of these factors—notably the construction costs of leaves, stems, and roots—may have an overwhelming effect on a plant's ecological compensation point. Of what significance is an instantaneous leaf compensation point of 1 μmol m^{-2} s^{-1}, or 0·001 μmol m^{-2} s^{-1}, if tissue construction costs raise the ecological compensation point by 100 μmol m^{-2} s^{-1}?

The foregoing analysis suggests the prime importance of support costs and, by inference, plant height in determining a plant's ability to persist in deep shade. The costs of leaf and root production may be reduced in shade-adapted leaves because they possess lower enzyme levels and transpire at lower rates. However, the support costs outlined above are essentially fixed as a function of plant height, no matter what are the mass and gas-exchange properties of the leaves. These support costs are likely to insure that the ecological compensation points of tall trees are usually greater than those of herbs, regardless of differences in leaf photosynthetic light response (Givnish 1982, 1984; Raven 1986). The fact that these costs are minimal in very short plants suggests that such plants may develop the lowest leaf respiration rates in shaded environments. This is because the photosynthetic light response of individual leaves would be most important as a determinant of shade tolerance in plants of the shortest stature. As shown quantitatively above, the height-dependent scaling of support costs may also limit maximum plant height, perhaps in concert with the effects of other height-dependent factors, such as reduced leaf water potential and/or increased susceptibility to drought.

A whole-plant perspective also suggests that plants in moister or more fertile sites may generally have lower ecological compensation points, based on reduced root costs (Givnish 1984), achieve higher maximum height and LAI at a given irradiance, and tolerate lower irradiance levels at a given height. Indeed, along gradients of increasing rainfall, forests generally increase in stature and LAI (e.g. Grier and Running 1977; Waring *et al.* 1978). The quantitative predominance of support and root costs in setting the ecological compensation point of *Liriodendron* strongly suggests that the investigation of shade adaptation requires a synthesis of studies aimed at roots, stems and leaves, and should incorporate the effects of water and nutrient supplies in determining the ability of a plant to persist at a given irradiance level.

Relative Crown Width

In addition to LAI, an important aspect of canopy geometry determining whole-plant carbon gain is mechanical efficiency, the fraction of energy allocated to foliage *v.* unproductive support tissue. Yet, because a plant's canopy is both an organ of energy capture and a direct instrument of competitive interference, analyses based purely on mechanical efficiency are inadequate. Among canopies having an adaptive LAI, the most mechanically efficient are extremely short and narrow, with horizontal and vertical lever arms of vanishingly small length and cost (Givnish 1982). Yet short plants are readily overtopped, and narrow canopies can incorporate little leaf area. A more important question, at least for woody plants of indeterminate growth, is which canopy geometries allow the greatest rate of height growth.

King (1981) addressed an important aspect of this question and presented a quantitative model for energy allocation between a tree's canopy (leaves and branches) and trunk that would maximise its overall rate of height growth. The fundamental tradeoff involved can be summarised as follows. If a tree allocates almost all of the photosynthetic return from its canopy into the trunk, it will produce little new foliage to fuel further height growth. As fractional allocation to the canopy increases, so does foliage area and canopy output, but with two costs. First, fractional allocation to the trunk is reduced; second, the more massive the leaf canopy, the more massive and costly the branches that must be replaced as the tree grows upward. At very high canopy allocations, these costs would reduce height growth to zero. Thus, an intermediate allocation to canopy *v.* trunk produces optimal height growth.

King (1981) tested this model using allometric data on branch and trunk mass in aspen (*Populus tremuloides*) growing in crowded stands undergoing self-thinning. The observed allocation between canopy and trunk was close to that predicted, and resulted in a calculated rate of growth nearly equal to the maximum possible rate. Perhaps not coincidentally, the ratio of canopy diameter to tree height observed in these optimally designed aspens (0·17) is very nearly the same as that seen across 19 single-species stands undergoing self-thinning, and across 79 mixed-species forest stands (Givnish 1986*g*). This suggests that selection, at least in crowded stands, does favour canopies with a geometry that maximises height growth; Givnish (1986*g*) used this canopy diameter/height ratio, in combination with constraints on LAI and the biomechanical properties of wood, to explain the slope and intercept of the $-3/2$ power law of self-thinning (White 1981).

How should the optimal width/height ratio of a tree's canopy vary with irradiance? In King's model, terms involving photosynthetic rate per unit canopy mass cancel, and the optimal allocation between trunk and canopy should not vary with irradiance. However, reductions in irradiance reduce the optimal density of leaves per unit area of ground occupied by the canopy (i.e. LAI), tending to reduce the mass of leaves and branches associated with a given canopy diameter. Thus, given that canopy mass should increase monotonically with canopy diameter, shady conditions should favour broader canopies.

There is another, more subtle reason why this should be true. On moist, relatively fertile sites, the canopies of understory shrubs or trees are likely to cover a lower proportion of the ground than are trees in the forest canopy: there are simply fewer microsites where ambient irradiance exceeds the ecological compensation point of such plants. Consequently, subcanopy plants will often find themselves surrounded by empty space. In the limiting case in which an understory tree—incapable of reaching the canopy—has *no* neighbours, growing upwards and slightly outwards in accordance with the canopy diameter/width ratio predicted by King's model will not increase energy capture as much as simply growing outwards. This is because vertical growth yields no benefit (at least in terms of light capture), whereas horizontal growth expands the area over which light can be captured. Thus, extremely uncrowded understory conditions should favour very broad, low canopies that maximise a plant's net rate of energy capture in the absence of competitors. Completely crowded conditions, on the other hand, place height growth at a premium, and narrower canopies that maximise height growth are favoured. At intermediate levels of subcanopy coverage, selection should favour trees with diameter/height ratios which are intermediate between these extremes, and which maximise net energy capture in the *presence* of competitors.

The evolutionarily stable diameter/height ratio for understory trees, which results in a net rate of energy capture that cannot be bettered by competitors with another ratio, must depend on subcanopy coverage and the resulting probability of a tree being adjacent to, and shaded by, a competitor. At very low subcanopy coverage, very broad canopies that maximise energy gain in the absence of competitors also maximise energy gain in the presence of competitors. At very high subcanopy coverage, narrow canopies with a diameter/height ratio in accord with King's (1981) model are the only ones not surely overtopped, and again maximise energy gain in the presence of competitors. Canopies that expand at an intermediate diameter/height ratio should maximise energy gain realised in subcanopies of intermediate coverage, reflecting a balance between the energetic costs of outgrowing all possible neighbours, the chance of encountering a neighbour, and the energetic benefits of a broad canopy. To the extent that stratal coverage is lower in shade than in sun, and lower in the subcanopy than the forest canopy, the diameter/height ratio of individual trees should be greater in subcanopy trees and shade-adapted canopy species.

Data to test these ideas were culled from the register of 'champion trees' for North America (Hunt 1986). Unusually large individuals of each species qualify as 'champions' based on a combination of their girth, height and canopy spread. Undoubtedly, most grow on sites that are, or have been, more open than is typical for their species; in that sense, the use of such individuals may reduce somewhat the effects of systematic differences between shade-tolerant and shade-intolerant species in the environmental conditions experienced during ontogeny. Canopy diameter/height ratios were calculated for all tree species occurring in eastern North America that could unequivocally be assigned to one of three classes: obligate subcanopy species, shade-tolerant canopy species, and shade-intolerant canopy species (Table 3). The expectation was that obligate subcanopy species, experiencing shady conditions throughout their lives, would have the greatest diameter/height ratios; that shade-tolerant species that achieve the canopy would have somewhat lower ratios; and that shade-intolerant canopy species would have the lowest ratios. This is exactly what is seen (Table 3): the mean diameter/height ratio for subcanopy species is $1 \cdot 09 \pm 0 \cdot 47$ ($n=8$); for shade-tolerant canopy

Table 3. Ratio of canopy diameter to height in selected 'champion' trees of eastern North America (see text)

Subcanopy species		Shade-tolerant canopy species		Shade-intolerant canopy species	
Acer pensylvanicum	0·49	*Acer saccharinum*	1·07	*Betula papyrifera*	0·70
Acer spicatum	0·55	*Acer saccharum*	0·87	*Liriodendron tulipifera*	0·98
Amelanchier arborea	1·30	*Betula allegheniensis*	1·20	*Nyssa sylvatica*	0·59
Carpinus caroliniana	1·55	*Fagus grandifolia*	0·58	*Platanus occidentalis*	0·81
Cercis canadensis	1·83	*Tilia americana*	0·66	*Populus grandidentata*	0·63
Cornus alternifolia	1·67			*Populus tremuloides*	0·43
Hamamelis virginica	0·86			*Prunus pensylvanica*	0·35
Lindera benzoin	0·96			*Robinia pseudoacacia*	0·96
Ostrya virginica	1·21			*Sassafras albidum*	0·49
				Sorbus americana	0·65

species, $0 \cdot 88 \pm 0 \cdot 26$ ($n=5$); and for shade-intolerant canopy species, $0 \cdot 66 \pm 0 \cdot 21$ ($n=10$). The distributions of the ratios for each group of species differs significantly in the direction expected ($P<0 \cdot 05$, Wilcoxon paired-rank test). These results confirm the trend predicted, and should be followed by further tests and a quantitative version of the model. It should be noted that the coverage-dependent argument for canopy diameter/height ratio in trees of indeterminate growth is, in many ways, analagous to the game-theory model for the evolution of leaf height in forest herbs of determinate growth, advanced and quantitatively tested by Givnish (1982, 1986*f*).

Orthotropy v. *Plagiotropy*

At a smaller scale of canopy geometry than overall canopy width and leaf area index, two modes of leaf arrangement, or phyllotaxis, are especially common. These involve the packing of leaves in spirals or whorls about erect twigs (spiral phyllotaxis on orthotropic axes), or in planar arrays along more or less horizontal axes (distichous phyllotaxis on plagiotropic axes) (Hallé *et al.* 1978). Spiral phyllotaxis on erect axes is common in sun plants, and distichous phyllotaxis on horizontal axes is common in shade plants (Leigh 1972, 1975; Hallé *et al.* 1978). Why should this be the case?

Givnish (1984) provides a qualitative explanation for this pattern, assessing the impact of phyllotaxis on twigs as organs of energy capture and as direct instruments of competitive interference. As organs of energy capture, orthotropic axes with a spiral phyllotaxis have an advantage in sunny environments because (i) erect lever arms can support more leaf mass per unit twig mass; (ii) greater self-shading relative to distichy would have a relatively small effect on carbon gain at irradiance levels near photosynthetic saturation; and (iii) such self-shading would substantially decrease heat load and transpirational costs in a high-radiation environment. Plagiotropic axes with a distichous phyllotaxis should be favoured in shady environments in spite of their mechanical inefficiency, because (i) their low degree of self-shading would have a large impact on net carbon gain at irradiance levels near the photosynthetic compensation point; and (ii) the additional transpirational costs imposed by direct exposure in a low-radiation environment would almost be nil. As organs of growth, orthotropic axes should be favoured in sun-adapted plants whose total carbon gain would be most strongly enhanced by vertical growth and attainment of the forest canopy, whereas plagiotropic axes should be favoured in shade-adapted plants whose total input would be most strongly enhanced by horizontal spread and an increase in canopy area.

Economics of Biotic Interactions

Irradiance level, by influencing the potential photosynthetic return from leaf tissue and the optimal allocation of energy to various photosynthetic compounds, affects the likely benefits and opportunity costs associated with different kinds and amounts of defensive measures aimed against folivores (Mooney and Gulmon 1982; Coley 1983; Coley *et al.* 1985; Givnish 1986*b*; Gulmon and Mooney 1986; Bazzaz *et al.* 1987). This section briefly reviews a general model for defensive allocation in sun and shade, and then discusses the possible significance of leaf mottling, an unusual feature of certain shade-adapted species, as an anti-herbivore defence.

Amounts and Kinds of Defensive Compounds as a Function of Irradiance

Angiosperms display an extraordinary range in the kind and amount of chemical, physical, and biological defences they deploy against folivores (Janzen 1966, 1983; Gilbert 1975, 1980; Feeny 1976; Barlow and Wiens 1977; Rausher 1978; Price *et al.* 1980; Berenbaum 1981; Schultz and Baldwin 1982; Denno and McClure 1983; Strong *et al.* 1984). Recently, attention has focused on attempts to predict the nature of such defences as a function of the ecological characteristics of the plants bearing them. Two classes of models have been developed, based on either (i) the effects of a plant species' abundance and/or longevity on the effectiveness of different defences (Feeny 1976; Rhoades and Cates 1976; Rhoades 1979); or (ii) the impact of different defences on a plant's net rate of energy capture as a function of environmental context (Janzen 1974; Mooney and Gulmon 1982; Coley 1983; Coley *et al.* 1985; Gulmon and Mooney 1986; Bazzaz *et al.* 1987). The latter set of cost–benefit models bears directly on the question of how irradiance level should influence the kind and amount of defences deployed against herbivores.

Coley (1983) predicted that shade-adapted species should allocate more to defence than sun-adapted species, because the effective cost of replacing a given amount of leaf tissue is larger in slow-growing shade plants, and because the opportunity cost associated with a given amount of defence is larger in fast-growing sun plants. That is, given the low photosynthetic rates characteristic of shaded leaves, it would take plants of a given size longer to replace an eaten leaf than it would under more productive sunny conditions. Similarly, given the high photosynthetic rates characteristic of sunlit leaves, a given allocation to defence instead of photosynthetic tissue would more greatly reduce a plant's potential growth than it would under less productive shady conditions. This analysis can be seen as a logical extension of Janzen's (1974) argument that the costs of replacing leaves eaten by herbivores would be greater in nutrient-poor habitats than in more fertile sites, given the longer time required to recoup the costs of leaf construction.

Coley *et al.* (1985) generalise the preceding arguments to conclude that, at least between species, defensive allocations should be greater in plants growing on less productive sites, regardless of the particular factor responsible for reduced productivity. In addition, they suggest that sunny conditions should favour carbon-based defensive compounds (e.g. tannins, phenols), whereas shady conditions should favour nitrogen-based defences (e.g. alkaloids). This is because nitrogen is more likely to limit photosynthesis under sunny conditions (Mooney and Gulmon 1979; Gulmon and Chu 1981), and because fixed carbon is likely to be in less abundant supply under shaded conditions. Finally, Coley *et al.* (1985), Bazzaz *et al.* (1987), and Coley (1987) suggest that 'quantitative' defences (e.g. tannins, phenols)—which reduce feeding or digestive efficiency, are effective only at relatively high concentrations, but are effective against almost all herbivores (Feeny 1976)—should be favoured in shade-adapted plants, and that 'qualitative' defences (e.g. alkaloids, cardiac glycosides)—which are toxins effective at low concentrations against most species, but which are ineffective against certain specialist herbivores—should be found in sun-adapted plants. They argue that the costs of construction are higher for quantitative defences, but have low continuing costs once deployed, so that they should be most cost-effective in long-lived leaves, typical of slow-growing plants (see Field 1983). Qualitative defences are effective at lower doses and thus have lower construction costs, but often turn over at high rates (Mooney and Gulmon 1982), implying that they might have substantial continuing costs and be cost-effective only in short-lived leaves, typical of faster-growing plants.

The predictions that result from the preceding arguments are in accord with many of the known facts regarding the nature and kind of chemical defences seen in sun- *v.* shade-adapted plants (Coley 1987), and the arguments might also apply to certain kinds of physical defences (e.g. thick cell walls, silica inclusions) and biological defences (e.g. extrafloral nectaries to attract ant bodyguards).

However, assessing the costs associated with visual biological defences—such as mimicry, aposematic coloration, or cryptic coloration (Gilbert 1975; Barlow and Wiens 1977; Rausher 1978)—can be more difficult. Ehleringer *et al.* (1986) provide a model for rigorously testing theories regarding visual defences, in their work on the degree of mimicry of host leaf shape in poorly defended mistletoes with high *v.* low levels of leaf nitrogen. Smith (1986) presents a hypothesis that mottled or variegated leaves may constitute an irradiance-dependent visual defence against leaf miners, in that they may (i) serve to mimic miner damage and thereby deter actual oviposition by leaf miners; (ii) reduce leaf temperature and transpiration in sunny, hot, and/or dry sites; but (iii) reduce light interception and photosynthesis, particularly in shady sites, given the masking or absence of chlorophyll over the mottled areas of the leaf. Consequently, mottled leaves should be favoured in sunny but not shady microsites.

Smith (1986) supports this conclusion with data on the distribution, growth, and damage by leaf miners of mottled and unmottled morphs of *Byttneria aculeata* Jacq.

(Sterculiaceae), a weedy tropical subcanopy vine growing in Panama. As predicted, mottled individuals are relatively more frequent in clearings than under closed canopies. In addition, leaf production by transplants in the absence of herbivory increases with the average amount of variegation per leaf in open habitats, and decreases with the average amount of variegation in shaded sites. Finally, mottled leaves are indeed less heavily attacked than unmottled leaves where they occur at similar frequencies. These results appear to provide the first experimental evidence for the adaptive significance of leaf variegation in sunny *v.* shady environments. Yet their applicability may be limited; Givnish (1988*b*) presents data showing that, at least in the flora of the north-eastern U.S.A., mottled leaves are far more common in herbs of shaded forest understories than in any other growth form, and are essentially absent in trees, shrubs, herbs or vines of sunny sites. He presents an alternative hypothesis that mottling serves to camouflage the foliage of certain, particularly vulnerable phenological groups of species (e.g. evergreens, spring ephemerals), by disrupting their outline as perceived by colour-blind vertebrate herbivores in sun-dappled understories. As Neger (1913) noted, the puzzling and oft-overlooked phenomenon of leaf mottling remains a rich field for further physiological and ecological studies.

Conclusions

This paper has illustrated the value of a whole-plant perspective in generating and testing hypotheses regarding adaptation to irradiance level, and in resolving certain paradoxes regarding the adaptive value of specific leaf and canopy traits. Three general conclusions emerge:

(1) To understand the adaptive significance of variations in leaf form, physiology, and arrangement, we must consider the functional integration of leaves with other plant parts, particularly roots and mechanical tissue. Many leaf and canopy traits that enhance whole-plant carbon gain have associated costs involving the uptake of water and nutrients, mechanical support and interactions with herbivores. The balance between these costs and associated energetic benefits create tradeoffs that underlie the economics of gas exchange, support and biotic interactions. Analysis of these tradeoffs is the key to understanding the response to irradiance of such key traits as photosynthetic light response, leaf nitrogen content, stomatal conductance, leaf area index, relative canopy width, whole-plant compensation point and allocation to antiherbivore defences.

(2) Two major themes of research on adaptations to irradiance level are likely to emerge over the next few years. First, approaches to the study of leaf, canopy and root adaptations are likely to merge, given that any completely quantitative theory for a trait influencing photosynthesis must incorporate the associated costs of transpiration and root function. Physiological ecologists—who heretofore have mainly been leaf physiologists—must collaborate more closely with root biologists if we are to achieve continued advances in our understanding of photosynthetic adaptations.

Second, the integration of support costs offers many opportunities for further study. This paper has shown that these costs profoundly affect the effective light compensation point of individual leaves as canopy height varies. Consequently, canopy height affects optimal leaf area index and the whole-plant compensation point. Irradiance level sets a limit on maximum canopy height, and canopy height defines the minimum irradiance level required for survival. In the case of the tree *Liriodendron tulipifera*, the maximum height observed accords fairly closely with the limits calculated from support costs. These results exemplify the kinds of conclusions that might be drawn by meshing traditional approaches to physiological ecology with biomechanical considerations, with the aim of determining whether particular growth forms confer a context-specific competitive advantage. The opportunities for such research—focusing on the significance of various aspects of canopy geometry, branching pattern and leaf arrangement—are, in many ways, as exciting as those facing physiological ecology in the 1960's and 1970's.

(3) Analysing the impact of a trait on whole-plant carbon gain is an important, if not essential, step toward understanding its adaptive value, but such studies alone are inadequate to demonstrate that such a value exists and is biologically relevant. Only appropriately controlled studies of competition between plants with different expressions of a trait, conducted under natural conditions, can demonstrate that such variants have adaptive significance. Natural within-population variation in traits that affect adaptation to irradiance level has been little studied, and the implications of such variation for whole-plant carbon gain under different conditions have been largely ignored. Ultimately, to demonstrate adaptation to irradiance level more rigorously, we must relate within-population variation to differences in whole-plant carbon gain, growth and reproductive success under different irradiance levels.

Acknowledgments

I wish to thank the Australian National University and the University of Wisconsin for providing travel funds, and Susan Knight for providing useful observations and references. I also wish to thank Ian Cowan for his kind hospitality and helpful suggestions regarding the manuscript.

References

Armond, P. A., and Mooney, H. A. (1978). Correlation of photosynthetic unit size and density with photosynthetic capacity. *Carnegie Inst. Wash. Year Book* **77**, 234–7.

Ball, M. C., and Farquhar, G. D. (1984*a*). Photosynthetic and stomatal responses of two mangrove species, *Aegiceras corniculatum* and *Avicennia marina*, to long term salinity and humidity conditions. *Plant Physiol.* **74**, 1–6.

Ball, M. C., and Farquhar, G. D. (1984*b*). Photosynthetic and stomatal responses of the grey mangrove, *Avicennia marina*, to transient salinity conditions. *Plant Physiol.* **74**, 7–11.

Barlow, B. A., and Wiens, D. (1977). Host–parasite resemblance in Australian mistletoes: the case for cryptic mimicry. *Evolution* **31**, 69–84.

Bazzaz, F. A., Chiariello, N. R., Coley, P. D., and Pitelka, L. F. (1987). Allocating resources to reproduction and defense. *BioScience* **37**, 58–67.

Berenbaum, M. (1981). Patterns of furanocoumarin distribution and insect herbivory in the Umbelliferae: plant chemistry and community structure. *Ecology* **62**, 1254–66.

Björkman, O. (1968*a*). Carboxydismutase activity in shade-adapted and sun-adapted species of higher plants. *Carnegie Inst. Wash. Year Book* **67**, 487–8.

Björkman, O. (1968*b*). Carboxydismutase activity in shade-adapted and sun-adapted species of higher plants. *Physiol. Plant.* **21**, 1–10.

Björkman, O. (1981). Responses to different quantum flux densities. In 'Physiological Plant Ecology. I. Responses to the Physical Environment'. (Eds O. L. Lange, P. S. Nobel, C. B. Osmond and H. Ziegler.) Encycl. Plant Physiol. New Ser., Vol. 12A, pp. 57–107. (Springer-Verlag: New York.)

Björkman, O., and Holmgren, P. (1963). Adaptability of the photosynthetic apparatus to light intensity in ecotypes from exposed and shaded habitats. *Physiol. Plant.* **16**, 889–914.

Björkman, O., and Powles, S. B. (1984). Inhibition of photosynthetic reactions under water stress: interaction with light level. *Planta* **161**, 490–504.

Björkman, O., Ludlow, M. M., and Morrow, P. A. (1972*a*). Photosynthetic performance of two rainforest species in their native habitat and analysis of their gas exchange. *Carnegie Inst. Wash. Year Book* **71**, 94–102.

Björkman, O., Boardman, N. K., Anderson, J. M., Thorne, S. W., Goodchild, D. J., and Pyliotis, N. A. (1972*b*). Effect of light intensity during growth of *Atriplex patula* on the capacity of photosynthetic reactions, chloroplast components and structure. *Carnegie Inst. Wash. Year Book* **71**, 115–35.

Blackman, G. E., and Black, J. N. (1959). Physiological and ecological studies in the analysis of plant environment. XI. A further assessment of the influence of shading on the growth of different species in the vegetative phase. *Ann. Bot. New Ser.* **23**, 51–63.

Boardman, N. K. (1977). Comparative photosynthesis of sun and shade plants. *Annu. Rev. Plant Physiol.* **28**, 355–77.

Boardman, N. K., Anderson, J. M., Thorne, S. W., and Björkman, O. (1972). Photochemical reactions of chloroplasts and components of the photosynthetic electron transport chain in two rainforest species. *Carnegie Inst. Wash. Year Book* **71**, 107–14.

Borchert, R., and Tomlinson, P. B. (1984). Architecture and crown geometry in *Tabebuia rosea* (Bignoniaceae). *Am. J. Bot.* **71**, 958–69.

Bunce, J. A., Patterson, D. T., and Peet, M. M. (1977). Light acclimation during and after leaf expansion in soybean. *Plant Physiol.* **60**, 255–8.

Chazdon, R. L. (1986). Light variation and carbon gain in an understory palm. *J. Ecol.* **74**, 995–1012.

Chazdon, R. L., and Pearcy, R. W. (1986*a*). Photosynthetic responses to light variation in rainforest species. I. Induction under constant and fluctuating conditions. *Oecologia* **69**, 517–23.

Chazdon, R. L., and Pearcy, R. W. (1986*b*). Photosynthetic responses to light variation in rainforest species. II. Carbon gain and photosynthetic efficiency during sunflecks. *Oecologia* **69**, 524–31.

Chew, F. S., and Rodman, J. E. (1979). Plant resources for chemical defense. In 'Herbivores: Their Interactions with Plant Secondary Metabolites'. (Eds G. A. Rosenthal and D. H. Janzen.) pp. 271–307. (Academic Press: New York.)

Clough, J. M., Terri, J. A., and Alberte, R. S. (1979). Photosynthetic adaptation of *Solanum dulcamara* L. to sun and shade environments. I. A comparison of sun and shade populations. *Oecologia* **38**, 13–21.

Coley, P. D. (1983). Herbivory and defensive characteristics of tree species in a lowland tropical forest. *Ecol. Monogr.* **53**, 209–33.

Coley, P. D. (1987). Interspecific variation in plant anti-herbivore properties: the role of habitat quality and rate of disturbance. *New Phytol.* **106** (Suppl.), 251–63.

Coley, P. D., Bryant, J. P., and Chapin, F. S. III. (1985). Resource availability and plant anti-herbivore defense. *Science* **230**, 895–9.

Coker, W. C. (1944). The woody smilaxes of the United States. *J. Elisha Mitchell Soc.* **60**, 27–69.

Cowan, I. R. (1977). Stomatal behaviour and the environment. *Adv. Bot. Res.* **4**, 1176–227.

Cowan, I. R. (1986). Economics of carbon fixation in higher plants. In 'On the Economy of Plant Form and Function'. (Ed. T. J. Givnish.) pp. 133–70. (Cambridge University Press: Cambridge.)

Cowan, I. R., and Farquhar, G. D. (1977). Stomatal function in relation to leaf metabolism and environment. *Symp. Soc. Exp. Biol.* **31**, 471–505.

Denno, R. F., and McClure, M. S. (Eds) (1983). 'Variable Plants and Herbivores in Natural and Managed Systems.' (Academic Press: New York.)

Ehleringer, J., and Björkman, O. (1977). Quantum yields for CO_2 uptake in C_3 and C_4 plants. *Plant Physiol.* **59**, 86–90.

Ehleringer, J., and Björkman, O. (1978). A comparison of photosynthetic characteristics of *Encelia* species possessing glabrous and pubescent leaves. *Plant Physiol.* **62**, 185–90.

Ehleringer, J., and Forseth, I. (1980). Solar tracking by plants. *Science* **210**, 1094–8.

Ehleringer, J., and Mooney, H. A. (1978). Leaf hairs: effects on physiological activity and adaptive value to a desert shrub. *Oecologia* **37**, 183–200.

Ehleringer, J. R., and Werk, K. S. (1986). Modification of solar radiation absorption patterns and implications for carbon gain at the leaf level. In 'On the Economy of Plant Form and Function'. (Ed. T. J. Givnish.) pp. 57–82. (Cambridge University Press: Cambridge).

Ehleringer, J. R., Ullmann, I., Lange, O., Farquhar, G. D., Cowan, I. R., Schulze, E.-D., and Ziegler, H. (1986). Mistletoes: a hypothesis concerning morphological and chemical avoidance of herbivory. *Oecologia* **70**, 234–7.

Erickson, R., George, A. S., Marchant, N. G., and Morcombe, M. K. (1979). 'Flowers and Plants of Western Australia.' (A. H. & A. W. Reed: Sydney.)

Farquhar, G. D. (1979). Carbon assimilation in relation to transpiration and fluxes of ammonia. In 'Photosynthesis and Plant Development'. (Eds R. Marcelle, H. Clijsters and M. van Poucke.) pp. 321–8. (Dr. Junk: The Hague.)

Farquhar, G. D., and Sharkey, T. D. (1982). Stomatal conductance and photosynthesis. *Annu. Rev. Plant Physiol.* **33**, 317–45.

Farquhar, G. D., Schulze, E.-D., and Küppers, M. (1980*a*). Responses to humidity by stomata of *Nicotiana glauca* L. and *Corylus avellana* L. are consistent with the optimization of carbon dioxide uptake with respect to water loss. *Aust. J. Plant Physiol.* **7**, 315–27.

Farquhar, G. D., von Caemmerer, S., and Berry, J. A. (1980*b*). A biochemical model of photosynthetic CO_2 assimilation in the leaves of C_3 species. *Planta* **149**, 78–90.

Feeny, P. (1976). Plant apparency and chemical defense. In 'Interactions between Plants and Insects'.

Vol. 10. (Eds J. W. Wallace and R. L. Mansell.) pp. 1-40. (Plenum Press: New York.)
Fernald, M. L. (1950). 'Gray's Manual of Botany'. (Van Nostrand: New York.)
Field, C. (1983). Allocating leaf nitrogen for the maximization of carbon gain: leaf age as a control on the allocation program. *Oecologia* **56**, 341-7.
Field, C. B. (1988). On the role of photosynthetic responses in constraining the habitat distribution of rainforest plants. *Aust. J. Plant Physiol.* **15**, 343-58.
Field, C., and Mooney, H. A. (1986). The photosynthesis-nitrogen relationship in wild plants. In 'On the Economy of Plant Form and Function'. (Ed. T. J. Givnish.) pp. 25-55. (Cambridge University Press: Cambridge.)
Field, C., Berry, J. A., and Mooney, H. A. (1982). A portable system for measuring carbon dioxide and water vapor exchange of leaves. *Plant Cell Environ.* **5**, 179-86.
Gilbert, L. E. (1975). Ecological consequences of a coevolved mutualism between butterflies and plants. In 'Coevolution of Animals and Plants'. (Eds L. E. Gilbert and P. H. Raven.) pp. 210-40. (University of Texas Press: Austin.)
Gilbert, L. E. (1980). Food web organization and the conservation of neotropical diversity. In 'Conservation Biology'. (Eds M. E. Soulé and B. A. Wilcox.) pp. 11-33. (Sinauer Associates: Sunderland, MA.)
Givnish, T. J. (1979). On the adaptive significance of leaf form. In 'Topics in Plant Population Biology'. (Eds O. T. Solbrig, S. Jain, G. B. Johnson and P. H. Raven.) pp. 375-407. (Columbia University Press: New York.)
Givnish, T. J. (1982). On the adaptive significance of leaf height in forest herbs. *Am. Nat.* **120**, 353-81.
Givnish, T. J. (1984). Leaf and canopy adaptations in tropical forests. In 'Physiological Ecology of Plants of the Wet Tropics'. (Eds E. Medina, H. A. Mooney and C. Vásquez-Yánes.) pp. 51-84. (Dr. Junk: The Hague.)
Givnish, T. J. (1986*a*). On the use of optimality arguments. In 'On the Economy of Plant Form and Function'. (Ed. T. J. Givnish.) pp. 3-9. (Cambridge University Press: Cambridge.)
Givnish, T. J. (1986*b*). Economics of biotic interactions. In 'On the Economy of Plant Form and Function'. (Ed. T. J. Givnish.) pp. 667-80. (Cambridge University Press: Cambridge.)
Givnish, T. J. (1986*c*). Economics of gas exchange. In 'On the Economy of Plant Form and Function'. (Ed. T. J. Givnish.) pp. 11-24. (Cambridge University Press: Cambridge.)
Givnish, T. J. (1986*d*). Optimal stomatal conductance, allocation of energy between leaves and roots, and the marginal cost of transpiration. In 'On the Economy of Plant Form and Function'. (Ed. T. J. Givnish.) pp. 171-213. (Cambridge University Press: Cambridge.)
Givnish, T. J. (1986*e*). Economics of support. In 'On the Economy of Plant Form and Function'. (Ed. T. J. Givnish.) pp. 413-20. (Cambridge University Press: Cambridge.)
Givnish, T. J. (1986*f*). Biomechanical constraints on crown geometry in forest herbs. In 'On the Economy of Plant Form and Function'. (Ed. T. J. Givnish.) pp. 525-83. (Cambridge University Press: Cambridge.)
Givnish, T. J. (1986*g*). Biomechanical constraints on self-thinning in plant populations. *J. Theor. Biol.* **119**, 139-46.
Givnish, T. J. (1987). Comparative studies of leaf form: assessing the relative roles of selective pressures and phylogenetic constraints. *New Phytol.* **106** (Suppl.), 131-60.
Givnish, T. J. (1988*a*). Ecology and evolution of carnivorous plants. In 'Plant-Animal Interactions'. (Ed. W. A. Abrahamson.) pp. 243-90. (Macmillan: New York.)
Givnish, T. J. (1988*b*). Leaf mottling: relation to growth form and leaf phenology, and possible role as camouflage. *Funct. Ecol.,* in press.
Givnish, T. J., and Vermeij, G. J. (1976). Sizes and shapes of liane leaves. *Am. Nat.* **100**, 743-78.
Givnish, T. J., Burkhardt, E. L., Happel, R. E., and Weintraub, J. D. (1984). Carnivory in the bromeliad *Brocchinia reducta*, with a cost/benefit model for the general restriction of carnivorous plants to sunny, moist, nutrient-poor habitats. *Am. Nat.* **124**, 479-97.
Grier, C. C., and Running, S. W. (1977). Leaf area of mature northwestern coniferous forests: relation to site water balance. *Ecology* **58**, 893-9.
Gulmon, S. L., and Chu, C. C. (1981). The effects of light and nitrogen on photosynthesis, leaf characteristics, and dry matter allocation in the chaparral shrub, *Diplacus aurantiacus. Oecologia* **49**, 207-12.
Gulmon, S. L., and Mooney, H. A. (1986). Costs of defense and their effect on plant productivity. In 'On the Economy of Plant Form and Function'. (Ed. T. J. Givnish.) pp. 681-98. (Cambridge University Press: Cambridge.)

Hall, A. E., and Schulze, E.-D. (1980). Stomatal response to environment and a possible interrelation between stomatal effects on transpiration and CO_2 assimilation. *Plant Cell Environ.* **3**, 467–74.
Hallé, F., Oldeman, R. A. A., and Tomlinson, P. B. (1978). 'Tropical Trees and Forests'. (Cambridge University Press: New York.)
Hirose, T., and Werger, M. J. A. (1987). Maximizing daily canopy photosynthesis with respect to the nitrogen allocation pattern in the canopy. *Oecologia* **72**, 520–6.
Honda, H., and Fisher, J. B. (1978). Tree branch angle: maximizing effective leaf area. *Science* **199**, 888–90.
Honda, H., and Fisher, J. B. (1979). Ratio of tree branch lengths: the equitable distribution of leaf clusters on branches. *Proc. Natl. Acad. Sci. U.S.A.* **76**, 3875–9.
Horn, H. S. (1971). 'The Adaptive Geometry of Trees'. (Princeton University Press: Princeton.)
Horn, H. S. (1979). Adaptation from the perspective of optimality. In 'Topics in Plant Population Biology'. (Eds O. T. Solbrig, S. Jain, G. B. Johnson and P. H. Raven.) pp. 48–61. (Columbia University Press: New York.)
Hunt, F. A. (1986). The complete updated national register of big trees. *Am. For.* **92**, 22–52.
Janzen, D. H. (1966). Coevolution of mutualism between ants and acacias in Central America. *Evolution* **20**, 249–75.
Janzen, D. H. (1974). Tropical blackwater rivers, animals, and the evolution of mast fruiting in the Dipterocarpaceae. *Biotropica* **6**, 69–103.
Janzen, D. H. (1983). Food webs: who eats what, why, how, and with what effects in a tropical forest? In 'Tropical Rain Forest Ecosystems: Structure and Function'. (Ed. F. B. Golley.) pp. 167–82. (Springer-Verlag: New York.)
Jurik, T. W., Chabot, J. F., and Chabot, B. F. (1979). Ontogeny of photosynthetic performance in *Fragaria virginiana* under changing light regimes. *Plant Physiol.* **63**, 542–7.
King, D. (1981). Tree dimensions: maximizing the rate of height growth in dense stands. *Oecologia* **51**, 351–6.
Leigh, E. G. (1972). The golden section and spiral leaf arrangement. In 'Growth by Intussusception'. (Ed. E. S. Deevey.) pp. 161–76. (Archon Press: Hamden, Conn.)
Leigh, E. G. (1975). Structure and climate in tropical rain forests. *Annu. Rev. Ecol. Syst.* **6**, 67–86.
Lincoln, D. E., Newton, T. S., Ehrlich, P. R., and Williams, K. S. (1982). Coevolution of the checkerspot butterfly *Euphydryas chalcedona* and its larval food plant *Diplacus aurantiacus*: larval response to protein and leaf resin. *Oecologia* **52**, 216–23.
Ludlow, M. M., and Wilson, G. L. (1971). Photosynthesis of tropical pasture plants. II. Photosynthesis and illuminance history. *Aust. J. Biol. Sci.* **24**, 1065–75.
Ludlow, M. M., and Björkman, O. (1984). Paraheliotropic movements in Siratro as a protective mechanism against drought-induced damage to primary photosynthetic reactions: damage by excessive light and heat. *Planta* **161**, 505–18.
Mangaly, J. K. (1968). A cytotaxonomic study of the herbaceous species of *Smilax*: section *Coprosmanthus*. *Rhodora* **70**, 55–82; 247–73.
McKey, D. B., Waterman, P. G., Msi, C. H., Gartlan, J. S., and Struhsaker, T. T. (1978). Phenolic content of vegetation in two African rain forests: ecological implications. *Science* **202**, 61–3.
Meinzer, F. C. (1982). The effect of vapour pressure on stomatal control of gas exchange in Douglas fir, *Pseudotsuga menziesii*. *Oecologia* **54**, 236–42.
Mooney, H. A. (1972). The carbon balance of plants. *Annu. Rev. Ecol. Syst.* **3**, 315–46.
Mooney, H. A., and Gulmon, S. L. (1979). Environmental and evolutionary constraints on the photosynthetic characteristics of higher plants. In 'Topics in Plant Population Biology'. (Eds. O. T. Solbrig, S. Jain, G. B. Johnson and P. H. Raven.) pp. 316–37. (Columbia University Press: New York.)
Mooney, H. A., and Gulmon, S. L. (1982). Constraints of leaf structure and function in reference to herbivory. *BioScience* **32**, 198–206.
Mooney, H. A., Ferrar, P. J., and Slatyer, R. O. (1978). Photosynthetic capacity and carbon allocation patterns in diverse growth forms of *Eucalyptus*. *Oecologia* **36**, 103–11.
Mooney, H. A., Field, C., Williams, W. E., Berry, J. A., and Björkman, O. (1983). Photosynthetic characteristics of plants of a Californian cool coastal environment. *Oecologia* **57**, 38–42.
Morley, B. (1973). Ecological factors of importance to *Columnea* taxonomy. In 'Taxonomy and Ecology'. (Ed. V. H. Heywood.) pp. 265–81. (Academic Press: New York.)
Munz, P. A. (1975). 'A California Flora, with Supplement.' (University of California Press: Berkeley.)
Neger, F. W. (1913). 'Biologie der Pflanzen.' (Enke: Stuttgart.)

Nobel, P. S. (1976). Photosynthetic rates of sun versus shade leaves of *Hyptis emoryi* Torr. *Plant Physiol.* **58**, 218–23.

Nobel, P. S. (1986). Form and orientation in relation to PAR interception by cacti and agaves. In 'On the Economy of Plant Form and Function'. (Ed. T. J. Givnish.) pp. 83–103. (Cambridge University Press: Cambridge.)

Orians, G. H., and Solbrig, O. T. (1977). A cost–income model of leaves and roots with special reference to arid and semi-arid areas. *Am. Nat.* **111**, 677–90.

Osmond, C. B. (1983). Interactions between irradiance, nitrogen nutrition, and water stress in the sun–shade responses of *Solanum dulcamara. Oecologia* **57**, 316–21.

Osmond, C. B., Björkman, O., and Anderson, D. J. (1980). 'Physiological Processes in Plant Ecology.' (Springer-Verlag: New York.)

Parkhurst, D. F. (1986). Internal leaf structure: a three-dimensional perspective. In 'On the Economy of Plant Form and Function'. (Ed. T. J. Givnish.) pp. 215–49. (Cambridge University Press: Cambridge.)

Patterson, D. T., Duke, S. O., and Hoagland, R. E. (1978). Effect of irradiance during growth on the adaptive photosynthetic characteristics of velvet leaf and cotton. *Plant Physiol.* **61**, 402–5.

Pearcy, R. W., Osteryoung, K., and Calkin, H. W. (1985). Photosynthetic responses to dynamic light environments by Hawaiian trees. The time course of CO_2 uptake during sunflecks. *Plant Physiol.* **79**, 896–902.

Preston, R. J. (1976). 'North American Trees'. (Iowa State University Press: Ames.)

Price, P. W., Bouton, C. E., Gross, P., McPheron, B. A., Thompson, J. N., and Weis, A. E. (1980). Interactions among three trophic levels: influence of plants on interactions between insect herbivores and natural enemies. *Annu. Rev. Ecol. Syst.* **11**, 41–66.

Rausher, M. D. (1978). Search image for leaf shape in a butterfly. *Science* **200**, 1071–3.

Rausher, M. D. (1980). Host abundance, juvenile survival, and oviposition preference in *Battus philenor. Evolution* **34**, 343–55.

Raven, J. A. (1986). Evolution of plant life forms. In 'On the Economy of Plant Form and Function'. (Ed. T. J. Givnish.) pp. 421–92. (Cambridge University Press: Cambridge.)

Rhoades, D. F. (1979). Evolution of plant chemical defenses against herbivores. In 'Herbivores: Their Interactions with Plant Secondary Metabolites'. (Eds G. A. Rosenthal and D. H. Janzen.) pp. 4–54. (Academic Press: New York.)

Rhoades, D. F., and Cates, R. G. (1976). A general theory of plant anti-herbivore chemistry. In 'Interactions between Plants and Insects'. Vol. 10. (Ed. J. W. Wallace and R. L. Mansell.) pp. 168–213. (Plenum Press: New York.)

Richards, P. W. (1952). 'The Tropical Rain Forest: an Ecological Study'. (Cambridge University Press: Cambridge.)

Rickett, H. W. (1965). 'Wild Flowers of the United States'. Vol. 1. (McGraw-Hill: New York.)

Rickett, H. W. (1970). 'Wild Flowers of the United States'. Vol. 6. (McGraw-Hill: New York.)

Schultz, J. C., and Baldwin, I. T. (1982). Oak leaf quality declines in response to defoliation by gypsy moth larvae. *Science* **217**, 149–51.

Schulze, E.-D., and Hall, A. E. (1982). Stomatal responses, water loss and CO_2 assimilation rates of plants in contrasting environments. In 'Physiological Plant Ecology II. Water Relations and Carbon Assimilation'. (Eds O. L. Lange, P. S. Nobel, C. B. Osmond and H. Ziegler) Encycl. Plant Physiol., New Ser., Vol. 12B, pp. 181–230. (Springer-Verlag: New York.)

Smith, A. P. (1986). Ecology of a leaf color polymorphism in a tropical forest species: habitat segregation and herbivory. *Oecologia* **69**, 283–7.

Strong, D. R., Lawton, J. H., and Southwood, R. (1984). 'Insects on Plants: Community Patterns and Mechanisms'. (Harvard University Press: Cambridge, Mass.)

Wallace, L. L., and Dunn, E. L. (1980). Comparative photosynthesis of three gap phase successional tree species. *Oecologia* **45**, 331–40.

Waring, R. H., Emmingham, W. H., Gholz, H. L., and Grier, C. C. (1978). Variation in maximum leaf area of coniferous forests in Oregon and its ecological significance. *For. Sci.* **24**, 131–40.

Wells, E. F. (1984). A revision of the genus *Heuchera* (Saxifragaceae) in eastern North America. *Syst. Bot. Monogr.* **3**, 45–121.

White, J. (1981). The allometric interpretation of the self-thinning rule. *J. Theor. Biol.* **89**, 475–500.

Whittaker, R. H., and Woodwell, G. M. (1968). Dimension and production relations of trees and shrubs in the Brookhaven Forest, New York. *Ecology* **56**, 1–25.

Whittaker, R. H., Cohen, N., and Olson, J. S. (1963). Net production relations of three tree species at Oak Ridge, Tennessee. *Ecology* **44**, 806–10.

Wilmot, A., and Moore, P. D. (1973). Adaptation to light intensity in *Silene alba* and *S. dioica*. *Oikos* **24**, 458–64.
Wong, S. C., Cowan, I. R., and Farquhar, G. D. (1979). Stomatal conductance correlates with photosynthetic capacity. *Nature* **282**, 424–6.
Wong, S. C., Cowan, I. R., and Farquhar, G. D. (1985*a*). Leaf conductance in relation to rate of CO_2 assimilation. I. Influence of nitrogen nutrition, phosphorus nutrition, photon flux density, and ambient partial pressure of CO_2 during ontogeny. *Plant Physiol.* **78**, 821–5.
Wong, S. C., Cowan, I. R., and Farquhar, G. D. (1985*b*). Leaf conductance in relation to rate of CO_2 assimilation. II. Effects of short-term exposure to different photon flux densities. *Plant Physiol.* **78**, 826–9.
Wong, S. C., Cowan, I. R., and Farquhar, G. D. (1985*c*). Leaf conductance in relation to rate of CO_2 assimilation. III. Influences of water stress and photoinhibition. *Plant Physiol.* **78**, 830–4.

Acclimation by the Thylakoid Membranes to Growth Irradiance and the Partitioning of Nitrogen between Soluble and Thylakoid Proteins

John R. Evans

Division of Plant Industry, CSIRO, G.P.O. Box 1600,
Canberra, A.C.T. 2601, Australia;
present address: Plant Environmental Biology Group, Research
School of Biological Sciences, Australian National University,
G.P.O. Box 475, Canberra, A.C.T. 2601, Australia.

Abstract

Three characteristics of shade plants are reviewed. Firstly, they have relatively more chlorophyll *b* and the associated light-harvesting chlorophyll *a/b*-protein complex (LHC). Two currently accepted reasons for this are not supported by quantitative analysis. Instead, the reduced protein cost of complexing chlorophyll in LHC and the turnover of the 32 kDa herbicide binding protein are considered.

Secondly, shade plants have low electron transport capacities per unit of chlorophyll. This is primarily related to a reduction in the amount of electron transport components such as the cytochrome *f* complex and the ATPase. The nitrogen cost of the thylakoid membranes per unit of light absorbed is thereby reduced, but the irradiance range over which light is used with high efficiency is also reduced.

Thirdly, shade plants have less RuP_2 carboxylase and other soluble proteins for a given amount of chlorophyll. However, while the ratio of RuP_2 carboxylase protein to thylakoid protein declined, the ratio of the RuP_2 carboxylase activity to electron transport activity increased. For several species, the relationship between the rate of CO_2 assimilation and leaf nitrogen content depends on the irradiance during growth.

Introduction

Characteristics of plants from shady and sunny habitats have frequently been compared and contrasted (Boardman 1977; Björkman 1981). Three of the characteristics of shade plants are relatively more chlorophyll *b*, low rates of electron transport per unit of chlorophyll and a reduction in soluble protein relative to chlorophyll. In this review, I shall discuss the consequences of these characteristics in terms of various photosynthetic properties and the possible benefits they confer.

Chlorophyll–Protein Complexes

Light Distribution Between the Two Photosystems

A shade environment obviously has a relatively low irradiance, but this is usually associated with an altered spectral distribution of light. The spectral quality is altered primarily by light absorption by chlorophyll in the leaf canopy. Shadelight, by which is meant the light within leaf canopies such as at ground level in dense forests, is different from sunlight in that it is relatively deficient in red light (600–680 nm) while being enriched in wavelengths greater than 700 nm (for examples of the different spectra, see Evans 1986). It has been suggested that shadelight is more effective in exciting photosystem I than sunlight and that, to redress this imbalance, shade plants have a larger ratio of photosystem II to photosystem I, in terms of both chlorophyll or reaction centres (Björkman 1981). Increasing the chlorophyll *b* content may also increase the proportion of blue-green light captured by photosystem II (Anderson 1982).

0310-7841 / 88 / 010093$03.00

These two suggestions can be questioned on two grounds. Firstly, the quantum yields of photosynthesis have been found to be similar for sun and shade plants and independent of growth irradiance, when measured with the same light source (Björkman *et al.* 1972; Boardman 1977; Björkman and Demmig 1987; Evans 1987*a*). The maximum quantum yield in red light (~600 nm) observed by several groups is 0·111 mol O_2 (or CO_2) mol^{-1} quanta (see Evans 1987*a*). This is the theoretical maximum in the absence of the *Q*-cycle, which implies that all absorbed light is used for photochemistry. The two photosystems are separated laterally, with photosystem II present in the appressed membranes of the grana and photosystem I in the stroma-exposed lamellae (Anderson and Boardman 1966; Sane *et al.* 1970; Anderson 1986; Anderson and Goodchild 1987), such that energy transfer between the two photosystems should be small. Given that all the light is used in photochemistry and energy transfer between photosystems I and II is negligible, the proportion of light absorbed by each photosystem must be 50%. Since the PAR light absorption is proportional to the chlorophyll content (Evans 1986) and virtually independent of the chlorophyll–protein complex (Evans and Anderson 1987), absorbing 50% of the light implies that half of the chlorophyll must be associated with each photosystem in both the sun and shade plants. If the *Q*-cycle operates then, in the absence of nitrate reduction, 11% of the light can be dissipated, i.e. one photosystem could not be associated with more than 56% of the chlorophyll.

The second reason for rejecting why shade plants have relatively more chlorophyll *b* is based on a quantitative analysis of light absorption by the chlorophyll–protein complexes. The analysis is based on a comparison between a shade plant *Alocasia* and a sun plant *Pisum*. The conclusion reached would have been the same if it was based on *Pisum* grown at low versus high irradiance (Evans 1986).

The consequences of altering the relative amounts of the chloroplyll–protein complexes is shown in Table 1. The *Alocasia* (which was grown in natural shade) has more light-harvesting chlorophyll *a*/*b* protein complex (LHC), hence the lower chlorophyll *a*/*b* ratio than *Pisum* grown under high irradiance. For the chlorophyll associated with each photosystem to remain equal, slightly less of the LHC is associated with photosystem II in *Alocasia* (68%) than *Pisum* (72%). The ratio of light absorbed by photosystem II to that by photosystem I is virtually identical in sunlight and shadelight. So, contrary to the suggestion of Björkman (1981) shadelight, despite being enriched in far-red light which is preferentially absorbed by photosystem I, is not deficient in light absorbed by photosystem II when the full PAR spectrum is analysed. Therefore, photosystem II does not require a relatively greater proportion of chlorophyll. The relative increase in the proportion of chlorophyll in LHC does not enhance absorption of either sunlight or shadelight by *Alocasia* relative to the *Pisum* chloroplast (Table 1). So, contrary to the suggestion of Anderson (1982), increasing the chlorophyll *b* content does not increase the proportion of light that is absorbed by photosystem II because a considerable proportion of the chlorophyll *b* is probably associated with photosystem I.

This conclusion relies on several assumptions which were discussed in greater detail by Evans (1986). Within both the leaf and chloroplast, the pigments are unevenly distributed. Light is scattered within the leaf, giving rise to wavelength-dependent path lengthening. While this increases the proportion of incident light absorbed, it does not influence the distribution of light between the two photosystems. The clumping of pigment within the chloroplast had a differential effect on the two photosystems because photosystem II is concentrated in the grana. For 680 nm light, the effectiveness of photosystem II was reduced relative to photosystem I by 15–20%. However, when white light was compared, photosystem II was only disadvantaged by 5%. Recently, Melis *et al.* (1987) argued that more than 50% of the chlorophyll needed to be associated with photosystem II to counteract poorer absorption by chlorophyll *b*, by pigments in grana and by photosystem II in shadelight. While that may well be true over the spectral range they considered (550–750 nm), it is certainly false when the whole PAR spectrum is considered. Chlorophyll *b* absorbs far more than chlorophyll *a* between 400 and 550 nm and the opposite conclusion would be reached for this waveband. Since photosynthesis uses light from these two wavebands about equally, the conclusion by Melis *et al.* (1987) is misleading.

In the earlier analysis (Evans 1986), it was noted that little was known about changes in the carotenoid composition during acclimation to different growth irradiances. The amounts of carotenoids determined by Eskins *et al.* (1983) appear to grossly understate their importance. More recent figures compiled by Siefermann-Harms (1985) suggest that the amounts of β-carotene : xanthophylls per 100 Chl *a* + *b* are 14 : 0 for photosystem I, 12 : 32 for LHC associated with photosystem I, 20 : 0 for photosystem II and 1 : 27 for LHC associated with photosystem II. A substantial fraction of these carotenoids would be retained in the complexes whose spectra were used in the calculations for Table 1. However, the conclusions reached in Table 1 are sensitive to the extent that carotenoid composition varies between species or with acclimation to growth irradiance, given that there are about 25 carotenoids per 100 chlorophylls.

Table 1. Absorption characteristics of *Alocasia* and *Pisum* chloroplasts in sunlight and shadelight

Calculated for 10 μM chlorophyll solutions in a 1 cm light path, assuming that the absorption spectra of the complexes from *Alocasia* and *Pisum* are the same as that for *Spinacia* (Evans and Anderson 1987). The proportion of chlorophyll associated with the complexes are the data of Chu and Anderson (1984) for *Alocasia* and of Leong and Anderson (1984*a*, 840 μmol quanta m^{-2} s^{-1} growth irradiance) for *Pisum*. PSI, PSII, photosystems I, II. Chl, chlorophyll

	Alocasia	*Pisum*
Chlorophyll *a*/*b* ratio	2·2	3·2
Proportion of total Chl in the three complexes:		
PS I : PS II : LHC	0·319 : 0·111 : 0·57	0·374 : 0·17 : 0·456
% LHC associated with PS II[A]	68	72
% Sunlight absorbed[B]	34·4	34·3
% Shadelight absorbed[B]	30·7	30·6
Ratio of light absorbed by PS II/PS I in		
Sunlight	1·01	1·01
Shadelight	0·99	0·99
Calculated nitrogen cost of the chlorophyll–protein complexes (mol N mol^{-1} Chl)	39·2	42·8

[A]Based on the assumption that 50% of the chlorophyll is associated with PS II.
[B]For the spectra, see Evans (1986).

Antenna Size

An increase in the proportion of LHC results from a concomitant decrease in photosystem II reaction centres and core chlorophyll, because the photosystem I reaction centre content is relatively stable (Leong and Anderson 1984*b*; Chow and Hope 1987; Evans 1987*b*; Terashima and Evans 1987). The photosystem II reaction centre content, equivalent to the number of atrazine-binding sites, *Q*, is thus linearly related to the chlorophyll *a*/*b* ratio (Fig. 1). In contrast to the model of Björkman (1981), for acclimation to low irradiance, lower chlorophyll *a*/*b* ratios are accompanied by fewer photosystem II reaction centres. It has been estimated that about 52 chlorophylls are bound to the photosystem II core complex (Evans 1987*b*), so that the assay for photosystem II reaction centres can be converted to a proportion of the total chlorophyll (left-hand *y*-axis, Fig. 1). The proportion of total chlorophyll present in LHC (□), determined by mild gel electrophoresis, has been added to the amount of chlorophyll in the photosystem II reaction centres. The resulting line is nearly independent of the chlorophyll *a*/*b* ratio (Fig. 1). From the discussion above, the high and constant quantum yield requires that 50% of the chlorophyll is associated with each photosystem (broken line). This suggests that approximately 20% of the LHC (the proportion of

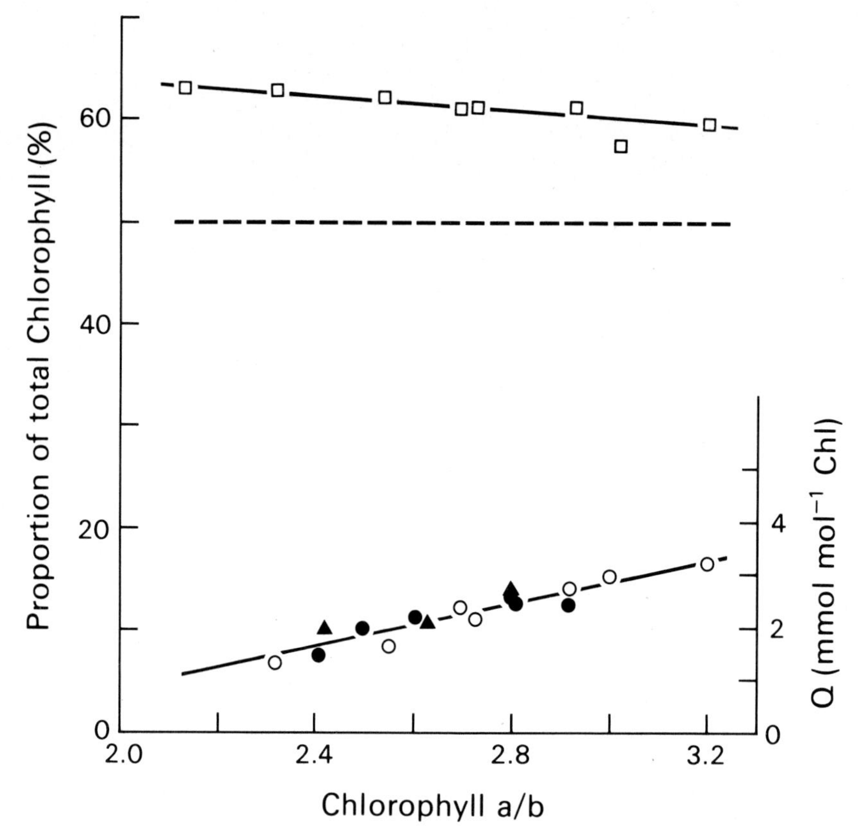

Fig. 1. Photosystem II core chlorophyll and light-harvesting chlorophyll *a*/*b*–protein complex versus the chlorophyll *a*/*b* ratio. *Q*, atrazine-binding sites. ○ *Pisum* (Leong and Anderson 1984*a*, 1984*b*). ● *Pisum* (Evans 1987*b*). ▲ *Spinacia* (Terashima and Evans 1988, assuming 52 Chl per PS II core). The proportion of chlorophyll in LHC is superimposed onto the PS II line. □ *Pisum* (Leong and Anderson 1984*a*). The broken line at 50% represents the proportion of chlorophyll associated with each photosystem.

LHC above the broken line) would be present in the stroma-exposed lamellae to transfer energy to photosystem I. This is in agreement with immunogold localisation and phase separation of the membranes (Anderson and Goodchild 1987; Andersson and Anderson 1980). Bassi (1985) has also shown that LHC is composed of several complexes, some of which may readily associate with photosystem I, as in the bundle sheath cells of maize.

Plants acclimated to low irradiance have relatively more LHC per photosystem II reaction centre (Leong and Anderson 1984*a*). The antenna size of *Pisum*, with chlorophyll a/b ratios of 2·2 and 3·2, can be calculated from Fig. 1. There are 500 chlorophylls per 1·24 and 3·27 Q, respectively, giving 404 and 153 photosystem II chlorophylls per Q.

The acclimation to lower irradiances is a combination of changing the antenna size of each photosystem II as well as changing the ratio of photosystem II to I reaction centres. If these changes cannot be justified on the basis of either redressing the imbalance of light distribution between the two photosystems, or enhancing total absorption, what is the benefit to the plant? One potential benefit arises from the protein cost of complexing chlorophyll. It has been estimated (Evans 1987*b*) that, if 11 chlorophylls

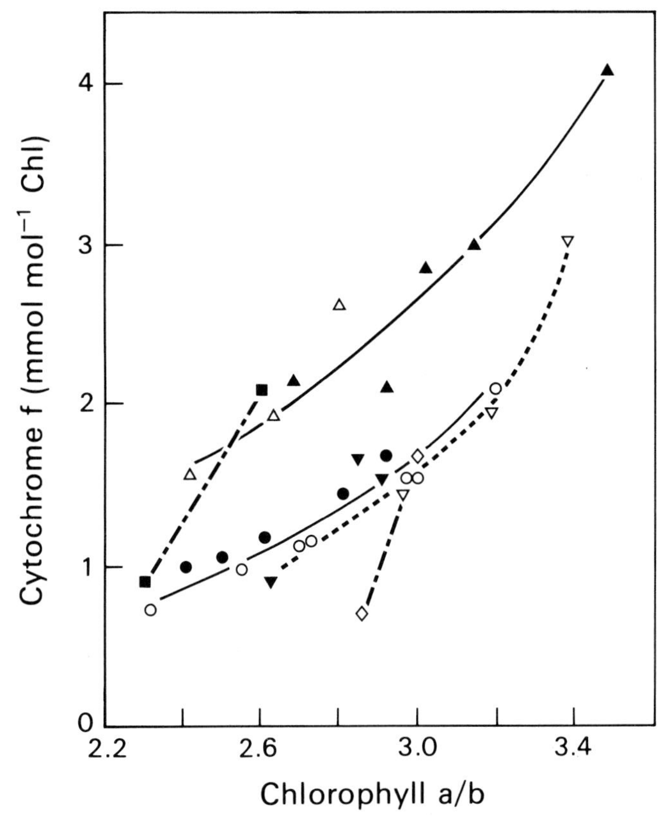

Fig. 2. The correlation between cytochrome f content and chlorophyll a/b ratio. ▲ *Spinacia* (Terashima and Inoue 1985). △ *Spinacia* (Terashima and Evans 1988). ○ *Pisum* (Leong and Anderson 1984*b*). ● *Pisum* (Evans 1987*b*). ▽ *Atriplex* (Björkman *et al.* 1972). ▼ *Atriplex* (Leong and Anderson 1984*c*). ◇ *Betula* (Oquist *et al.* 1982). ■ *Sinapis* (Wild *et al.* 1984).

are bound per 24·5 kDa LHC polypeptide, this complex costs 25·5 mol N mol^{-1} Chl. For the photosystem II core complex, 52 chlorophylls are bound per 318 kDa protein complex equivalent to 69·8 mol N mol^{-1} Chl. If light harvesting is of paramount importance, then the plant should maximise the amount of pigment bound per unit of protein. The nitrogen cost of the chlorophyll–protein complexes per unit of chlorophyll for *Alocasia* was only 92% of that for *Pisum* (39·2 cf. 42·8 mol N mol^{-1} Chl, Table 1). The benefit to peas by increasing the proportion of chlorophyll in LHC, when grown at low irradiance, was only 4% (Evans 1987*b*).

Another reason for reducing the number of photosystem II reaction centres when acclimated to low irradiance could be the turnover of the herbicide-binding protein, Q_B. The herbicide-binding protein turns over more rapidly than other thylakoid proteins and the turnover is greater at higher irradiances (see Anderson 1986). Presumably, the respiratory cost associated with the turnover of this polypeptide would be proportional to both the amount of Q_B protein and the irradiance. By having fewer photosystem II reaction centres, the maintenance cost of this complex could be reduced. However, the concomitant increase in antenna size would result in relatively more electron transfers and consequently could increase the protein turnover. Since fewer photosystem II reaction centres may be associated with some decline in the maximum photosystem II electron transport capacity, the efficiency of utilisation of moderate irradiances may also decline.

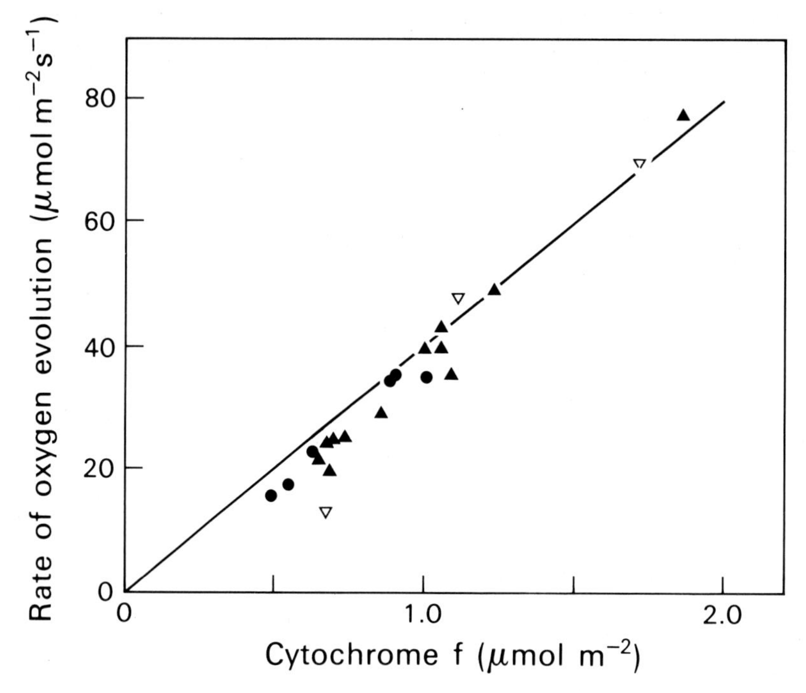

Fig. 3. Relationship between the CO_2-saturated rate of oxygen evolution and the cytochrome *f* content. The irradiance was 2·7 mol absorbed quanta mol^{-1} Chl s^{-1}. ▲ *Spinacia* (Terashima and Evans 1988). ● *Pisum* (Evans 1987*b*). ▽ *Atriplex* (Björkman *et al.* 1972, photosynthetic rate multiplied by 1·75 to approximately convert the rate of CO_2 assimilation at 320 p.p.m. external to a CO_2 saturated rate of oxygen evolution).

The relative constancy of chlorophyll associated with photosystem I with changing chlorophyll *a*/*b* ratios (Leong and Anderson 1984*b*; Shmeleva and Ivanov 1985; Chow and Hope 1987; Evans 1987*b*; Fig. 1) contrasts with the observed association of thylakoid number per granum (Terashima and Inoue 1985; Terashima *et al.* 1986) and proportion of appressed membrane (Aro *et al.* 1986) with the chlorophyll *a*/*b* ratio. The proportion of non-appressed membranes may be underestimated when the chlorophyll *a*/*b* ratio is low, or the surface area of the membrane may not be directly related to its chlorophyll content. Inferring the proportion of chlorophyll associated with each photosystem from either thylakoid number per granum or proportion of appressed membrane may, therefore, be misleading.

Electron Transport Capacity

The changes noted in the relative amounts of the chlorophyll–protein complexes cannot be separated from changes which also occur in the electron transport components. One of the most striking is the cytochrome *f* complex (Fig. 2). Growth under low irradiance results in a reduction in both the chlorophyll *a*/*b* ratio and cytochrome *f* content. A similar relationship holds for the ATPase (Leong and Anderson 1984*b*; Davies *et al.* 1986; Evans 1987*b*). Both cytochrome *f* content and ATPase activity are almost directly proportional to the electron transport capacity of leaves when measured with saturating CO_2 and a constant amount of absorbed light per unit of chlorophyll (Fig. 3). This is not the case for photosystem I reaction centre content, and only a weak relationship was evident with photosystem II reaction centre content (Evans 1987*b*).

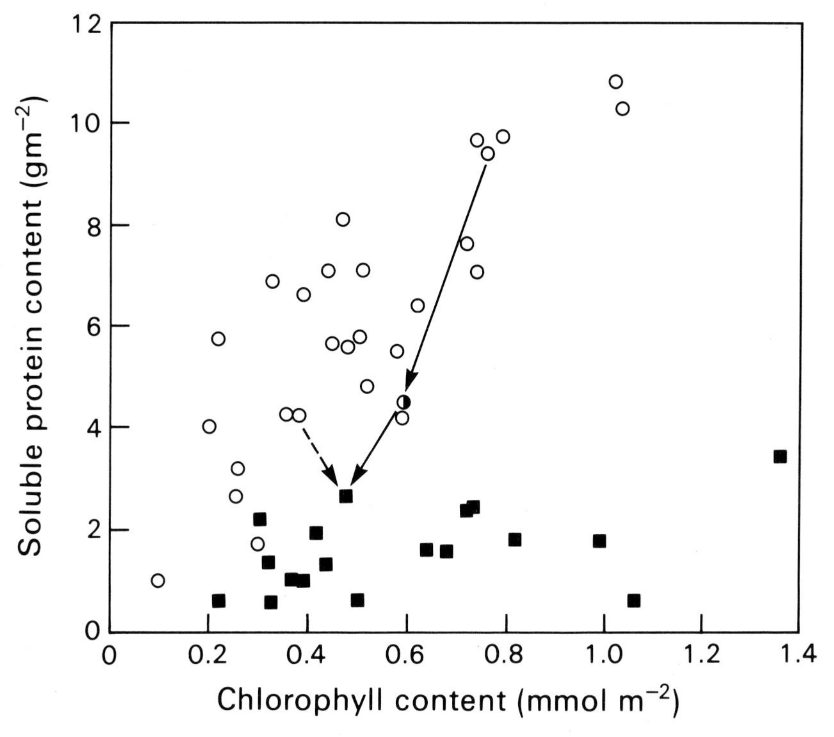

Fig. 4. Relationship between soluble protein content and chlorophyll content per unit leaf area. ○ Sunny habitats. ■ Shady habitats. ◑ 30% sunlight. Data are from Goodchild *et al.* (1972), Boardman (1977) and Terashima and Evans (1988). The solid line joins spinach leaves grown with 12 mM nitrate at 100, 30 and 15% sunlight; the broken line joins two spinach leaves with the same nitrogen content per unit leaf area.

Acclimation to low irradiance has been suggested to involve reducing the protein cost of complexing chlorophyll (Evans 1986, 1987*b*). Low irradiance also reduces the need for a large electron transport capacity. The protein saved by reducing the amount of electron transport components can then be used to greater advantage in complexing chlorophyll. For example, the protein cost associated with just the electron transport chain and ATPase estimated for *Alocasia* (with a cytochrome *f* content of 0·89 mmol mol^{-1} chlorophyll, Boardman *et al.* 1972) and *Pisum* (2·12 mmol cytochrome *f* mol^{-1} chlorophyll, Leong and Anderson 1984*b*) would be 5·6 and 13·3 mol N mol^{-1} chlorophyll. Combining this with the nitrogen cost associated with the chlorophyll–protein complexes and reaction centres (Table 1) means that the total thylakoid protein costs would be 44·9 and 56·1 mol N mol^{-1} chlorophyll, for *Alocasia* and *Pisum* respectively. In terms of light capture, the *Alocasia* thylakoids required only 80% as much protein as *Pisum*. Improved performance for high irradiance involves a different approach. Since the chlorophyll–proteins are the major investment in the thylakoids, containing 75–90% of the thylakoid nitrogen, plants should maximise the electron transport capacity of that chlorophyll by increasing the relative amounts of cytochrome *f* and ATPase. The additional 7·7 mol N mol^{-1} chlorophyll in the electron transport chain and ATPase that *Pisum* invested relative to *Alocasia* increased the electron transport capacity of its chlorophyll by 238%. Put another way, the nitrogen costs in terms of

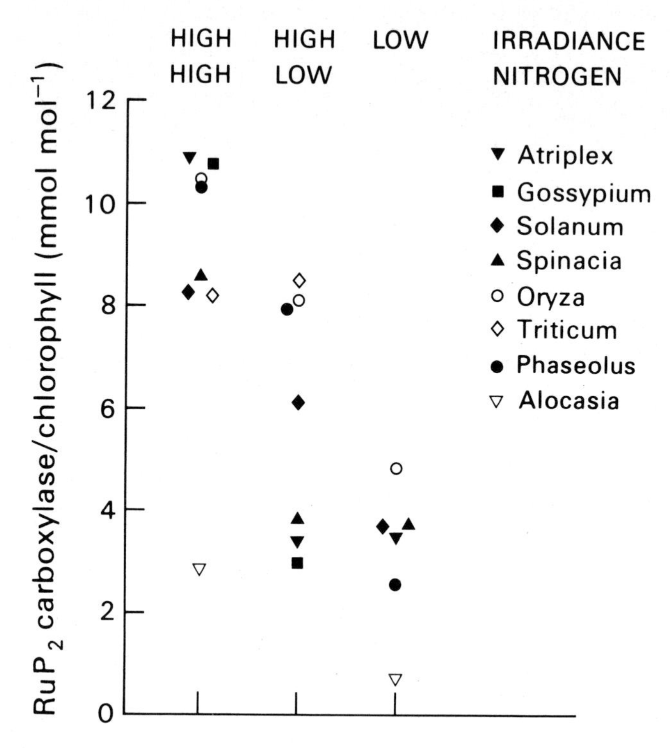

Fig. 5. The effect of nitrogen nutrition or irradiance during growth on the ratio of RuP_2 carboxylase to chlorophyll. ▼ *Atriplex* (Medina 1971). ■ *Gossypium* (Wong 1979). ● *Phaseolus* (Seemann *et al.* 1987). ◆ *Solanum* (Ferrar and Osmond 1986). ▲ *Spinacia* (Terashima and Evans 1988). ○ *Oryza* (Uchida *et al.* 1980). ◇ *Triticum* (Evans 1983). ▽ *Alocasia* (Seemann *et al.* 1987).

electron transport capacity are 26·5 and 50·4 mol N mol^{-1} cytochrome *f* for *Pisum* and *Alocasia*, respectively. *Pisum* requires only 53% as much thylakoid protein as *Alocasia* to achieve the same electron transport capacity.

Nitrogen Partitioning Between Soluble and Thylakoid Protein

Leaves from shady environments were found to have far less soluble protein for a given amount of chlorophyll per unit leaf area (Fig. 4). Acclimation by spinach to different irradiances suggested that a continuum existed between the sun and shade extremes. Surprisingly, the proportion of total leaf nitrogen associated with the thylakoids was independent of the growth irradiance, being 27% for peas (Evans 1987*b*) and 24% for spinach (Terashima and Evans 1987). By contrast, for *Alocasia* this proportion increased at reduced growth irradiance (Seemann *et al.* 1987), which may be a property of shade-tolerant species. The proportion of total nitrogen in soluble protein increased with leaf nitrogen content and was also greater in leaves grown under higher irradiances with sufficient nitrogen. In spinach, the proportion of soluble protein that was RuP_2 carboxylase was independent of growth irradiance.

The changing ratio of soluble protein to chlorophyll is also evident for RuP_2 carboxylase. In general, with nitrogen stress or reduced irradiance during growth, the ratio of RuP_2 carboxylase to chlorophyll declines considerably (Fig. 5). Two exceptions to this are wheat and rice, where nitrogen deficiency did not alter the ratio. The functional significance of this changing ratio differs between the two treatments. Nitrogen deficiency does not alter the electron transport capacity per unit of chlorophyll (Evans and Terashima 1987), whereas growth irradiance does. Unfortunately, there are few

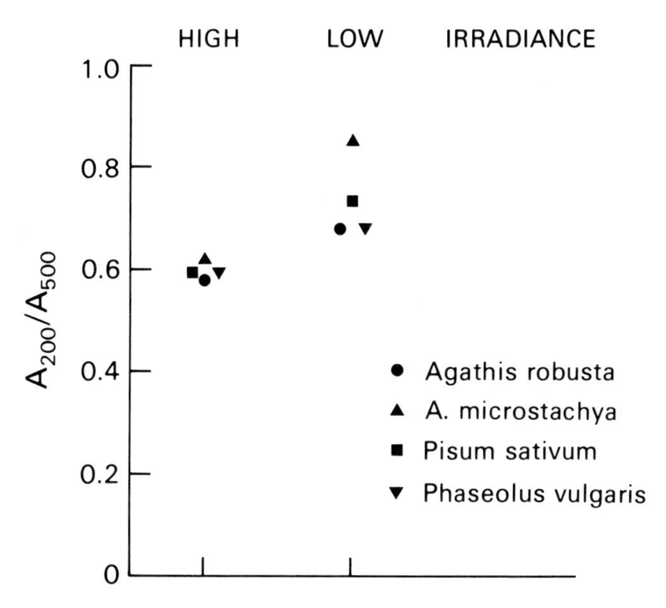

Fig. 6. The effect of irradiance during growth on the ratio of the rates of CO_2 assimilation with intercellular $p(CO_2)$ of 200 and 500 μbar. ● *Agathis robusta*. ▲ *A. microstachya* (Langenheim *et al.* 1984). ■ *Pisum* (Evans 1987*b*). ▼ *Phaseolus* (Seemann *et al.* 1987).

data available where both RuP_2 carboxylase activity and electron transport activity have been determined for the material grown under different irradiances.

This relationship can be deduced from the rate of CO_2 assimilation determined as a function of intercellular $p(CO_2)$, p_i (Farquhar and Caemmerer 1982). At high p_i ($\sim 500\ \mu$bar), the rate of CO_2 assimilation reflects the electron transport capacity if Calvin cycle enzymes do not limit RuP_2 regeneration (Caemmerer and Farquhar 1981; Evans and Terashima 1988). At low p_i, it reflects the RuP_2 carboxylase activity (Caemmerer and Farquhar 1981, 1984; Evans 1983, 1986; Evans and Seemann 1984; Brooks 1986; Evans and Terashima 1988), although the activation state of RuP_2 carboxylase can complicate the interpretation (Caemmerer and Edmondson 1986). The ratio of the rates of CO_2 assimilation at 200 and 500 μbar may therefore reflect the relative capacities of RuP_2 carboxylase and electron transport (Fig. 6). In contrast to the decline in the ratio of RuP_2 carboxylase to chlorophyll, the ratio of RuP_2 carboxylase to electron transport capacity actually increases following acclimation to lower irradiances. This can be seen in the curves relating A to p_i (Fig. 7A; see also Walters and Field 1987). When a plant is grown at low irradiance, the CO_2 saturated part of the curve is relatively lower than that expected from the $A : p_i$ curve obtained from a plant grown at high irradiance (denoted by the arrow). While the RuP_2 carboxylase

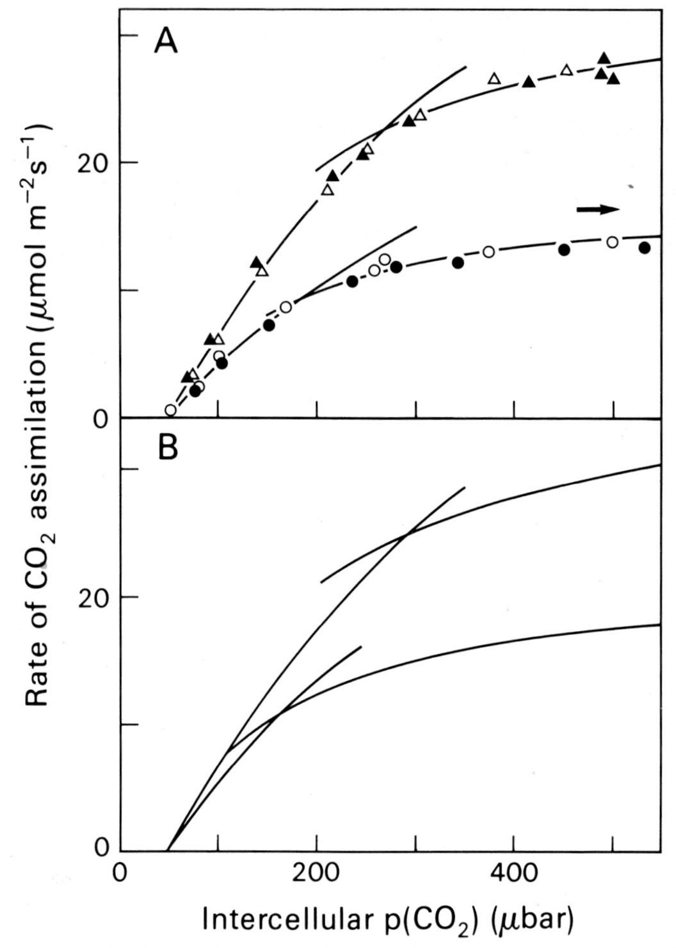

Fig. 7. The rate of CO_2 assimilation versus intercellular $p(CO_2)$. (A) Pea leaves grown under 1000 (△▲) or 130 (○●) μmol quanta m^{-2} s^{-1}. Lines are modelled according to Farquhar and Caemmerer (1982) with $V_{c,max}$, J and R_d of 82, 165, 1·5 and 50, 85, 1 μmol m^{-2} s^{-1} for the high and low growth irradiances, respectively. The arrow shows where the curve would reach if the ratio of $V_{c,max}/J$ for the low growth irradiance curve was the same as that for the high growth irradiance. (B) Modelled $A : p_i$ response curves for high and low growth irradiances with a leaf nitrogen content of 100 mmol m^{-2}. $V_{c,max}$ and (J) were 84, (175) and 64, (105) for the high and low irradiances respectively. For derivation, see text.

capacity is less for plants grown at the lower irradiance, the decline is less than in the electron transport capacity. Further examples need to be obtained to see whether this holds true in a more general sense.

The functional consequences of changes in all of the parameters described above can be illustrated by the following calculations. Consider two pea leaves acclimated to a low or high irradiance during growth, each with 100 mmol N m^{-2} and chlorophyll *a/b* ratios of 2·5 and 2·9, respectively. They contain either 1 or 2 mmol cytochrome *f* mol^{-1} chlorophyll (Leong and Anderson 1984*b*). Given the nitrogen costs of 45 and 52·3 mol N mol^{-1} chlorophyll and 25 mmol thylakoid N m^{-2} (Evans 1987*b*), there are 0·555 and 0·478 mmol chlorophyll m^{-2} and 0·55 and 0·96 mmol cytochrome *f* m^{-2}. The different chlorophyll contents result in leaf absorptances of 0·877 and 0·86. Assuming that electron transport capacity, J (μmol e^- m^{-2} s^{-1}) = 200 × cytochrome *f* (mmol m^{-2}), the electron transport rates at 2000 μmol quanta m^{-2} s^{-1} are 105 and 175 μmol e^- m^{-2} s^{-1}, for the low- and high-light leaves. From the relationships between mesophyll conductance, g_m, and electron transport (Evans 1987*b*), this predicts mesophyll conductances of 0·099 and 0·130 mol m^{-2} s^{-1} bar^{-1}. These are equivalent to RuP_2 carboxylase activities of 63·7 and 83·6 μmol CO_2 m^{-2} s^{-1}, using C = 75 μbar, K_c = 252 μbar, K_o = 192 mbar, O = 210, Γ_* = 37 μbar at 23°C, $V_c = g_m (C + K_c(1 + O/K_o))^2/(\Gamma_* + K_c(1 + O/K_o))$; Caemmerer and Farquhar (1981). The RuP_2

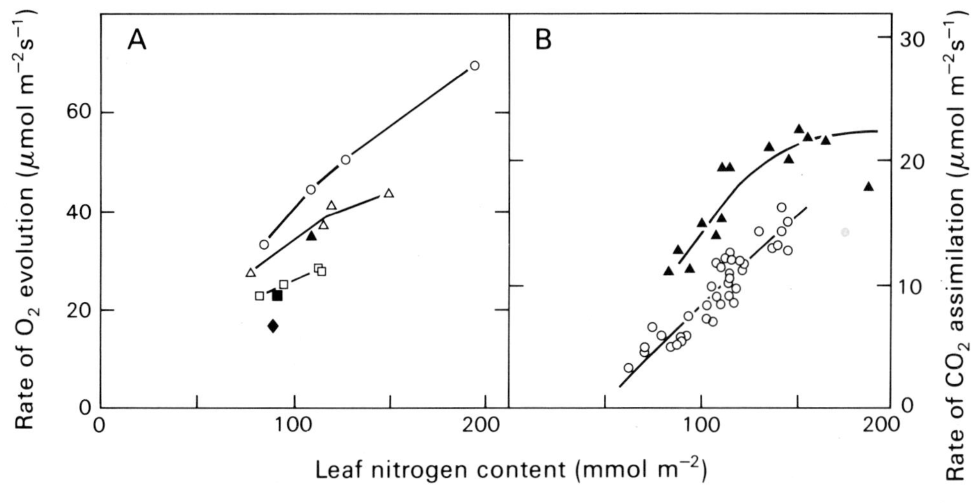

Fig. 8. The relationship between the rate of CO_2 or O_2 exchange and the leaf nitrogen content. (*A*) ○ 100%, △ 30% and □ 15% sunlight of spinach grown with different levels of nitrogen nutrition (Terashima and Evans 1988), ▲ >250, ■ 80 and ♦ <50 μmol quanta m^{-2} s^{-1} for peas (Evans 1987*b*). Oxygen evolution was measured with 1% CO_2 and 2·7 mol absorbed quanta mol^{-1} Chl s^{-1}. (*B*) ▲ *Diplacus* (Gulmon and Chu 1981) and ○ *Solidago* (Hirose and Werger 1987) measured with normal ambient $p(CO_2)$ and 1200 or 1000 μmol quanta m^{-2} s^{-1}, respectively.

carboxylase to chlorophyll ratios are then 5·1 and 7·8 mmol mol^{-1}, using a turnover number of 22·4 mol CO_2 mol^{-1} enzyme s^{-1} (Evans and Terashima 1987). These parameters have been used to predict the response of the rate of CO_2 assimilation to intercellular $p(CO_2)$ (Fig. 7*B*). Despite similar leaf nitrogen contents, the two irradiance treatments result in considerable differences in photosynthetic capacity. The higher rate of CO_2 assimilation is due to both a greater amount of electron transport components such as cytochrome *f* and ATPase and a larger pool of soluble protein. While the nitrogen contents of the leaves whose $A : p_i$ responses are shown in Fig. 7*A* were not determined, the chlorophyll contents were 0·56 and 0·59 mmol m^{-2} for the low and high light treatments.

The strong relationships between photosynthetic capacity and leaf nitrogen content have been used to argue that nitrogen is distributed within a canopy in proportion to the irradiance at each leaf position (Gulmon and Chu 1981; Hirose and Werger 1987;

Fig. 8*B*). However, from the preceding example, the relationship between photosynthetic capacity and leaf nitrogen content depends on the irradiance during growth. This is illustrated with data from spinach which was grown at three different irradiances and with four levels of nitrogen nutrition (Terashima and Evans 1988; Fig. 8*A*). A similar effect was found for pea leaves grown at different irradiances: when grown below 50 μmol quanta m^{-2} s^{-1}, the reduction by 60% of the photosynthetic capacity of leaves grown above 250 μmol quanta m^{-2} s^{-1} was accompanied by only a 20% reduction in the leaf nitrogen content. The simple relationships observed in Fig. 8*B* conceal the separate effects of leaf nitrogen content and growth irradiance seen in Fig. 8*A*.

Taken at face value, leaves developed under higher irradiance have a greater photosynthetic capacity for a given nitrogen content. In spinach, this was associated with both a higher specific leaf weight and larger dark respiration rate. The dependence upon the irradiance during growth is evident whether both axes are expressed on an area or dry weight basis. For example, when grown with 12 mM nitrate, the nitrogen contents for the 100, 30 and 15% sunlight treatments were all 4·1 mmol N g^{-1}, whereas the light and CO_2 saturated rate of oxygen evolution declined from 1·62 to 1·16 and 0·89 μmol g^{-1} s^{-1}, respectively. This contrasts with the assertion of Field and Mooney (1986) that a unique relationship across C_3 species exists in the relationship between photosynthetic rate and leaf nitrogen content. Presumably the acclimation to low irradiance is beneficial to the plant, and to demonstrate this might require daily integrated carbon fixation rather than measurements of the maximum photosynthetic rate (Hirose and Werger 1987).

Acknowledgments

The discussions and collaboration with Dr I. Terashima are gratefully acknowledged. This work was supported in part by an ANU–CSIRO collaborative grant.

References

Anderson, J. M. (1982). The role of chlorophyll–protein complexes in the function and structure of chloroplast thylakoids. *Mol. Cell. Biochem.* **46**, 161–72.

Anderson, J. M. (1986). Photoregulation of the composition, function and structure of thylakoid membranes. *Annu. Rev. Plant Physiol.* **37**, 93–136.

Anderson, J. M., and Boardman, N. K. (1966). Fractionation of the photochemical systems of photosynthesis. I. Chlorophyll contents and photochemical activities of particles isolated from spinach chloroplasts. *Biochim. Biophys. Acta* **112**, 403–21.

Anderson, J. M., and Goodchild, D. J. (1987). Lateral distribution of the photosystem I complex between the appressed and non-appressed regions of spinach thylakoid membranes: an immunocytochemical study. In 'Progress in Photosynthesis Research'. (Ed. J. Biggins.) Vol. 2, pp. 301–4. (Martinus Nijhoff: Dordrecht.)

Andersson, B., and Anderson, J. M. (1980). Lateral heterogeneity in the distribution of chlorophyll–protein complexes of the thylakoid membranes of spinach chloroplasts. *Biochim. Biophys. Acta* **593**, 427–40.

Aro, E. M., Rintamäki, E., Korhonen, P., and Mäenpää, P. (1986). Relationship between chloroplast structure and O_2 evolution rate of leaf discs in plants from different biotypes in south Finland. *Plant Cell Environ.* **9**, 87–94.

Bassi, R. (1985). Spectral properties and polypeptide composition of the chlorophyll–proteins from thylakoids of granal and agranal chloroplasts of maize (*Zea mays* L.). *Carlsberg Res. Commun.* **50**, 127–43.

Björkman, O. (1981). Responses to different quantum flux densities. In 'Physiological Plant Ecology I. Responses to the Physical Environment'. (Eds O. L. Lange, P. S. Nobel, C. B. Osmond and H. Ziegler.) Encycl. Plant Physiol. New Ser., Vol. 12*A*, pp. 57–107. (Springer-Verlag: Berlin.)

Björkman, O., Boardman, N. K., Anderson, J. M., Thorne, S. W., Goodchild, D. J., and Pyliotis, N. A. (1972). Effect of light intensity during growth of *Atriplex patula* on the capacity of photosynthetic reactions, chloroplast components and structure. *Carnegie Inst. Wash. Year Book* **71**, 115–35.

Björkman, O., and Demmig, B. (1987). Photon yield of O_2 evolution and chlorophyll fluorescence characteristics at 77K among vascular plants of diverse origins. *Planta* **170**, 489–504.

Boardman, N. K. (1977). Comparative photosynthesis of sun and shade plants. *Annu. Rev. Plant Physiol.* **28**, 355–77.

Boardman, N. K., Anderson, J. M., Thorne, S. W., and Björkman, O. (1972). Photochemical reactions of chloroplasts and components of the photosynthetic electron transport chain in two rainforest species. *Carnegie Inst. Wash. Year Book* **71**, 107–14.

Brooks, A. (1986). Effect of phosphorus nutrition on ribulose 1,5-bisphosphate carboxylase activation, photosynthetic quantum yield and amount of some Calvin cycle metabolites in spinach leaves. *Aust. J. Plant Physiol.* **13**, 221–37.

Caemmerer, S. von, and Edmondson, D. L. (1986). Relationship between steady-state gas exchange, *in vivo* ribulose bisphosphate carboxylase activity and some carbon reduction cycle intermediates in *Raphanus sativus*. *Aust. J. Plant Physiol.* **13**, 669–88.

Caemmerer, S. von, and Farquhar, G. D. (1981). Some relationships between the biochemistry of photosynthesis and the gas exchange of leaves. *Planta* **153**, 376–87.

Chow, W. S., and Hope, A. B. (1987). The stoichiometries of supramolecular complexes in thylakoid membranes from spinach chloroplasts. *Aust. J. Plant Physiol.* **14**, 21–8.

Chu, Z. X., and Anderson, J. M. (1984). Modulation of the light harvesting assemblies in chloroplasts of a shade plant, *Alocasia macrorrhiza*. *Photobiochem. Photobiophys.* **8**, 1–10.

Davies, E. C., Chow, W. S., and Jordan, B. R. (1986). A study of factors which regulate the membrane appression of lettuce thylakoids in relation to irradiance. *Photosynth. Res.* **9**, 359–70.

Eskins, K., Duysen, M. E., and Olson, L. (1983). Pigment analysis of chloroplast pigment-protein complexes in wheat. *Plant Physiol.* **71**, 777–9.

Evans, J. R. (1983). Nitrogen and photosynthesis in the flag leaf of wheat (*Triticum aestivum* L.). *Plant Physiol.* **72**, 297–302.

Evans, J. R. (1986). A quantitative analysis of light distribution between the two photosystems, considering variation in both the relative amounts of the chlorophyll–protein complexes and the spectral quality of light. *Photobiochem. Photobiophys.* **10**, 135–47.

Evans, J. R. (1987*a*). The dependence of quantum yield on wavelength and growth irradiance. *Aust. J. Plant Physiol.* **14**, 69–79.

Evans, J. R. (1987*b*). The relationship between electron transport components and photosynthetic capacity in pea leaves grown at different irradiances. *Aust. J. Plant Physiol.* **14**, 157–70.

Evans, J. R., and Anderson, J. M. (1987). Absolute absorption spectra for the five major chlorophyll–protein complexes and their 77°K fluorescence excitation spectra. *Biochim. Biophys. Acta* **892**, 75–82.

Evans, J. R., and Seemann, J. R. (1984). Differences between wheat genotypes in specific activity of ribulose 1,5-bisphosphate carboxylase and the relationship to photosynthesis. *Plant Physiol.* **74**, 759–65.

Evans, J. R., and Terashima, I. (1987). Effects of nitrogen nutrition on electron transport components and photosynthesis in spinach. *Aust. J. Plant Physiol.* **14**, 59–68.

Evans, J. R., and Terashima, I. (1988). Photosynthetic characteristics of spinach leaves grown with different nitrogen treatments. *Plant Cell Physiol.* (in press).

Farquhar, G. D., and Caemmerer, S. von (1982). Modelling of photosynthetic response to environmental conditions. In 'Physiological Plant Ecology II. Water Relations and Carbon Assimilation'. (Eds O. L. Lange, P. S. Nobel, C. B. Osmond and H. Ziegler.) Encycl. Plant Physiol. New Ser., Vol. 12B, pp. 550–87. (Springer-Verlag: Berlin.)

Ferrar, P. J., and Osmond, C. B. (1986). Nitrogen supply as a factor influencing photoinhibition and photosynthetic acclimation after transfer of shade-grown *Solanum dulcamara* to bright light. *Planta* **168**, 563–70.

Field, C., and Mooney, H. A. (1986). The photosynthesis–nitrogen relationship in wild plants. In 'On the Economy of Form and Function'. (Ed. T. J. Givnish.) pp. 25–55. (Cambridge University Press: Cambridge.)

Goodchild, D. J., Björkman, O., and Pyliotis, N. A. (1972). Chloroplast ultrastructure, leaf anatomy and content of chlorophyll and soluble protein in rainforest species. *Carnegie Inst. Wash. Year Book* **71**, 102–7.

Gulmon, S. L., and Chu, C. C. (1981). The effects of light and nitrogen on photosynthesis, leaf characteristics and dry matter allocation in chaparral shrub, *Diplacus aurantiacus*. *Oecologia (Berlin)* **49**, 207–12.

Hirose, T., and Werger, M. J. A. (1987). Nitrogen use efficiency in instantaneous and daily photosynthesis of leaves in the canopy of a *Solidago altissima* stand. *Physiol. Plant.* **70**, 215–22.

Langenheim, J. H., Osmond, C. B., Brooks, A., and Ferrar, P. J. (1984). Photosynthetic responses to light in seedlings of selected Amazonian and Australian rainforest tree species. *Oecologia (Berlin)* **63**, 215–24.

Leong, T., and Anderson, J. M. (1984*a*). Adaptation of the thylakoid membranes of pea chloroplasts to light intensities. I. Study of the distribution of chlorophyll–protein complexes. *Photosynth. Res.* **5**, 105–15.

Leong, T., and Anderson, J. M. (1984*b*). Adaptation of the thylakoid membranes of pea chloroplasts to light intensities. II. Study of electron transport capacities, coupling factor (CF_1) activity and rates of photosynthesis. *Photosynth. Res.* **5**, 117–28.

Leong, T., and Anderson, J. M. (1984*c*). The effect of light quality on the composition and function of thylakoid membranes in *Atriplex triangularis*. *Biochim. Biophys. Acta* **766**, 533–41.

Medina, E. (1971). Effect of nitrogen supply and light intensity during growth on the photosynthetic capacity and carboxydismutase activity of leaves of *Atriplex patula* ssp. *hastata*. *Carnegie Inst. Wash. Year Book* **70**, 551–9.

Melis, A., Spangfort, M., and Andersson, B. (1987). Light-absorption and electron transport balance between photosystem II and photosystem I in spinach chloroplasts. *Photochem. Photobiol.* **45**, 129–36.

Oquist, G., Brunes, L., and Hallgren, J. E. (1982). Photosynthetic efficiency of *Betula pendula* acclimated to different quantum flux densities. *Plant Cell Environ.* **5**, 9–15.

Sane, P. V., Goodchild, D. J., and Park, R. B. (1970). Characterization of chloroplast photosystems 1 and 2 separated by a non-detergent method. *Biochim. Biophys. Acta* **216**, 162–78.

Seemann, J. R., Sharkey, T. D., Wang, J. L., and Osmond, C. B. (1987). Environmental effects on photosynthesis, nitrogen-use efficiency and metabolite pools in leaves of sun and shade plants. *Plant Physiol.* **84**, 796–802.

Shmeleva, V. L., and Ivanov, B. N. (1985). Effect of irradiance during plant growth on electron transport chain in pea chloroplasts: state of cytochromes. *Photosynthetica* **19**, 402–10.

Siefermann-Harms, D. (1985). Carotenoids in photosynthesis. I. Location in photosynthetic membranes and light-harvesting function. *Biochim. Biophys. Acta* **811**, 325–55.

Terashima, I., and Evans, J. R. (1988). Effects of light and nitrogen nutrition on the organization of the photosynthetic system in spinach leaves. *Plant Cell Physiol.* (in press).

Terashima, I., and Inoue, Y. (1985). Vertical gradient in photosynthetic properties of spinach chloroplasts dependent on intraleaf light environment. *Plant Cell Physiol.* **26**, 781–5.

Terashima, I., Sakaguchi, S., and Hara, N. (1986). Intraleaf and intracellular gradients in chloroplast ultrastructure of dorsiventral leaves illuminated from the adaxial or abaxial side during their development. *Plant Cell Physiol.* **27**, 1023–31.

Uchida, N., Itoh, R., and Murata, Y. (1980). Studies on the changes in the photosynthetic activity of a crop leaf during its development and senescence. I. Changes in the developmental stage of a rice leaf. *Jpn J. Crop Sci.* **49**, 127–34.

Walters, M. B., and Field, C. B. (1987). Photosynthetic light acclimation in two rainforest *Piper* species with different ecological amplitudes. *Oecologia (Berlin)* **72**, 449–56.

Wild, A., Nies, E., and Ewen, J. (1984). Correlative analysis of the photosynthetic capacity and different components of the photosynthetic apparatus. In 'Advances in Photosynthesis Research'. (Ed. C. Sybesma.) Vol. 4, pp. 333–6. (Martinus Nijhoff/Dr W. Junk: The Hague.)

Wong, S. C. (1979). Elevated atmospheric partial pressure of CO_2 and plant growth. I. Interactions of nitrogen nutrition and photosynthetic capacity in C_3 and C_4 plants. *Oecologia (Berlin)* **44**, 68–74.

Photosynthetic Acclimation of *Alocasia macrorrhiza* (L.) G. Don to Growth Irradiance: Structure, Function and Composition of Chloroplasts

W. S. Chow[A], *Luping Qian*[B], *D. J. Goodchild*[A] *and Jan M. Anderson*[A]

[A]Division of Plant Industry, CSIRO, G.P.O. Box 1600, Canberra, A.C.T. 2601, Australia.
[B]Shanghai Institute of Plant Physiology, Academia Sinica, Shanghai, China.

Abstract

The photosynthetic acclimation of *Alocasia macrorrhiza* (L.) G. Don, a species naturally occurring in deep shade in rainforests, has been studied in relation to a wide range of controlled irradiances during growth (~3–780 μmol photons m^{-2} s^{-1} of fluorescent or incandescent light, 10 h light/14 h dark). At the maximum growth irradiances, the light- and CO_2-saturated rates of O_2 evolution per unit leaf area were ~4 times as high as at low irradiance, and approached those of glasshouse-grown spinach. Growth at maximum irradiances reduced the quantum yield of O_2 evolution only slightly.

Changes in the anatomy of leaf tissue, the ultrastructure of chloroplasts and the composition of chloroplast components accompanied the changes in photosynthetic functional characteristics. At low growth irradiance, palisade cell chloroplasts were preferentially located adjacent to the distal periclinal cell walls and had large granal stacks, and the destacked thylakoids had a very low surface charge density. In contrast, at higher growth irradiance, palisade cell chloroplasts were preferentially located adjacent to the anticlinal cell walls; they had small granal stacks, large stromal space, and a high surface charge density on the destacked thylakoids. The number of chloroplasts per unit section length increased with growth irradiance. Ribulosebisphosphate carboxylase activity per unit leaf area increased markedly with irradiance. Photosystem II, cytochrome *f* and latent ATPase activity per unit chlorophyll increased to a lesser extent. While the chlorophyll *a*/chlorophyll *b* ratio increased substantially with growth irradiance, the chlorophyll content per unit leaf area declined slightly.

Our results show that coordinated changes in the structure of leaf tissue, and the organisation and composition of chloroplast components are responsible for *Alocasia* being capable of acclimation to high as well as low irradiance.

Introduction

From a number of studies, it is evident that the light environment has a profound influence on the structure and composition of the photosynthetic apparatus, and thus on the capacity for photosynthetic performance (Boardman 1977; Wild 1979; Björkman 1981; Anderson 1986; Anderson and Osmond 1987; Anderson *et al.* 1988). The dynamic extent of photosynthetic adaptation to growth irradiance may, however, differ between species. For example, extreme shade species can survive at much lower irradiances than sun species, but may possess only limited adaptability to high light environments (Boardman 1977; Anderson and Osmond 1987). However, the adaptability of a species to low light environment does not necessarily preclude its sustained growth in high light environments, particularly if other conditions such as nutrient and water supply are favourable. Indeed, since different strategies may be available to the photosynthetic apparatus for coping with high irradiance during growth, any attempts to categorise plants into sun and shade species are fraught with difficulties (Anderson and Osmond 1987).

0310-7841/88/010107$03.00

Alocasia macrorrhiza is capable of sustained growth on the floor of a rainforest where the average irradiance is less than 0·5% of full sunlight (Björkman and Ludlow 1972). Leaves obtained from such an extreme shade habitat exhibit low rates of CO_2 assimilation at light saturation (Björkman *et al.* 1972) and the *in vitro* rates of electron transport are correspondingly low (Boardman *et al.* 1975). Because of its tolerance to extreme shade, *Alocasia* is sometimes assumed to be an archetypal obligate shade species. Yet it can also grow in open sunny areas, although apparently with reduced quantum efficiency (Pearcy 1987). In this study, we investigate the extent to which *Alocasia* is able to acclimate to a range of controlled irradiances. Our study covers a wide range of responses at the functional, compositional, anatomical and ultrastructural levels. Its aims are (1) to define the extent of photosynthetic acclimation to growth irradiance and (2) to probe the mechanisms whereby *Alocasia* acclimates to low and high irradiances.

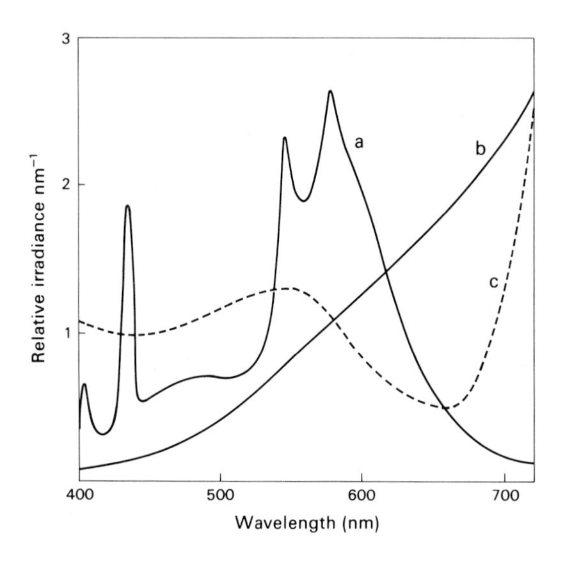

Fig. 1. Spectral distribution of (*a*) fluorescent light with an incandescent supplement; (*b*) incandescent light alone used to grow *Alocasia* in controlled environment cabinets; and (*c*) light climate on a rainforest floor where *Alocasia* was studied in its 'native' habitat (Björkman and Ludlow 1972). The spectra are normalised to give equal irradiance in the range 400–720 nm.

Materials and Methods

Plant Material

Alocasia macrorrhiza (L.) G. Don plants, naturally occurring in south-eastern Queensland, were obtained from the Australian National Botanic Gardens, Canberra. The plants were grown in a mixture of vermiculite and perlite in a growth cabinet at 24°C, and watered twice daily, once with Hoagland's solution and once with demineralised water. Either incandescent light alone or fluorescent light supplemented by incandescent light was supplied, with a regime of 10 h day/14 h night. The spectral distribution of the lights, determined with a spectroradiometer (SR 3000A, Macam Photometrics Ltd, Livingston, Scotland), is shown in Fig. 1, together with that of the diffuse light on a tropical rainforest floor reported by Björkman and Ludlow (1972). Total irradiance in the wavelength range 400–700 nm, at a particular portion of leaf to be sampled, was measured with a Li-Cor (LI-185A) quantum photometer. At intermediate and low irradiance of fluorescent light, the spectral distribution of the light sensed by a particular portion of a leaf would have deviated somewhat from that of the light source because of partial shading by other leaves which absorb maximally in the blue and the red regions.

In addition to the plants grown in cabinets, *Alocasia* leaves were also sampled from the Australian National Botanic Gardens and used directly.

Measurements of Photosynthesis

Leaf photosynthesis was measured as the steady-state rate of O_2 evolution in white light from a slide projector, using a Hansatech (King's Lynn, England) leaf disc O_2 electrode. Measurements were made at a leaf temperature of 25°C in saturating CO_2 conditions (~1% CO_2 from a 1 M carbonate/bicarbonate buffer solution at pH 9). Quantum yield was determined as the oxygen evolved per absorbed photon over the range of irradiance where the relationship was linear (Evans 1987*a*). Photosynthetic capacity was measured as the light- and CO_2-saturated rate of O_2 evolution per unit leaf area.

Determination of Chloroplast Components

Chloroplasts were isolated by slicing leaf tissue immersed in the isolation medium at 0°C and then homogenising the slices as described previously (Chow and Hope 1987), with the isolation medium modified to include sodium ascorbate (0·2%, w/v) and polyvinylpyrrolidone (PVP 40T, 1% w/v). Chlorophyll was determined in 80% acetone with either a Hitachi model U-3200 or Cary 14 spectrophotometer. Both instruments gave very similar values of total Chl* or Chl *a/b* ratios.

The *in vitro* activity of RuP_2 carboxylase (EC 4.1.1.39) was assayed at 25°C as described by Evans (1983). Leaf discs were collected and stored in liquid nitrogen until used. Each leaf disc was ground immediately upon thawing, and the enzyme fully activated by CO_2 and Mg^{2+}.

PS II reaction centres were assayed as atrazine-binding sites according to the general method of Tischer and Strotmann (1977) with modifications as described by Chow and Hope (1987). The number of functional PS II centres *in vivo* was determined from the O_2 yield per single-turnover xenon flash given to a leaf disc at 4 Hz. Background far-red light was supplied at about 17 μmol photons $m^{-2} s^{-1}$ in the wavelength range 700–730 nm to ensure no limitation by PS I turnover.

Cyt *f* was estimated from hydroquinol-reduced minus ferricyanide-oxidised difference spectra (Bendall *et al.* 1971), using an Hitachi 557 double-beam spectrophotometer.

PS I reaction centres were assayed by the light-induced absorbance change at 703 nm, using a flash photometer constructed by Professor W. Haehnel as reported previously (Chow and Hope 1987).

The Mg^{2+}-specific ATPase activity in chloroplast CF_1 was assayed in the presence of octyl glucoside (Pick and Bassilian 1981), with details as given in Chow and Hope (1987).

The overall surface charge density of destacked thylakoids was determined from the sequestering of a screening divalent cation (methyl viologen) in the diffuse layer adjacent to the thylakoid surface, as described by Chow and Barber (1980), with the modification that the upper distance limit of integration, *d*, was 50 nm. Thylakoid membrane components were first randomised by destacking in 20 mM KCl (Chow 1984*a*) so as to yield a more homogeneous distribution of surface charge. Destacked thylakoids were resuspended in a medium containing 100 mM sorbitol, 0·1 mM Na-EDTA and 0·1 mM HEPES (pH ~7·5, KOH), together with various amounts of methyl viologen. After incubation at room temperature for several minutes, the thylakoids were centrifuged through a silicone oil layer. The concentration of methyl viologen remaining in the clear supernatant was taken as the bulk concentration. The background concentration of monovalent cations (Na^+, K^+) was estimated and taken into account in applying the Gouy-Chapman theory of the electric double layer.

Microscopy

For microscopy, tissue pieces approx. 1 mm × 1 mm were cut from leaves at the different irradiances and immediately fixed in 3% glutaraldehyde in 25 mM Na-phosphate buffer (pH 7·2) for 1·5 h. To ensure uniform conditions during fixation, vials were placed in the dark. After several buffer washes the tissue pieces were post-fixed in 2% OsO_4 in the phosphate buffer for 1·5 h, washed, dehydrated through an alcohol series and embedded in Spurr's resin (Spurr 1969). For light microscopy, sections approx. 1 mm long and 0·5 μm thick were stained with toluidine blue, observed and photographed. Photographic montages of each section were prepared at known magnifications of approx. 550×, the cell numbers were counted, and the areas of the cell layers, upper and lower epidermis, palisade, mesophyll and vascular tissue were determined using a digitiser tablet (Summagraphics, Fairfield, Conn., U.S.A.). Chloroplast numbers in the palisade and mesophyll cells were also determined from the montages and the results normalised to 1 mm section length.

*Abbreviations used: CF_1, coupling factor 1; Chl, chlorophyll; cyt, cytochrome; HEPES, *N*-2-hydroxyethylpiperazine-*N'*-2-ethanesulfonic acid; LHCII, light-harvesting chlorophyll *a/b*-protein complexes of photosystem II; PS, photosystem; Q_A, Q_B, first and second PS II quinone-type acceptors, respectively; RuP_2, ribulose 1,5-bisphosphate.

For electron microscopy, thin sections were prepared from the same embedded tissue, stained with uranyl acetate and lead citrate and examined in a JEOL 100S electron microscope at 80 kV. Chloroplast semi-axes from the palisade and mesophyll cells were measured directly from the image on the electron microscope screen at an instrument magnification of approx. 1000× using an ocular graticule in the binocular viewing microscope at a further magnification of approx. 10×. A grating replica was used in place of the specimen to calibrate the graticule. For microscopic comparison between the artificial irradiances and natural sunlight, *Alocasia* tissue embedded for a previous experiment (Anderson *et al.* 1973) was used. This material had been collected from an extreme shade habitat on the floor of a rainforest in Lamington National Park of SE. Queensland where the average daily radiation in the wavelength range 400–700 nm was 0·21 mol photons m^{-2} day^{-1} (Björkman and Ludlow 1972). The spectral distribution of this irradiation is shown in Fig. 1, curve *c*. The area of each chloroplast was calculated and the mean and standard error for each light treatment determined from these individual areas. All measurable chloroplasts in a single section from each treatment were used resulting in sample sizes from 50 to 180. Despite the variation in sample size and variation due to randomness in sectioning of chloroplasts, the method produced small standard errors in comparison to the differences between means. To investigate chloroplast membrane characteristics between light treatments, electron micrograph prints were prepared.

Chlorophyll per Chloroplast

For this determination, chloroplasts from plants grown at about 5 and 350 µmol photons m^{-2} s^{-1} (fluorescent light) were used; higher irradiances produced chloroplasts engorged with starch grains that fragmented on isolation and this precluded their use for counting. Chloroplasts isolated as described above were further fractionated on a Percoll gradient with three steps (Robinson 1983). This procedure produced the very high proportion of class I chloroplasts, with intact envelopes, that are required to accurately determine numbers. A Petroff-Hauser bacteria counter was used to count the chloroplasts (Anderson *et al.* 1973).

Results

Photosynthetic Characteristics

The overall photosynthetic response of a leaf to the growth irradiance is reflected in the photosynthetic capacity, measured as the CO_2- and light-saturated rate of O_2 evolution per unit leaf area. Fig. 2*a* shows the photosynthetic capacity of *Alocasia* leaves as a function of growth irradiance (wavelength range 400–700 nm). The low photosynthetic capacity of leaves from plants grown under low incandescent or fluorescent light is comparable to that reported by Björkman *et al.* (1972) for *Alocasia* grown under a Queensland rainforest canopy, where the daily average irradiance is indicated by the arrow (Fig. 2*a*). Growth in incandescent light (closed circles) gave rise to a lower photosynthetic capacity than in fluorescent light (open circles) of equivalent irradiance. As the growth irradiance increased, the photosynthetic capacity of *Alocasia* leaves approached that of spinach grown in a glasshouse (squares), or that of *Alocasia* growing in the open in late spring at the Australian National Botanic Gardens (triangles). Whether expressed on a leaf area basis or a chlorophyll basis, the relative changes in the maximum photosynthetic rate as a function of growth irradiance were comparable. Thus, *Alocasia* grown under these varying conditions showed considerable adaptability to growth irradiance.

To investigate whether the increase in photosynthetic capacity might have occurred at the expense of quantum efficiency, we measured the quantum yield of O_2 evolution (Fig. 2*b*). The quantum yield of *Alocasia* leaves grown in low fluorescent or incandescent light, measured with white projector light, was approximately 0·10 mol O_2 (mol absorbed photons)$^{-1}$, similar to that of glasshouse-grown spinach (squares). Only at the maximum irradiances in a growth cabinet did the quantum yield decline to ~0·09 mol O_2 (mol absorbed photons)$^{-1}$. Interestingly, *Alocasia* grown in an open, sunny area in spring maintained its quantum yield at ~0·085 mol O_2 (mol absorbed photons)$^{-1}$.

Chloroplast Components

The chlorophyll content per unit leaf area declined with increasing irradiance, with ~30% difference between extremes (Fig. 2*c*). In contrast, the Chl *a/b* ratio increased markedly with growth irradiance (Fig. 2*d*). Incandescent light (closed circles) gave rise

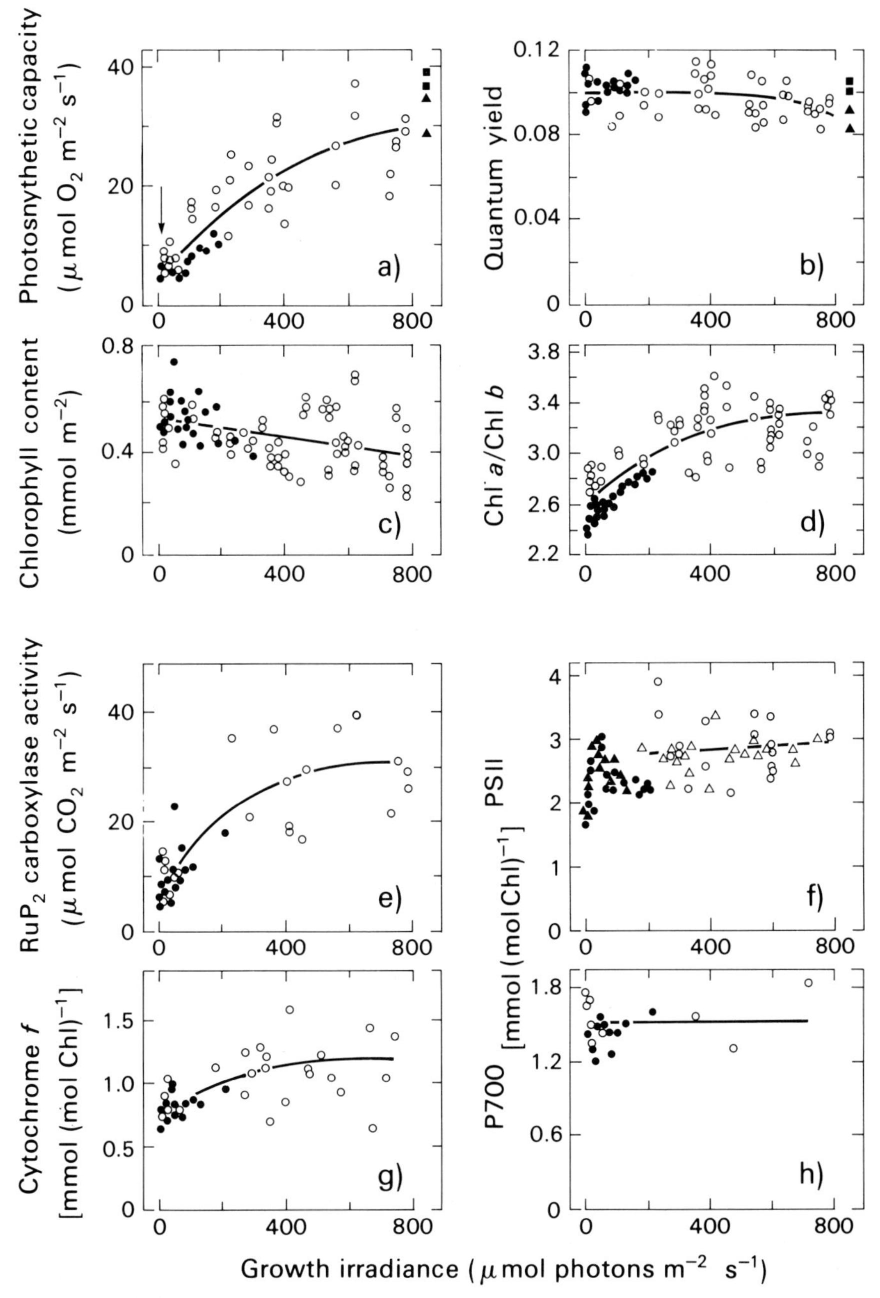

Fig. 2. Photosynthetic parameters and components of *Alocasia* leaves grown under different irradiances of incandescent light alone (●) or fluorescent with an incandescent supplement (○). The arrow in (*a*) shows the approximate irradiance equivalent to the average irradiance on a rainforest floor. The unit for quantum yield is mol O_2 (mol photons absorbed)$^{-1}$. Lines have been drawn to indicate trends.

to a lower ratio than did fluorescent light (open circles) of equivalent irradiance. The lowest Chl *a/b* ratios reached in the growth cabinet were similar to that of *Alocasia* from the rainforest floor (Anderson *et al.* 1973).

As a representative stromal enzyme, RuP_2 carboxylase was assayed for its maximum activity. There was a difference of approximately 5-fold between the extreme irradiances, suggesting large increases in the amounts of this enzyme at high irradiance (Fig. 2*e*).

To determine the number of PS II reaction centres, two methods were used (Fig. 2*f*). In one case, the number of functional PS II centres was determined *in vivo* from the O_2 yield per single turnover flash in the presence of background far-red light (circles). In the other case, the number of PS II centres obtained by atrazine-binding (triangles) was about 20% lower; these values were therefore multiplied by 1·20 to superimpose the two sets of data. Over a large range of irradiance, the PS II content was, on average, slightly under 3 mmol (mol Chl)$^{-1}$, comparable to that of spinach (see Chow and Hope 1987). In incandescent light, a peak value of [PS II] seemed to occur at about 40 μmol photons m^{-2} s^{-1}. At extremely low irradiance, incandescent light gave rise to a low value of [PS II] which was below 2 mmol (mol Chl)$^{-1}$.

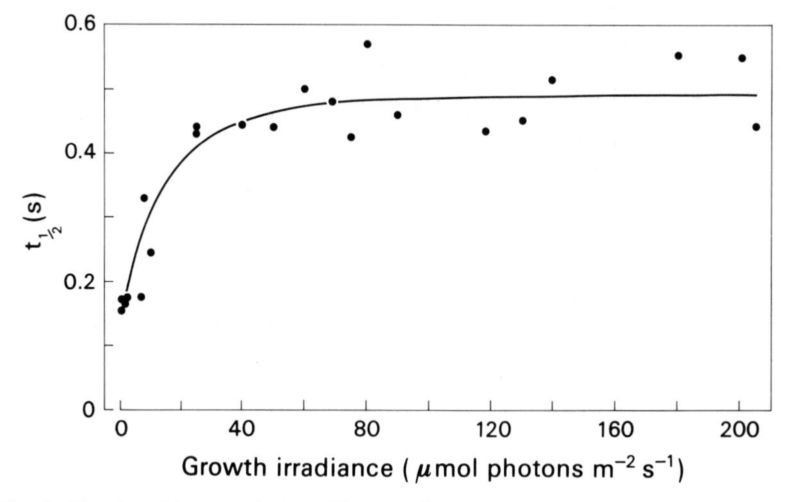

Fig. 3. The time ($t_{\frac{1}{2}}$) to reach the mid-way point between F_o and F_p in the chlorophyll fluorescence induction curve for discs of *Alocasia* leaves grown under different irradiances of incandescent light. Chl fluorescence was excited by blue-green light transmitted by a Corning 4-72 filter (~100 μmol photons m^{-2} s^{-1}) and detected at 685 nm through a light guide using a digital storage oscilloscope.

The cyt *f* content of thylakoids isolated from leaves grown in low irradiance was comparable to that reported for rainforest *Alocasia* (Boardman *et al.* 1972). At maximum irradiances, there was about 50% more cyt *f*, although the values were scattered (Fig. 2*g*). In contrast, there was little consistent variation of P700 with growth irradiance (Fig. 2*h*).

The latent ATPase activity of thylakoids was assayed as a measure of the amount of CF_1. Thylakoids isolated from *Alocasia* leaves grown in a mean fluorescent irradiance of 606 μmol photons m^{-2} s^{-1} had approximately 70% higher ATPase activity than from leaves grown in a mean irradiance of 9 μmol photons m^{-2} s^{-1}.

A useful *in vivo* indicator for the adaptation of thylakoids is given by the chlorophyll fluorescence induction signal from leaf discs (Davies *et al.* 1986*a*). When the excitation light is turned on, the fluorescence rises instantaneously to the level F_o, and subsequently more gradually to the peak F_p. The time ($t_{\frac{1}{2}}$) for the fluorescence to reach the average value of F_o and F_p is a measure of the speed of the net reduction of the primary quinone acceptor (Q_A) in PS II. At a fixed excitation irradiance, $t_{\frac{1}{2}}$ depends, among other factors, on the amount of electron carriers (particularly plastoquinone) which drain electrons from PS II. For *Alocasia* thylakoids *in vivo* (after an incubation in the dark for ~20 min), $t_{\frac{1}{2}}$ increased 3-fold over the range of incandescent irradiance examined (Fig. 3).

Anatomical and Ultrastructural Features

Microscopy

Figs 4*A* and 4*B* are light micrographs of *Alocasia* sections from the high fluorescent and low incandescent light treatments, respectively, which represent two extremes. Palisade cell chloroplasts from the high light treatment are preferentially located adjacent to the anticlinal cell walls; they have a length / width ratio larger than those for all other treatments (Table 1). There are also fewer grana and fewer thylakoids in the grana stacks (cf. Figs 4*C* and 4*D*). In addition, a higher percentage of the high light chloroplasts is occupied by stroma, which is reflected in the higher levels of RuP_2 carboxylase activity reported above. Palisade cell chloroplasts from the low irradiance treatment are preferentially located adjacent to the periclinal cell walls with a higher proportion adjacent to the distal periclinal cell walls and relatively few along the anticlinal cell walls (Fig. 4*B*). To some extent this distribution persists in the mesophyll cells adjacent to the palisade, but a more random distribution in chloroplast location is seen in the mesophyll cells adjacent to the lower epidermis. This effect of low irradiance on chloroplast position was also evident in sections from *Alocasia* grown in the rainforest (data not shown).

Although measurements were not made, mesophyll chloroplasts in high, and to a lesser degree in low irradiance-adapted leaves, possessed larger granal stacks than palisade chloroplasts. This is consistent with the intraleaf gradient of thylakoid stacking and photosynthetic characteristics reported by Terashima and Inoue (1985). In high irradiance, palisade chloroplasts possessed few starch grains, whereas mesophyll chloroplasts contained more starch grains (data not shown).

Cell Number

Cell number increased with increasing fluorescent irradiance and the increase in the number of mesophyll cells reflects an additional cell layer in some places (Table 1). Palisade cells were packed more closely as shown by their increase in number. These trends were not apparent in incandescent irradiance but may reflect the comparatively low maximum incandescent irradiance employed.

Photosynthetic Tissue

The high fluorescent irradiance treatment produced the largest number of chloroplasts per section length and these chloroplasts had a markedly elongated shape by comparison with other treatments. Chloroplast number per palisade cell was, however, highest in the higher incandescent irradiance treatment. Increasing fluorescent or incandescent irradiance significantly increased chloroplast area in palisade cell chloroplasts and the trend was also evident in the mesophyll. There is an additional effect of incandescent light on chloroplast area per chloroplast that results in the marked increases at high incandescent irradiance. The differences seen in chloroplast areas for palisade and mesophyll cell chloroplasts are sustained in the values for total chloroplast area per millimetre of section (Table 1).

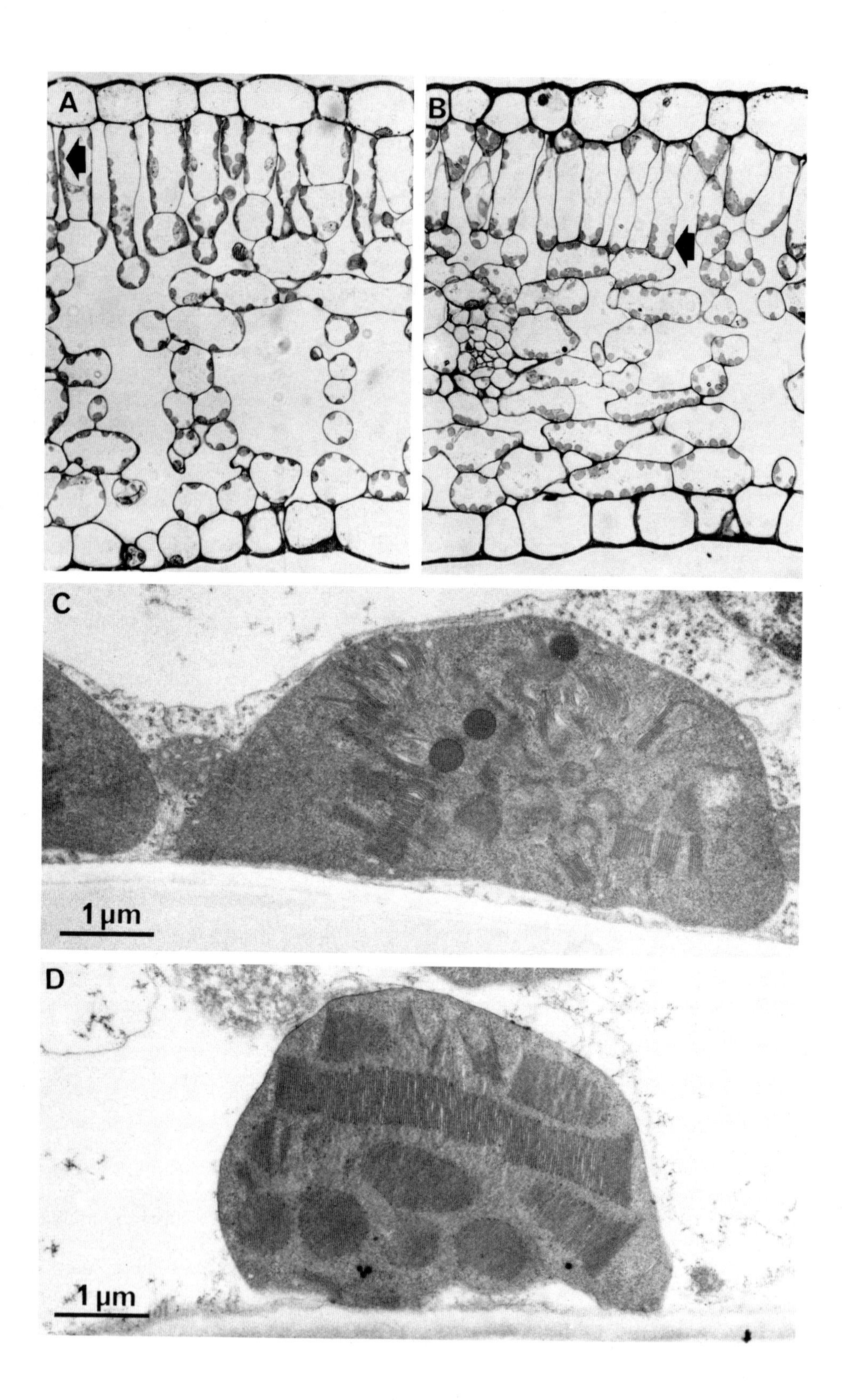
A
B
C
1 µm
D
1 µm

The results for chlorophyll content per chloroplast were limited by the ability to prepare intact chloroplasts for counting when large starch grains are present. The differences in Chl per chloroplast between the low and high fluorescent irradiance treatments studied are of a similar proportion and opposite to the differences measured in

Table 1. Morphometric analysis of leaf sections from plants of *Alocasia macrorrhiza* grown in four artificial irradiance conditions and a natural Queensland rainforest environment

Chlorophyll content per chloroplast under fluorescent irradiance of 5 and 350 μmol photons $m^{-2} s^{-1}$ was 1·73 and 1·27 fmol, respectively. 'Area' refers to chloroplast cross-sectional area

	Irradiance treatment (μmol photons $m^{-2} s^{-1}$)				
	Fluorescent		Incandescent		Rainforest
	18	780	10	75	5[A]
Leaf section thickness (μm)	258	265	205	220	176
Palisade					
No. of cells	67	97	70	79	44
Chloroplasts					
No. per cell	4·9	4·8	4·9	6·0	4·7
No. per mm section	325	458	342	407	207
Length/width	1·5	3·0	1·5	1·8	2·2
Area ± s.e. (μm^2)	37·9±1·5	51·5±1·8	44·4±1·7	92·3±2·8	122·5±5·9
10^{-4} × Area/mm section (μm^2)	1·42	2·36	1·52	3·76	2·54
Mesophyll					
No. of cells	175	201	179	131	88
Chloroplasts					
No. per cell	3·3	3·4	2·6	3·2	2·6
No. per mm section	462	705	472	421	227
Length/width	1·4	1·7	1·5	1·4	1·3
Area ± s.e. (μm^2)	38·1±1·8	40·0±1·6	38·1±1·6	62·2±2·7	81·8±5·7
10^{-4} × Area/mm section (μm^2)	1·76	2·82	1·80	2·62	1·86
Total area as % of section area	12	19	16	29	25

[A]Average irradiance for 12 h day calculated from integrated daily amount including light flecks (400–700 nm).

Fig. 4. (*A*) Section of *Alocasia* leaf from a plant grown at high fluorescent irradiance (780 μmol photons $m^{-2} s^{-1}$). Arrow shows disposition of chloroplasts along the anticlinal palisade cell walls. A proportion of mesophyll cell chloroplasts of the subpalisade layer also tend to be located along the anticlinal cell walls. Palisade cell chloroplasts are predominantly elongated in shape. Light micrograph magnification is ×285.

(*B*) Section of *Alocasia* leaf from a plant grown at low incandescent irradiance (10 μmol photons $m^{-2} s^{-1}$). Arrow shows disposition of chloroplasts along the periclinal cell walls, a pattern also seen in the mesophyll cells. Chloroplasts from both cell layers tend to be globular in shape. Light micrograph magnification is ×285.

(*C*) Electron micrograph of a chloroplast from a plant grown at high fluorescent irradiance (780 μmol photons $m^{-2} s^{-1}$). Chloroplast grana are randomly oriented but with fewer thylakoids and a larger cross-sectional area of stroma than seen in low irradiance-adapted chloroplasts (*D*).

(*D*) Electron micrograph of a chloroplast from a plant grown at low incandescent irradiance (10 μmol photons $m^{-2} s^{-1}$). The chloroplast is more globular than that in (*C*), the randomly oriented grana are composed of large numbers of thylakoids, and a much smaller area of stroma is evident.

total chloroplast area as a percentage of section area (Table 1). This result is in accord with the small differences in chlorophyll per unit leaf area for these treatments (Fig. 2*c*), and may reflect changes necessary to maintain chlorophyll per unit leaf area at an approximately constant level.

The rainforest *Alocasia* differs from the artificial light treatments in having considerably thinner leaves, fewer leaf cells and the largest chloroplast areas for both palisade and mesophyll chloroplasts. These large areas resulted in a total chloroplast area per millimetre of section similar to that of the high incandescent light treatment.

In addition to the results shown in Table 1, for the *Alocasia* grown under fluorescent and incandescent light conditions, the upper and lower epidermis each accounted for 9–13% of the section area, the palisade tissue 15–17%, mesophyll 24–29% and air space

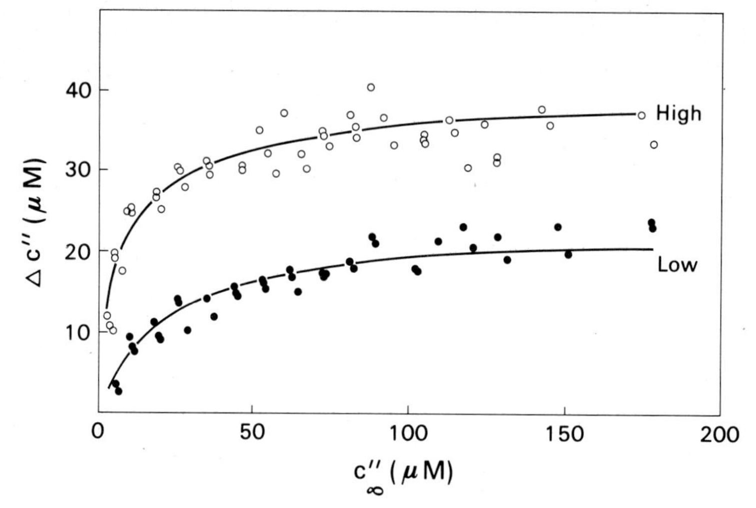

Fig. 5. The sequestering of a divalent cation, methyl viologen, in the diffuse layer adjacent to charged thylakoid surfaces. The difference ($\Delta c''$) between the total concentration of divalent cation added and the bulk concentration (c_∞'') represents the amount sequestered into the diffuse layer. ○ *Alocasia* grown in high fluorescent light with an incandescent supplement, Chl *a*/*b* = 3·05, background monovalent cation concentration ~0·4 mM, fitted curve calculated by taking surface charge density = $-14\cdot5$ mC m^{-2}, and effective surface area = $7\cdot8 \times 10^6$ m^2 $(mol\ Chl)^{-1}$. ● *Alocasia* grown in low incandescent light, Chl *a*/*b* = 2·36, background monovalent cation concentration ~0·15 mM, fitted curve calculated by taking surface charge density = $-1\cdot08$ mC m^{-2} and effective surface area = $8\cdot2 \times 10^7$ m^2 $(mol\ Chl)^{-1}$.

30–33%. For the *Alocasia* from the rainforest environment, however, the upper epidermis accounted for 9% and the lower epidermis 6% of these sections. The results (Table 1) also show that the sections from *Alocasia* growing in the rainforest environment were significantly thinner than those grown in the artificial light conditions although the percentages of palisade, mesophyll, and air spaces were similar.

Surface Charge Density

The ultrastructures of high and low light-adapted *Alocasia* chloroplasts exhibit striking differences, particularly in granal size (Figs 4*C* and 4*D*). The extent of appression of thylakoid surfaces under the influence of van der Waals attraction is strongly determined by the electrostatic properties of the membranes (Barber 1980). Hence, we compared the surface charge densities of destacked thylakoids of *Alocasia* grown at low and moderately high irradiances. As seen in Fig. 5, the surface charge densities of thylakoids from a low incandescent ($\sim 10\ \mu$mol $m^{-2} s^{-1}$) and a moderately high fluorescent ($\sim 400\ \mu$mol photons $m^{-2} s^{-1}$) irradiance environment were very different, namely, $-1 \cdot 1$ and $-14 \cdot 5$ mC m^{-2}, respectively. The extremely low negative surface charge density of low light-adapted thylakoids suggests very low electrostatic repulsion between these thylakoid membranes. This may be the main factor giving rise to the large granal stacks of thylakoids in low irradiance *Alocasia*.

Discussion

The results of this study show that it is no longer valid to regard *Alocasia* as an archetypal shade plant. Rather, *Alocasia* plants are capable of adapting not only to low but also to fairly high irradiances. Indeed its luxuriant growth in a controlled environment with adequate water and nutrient supply is evidence in support of its adaptability to high irradiance conditions. The process of photosynthetic acclimation to irradiance is complex, and is brought about by changes at various levels of chloroplast and leaf organisation, the net result being reflected in functional characteristics which are presumably a compromise between the different constraints.

Photosynthetic Characteristics

The photosynthetic capacity of leaves from plants grown at low irradiance (Fig. 2*a*) was comparable to that reported for *Alocasia* in the rainforest (Björkman *et al.* 1972). The photosynthetic capacity of *Alocasia* leaves grown at the maximum irradiances in a cabinet (approx. 40% of full sunlight) was high, and approached that of glasshouse spinach, in contrast to low values reported by Seeman *et al.* (1987) for *Alocasia* grown in sunlight in a glasshouse. Further evidence for the ability of *Alocasia* to acclimate in high light is provided by measurements of quantum yield, which declined only slightly at the highest irradiance attained in a growth cabinet (Fig. 2*b*), indicating that little photoinhibition had taken place. Although it has been reported that *Alocasia* plants growing in the open were severely photoinhibited (Pearcy 1987) it is likely that, in full sunlight in the open, *Alocasia* is more susceptible to photoinhibition particularly if nitrogen nutrition (Ferrar and Osmond 1986) and water are limiting. Interestingly, *Alocasia* growing in an open, sunny position in spring at the Australian National Botanical Gardens was not severely photoinhibited (Figs 2*a* and 2*b*).

Chloroplast Components

The Chl *a*/*b* ratio has been shown to be a good indicator of adaptation to irradiance occurring at the thylakoid level (Leong and Anderson 1984*a*, 1984*b*; Anderson *et al.* 1988). In *Alocasia* grown at the different irradiances the variation in the ratio between 2·4 and 3·5 represents a substantial variation in the relative amounts of core Chl a–proteins and light-harvesting Chl *a*/*b*–proteins. With more light harvesting chlorophyll of PS II (LHCII) and fewer PS II reaction centres in the low-light *Alocasia*, the photosynthetic unit size of PS II is larger than at high irradiance. Despite these modulations in Chl–proteins, the total Chl content per unit leaf area declined only slightly with increase in irradiance. The Chl content per unit leaf area in Fig. 2*c* is considerably lower than that reported by Seemann *et al.* (1987) for *Alocasia* grown in low-light, high-nitrogen conditions.

We observed that the Chl *a*/*b* ratio was lower in incandescent light than fluorescent light of equivalent irradiance (Fig. 2*d*). A similar observation was made in pea (Glick *et al.* 1985). Light quality, especially the enrichment (incandescent) or depletion (fluorescent) of far-red light is thought to influence the composition of Chl–proteins

(Glick *et al.* 1985). With the far-red-enriched incandescent irradiance favouring light absorption by PS I, *Alocasia* chloroplasts could compensate by increasing the PS II chlorophyll content so as to maintain a balanced light-harvesting capacity in both photosystems. The increase in PS II chlorophyll content could be brought about by increasing the PS II Chl *a/b*-proteins, as indicated by the lower overall Chl *a/b* ratios.

Of the *Alocasia* chloroplast components examined, the RuP_2 carboxylase activity, and presumably the amount of the enzyme, exhibits the greatest increase with irradiance (Fig. 2*e*). Substantial increases in RuP_2 carboxylase activity have also been reported in other species following their transfer from low to high light (Besford 1986; Chow and Anderson 1987*a;* Prioul and Reyss 1987). To a lesser extent, the amounts of PS II and cyt *f* (Fig. 2) per unit Chl also increase with growth irradiances. Because the Chl content per unit leaf area declined slightly with irradiance, it follows that the amount of PS II and cyt *f* per unit leaf area increased only slightly with growth irradiance. Thus it appears that the substantial increase in photosynthetic capacity of *Alocasia* with increased growth irradiance was achieved without large increases in cyt *f* or PS II. If so, *Alocasia* would not have the same linear relationship between leaf photosynthetic capacity and cyt *f* as that reported for pea (Evans 1987*b*). The content of P700 in *Alocasia*, on a chlorophyll basis, was independent of growth irradiance (Fig. 2*h*) as has been observed for many species: mustard (Grahl and Wild 1975), pea (Chow and Anderson 1987*a*) and spinach (Chow and Hope 1987). These results suggest that PS I does not limit electron transport.

The half-time for rise in the variable Chl fluorescence increased with incandescent irradiance (Fig. 3), suggesting that the net reduction of Q_A required more time as growth irradiance increased. One explanation is that the plastoquinone content was greater in leaves grown at high irradiance, thereby accelerating the oxidation of Q^-_B and hence of Q^-_A; at a given rate of excitation of PS II, a faster reoxidation of Q^-_A would lead to a slower net reduction of Q_A. An increase in the plastoquinone content in high irradiance has been observed in other species, e.g. mustard (Wild *et al.* 1975) and pea (Leong and Anderson 1984*b;* Chow and Anderson 1987*b*). A second explanation is that higher growth irradiance leads to a smaller antenna size of each PS II.

Anatomical and Ultrastructural Features

Alocasia grown in high fluorescent irradiance has a large number of chloroplasts per unit section length in both palisade and spongy mesophyll tissue. This increase in chloroplast number is mainly due to an increase in cell number rather than the number of chloroplasts per cell. The chloroplast number per unit section length (and by inference, per unit leaf area) could partly determine the photosynthetic capacity, as has been suggested for other 'shade' species (Araus *et al.* 1986). In contrast, the chloroplast section area as a proportion of total section area and chloroplast section area per mm section length would seem to be poor indicators of photosynthetic capacity, since both parameters were high in incandescent irradiance (75 μmol photons $m^{-2} s^{-1}$), even though the photosynthetic capacity was still low at this irradiance.

Ultrastructural differences between sun and shade chloroplasts were seen in *Alocasia* adapted to high and low irradiances. In high fluorescent irradiance (Fig. 4*C*), a large proportion of the chloroplast area is stroma, consistent with a high content of RuP_2 carboxylase. At the other extreme, in low incandescent irradiance where the lowest Chl *a/b* ratios were found, chloroplasts possessed large granal stacks.

Factors which influence the stacking of lettuce thylakoids (Davies *et al.* 1986*b*) also seem to apply in *Alocasia*. The predominant factor is probably the surface charge density of thylakoids. The results show that the average surface charge density of destacked thylakoids was an order of magnitude lower for low growth irradiance than for high irradiance (Fig. 5). This implies that the low-light *Alocasia* thylakoid membranes would readily adhere together under the influence of van der Waals attraction. Barber and Chow (1979) proposed that membrane appression and formation of the PS II–LHCII domain is favoured in grana partitions where the net surface charge density is low (minimum electrostatic repulsion) and van der Waals attraction is high, presumably mainly due to LHCII. Low light-adapted *Alocasia* thylakoids possess more Chl *a/b*-protein (LHCII) which thus favours thylakoid stacking. The lower amount of CF_1, which is known to be segregated into non-appressed membrane regions (Miller and Staehelin 1976), presumably for steric reasons, would also favour stacking in low light *Alocasia* chloroplasts. Taken together, these factors could explain the formation of the prominent, large granal stacks in low light *Alocasia* chloroplasts.

Our results in Table 1 clearly show that incandescent irradiance is affecting chloroplast section area in *Alocasia*. Comparing 10 and 75 μmol photons $m^{-2} s^{-1}$ of incandescent irradiance, there was a clear increase in chloroplast section area with the rise in irradiance in both palisade and spongy mesophyll chloroplasts. As the spectral distribution is the same, the difference associated with the two levels of incandescent irradiance must be related to the absolute irradiance in a critical wavelength region. The rainforest-grown *Alocasia* exhibits an equally large effect but at a very low average irradiance in the visible wavelength range. While the exact cause of the light quality or quantity effect on chloroplast section area remains to be investigated, it could be that elevated levels of far-red light (>700 nm), a feature common to the rainforest floor and the higher incandescent light environment (Fig. 1), were responsible for the increase in chloroplast section area.

Successful Adaptation to High Irradiance

The mechanisms by which *Alocasia* acclimates to high growth irradiance are probably related to both the amount and composition of its chloroplast components, as well as the spatial distribution of the chloroplasts within the leaves. In response to an increase in irradiance, *Alocasia* chloroplasts increase a number of components (particularly RuP_2 carboxylase, cyt *f/b* complex and ATP synthase) in a coordinated, integrated manner, while the photosynthetic unit size of PS II was decreased. The accompanying increase in photosynthetic capacity not only ensures the effective utilisation of high light, but also minimises damage due to excessive irradiation, thus helping to maintain high quantum efficiency (Fig. 2*b*). In palisade cells, chloroplasts from the high irradiance treatment are preferentially located adjacent to the anticlinal palisade cell walls (Fig. 4*A*). This distribution should minimise the direct radiation reaching the chloroplasts, and may ensure a more even distribution of light among the chloroplasts. Another mechanism whereby damage due to excessive irradiation is minimised is the alteration of the leaf angle with respect to the on-coming radiation; *Alocasia* leaves growing in open, sunny areas sometimes assume a more vertical orientation.

In our study we have not dealt with transient responses of *Alocasia* to sudden transfers from low to high irradiance; the speed of response in *Alocasia* remains to be investigated. In pea plants grown in low irradiance (60 μmol photons $m^{-2} s^{-1}$) and then transferred to higher irradiance (390 μmol photons $m^{-2} s^{-1}$), there is a temporary inhibition of PS II as assayed by atrazine-binding (Chow and Anderson 1987*b*) and a temporary loss of quantum efficiency (Chow and Anderson, unpublished data) before the chloroplasts finally acclimate to the higher irradiance. In *Solanum*, the temporary photoinhibition which occurs on transferring plants to bright light has been shown to be more pronounced in nitrogen-deficient conditions (Ferrar and Osmond 1986). Clearly both the extent and rate of acclimation are important factors underlying successful adaptation to a varying light environment.

Successful Adaptation to Low Light

Alocasia is able to grow at a low irradiance at which many species would not survive (Björkman *et al.* 1972). For example, at irradiances lower than about 15 μmol photons $m^{-2} s^{-1}$, maize (Bennett *et al.* 1987) and pea (J. Bennett, personal communication) have only low rates of chlorophyll synthesis and hence minimal amounts of Chl–protein complexes, and their growth is poor. If a species is to grow in extremely low irradiance conditions, it must be able to harvest light, including diffuse light, efficiently. In *Alocasia*, the highly randomised orientation of the granal stacks in chloroplasts from a low light environment (Fig. 4*D*) could help in the absorption of diffuse light. It is also interesting to note that the low-light *Alocasia* chloroplasts, particularly in palisade cells, were located adjacent to the periclinal cell walls, with a higher proportion adjacent to the distal periclinal cell walls. It is possible that such a spatial distribution helps in the absorption of direct radiation, for example in sunflecks.

The development of large granal stacks is a prominent feature of natural shade chloroplasts. It helps to pack a large quantity of pigments into a small volume. Apart from this function, the intriguing question is whether the formation of the large granal stacks is just an inevitable consequence of the physical factors discussed above (enhanced van der Waals attraction, low electrostatic repulsion and decreased steric hindrance), or whether it serves useful functions. The formation of grana is not essential for photosynthesis, but several possible useful functions have been proposed (Anderson 1982; Larkum and Barrett 1983). In *in vitro* studies, granal formation enhances light-saturated non-cyclic photophosphorylation without adversely affecting non-cyclic electron transport capacity (Chow 1984*b*). Further, *in vitro* ageing of thylakoids is slower in the stacked condition than in the destacked condition (Chow, unpublished data). It could be that the presence of large granal stacks confers structural stability to the membrane system, resulting for example in a slower degradation of membrane components. If so, the maintenance energy required would be lower, consistent with the known low rate of dark respiration, thus favouring the tolerance by *Alocasia* of low-light conditions.

On the rainforest floor, where *Alocasia* successfully acclimates to the very low average irradiance, the leaves are intermittently exposed to potentially damaging sunflecks. However, the increased amount of Chl *a/b*-protein in low-light-acclimated *Alocasia* chloroplasts should be accompanied by a greater amount of carotenoids bound in the pigment–protein complexes. Carotenoid-enriched chloroplasts may be able to mitigate the potentially adverse effects of sunflecks, for example, through the operation of a carotenoid cycle (Demmig *et al.* 1987).

Conclusion

Alocasia is a shade-tolerant but not, as sometimes assumed, an obligate shade species. Given sufficient nutrients and water, it can acclimate photosynthetically to relatively high growth irradiances without suffering serious loss of quantum efficiency. The acclimation of *Alocasia* chloroplasts to high or very low light conditions occurs via coordinated changes at the biochemical, anatomical and ultrastructural levels.

Acknowledgments

We thank Miss Stephanie Hossack-Smith and Ms C. Miller for skilled technical assistance, Dr J. Armstrong of the Australian National Botanic Gardens for kindly providing the *Alocasia* plants, and Mr I. Dawson and Dr R. W. King of the Ceres Phytotron for providing growth facilities. Mr W. Müller of the Division of Mathematics and Statistics, CSIRO, carried out the statistical analysis of chloroplast areas. L.P.Q. gratefully acknowledges financial support from CSIRO and Academia Sinica.

References

Anderson, J. M. (1982). The significance of grana stacking in chlorophyll *b*-containing chloroplasts. *Photobiochem. Photobiophys.* **3**, 225–41.

Anderson, J. M. (1986). Photoregulation of the composition, function and structure of thylakoid membranes. *Annu. Rev. Plant Physiol.* **37**, 93–136.

Anderson, J. M., Chow, W. S., and Goodchild, D. J. (1988). Thylakoid membrane organisation in sun/shade acclimation. *Aust. J. Plant Physiol.* **15**, 11–26.

Anderson, J. M., Goodchild, D. J., and Boardman, N. K. (1973). Composition of the photosystems and chloroplast structure in extreme shade plants. *Biochim. Biophys. Acta* **325**, 573–85.

Anderson, J. M., and Osmond, C. B. (1987). Shade–sun responses: compromises between acclimation and photoinhibition. In 'Photoinhibition'. (Eds D. J. Kyle, C. B. Osmond and C. J. Arntzen.) Topics in Photosynthesis, Vol. 9, pp. 1–38. (Elsevier: Amsterdam.)

Araus, J. L., Alegre, L., Tapia, L., Calafell, R., and Serret, M. D. (1986). Relationship between photosynthetic capacity and leaf structure in several shade plants. *Am. J. Bot.* **73**, 1760–70.

Barber, J. (1980). Membrane surface charges and potential in relation to photosynthesis. *Biochim. Biophys. Acta* **594**, 253–308.

Barber, J., and Chow, W. S. (1979). A mechanism for controlling the stacking and unstacking of chloroplast thylakoid membranes. *FEBS Lett.* **105**, 5–10.

Bendall, D. S., Davenport, H. E., and Hill, R. (1971). Cytochrome components in chloroplasts of the higher plants. *Methods Enzymol.* **23**, 327–44.

Bennett, J., Schwender, J. R., and Shaw, E. K. (1987). Protochlorophyllide reductase and the failure of corn to acclimate to low irradiances. In 'Progress in Photosynthesis Research'. (Ed. J. Biggins.) Vol. 4, pp. 535–8. (Martinus Nijhoff: Dordrecht.)

Besford, R. T. (1986). Changes in some Calvin cycle enzymes of the tomato during acclimation to irradiance. *J. Exp. Bot.* **37**, 200–10.

Björkman, O. (1981). Responses to different quantum flux densities. In 'Physiological Plant Ecology I. Responses to the Physical Environment'. (Eds O. L. Lange, P. S. Nobel, C. B. Osmond and H. Ziegler.) Encycl. Plant Physiol. New Ser., Vol. 12A, pp. 57–107. (Springer-Verlag: Berlin.)

Björkman, O., and Ludlow, M. M. (1972). Characterization of the light climate on the floor of a Queensland rainforest. *Carnegie Inst. Wash. Year Book* **71**, 85–94.

Björkman, O., Ludlow, M. M., and Morrow, P. A. (1972). Photosynthetic performance of two rainforest species in their native habitat and analysis of their gas exchange. *Carnegie Inst. Wash. Year Book* **71**, 94–102.

Boardman, N. K. (1977). Comparative photosynthesis of sun and shade plants. *Annu. Rev. Plant Physiol.* **28**, 355–77.

Boardman, N. K., Anderson, J. M., Thorne, S. W., and Björkman, O. (1972). Photochemical reactions of chloroplasts and components of the photosynthetic electron transport chain in two rainforest species. *Carnegie Inst. Wash. Year Book* **71**, 107–14.

Boardman, N. K., Björkman, O., Anderson, J. M., Goodchild, D. J., and Thorne, S. W. (1975). Photosynthetic adaptation of higher plants to light intensity: relationship between chloroplast structure, composition of the photosystems and photosynthetic rates. In 'Proc. 3rd Int. Congr. Photosynthesis'. (Ed. M. Avron.) pp. 1809–27. (Elsevier: Amsterdam.)

Chow, W. S. (1984*a*). The extent to which the spatial separation between photosystems I and II associated with granal formation limits noncyclic electron flow in isolated lettuce chloroplasts. *Arch. Biochem. Biophys.* **232**, 162–71.

Chow, W. S. (1984*b*). Electron transport, photophosphorylation and thylakoid stacking. In 'Advances in Photosynthesis Research'. (Ed. C. Sybesma.) Vol. 3, pp. 83–6. (Martinus Nijhoff/Dr W. Junk: The Hague.)

Chow, W. S., and Anderson, J. M. (1987*a*). Photosynthetic responses of *Pisum sativum* to an increase in irradiance during growth. I. Photosynthetic activities. *Aust. J. Plant Physiol.* **14**, 1–8.

Chow, W. S., and Anderson, J. M. (1987*b*). Photosynthetic responses of *Pisum sativum* to an increase in irradiance during growth. II. Thylakoid membrane components. *Aust. J. Plant Physiol.* **14**, 9–19.

Chow, W. S., and Barber, J. (1980). Salt-dependent changes of 9-aminoacridine fluorescence as a measure of charge densities of membrane surfaces. *J. Biochem. Biophys. Methods* **3**, 173–85.

Chow, W. S., and Hope, A. B. (1987). The stoichiometries of supramolecular complexes in thylakoid membranes from spinach chloroplasts. *Aust. J. Plant Physiol.* **14**, 21–8.

Davies, E. C., Chow, W. S., LeFay, J. M., and Jordan, B. R. (1986*a*). Acclimation of tomato leaves to changes in light intensity: effects on the function of the thylakoid membrane. *J. Exp. Bot.* **37**, 211–20.

Davies, E. C., Chow, W. S., and Jordan, B. R. (1986*b*). A study of factors which regulate the membrane appression of lettuce thylakoids in relation to irradiance. *Photosynth. Res.* **9**, 359–70.

Demmig, B., Winter, K., Krüger, A., and Czygan, F.-C. (1987). Photoinhibition and zeaxanthin formation in intact leaves. A possible role of the xanthophyll cycle in the dissipation of excess light energy. *Plant Physiol.* **84**, 218–24.

Evans, J. R. (1983). Nitrogen and photosynthesis in the flag leaf of wheat (*Triticum aestivum* L.). *Plant Physiol.* **72**, 297–302.

Evans, J. R. (1987*a*). The dependence of quantum yield on wavelength and growth irradiance. *Aust. J. Plant Physiol.* **14**, 69–79.

Evans, J. R. (1987*b*). The relationship between electron transport components and photosynthetic capacity in pea leaves grown at different irradiances. *Aust. J. Plant Physiol.* **14**, 157–70.

Ferrar, P. J., and Osmond, C. B. (1986). Nitrogen supply as a factor influencing photoinhibition and photosynthetic acclimation after transfer of shade-grown *Solanum dulcamara* to bright light. *Planta* **168**, 563–70.

Glick, R. E., McCauley, S. W., and Melis, A. (1985). Effect of light quality on chloroplast-membrane organization and function in pea. *Planta* **164**, 487–94.

Grahl, H., and Wild, A. (1975). Studies on the content of P700 and cytochromes in *Sinapis alba* during growth under two different light intensities. In 'Environmental and Biological Control of Photosynthesis'. (Ed. R. Marcelle.) pp. 107–13. (Dr W. Junk: The Hague.)

Larkum, A. W. D., and Barrett, J. (1983). Light-harvesting processes in algae. *Adv. Bot. Res.* **10**, 1–219.

Leong, T. Y., and Anderson, J. M. (1984*a*). Adaptation of the thylakoid membranes of pea chloroplasts to light intensities. I. Study on the distribution of chlorophyll–protein complexes. *Photosynth. Res.* **5**, 105–15.

Leong, T. Y., and Anderson, J. M. (1984*b*). Adaptation of the thylakoid membranes of pea chloroplasts to light intensities. II. Regulation of electron transport capacities, electron carriers, coupling factor (CF_1) activity and rates of photosynthesis. *Photosynth. Res.* **5**, 117–28.

Miller, K. R., and Staehelin, L. A. (1976). Analysis of the outer thylakoid surface: coupling factor is limited to unstacked membranes. *Cell Biol.* **68**, 30–47.

Pearcy, R. W. (1987). Photosynthetic responses of tropical forest trees. In 'Ecophysiology of Tropical Plants'. (Eds J. Dooley, C. B. Osmond and W. C. Wongkaew.) Biotrop Special Publication No. 24. (Biotrop: Bogor.) (In press).

Pick, U., and Bassilian, S. (1981). Octyl glucoside stimulates a Mg^{++}-specific ATPase activity in chloroplast CF_1. In 'Energy Coupling in Photosynthesis'. (Eds B. R. Selman and S. Selman-Reimer.) pp. 251–60. (Elsevier: Amsterdam.)

Prioul, J. L., and Reyss, A. (1987). Acclimation of Rubisco activity and quantity to different irradiances within the same tobacco leaf. Differential expression in SSU and LSU mRNA. In 'Progress in Photosynthesis Research'. (Ed. J. Biggins.) Vol. 4, pp. 553–6. (Martinus Nijhoff: Dordrecht.)

Robinson, S. P. (1983). Isolation of intact chloroplasts with high CO_2 fixation capacity from sugarbeet leaves containing calcium oxalate. *Photosynth. Res.* **4**, 281–7.

Seemann, J. R., Sharkey, T. D., Wang, J.-L., and Osmond, C. B. (1987). Environmental effects on photosynthesis, nitrogen-use efficiency and metabolite pools in leaves of sun and shade plants. *Plant Physiol.* **84**, 796–802.

Spurr, A. R. (1969). A low-viscosity epoxy resin embedding medium for electron microscopy. *J. Ultrastruct. Res.* **26**, 31–43.

Terashima, I., and Inoue, Y. (1985). Vertical gradient in photosynthetic properties of spinach chloroplasts dependent on intra-leaf light environment. *Plant Cell Physiol.* **26**, 781–5.

Tischer, W., and Strotmann, H. (1977). Relationship between inhibitor binding by chloroplasts and inhibition of photosynthetic electron transport. *Biochim. Biophys. Acta* **460**, 113–25.

Wild, A. (1979). Physiologie der Photosynthese höherer Pflanzen. Die Anpassung an Lichtbedingungen. *Ber. Dtsch. Bot. Ges.* **92**, 341–64.

Wild, A., Rühle, W., and Grahl, H. (1975). The effect of light intensity during growth of *Sinapis alba* on the electron-transport and the noncyclic photophosphorylation. In 'Environmental and Biological Control of Photosynthesis'. (Ed. R. Marcelle.) pp. 115–21. (Dr W. Junk: The Hague.)

Photosynthetic Acclimation and Photoinhibition of Terrestrial and Epiphytic CAM Tissues Growing in Full Sunlight and Deep Shade

William W. Adams III

Plant Environmental Biology Group, Research School of Biological Sciences, Australian National University, G.P.O. Box 475, Canberra, A.C.T. 2601, Australia; present address: Lehrstuhl für Botanik II der Universität, Mittlerer Dallenbergweg 64, 8700 Würzburg, West Germany.

Abstract

Light response curves of O_2 exchange and fluorescence from photosystem II at 77K were examined in several species possessing crassulacean acid metabolism (CAM) growing under natural conditions of varying irradiance. All of the species exhibited considerable acclimation to low irradiance, but the epiphytes *Dendrobium speciosum* and *Pyrrosia confluens* were unable to acclimate fully to full sunlight. Although *Opuntia stricta* possessed a reduced quantum yield under full sunlight conditions, its photosynthetic capacity, unlike that of *D. speciosum* and *P. confluens*, was very high, and was probably due, at least in part, to the higher level of CAM activity present in this species. Shade tissues of *O. stricta*, however, were quite susceptible to photoinhibition by high light, as indicated by reductions in 77K fluorescence.

Fronds of *P. confluens* growing in full sunlight were clearly photoinhibited. However, the light environment in the epiphytic habitat is rarely static, and considerable recovery from photoinhibition in a sun population of *P. confluens* fronds was observed near the summer solstice when the canopy shielded this population from direct sunlight for most of the day. Such seasonal and diurnal changes in irradiance may be important to the survival of epiphytic species with a limited capacity for CAM activity.

Introduction

It has become increasingly clear that plants possessing crassulacean acid metabolism (CAM) have the ability to acclimate and grow under conditions of low irradiance (Martin *et al.* 1986; Winter *et al.* 1986; Adams *et al.* 1987*a*). Interestingly, however, the major limitations appear to be in photosynthetic acclimation to higher irradiance, where reductions in photosynthetic efficiency (as measured by quantum yield and PS II* fluorescence at 77K), photosynthetic capacity, and nocturnal acid accumulation have been observed (Adams *et al.* 1987*a*, 1987*b*).

This investigation was carried out in order to ascertain whether several CAM species growing in Australia exhibited responses to growth irradiance similar to those observed in glasshouse grown plants and to determine if photoinhibition was as prevalent in CAM tissues growing in high light as suggested by previous findings (Adams *et al.* 1987*a*, 1987*b*).

Materials and Methods

Field Sites

The terrestrial cactus *Opuntia stricta* Haw. was studied *in situ* in Warrumbungle National Park, N.S.W. (31°25′ S., 149°00′ E., 550 m elevation), where it is found growing in a mixed woodland of *Eucalyptus, Acacia* and *Callitris*. The epiphytic orchid *Dendrobium speciosum* Smith was collected from an area of subtropical rainforest remnants at Long Point, N.S.W. (30°25′ S., 151°55′ E., 850 m

*Abbreviations used: CAM, crassulacean acid metabolism; PS II, photosystem II.

0310-7841/88/010123$03.00

elevation) and returned to Canberra where light response curves were immediately obtained. The epiphytic fern *Pyrrosia confluens* (R.Br.) Ching was studied *in situ* on the northern (sun exposure) edge of a subtropical rainforest at Dorrigo National Park, N.S.W. (30°22′ S., 152°45′ E., 720 m elevation), as well as on the eastern coast at Middle Head, N.S.W. (30°22′ S., 153°00′ E., 20 m elevation) where it grew as a lithophyte. Irradiance was measured with a quantum sensor (Li-Cor 190S).

Light Response Curves

Light response curves of O_2 exchange at 5% CO_2 and 25°C were obtained with a Hansatech leaf disc O_2 electrode (Delieu and Walker 1981) as described previously (Adams *et al.* 1986). Quantum yields (ϕ) were calculated from a linear regression of the linear portion of the light response curve near and above the light compensation point. Absorptances were determined using a small integrating sphere and three standards of known reflectance.

Chlorophyll Fluorescence from PS II at 77K

Fluorescence at 690 nm from tissues frozen to 77K was obtained with a system similar to that described in Adams *et al.* (1987*b*). However, the excitation light (1·3 μmol quanta $m^{-2} s^{-1}$ at the sample surface) passed through a broad band blue cut-off filter (Corning 4–96) and the fluorescence at 690 nm was monitored with a photodiode (EG & G model HUV-4000B). Samples were placed in the dark for 7 min at ambient temperature prior to being frozen in liquid nitrogen for a minimum of 4 min.

Chlorophyll Analyses and Titratable Acidity

Chlorophyll concentrations were ascertained as described previously (Adams *et al.* 1987*b*). Samples for determination of nocturnal acid accumulation were collected at dusk and dawn, stored in 80% ethanol, and returned to the laboratory for extraction and titration to pH 7·0 with 0·01 N NaOH.

Results and Discussion

Given that the plants in this study were generally examined in habitats of either full sunlight or deep shade, it is perhaps instructive to briefly review the results of a previous study (Adams *et al.* 1987*a*) in which two species possessing CAM (*Kalanchoë daigremontiana* and *Hoya carnosa*) were grown under a range of light regimes and characterised (Figs 1 and 2). Acclimation to low irradiance was evident in reduced leaf thickness, light compensation point, photosynthetic capacity, dark respiration, and nocturnal acid accumulation, as well as increased chlorophyll content. All of these parameters exhibited considerable adjustment as growth irradiance increased, with increases in all of them, except chlorophyll content which decreased with increased irradiance. These trends held true in all cases except for *H. carnosa* at the highest irradiance, where it exhibited reduced photosynthetic capacity and nocturnal acid accumulation (Fig. 2) accompanied by reduced levels of F_v/F_m fluorescence at 690 nm and reduced quantum yields (Adams *et al.* 1987*a*). These all indicate that *H. carnosa* was incapable of acclimating to full sunlight and experienced photoinhibition under such conditions. With this characterisation of two CAM species growing across a range of light environments, a better interpretation of similar characteristics of CAM plants growing in extreme light environments in the field is possible.

Opuntia stricta

The light response curves for the sun-exposed and sun-protected faces from the cladode of an *O. stricta* plant growing in full sunlight are shown in Fig. 3*A*. The shade face (which never received direct sunlight: see Fig. 4*a*) exhibited a lower light compensation point and a higher quantum yield, but a much lower photosynthetic capacity than did the sun face. This contrasts sharply with the light response curves obtained from the desert cactus *O. basilaris* (shown in Fig. 3*B* for comparison), where both the quantum yield and photosynthetic capacity were lower on the sun-exposed surface than on the sun-protected surface. This capacity to maintain high photosynthetic rates at high irradiance despite reductions in quantum efficiency is probably related to the high levels of CAM activity present in *O. stricta*. While the level of nocturnal acid accumulation in *O. stricta* was 134 μequiv. acid g^{-1} fresh weight, it was less than 40 μequiv. acid g^{-1} fresh weight in *O. basilaris*. The moister climate of the Warrumbungles relative to

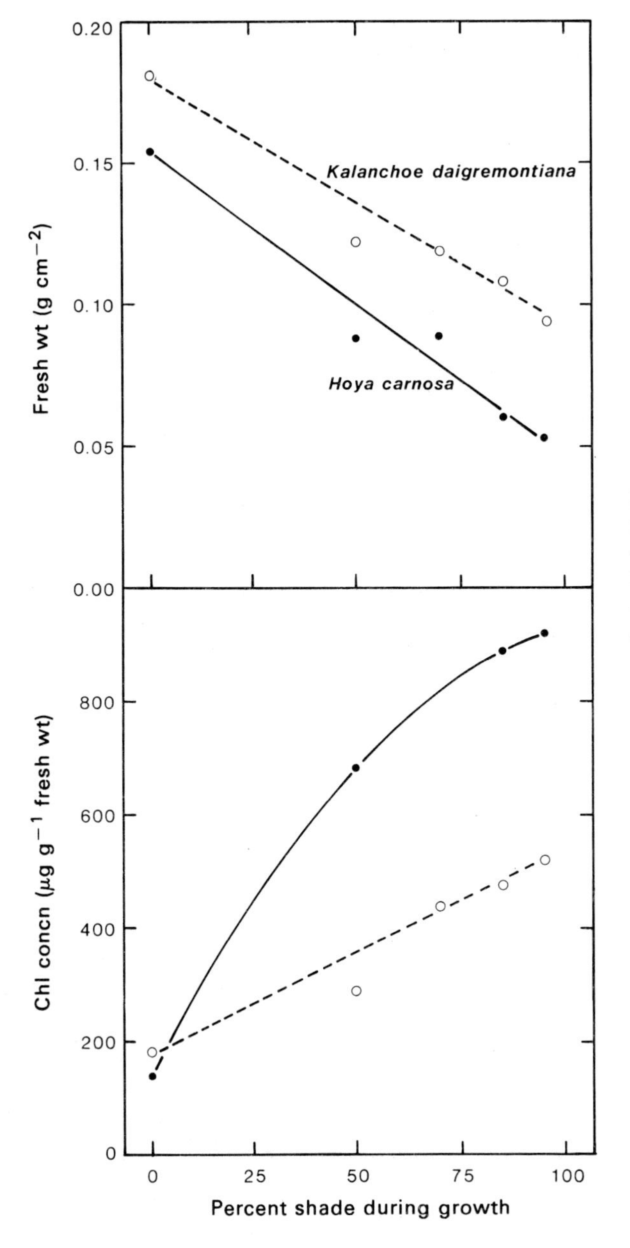

Fig. 1. Responses of fresh weight per unit leaf area and chlorophyll concentration to growth under a range of irradiances in *K. daigremontiana* (○) and *H. carnosa* (●) (after Adams *et al.* 1987*a*).

Death Valley is probably largely responsible for the difference in CAM activity between the two *Opuntia* species (Osmond *et al.* 1979*a*; Adams *et al.* 1987*b*), although the possibility of high temperature effects on both CAM activity and photosynthesis of *O. basilaris* growing in Death Valley cannot be excluded.

The capacity for relatively high photosynthetic rates (above 30 μmol O_2 m^{-2} s^{-1} far below light saturation, Fig. 3*A*) in *O. stricta* confirm those predicted on the basis of rates of deacidification (Osmond *et al.* 1979*b*). Similarly high rates have been observed in a number of other CAM species (Adams *et al.* 1986, 1987*a*; and unpublished data) under conditions of high CO_2 concentrations supplied internally (via malic acid

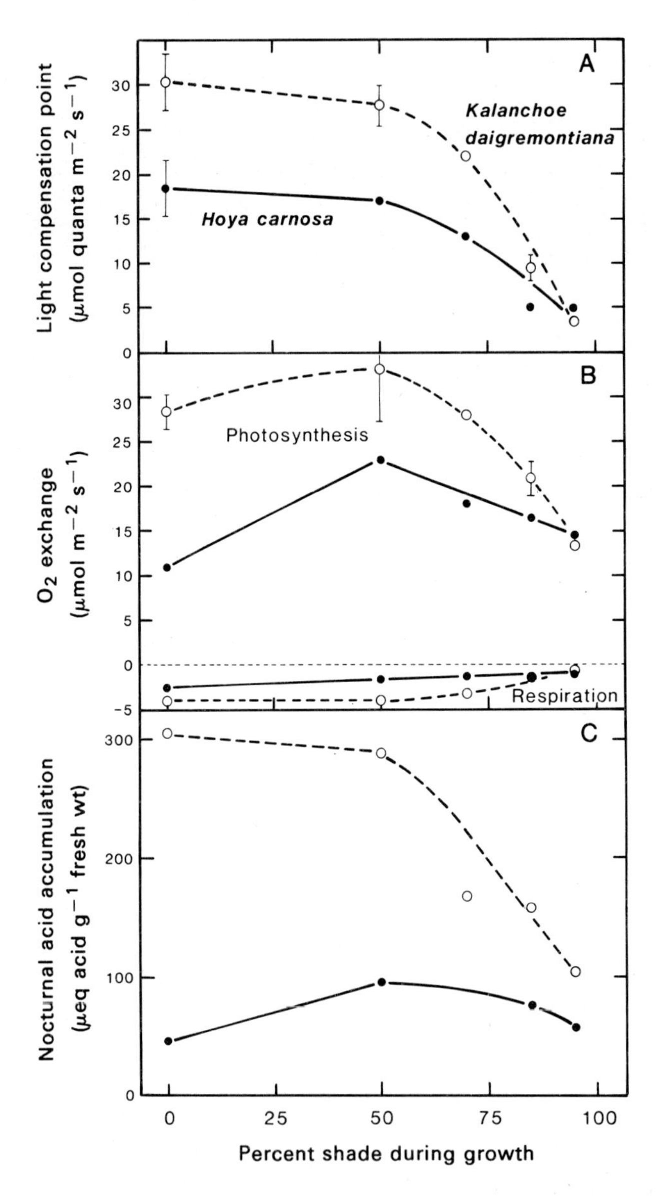

Fig. 2. Responses of photosynthetic characteristics (*A*, light compensation point and *B*, photosynthesis near light saturation), respiration rate (*B*), and nocturnal acid accumulation (*C*) to growth under a range of irradiances in *K. daigremontiana* (○) and *H. carnosa* (●) (after Adams *et al.* 1987*a*).

decarboxylation during deacidification) or supplied externally (5% CO_2). Thus the lower rates of CO_2 fixation generally reported for CAM plants at ambient CO_2 concentrations during phase IV (including *O. stricta*: Osmond *et al.* 1979*b*) must be attributable to CO_2 limitations, despite the fact that CO_2 response curves for *O. stricta* indicated saturation near ambient CO_2 levels (Osmond *et al.* 1979*a*).

Fluorescence from PS II at 77K was monitored from both faces of the same cladode examined above concomitantly with changes in incident irradiance throughout the day on 13 October 1986 (Fig.4*A*). The shade face did not receive light in excess of 140 μmol quanta m^{-2} s^{-1}, while the sun face received an irradiance parallel to, and just slightly lower than, that received on a horizontal plane. The shade face exhibited much higher levels of absolute fluorescence (F_o and F_m) relative to the sun face, and values of F_v/F_m indicative of healthy non-photoinhibited tissue throughout the day (around 0·8). The sun face, however, exhibited a predawn F_v/F_m of 0·52 which declined slightly during midday as incident irradiance reached its maximum and then returned to

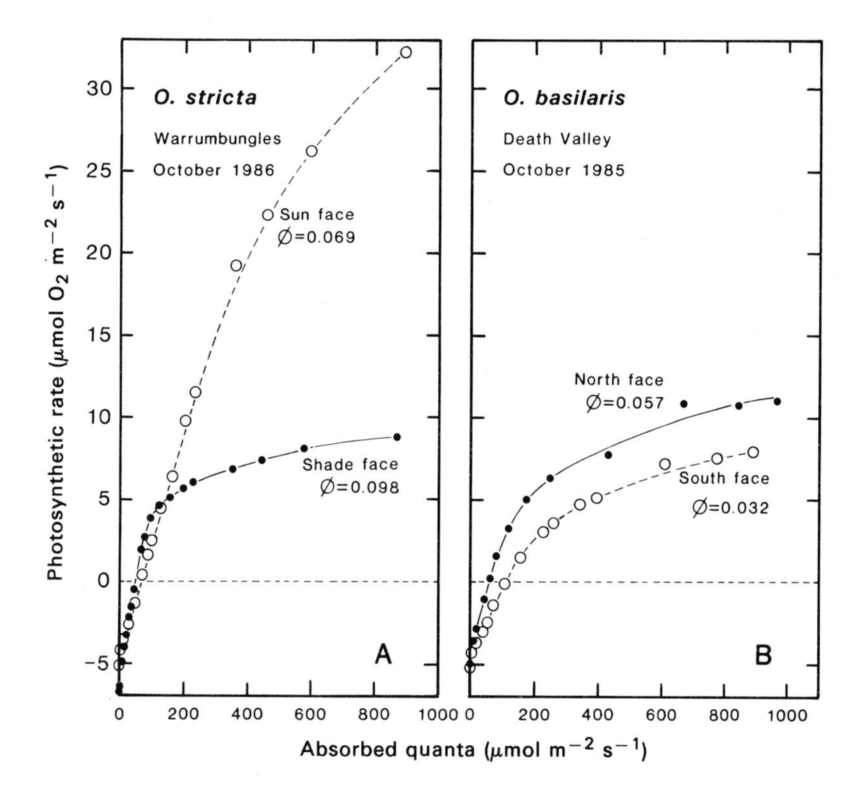

Fig. 3. (*A*) Light response curves from the exposed (○) and shaded (●) surface of an *O. stricta* cladode (see Fig. 4*A* for incident irradiance on each surface) growing in Warrumbungle National Park, N.S.W. on 13 October 1986. (*B*) Light response curves from the north (sun-protected, ●) and south (sun-exposed, ○) faces of an *O. basilaris* cladode sampled in Death Valley, California in October 1985.

predawn levels by the end of the day (incident irradiance is low during the last few hours of the day after the sun drops behind trees and a nearby hill). This slight depression of F_v/F_m during the midday peak of irradiance in *O. stricta* (Fig. 4*A*) is similar to that observed in *O. basilaris* (Adams *et al.* 1987*b*), and is probably indicative of increased levels of non-radiative energy dissipation during direct illumination (Demmig and Björkman 1987; Demmig *et al.* 1987). However, the low levels of F_o and F_m preclude any detailed analysis of changes in F_o and F_m which might confirm this possibility.

The capacity for high rates of photosynthesis despite reductions in quantum efficiency in a plant such as *O. stricta* growing in full sunlight (Fig. 3*A*) also highlights the potential differences between photoinhibition resulting from a photoinhibitory event, such as the exposure of a shade leaf to high light which yields reductions in both quantum yield and photosynthetic capacity (Powles and Björkman 1982; Winter *et al.* 1986; Adams *et al.* 1987*a*), and that resulting from growth under high light conditions. Although the fluorescence data in this study are not sufficient to distinguish between photoinhibitory damage and photoinhibition with a potentially protective function, it is conceivable that the majority of the reduced quantum yield and F_v/F_m values in the exposed face of

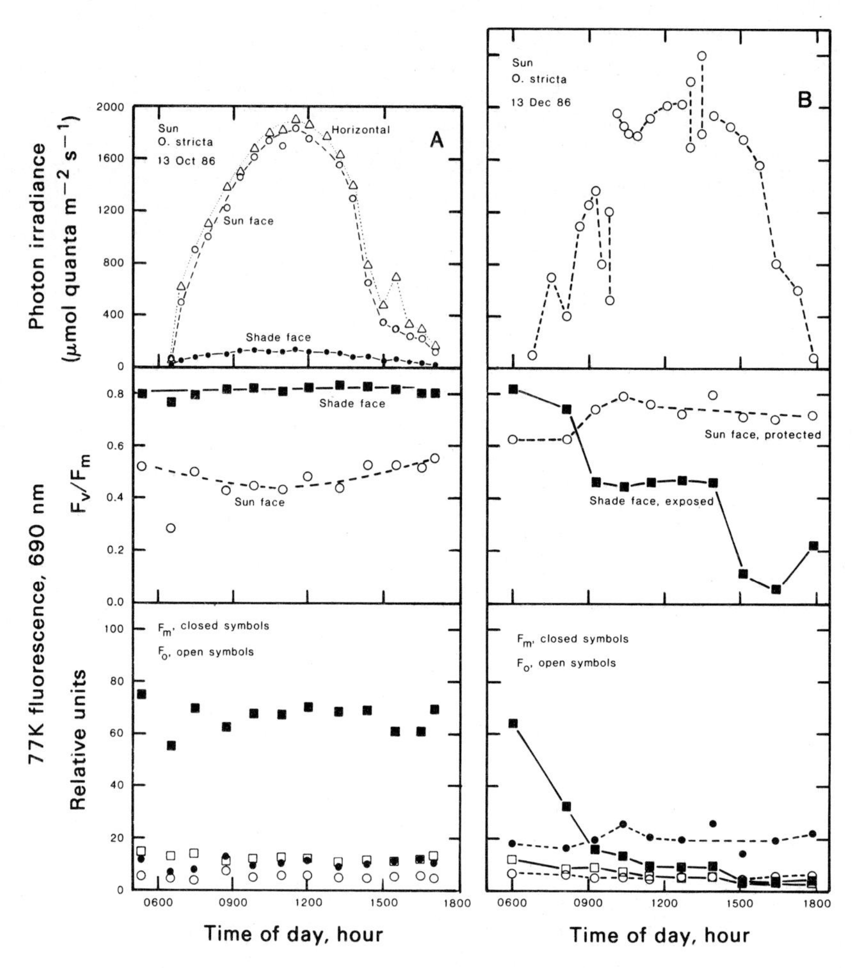

Fig. 4. (*A*) Diurnal measurements of irradiance and PS II fluorescence at 77K from the exposed (sun) and protected (shade) faces of an *O. stricta* cladode growing in Warrumbungle National Park, N.S.W. on 13 October 1986. (*B*) Diurnal measurements of irradiance and PS II fluorescence at 77K from the same cladode which was removed from the plant and placed on the ground with its shade face exposed to direct sunlight on 13 December 1986.

O. stricta results from high levels of non-radiative energy dissipation (Demmig and Björkman 1987; Demmig *et al.* 1987). As this serves the function of diverting energy which is in excess of that needed to run photosynthesis, it may have no effect on photosynthesis at higher light levels. Any reduction in the quantum yield resulting from such a protective mechanism would be of little consequence to tissues which experience high irradiance for much of the day (Fig. 4). The possibility of even greater photosynthetic capacity in tissues growing under irradiances slightly below full sunlight cannot, of course, be excluded.

Prior to sunrise on 13 December 1986, the same cladode examined in Figs 3*A* and 4*A* was removed from the plant and placed on the ground with the shade face exposed to direct sunlight. Both F_o and F_m declined in the exposed shade face during the day, resulting in a decrease in F_v/F_m from the predawn level of 0·82 to a minimum of 0·12 before climbing slightly to 0·22 at the very end of the day (Fig. 4*B*). The decline in F_o, F_m, and F_v/F_m was rather abrupt initially, and appeared to halt as the rise in incident irradiance was interrupted by clouds. The protected sun face, however, did not exhibit any decline in F_v/F_m, and may have even experienced some recovery. The reductions in F_v/F_m during exposure of the shade face of *O. stricta* to full sunlight are clearly indicative of photoinhibition. This appeared to be due to a quenching of both F_m and F_o (Fig. 4*B*). However, as pointed out by Demmig and Björkman (1987), most plants

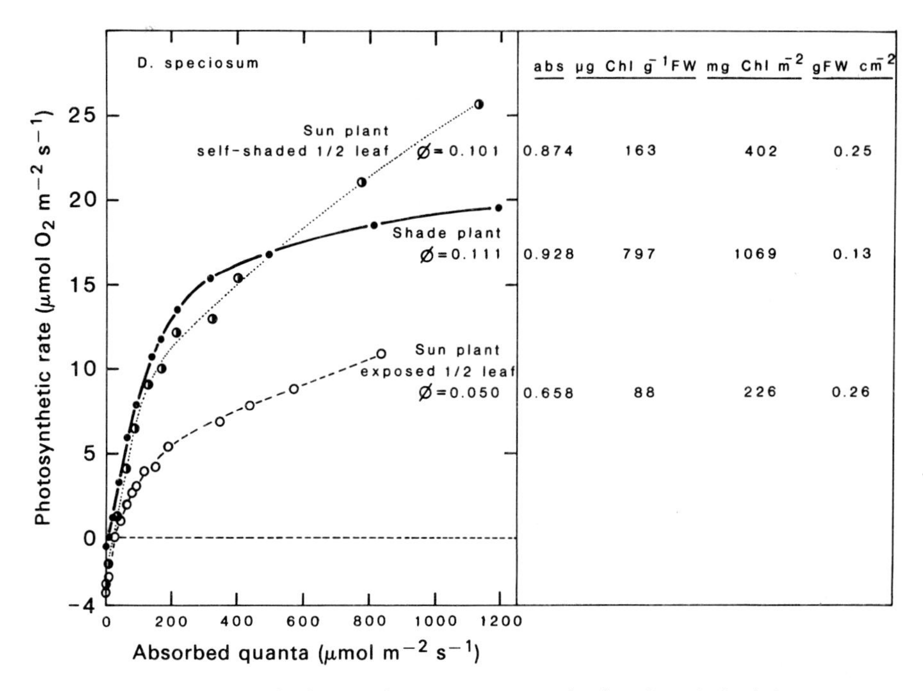

Fig. 5. Light response curves for leaves of *D. speciosum* growing in subtropical rainforest remnants at Long Point, N.S.W., which had experienced different levels of irradiance during growth. Absorptances, chlorophyll contents, and the fresh weight per unit leaf area are also given for each leaf.

are likely to exhibit a combination of a general quenching of fluorescence, indicative of an increase in non-radiative energy dissipation, and an increase in F_o indicative of damage to PS II (both of which will result in a decrease in F_v/F_m). Thus it is quite probable that there was some damage to PS II, but the increase in F_o which might indicate this was masked by the general quenching phenomenon which occurred during the exposure. This contrasts with the response of shade leaves of the CAM species *Hoya australis* which did exhibit an increase in F_o during exposure to full sunlight (unpublished data).

Dendrobium speciosum

Light response curves for tissues of *D. speciosum* which grew under full sunlight, medium shade (self-shaded leaf), and deep shade are shown in Fig. 5. Respiration rates and light compensation points were greatest in the sun-exposed leaf and lowest in the shade leaf, while the quantum yield of the sun leaf was approximately half that of the medium shade and deep shade leaves. The shade leaf possessed the greatest absorptance, highest chlorophyll content (on either an area or fresh weight basis), and was thinner than those which developed under higher light. The light response curve of the self-shaded leaf from the sun plant (Fig. 5) is similar to that of a *D. speciosum* leaf reported to be growing in intermediate to full sunlight (Winter *et al.* 1986).

Table 1. Measurements of nocturnal acid accumulation and fresh weight per unit leaf area for sun and shade populations of *P. confluens* fronds during winter, spring, and summer in a subtropical rainforest at Dorrigo National Park, N.S.W.

The levels of nocturnal acid accumulation represent the means of two determinations, each of which was garnered from three to seven leaves

Date	Population	Acid accumulated (μequiv. g^{-1} fresh wt)	Fresh wt per unit area (g cm^{-2}) (mean ± s.e. (*n*))
31 July 1986	Sun	14·0	0·111 ± 0·004 (6)
	Shade	13·5	0·071 ± 0·007 (6)
11 Oct. 1986	Sun	5·2	0·086 (2)
	Shade	23·9	0·076 ± 0·001 (4)
11 Dec. 1986	Sun	21·1	0·105 ± 0·006 (4)
	Shade	23·9	0·090 ± 0·004 (4)

Table 2. Frond length and percentage of fronds with sporangia for sun and shade populations of *P. confluens* growing on the coast at Middle Head, N.S.W.

Values for frond length are means ± s.e., with number of samples in parentheses

Population	Frond length (cm)	Sporangia present (%)
Sun	6·6 ± 2·1 (123)	62·6
Shade	12·1 ± 0·4 (111)	10·8

Pyrrosia confluens

The shade fronds of *P. confluens*, which were a dark green in colour, were thinner (Table 1) and longer (Table 2) than sun fronds, which were generally somewhat yellow in appearance. This was also recognised by Winter *et al.* (1983). However, reproduction was much greater in the sun fronds (Table 2). The levels of nocturnal acid accumulation in both the sun and shade fronds of *P. confluens* were relatively low and comparable to one another in July and December (Table 1). Although the species clearly exhibits CAM, it is only of limited capacity (Table 1). It may, in fact, be inhibited in strong light (see October, Table 1), or under conditions of extreme cold (the minimum temperature on 1 August 1986 was −5°C).

Light response curves for several fronds of *P. confluens* growing in full sunlight, medium shade, and deep shade reveal similar responses to those seen in *D. speciosum* (Fig. 6). Most specifically, the shade plant possessed a high quantum yield while the sun plants exhibited greatly reduced quantum yields indicative of photoinhibition. Light response curves from a closely related species, *P. longifolia*, did not reveal any differences in the quantum yields of fronds growing in sun-exposed and shaded habitats, although the light compensation point was higher in the sun fronds (Winter *et al.* 1986). They attributed the absence of acclimation of photosynthetic capacity to high light in this species to nutrient limitations. As in any epiphytic habitat, this is probably also important in the inability of *P. confluens* to acclimate to full sunlight.

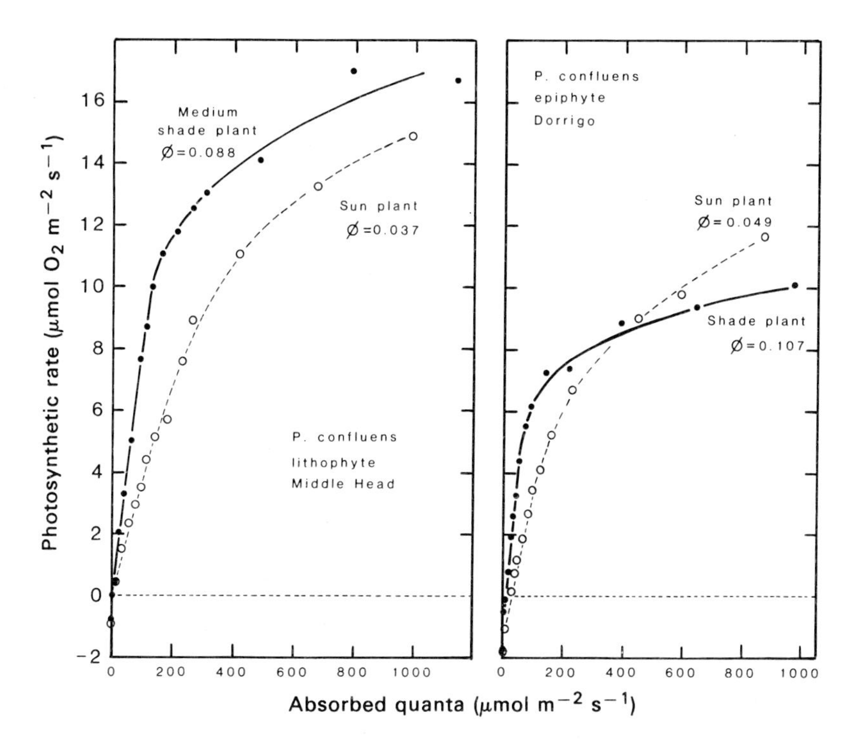

Fig. 6. Light response curves for fronds of *P. confluens* growing as a lithophyte on the coast at Middle Head, N.S.W., and as an epiphyte in a subtropical rain forest at Dorrigo National Park, N.S.W.

Measurements of fluorescence from PS II at 77K from a sun population of fronds on the east side of a tree and a shade population of fronds on 11 October 1986 revealed that the shade fronds exhibited higher levels of fluorescence (F_o and F_m) as well as F_v/F_m (Fig. 7*A*). F_v/F_m in the sun fronds appeared to be lower in the morning during exposure to high light and to rise in the afternoon when the irradiance was only slightly greater than that incident on the shade fronds. This was due primarily to higher levels of F_m in the afternoon, and may again be indicative of the operation of a protective mechanism whereby excess energy is dissipated via an increase in non-radiative energy transfer (Demmig and Björkman 1987; Demmig *et al.* 1987). However, as each point in Fig. 7*A* represents a sample from a different frond, this remains as speculation only.

In December the sampled shade population actually received some fairly high irradiance (much of it sunflecks) during the morning, and F_v/F_m rose slightly in the afternoon during the low light (Fig. 7*B*). The sampled sun population, however, received very little direct light at all. This was near the summer solstice, and the height of the sun was such that the canopy shielded these west-facing fronds from direct sunlight for most of the day. Consequently the levels of F_m were elevated in the sun fronds relative to those sampled in October, and F_v/F_m was comparable to that of the shade population. These results emphasise the dynamic nature of the light environment in the epiphytic habitat and the potential for recovery from photoinhibition in sun-exposed tissues. Unlike exposed, terrestrial habitats, where changes in irradiance could only result from seasonal changes in the path of the sun or transient weather phenomena, the epiphytic habitat is one in which it is unlikely that plant tissues will be totally

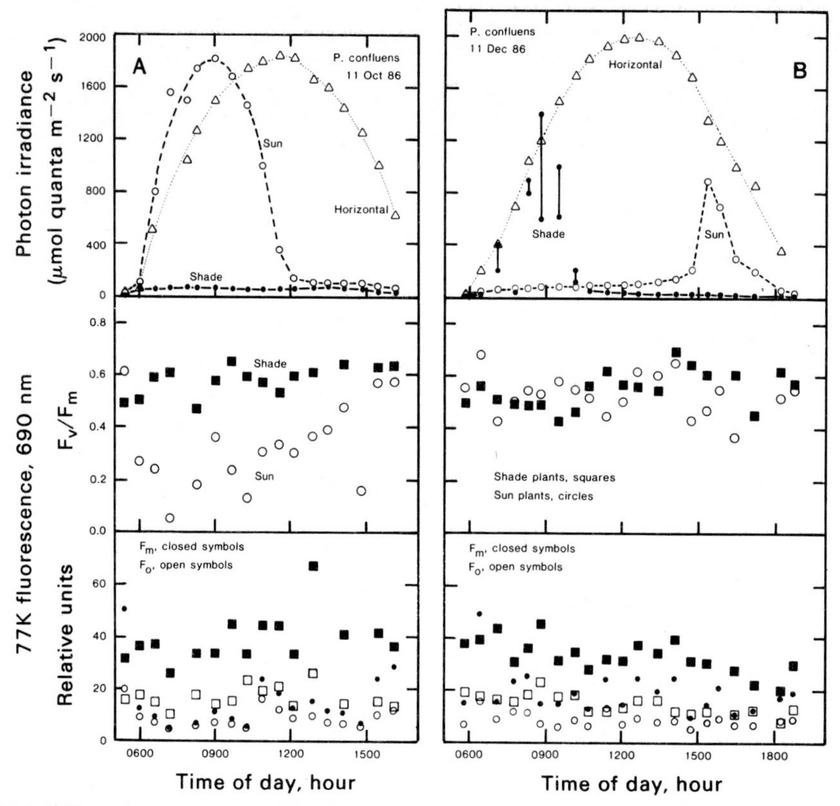

Fig. 7. (*A*) Diurnal measurements of irradiance and PS II fluorescence at 77K from sun and shade fronds of *P. confluens* at Dorrigo National Park on 11 October 1986. (*B*) Diurnal measurements of irradiance and PS II fluorescence at 77K from sun and shade fronds of *P. confluens* at Dorrigo National Park on 11 December 1986.

exposed to direct sunlight for the entire day. Given the evergreen nature of most CAM tissues, the dynamic aspects of the epiphytic light environment, both diurnally and seasonally (particularly important in deciduous forests), is undoubtedly instrumental in the survival of these plants.

Conclusions

It is fairly evident from the photosynthetic light response curves and measurements of fluorescence from PS II at 77K that the CAM species examined in this study are capable of acclimation to low irradiance. The low light compensation points, high quantum yields, and relatively low light levels required for saturation of photosynthesis by the shade acclimated tissues (Figs 3, 5, and 6) ensure the efficient utilisation of low irradiance levels. Under conditions of full sunlight, however, a considerable degree of photoinhibition is experienced by these plants. This is evident from the reduced levels of photosynthetic efficiency (lower quantum yields and F_v/F_m fluorescence at 690 nm) and capacity. The one exception to the latter, where the quantum yield of photosynthesis was reduced, but photosynthetic capacity was quite elevated in the sun-exposed face of *O. stricta* (Fig. 3*A*), suggests that plants possessing a high level of CAM activity may be quite capable of acclimation to high irradiance.

The recovery of F_v/F_m in the shaded sun population of *P. confluens* (Fig. 7*B*) deserves some further consideration. In this case F_v/F_m was essentially at the same level found in the leaves from the shade population, and this appeared to be due primarily to a relaxation of F_m quenching with no change in F_o relative to their levels from sun leaves on previous occasions (compare Fig. 7*A* with Fig. 7*B*, and unpublished data). This suggests that recovery in the previously sun-exposed leaves was due to a decrease in non-radiative energy dissipation rather than any repair of damage to PS II. Björkman (1987) reported that reduced levels of photochemical efficiency of PS II in leaves of mangroves growing in full sunlight, relative to those in shade, were largely due to much higher levels of non-radiative energy dissipation, and that it required 7 days of shading for this protective mechanism to relax sufficiently for the photochemical efficiency of PS II in the sun leaves to approach that found in the shade leaves. It seems entirely conceivable that the majority of photoinhibition observed in the CAM species growing in full sunlight is due to a high degree of non-radiative energy dissipation rather than any damage to PS II. Even *O. stricta*, which had relatively high levels of nocturnal acid accumulation and a very high capacity for photosynthetic O_2 evolution at high irradiance (Fig. 3*A*), exhibited evidence of reduced photochemical efficiency (reduced quantum yield and level of F_v/F_m; Figs 3*A* and 4*A*) from its sun-exposed surface. Given its capacity for high rates of photosynthesis at high irradiance, it seems unlikely that the reductions in photochemical efficiency are due to damaged PS II reaction centres.

Although most of the plants examined in this study were growing in either full sunlight or deep shade, it should be emphasised that in most habitats a species, or even the photosynthetic organs of a given plant, will be subject to a range of light environments which may change with time. Thus, as in the controlled experiments depicted in Figs 1 and 2, one could anticipate a range of acclimation phenomena in natural populations as well. This is certainly suggested by the light response curves in this study (Figs 3, 5, and 6), as well as the chlorophyll and morphological data (Fig. 5 and Tables 1 and 2). The degree of incomplete acclimation to high light (photoinhibition) for a given species will undoubtedly be found to be dependent on the age and genetic potential of that species (or photosynthetic organ) responding to a combination of environmental factors such as temperature, water status, and nutrient availability.

Acknowledgments

I thank S. C. Wong, W. Coupland, and P. Groeneveld for constructing the 77K fluorescence system used in this study, B. Wallace for suggesting some of the field sites, I. Terashima, E. Brugnoli, and S. Prober for assistance in the field, the National Parks and Wildlife Service of New South Wales for permission to work in Warrumbungle and Dorrigo National Parks, and C. B. Osmond for critically reading the manuscript.

References

Adams, W. W. III, Nishida, K., and Osmond, C. B. (1986). Quantum yields of CAM plants measured by photosynthetic O_2 exchange. *Plant Physiol.* **81**, 297–300.

Adams, W. W. III, Osmond, C. B., and Sharkey, T. D. (1987*a*). Responses of two CAM species to different irradiances during growth and susceptibility to photoinhibition by high light. *Plant Physiol.* **83**, 213–18.

Adams, W. W. III, Smith, S. D., and Osmond, C. B. (1987*b*). Photoinhibition of the CAM succulent *Opuntia basilaris* growing in Death Valley: evidence from 77K fluorescence and quantum yield. *Oecologia (Berlin)* **71**, 221–8.

Björkman, O. (1987). Low temperature chlorophyll fluorescence in leaves and its relationship to photon yield of photosynthesis in photoinhibition. In 'Photoinhibition'. (Eds D. J. Kyle, C. B. Osmond and C. J. Arntzen.) Topics in Photosynthesis, Vol. 9, pp. 123–44. (Elsevier: Amsterdam.)

Delieu, T., and Walker, D. A. (1981). Polarographic measurement of oxygen evolution by leaf discs. *New Phytol.* **89**, 165–75.

Demmig, B., and Björkman, O. (1987). Comparison of the effect of excessive light on chlorophyll fluorescence (77K) and photon yield of O_2 evolution in leaves of higher plants. *Planta* **171**, 171–84.

Demmig, B., Winter, K., Krüger, A., and Czygan, F.-C. (1987). Photoinhibition and zeaxanthin formation in intact leaves. A possible role of the xanthophyll cycle in the dissipation of excess light energy. *Plant Physiol.* **84**, 218–24.

Martin, C. E., Eades, C. A., and Pitner, R. A. (1986). Effects of irradiance on crassulacean acid metabolism in the epiphyte *Tillandsia usneoides* L. (Bromeliaceae). *Plant Physiol.* **80**, 23–6.

Osmond, C. B., Nott, D. L., and Firth, P. M. (1979*a*). Carbon assimilation patterns and growth of the introduced CAM plant *Opuntia inermis* in Eastern Australia. *Oecologia (Berlin)* **40**, 331–50.

Osmond, C. B., Ludlow, M. M., Davis, R., Cowan, I. R., Powles, S. B., and Winter, K. (1979*b*). Stomatal responses to humidity in *Opuntia inermis* in relation to control of CO_2 and H_2O exchange patterns. *Oecologia (Berlin)* **41**, 65–76.

Powles, S. B., and Björkman, O. (1982). Photoinhibition of photosynthesis: effect on chlorophyll fluorescence at 77K in intact leaves and in chloroplast membranes of *Nerium oleander*. *Planta* **156**, 97–107.

Winter, K., Wallace, B. J., Stocker, G. C., and Roksandic, Z. (1983). Crassulacean acid metabolism in Australian vascular epiphytes and some related species. *Oecologia (Berlin)* **57**, 129–41.

Winter, K., Osmond, C. B., and Hubick, K. T. (1986). Crassulacean acid metabolism in the shade. Studies on an epiphytic fern, *Pyrrosia longifolia*, and other rainforest species from Australia. *Oecologia (Berlin)* **68**, 224–30.

Molecular Events of Photoinhibitory Inactivation in the Reaction Centre of Photosystem II

Robyn E. Cleland

Botany Department, Australian National University,
G.P.O. Box 4, Canberra, A.C.T. 2601, Australia;
present address: Research Institute for Photosynthesis,
University of Sheffield, Sheffield, S10 2TN, England.

Abstract

The events of photoinhibition were examined at the molecular level. The evidence presented here suggests that the primary event of photoinhibition involves inactivation of photosystem II reaction centre function. Chlorophyll fluorescence emission and excitation spectra and fluorescence induction transients were suppressed after photoinhibitory treatment. Such fluorescence is directly dependent on photosystem II photochemistry. In addition, photoinhibition caused a decline in charge separation measured by the absorbance change at 320 nm arising from the light-induced reduction of the primary acceptor of photosystem II, Q_A. Inhibition of this parameter is indicative of damage to a component involved in primary photochemistry. That the effect of high light treatment could not be correlated with any loss of the D1 protein supports the suggestion that a cofactor involved in primary photochemistry is the initial site of photoinhibition. It is possible that D1 may eventually be lost as a result of such damage. The reaction centre chlorophyll of photosystem II, P680, is suggested to be the cofactor involved. Possible mechanisms of damage are discussed with reference to the prosthetic components of the reaction centre of photosynthetic bacteria.

Introduction

The interactions between light and the photosynthetic membrane appear more and more complex as we endeavour to understand the processes of capture and conversion of that light energy to a chemical form. In seeking to elucidate the paradoxical inhibition of photosynthesis by light (photoinhibition) it is these interactions that must be considered, both at the level of the exciton within the antenna and at the level of an electron passing down the photosynthetic electron transport chain. An understanding of the phenomenon of photoinhibition at the molecular level will contribute to a better appreciation of its role in photosynthesis and productivity as a whole.

Not only visible light but also ultraviolet light and heat can inhibit photosynthesis. The term photoinhibition was defined by Powles (1984) to refer to the inhibition of photosynthesis induced by visible light and was thought to arise from physical damage to PS* II. It is this definition which is applied here. Photoinhibition within a leaf

*Abbreviations used: ΔA_{320}, absorbance change at 320 nm due to the light-induced reduction of Q_A; Bchl, bacteriochlorophyll; Bpheo, bacteriopheophytin; BSA, bovine serum albumin; C550, absorbance change at 550 nm attributed to Pheo; Chl, chlorophyll; D1, the herbicide-binding 32 kDa protein encoded on psb A; DCMU, dichlorophenyldimethylurea; DMQ, dimethylbenzoquinone; DTT, dithiothreitol; F_{685}, F_{695}, F_{735}, fluorescence emission peaks at 685, 695 and 735 nm respectively; F_0, F_M, F_V, initial, maximum and variable ($F_M - F_0$) fluorescence yield respectively; Hepes, *N*-2-hydroxyethylpiperazine *N'*-2-ethanesulfonic acid; K_d, rate constant for non-radiative decay in the reaction centre; K_D, rate constant for non-radiative decay from the antenna; K_P, rate constant for photochemistry; LDS, lithium dodecylsulfate; P680, reaction centre Chl of PS II; P700, reaction centre Chl of PS I; P_{bac}, reaction centre Bchl of photosynthetic bacteria; PAGE, polyacrylamide gel electrophoresis; Pheo, the initial acceptor of PS II; PQ, plastoquinone; PS I and II, photosystems I and II; Q_A, first stable acceptor of PS II; Q_B, secondary acceptor of PS II; tricine, N-tris(hydroxymethyl)methylglycine; tris, 2-amino-2-hydroxymethylpropane-1,3-diol.

0310-7841/88/010135$03.00

cannot be simply regarded as a single event that damages a component of the photosynthetic apparatus and results in the decline of quantum yield, light saturated photosynthesis and Chl fluorescence that is characteristic of photoinhibition. Recent studies (Demmig and Björkman 1987) suggest that depression of quantum yield and of Chl fluorescence may result not only from damage to the photosynthetic apparatus, but also from an increase in processes responsible for the non-radiative dissipation of excess energy whose function is actually protective in nature. The relative contributions of this protective mechanism and actual damage to the overall inhibition of photosynthetic capacity will be different under different conditions and, at present, can be assessed only qualitatively. This work will address inhibition due primarily to damage to the photosynthetic apparatus. The term damage will be applied to inhibition of photosynthesis due to injury in order to distinguish such inhibition from that due to possible protective mechanisms. In addition, in intact photosynthetic tissues there is the complicating factor of repair occurring concurrently with damage (Myers and Burr 1940; Kok 1956; Ohad *et al.* 1984; Greer *et al.* 1986). The observed inhibition of photosynthetic activity seen under conditions of excess excitation energy is therefore the net result of dissipative, damaging and repair processes.

In isolated photosynthetic membranes complicating factors such as concurrent repair are eliminated and the initial events of primary inhibition can be examined. Yet it is important to ensure that the phenomenon being examined is that which also occurs *in vivo*. The similarity between *in vivo* and *in vitro* photoinhibitory treatment in the effects they induce on defined photosynthetic parameters, including light-limited and light-saturated rates of photosynthesis and Chl fluorescence, suggests that the primary cause of damage is identical in both systems. *In vivo* studies of photoinhibition have delineated the site of damage within PS II (Critchley 1981) although *in vitro* both PS II and PS I show inhibition of electron transport activity (Kok *et al.* 1965; Satoh 1970; Barényi and Krause 1985). In this work both *in vivo* and *in vitro* techniques were employed in order to identify the site of photoinhibitory damage within PS II. The results of these experiments are discussed in terms of possible hypotheses concerning the initial event of photoinhibition.

Chlorophyll Fluorescence

Chl *a* fluorescence arises primarily from PS II. The first stable PS II acceptor, Q_A, can accumulate in the reduced form as a result of the limitation of electron turnover imposed by the PQ pool. The presence of Q_A^- prevents energy dissipation through photochemistry, causing fluorescence to increase. The kinetics of fluorescence induction upon a dark–light transition reflect the reduction of Q_A^- (Duysens and Sweers 1963). DCMU, which blocks electron transport into the PQ pool, causes an accelerated rise in Chl fluorescence by preventing the quenching that occurs in membranes where reoxidaiion of Q_A is permitted. A similar effect is obtained at 77K where low temperature prevents the secondary events of electron transfer out of Q_A. Hence these two types of Chl fluorescence are equivalent. PS I fluoresces only at low temperature; since its primary acceptor does not accumulate in the reduced form at physiological temperatures, excitation energy can be quenched continually by photochemistry. Fluorescence has been assumed to arise from the immediate antenna associated with P680 (Kitajima and Butler 1975), although Breton (1982, 1983) has suggested it results initially from a recombination between $P680^+$ and $Pheo^-$ resulting in an ultrafast luminescence (see also Critchley 1988).

Kok *et al.* (1965) first used Chl fluorescence as a tool to investigate the effects of photoinhibition. These workers observed a quenching of fluorescence at both room temperature and at 77K, and the abolition of the variable component of the induction transient. These effects have since been shown to be characteristic of high light exposure of intact and isolated photosynthetic systems (Malkin and Jones 1968; Powles *et al.* 1979; Critchley and Smillie 1981; Powles and Björkman 1982; Cleland and Critchley 1985; Bradbury and Baker 1986).

In a 77K fluorescence emission spectrum, preillumination of uncoupled spinach thylakoids with strong light caused a decline in fluorescence yield (Fig. 1). The emission peaks at 685 nm (F_{685}) and 695 nm (F_{695}) have been assigned to PS II antenna Chl and the PS II reaction centre, respectively (Breton 1982, 1983) while the 735 nm emission (F_{735}) has been attributed to PS I. Initial exposure of spinach thylakoids to strong light preferentially quenched F_{685} and F_{695} (Fig. 1*a*). As preillumination time increased F_{735} became affected (Figs 1*b* and 1*c*), indicating inhibition of PS I. The overall characteristics of the spectra were unaltered by the treatment (no peak shifts or extraneous features were observed), suggesting that a normal light-trapping apparatus connected to

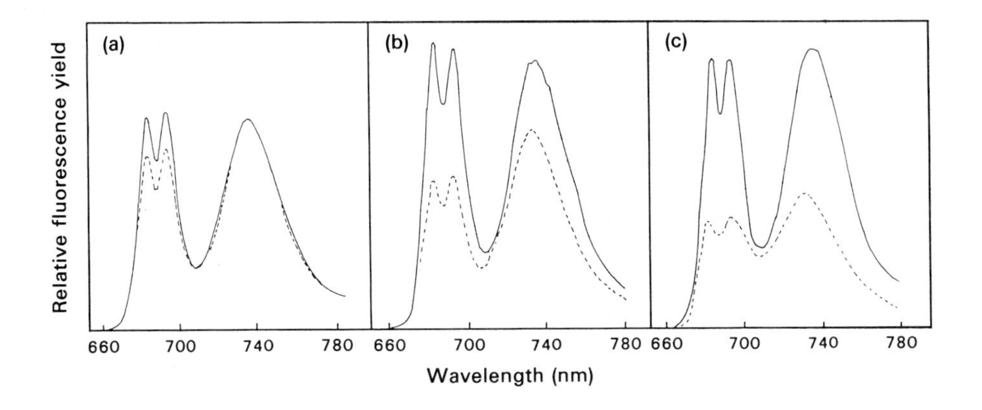

Fig. 1. The Chl *a* fluorescence emission spectra at 77K of control (———) and photoinhibited (----) spinach thylakoids (3 μg Chl ml^{-1}) normalised at 540 nm to sodium fluorescein. (*a*) 2 min preillumination with 6000 μmol quanta m^{-2} s^{-1} white light and 25°C. (*b*) 4 min preillumination. (*c*) 8 min preillumination. Thylakoids were isolated from glasshouse-grown plants and subjected to photoinhibitory treatment at 10 μg ml^{-1} in 50 mM Hepes–NaOH, pH 7·6, 20 mM NaCl, 10 mM NH_4Cl, 5 mM $MgCl_2$, according to the method of Cleland and Critchley (1985). Electron transport activity was measured polarographically using FeCN and DMQ according to Cleland and Critchley (1985) and was 46%, 27% and 6% of the control for (*a*), (*b*) and (*c*), respectively. Fluorescence spectra were measured using a Perkin-Elmer MPF-44B fluorescence spectrophotometer as described by Cleland and Critchley (1985) which incorporated automatic correction for photomultiplier and monochromator responses and variation in the energy output of the light source. The excitation wavelength was 440 nm, obtained with Corning CS 4-96 and Schott BG filters, the photomultiplier was shielded with a Schott OG 4 (yellow) filter.

the reaction centre was maintained. This quenching cannot be ascribed to the non-radiative quenching observed by Demmig and Björkman (1987) since the thylakoids were uncoupled and therefore incapable of generating the acidity within the lumen required to generate the quenching pigment (Demmig *et al.* 1987). If photoinhibition caused dissociation of the light-harvesting complex from the PS II reaction centre one might expect to observe the appearance of a shoulder at 680 nm, and clearly this was not the case. Photoinhibited centres have been shown to trap about 20% less excitation energy than the remaining active centres (Cleland *et al.* 1986), although the reason for this is unclear.

Excitation spectra also showed an overall quenching of emission after photoinhibitory treatment although the characteristics of the spectra measured at

695 nm and 735 nm were not altered (Fig. 2). If F_{695} does indeed arise from the reaction centre (Breton 1982, 1983), the overall quenching of both emission from and excitation of this peak rather than the appearance of abnormal features suggests that the reaction centre does not undergo modification observable by this technique but continues to function as a sink for excitation energy. However the captured light energy cannot be utilised in photochemistry but must be dissipated in some other manner.

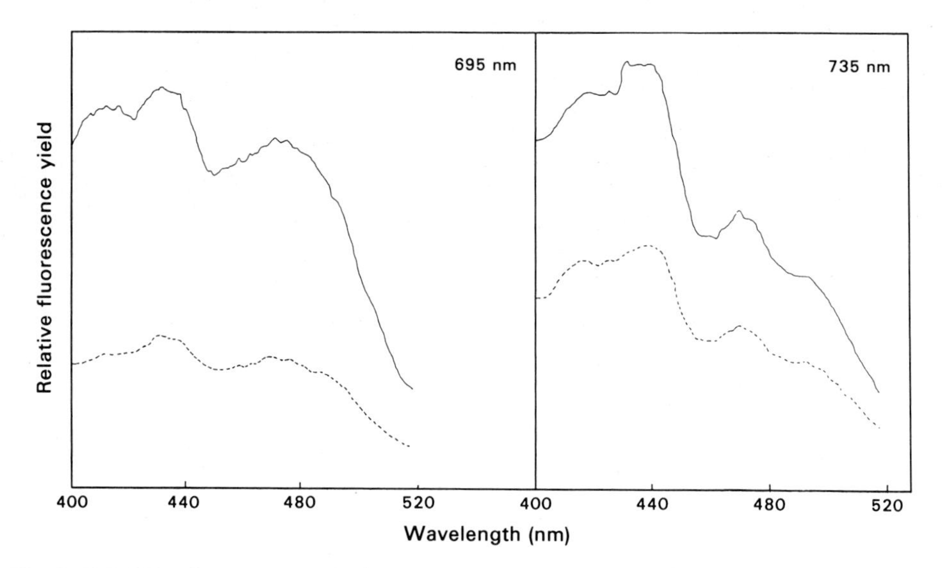

Fig. 2. 77K Chl *a* fluorescence excitation spectra, measured at 695 and 735 nm, of isolated spinach thylakoids. Control (———), 4 min preillumination at 6000 μmol quanta m^{-2} s^{-1} and 25°C (----). Experimental conditions and instrument configuration as described in Fig. 1, except that a red cut-off filter shielded the photomultiplier.

Primary Photochemistry

In Fig. 3, a series of Chl fluorescence induction transients measured at room temperature in the presence of DCMU is presented for isolated barley thylakoids exposed to strong actinic light (2500 μmol quanta m^{-2} s^{-1} at 4°C). The higher Chl concentration, lower light intensity and lower temperature used in this photoinhibitory treatment resulted in a longer time-course for inhibition compared to that in Fig. 1. The variable component of the induction curve was drastically affected by the preillumination treatment. The effect of this treatment on each component of the transient, F_0, the initial level of fluorescence, F_M, the maximum fluorescence and F_V, the variable component ($F_M - F_0$) is shown in Fig. 4. Both F_M and F_V were quenched in an exponential manner with time of incubation. F_0 rose initially, before undergoing a gradual decline during the remaining period of treatment. Such an increase in F_0 has been suggested to be characteristic of photoinhibitory damage and to distinguish such damage from quenching of fluorescence due to an increase in non-radiative dissipation (Björkman 1987; Demmig and Björkman 1987).

The characteristic quenching effect of photoinhibition on Chl fluorescence has been extensively used to determine whether photoinhibition has taken place. The model of Kitajima and Butler (1975) describes fluorescence in terms of a series of rate constants associated with energy-dissipating events in the reaction centre and antenna, of which fluorescence is one. This model assumes that both F_0 and F_V originate from antenna

Chls, F_0 reflecting the initial number of open centres while F_V reflects the closure of those centres. In explaining the fluorescence quenching observed under conditions of excess excitation (not necessarily photoinhibitory), Demmig and Björkman (1987) sug-

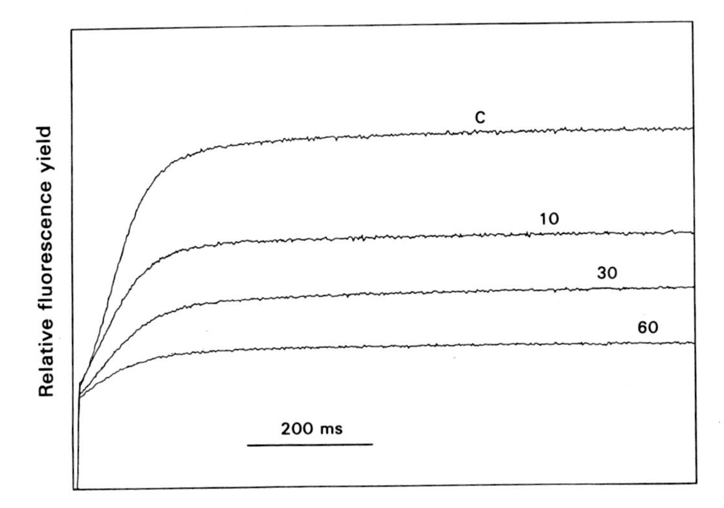

Fig. 3. Chl *a* fluorescence induction kinetics from DCMU-poisoned (12 μM) barley thylakoids (180 μg Chl ml^{-1}), after exposure to 0, 10, 30 and 60 min strong actinic light (2500 μmol quanta m^{-2} s^{-1}, 4°C). Barley plants were grown in a growth chamber at 25°C and 50 μmol quanta m^{-2} s^{-1}, thylakoids were isolated according to Cleland and Melis (1987) and subjected to photoinhibitory treatment in 0·4 M sucrose, 50 mM tricine-NaOH, pH 8·0, 10 mM NaCl, 5 mM $MgCl_2$, following the procedure of Cleland *et al.* (1986). Fluorescence was measured in a laboratory-constructed split-beam absorbance difference spectrophotometer (Melis and Brown 1980) as described by Cleland *et al.* (1986). The optical pathlength of the cuvette for the measuring beam was 2·06 mm and for the actinic beam 1·46 mm. Actinic excitation was provided in the green region by a combination of Corning CS 4-96 and CS 3-69 filters.

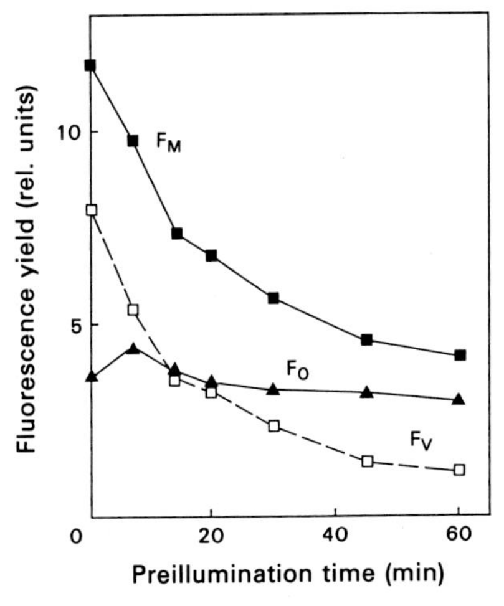

Fig. 4. The fluorescence yield of each component of the induction transients from barley thylakoids shown in Fig. 3, F_0 (▲), F_M (■) and F_V (□) as a function of incubation time in strong light.

gested that an increase in K_D, the rate constant for non-radiative decay in the antenna, caused the observed quenching in F_V and F_0 and hence the decline in F_V/F_M. A decrease in K_P, the rate constant for photochemistry, was suggested to explain the rise of F_0 (Björkman 1987). The increase in F_0 has been observed by a number of other workers during or following photoinhibition (Powles and Björkman 1982; Bradbury and Baker 1986; Krause and Behrend 1986) and was also noted here.

The initial rise in F_0 and decrease in F_M, and the subsequent decrease in F_0 observed in Fig. 4, suggest that two quenching processes may be operative during photoinhibition. The initial process can be described by a decrease in K_P as suggested above, which causes the initial rise in F_0. Mechanistically this could reflect the actual event of damage. Superimposed on this is a second process which could be caused by an increase in either K_d, non-radiative decay from the reaction centre and/or K_D, and is responsible for the quenching of F_V and F_0. An increase in K_d may reflect the dissipative characteristics of the damaged molecule. Kok *et al.* (1965) suggested that 'pre-illumination leaves these traps (centres) in a state capable of draining light from sensitising pigments but unable to perform useful photochemistry', i.e. after photoinhibition centres become permanently open, but photochemically inactive sinks for excitation energy.

Charge separation gives rise to the formation of the stable semiquinone anion Q_A^-. The light-induced absorbance change at 320 nm is quantitatively related to the number of PS II reaction centres (Melis and Brown 1980; McCauley and Melis 1986). As such it provides a reliable measure of the integrity of PS II. Fig. 5 shows the relative ampli-

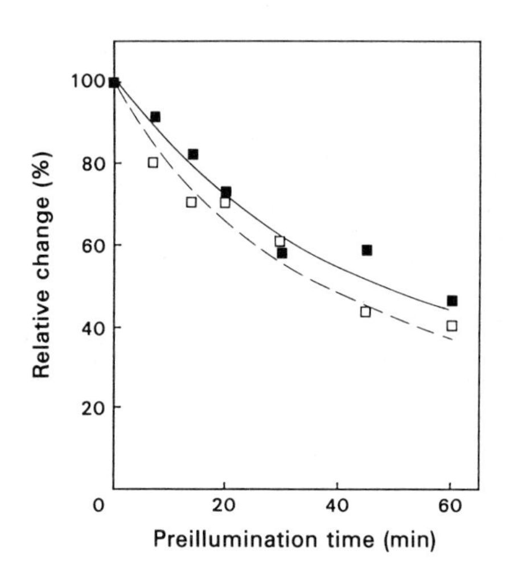

Fig. 5. The relative amplitude of the light-induced absorbance change arising from Q_A^- at 320 nm, ΔA_{320} (■), and the fluorescence ratio, F_V/F_M (□), from barley thylakoids as a function of preillumination time in strong light. Instrument configuration and photoinhibitory treatment as described in Fig. 3; conditions of measurement were after Cleland *et al.* (1986).

tude of the ΔA_{320} signal attributed to Q_A^- as a function of preillumination in strong actinic light. As the period of exposure to strong light increased, the extent of Q_A reduction was depressed in an exponential manner. This decline was slightly slower than that of F_V/F_M. Inhibition of charge separation after high light treatment indicates an effect of photoinhibition on the primary photochemistry of PS II (Cleland *et al.* 1986) and it suggests fundamental damage to a component within the reaction centre of PS II, either P680, Pheo or Q_A.

Secondary Events

The work of Kyle *et al.* (1984) and Ohad *et al.* (1984) has suggested a link between degradation of D1 (the 32 kDa, herbicide-binding, Q_B protein) and photoinhibitory damage. This possibility was investigated in the shade-adapted species *Oxalis oregona*. D1 can be readily visualised by separation and subsequent autoradiography of thylakoid proteins after incubation of photosynthetic material capable of protein synthesis (leaves, algae, intact chloroplasts) with $[^{35}S]$methionine. Labelling of *O. oregona* leaves with $[^{35}S]$met at low light resulted in the rapid appearance of the heavily labelled D1 protein within 1 h. Prelabelling had no effect on photosynthetic activity measured as Chl fluorescence (F_V/F_M measured at 77K) or on leaf oxygen evolution activity (Fig. 6). In contrast, exposure of the leaves to 3500 μmol quanta m^{-2} s^{-1} dramatically reduced both F_V/F_M and the rate of oxygen evolution. The decline in F_V/F_M was due to the combined effect of a quenching of F_M and an initial rise in F_0 followed by a gradual

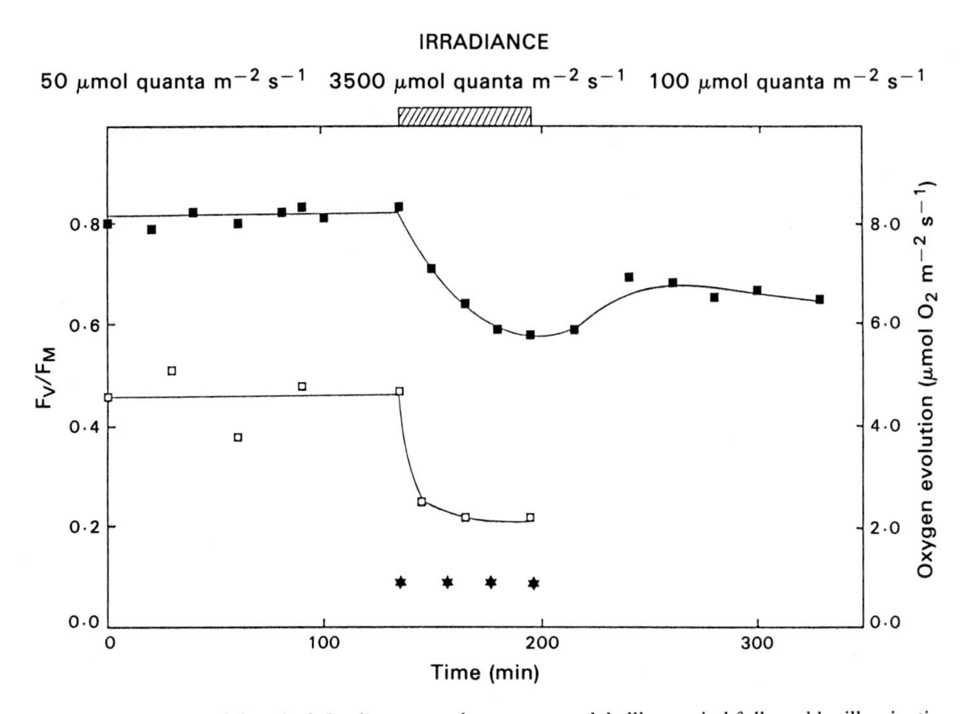

Fig. 6. The response of detached *Oxalis oregona* leaves to a prelabelling period followed by illumination at an irradiance of 3600 μmol quanta m^{-2} s^{-1}. Rates of leaf O_2 evolution (□) and the fluorescence ratio, F_V/F_M (■) were monitored throughout the experiment and recovery was followed by the response of F_V/F_M. The shaded region indicates the period of photoinhibitory treatment. The time points sampled in Fig. 7 are shown by stars. *O. oregona* was grown in a shaded situation in soil at Stanford, California during January–March 1985 and did not at any stage experience full sunlight. The petioles of trifoliate leaves were cut under water before being placed in Eppendorf tubes containing 39 μCi of [^{35}S]methionine (Amersham) in 0·2 ml of water. Leaves were exposed to 50 μmol quanta m^{-2} s^{-1} (sunlight attenuated through several layers of shade cloth) in the presence of label for 135 min. After transfer to damp filter paper, adaxial surface uppermost, leaves were then exposed to 3500 μmol quanta m^{-2} s^{-1} provided by a multivapour lamp (GTE Sylvania 'Metalarc', 1000 W) equipped with a diffusor and water filter in a chamber maintained at 20°C for the desired period of time. Humidified air was passed over the leaf surface during the treatment and the petioles were immersed in water. Two leaflets from separate leaves were removed for measurements at the times indicated. The irradiance was reduced for the recovery period by interrupting the light beam with wire screens until the irradiance was reduced to 100 μmol quanta m^{-2} s^{-1}. O_2 evolution from leaf discs was measured in a leaf oxygen electrode (Hansatech, U.K.) according to the procedure of Björkman and Demmig (1987) at 25°C and saturating CO_2 using 55 μmol quanta m^{-2} s^{-1} which was within the quantum yield region for *O. oregona*. Chl fluorescence (77K) transients were measured with a laboratory-built apparatus (Björkman and Demmig 1987). The excitation beam was passed through a 480 nm, 10 nm half-band width interference filter (Ditric Optical Co., MA), providing an irradiance of 1·0 μmol quanta m^{-2} s^{-1} at the surface of the leaf sample during measurements. The irradiance was raised to 10 μmol quanta m^{-2} s^{-1} for 3 min toward the end of each measurement to ascertain that maximum fluorescence had been obtained. Emission from the sample was measured through a 692 nm, 12 nm half-bandwidth interference filter (Ditric). Leaf discs were dark-adapted for 5 min prior to freezing with liquid N_2 and allowed to equilibrate for a further 2 min before measurement.

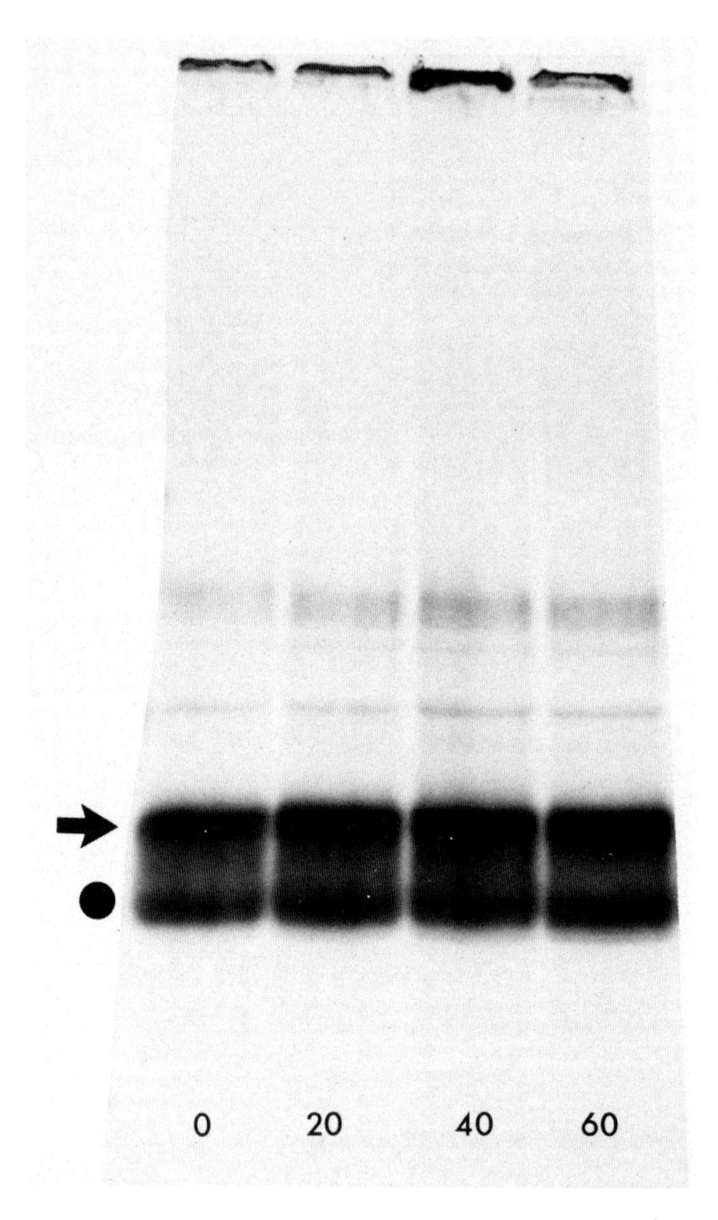

Fig. 7. Autoradiogram of thylakoids separated by LDS–PAGE after isolation from *O. oregona* leaves prelabelled *in vivo* with [^{35}S]met and subsequently exposed to 3500 μmol quanta m^{-2} s^{-1} for 0, 20, 40 or 60 min. The circle indicates LHC at 30 kDa and the arrow points to D1 at 35 kDa. Thylakoid membranes non-functional in O_2 evolution were isolated for electrophoresis by grinding the leaflets in 200 mM Hepes–KOH, pH 7·2, 15 mM NaCl, 2 mM EDTA, 0·2% w/v BSA. The homogenate was filtered through two layers of Miracloth and spun at full speed in an Eppendorf centrifuge for 1·5 min. The pellet was washed and resuspended in a small volume of the same buffer. *O. oregona* thylakoids were solubilised in 100 mM tris–HCl, pH 6·8, 13% v/v glycerol, DTT and LDS to give an LDS:DTT:Chl ratio of 50:50:1 and heated for 90 s at 100°C. 10 μg Chl was loaded per lane on a stacking gel of 7% v/v acrylamide/bisacrylamide (39:1) which overlaid a resolving gel of 10–22·5% v/v polyacrylamide. 0·1% v/v LDS was present in the gel and in the upper reservoir buffer. The gel was run overnight at 4°C using a constant current in a laboratory-built apparatus.

quenching of F_0. The recovery of F_V/F_M upon lowering the irradiance to 100 μmol quanta m^{-2} s^{-1} was small. Recovery of oxygen evolution was not measured.

When thylakoids were isolated from prelabelled leaves which had been exposed to photoinhibitory treatment, no change could be observed in the levels of D1 (Fig. 7) or in the labelled band below D1 which corresponded exactly with the LHCII polypeptide observed by staining with coomassie brilliant blue. A period of dark incubation had no effect on the intensity of label. The level of label present after 60 min preillumination was virtually identical to that observed at the onset of treatment. Clearly photoinhibition of photosynthetic activity measured either by O_2 evolution or Chl fluorescence could not be correlated with a loss of this protein in this experiment. The suppression of oxygen evolution and F_V/F_M could not be explained by an increase in non-radiative decay since F_0 rose (Björkman 1987) in the early stages of the treatment. In addition, when the protein was labelled with [^{35}S]met *in organello* and the ruptured thylakoids exposed to photoinhibitory light, no loss of D1 was observed (Cleland and Critchley 1985). This leads to the conclusion that D1 does not have a primary role in photoinhibitory damage although subsequent degradation may occur as a result of that primary event.

Lessons from Photosynthetic Bacteria

In endeavouring to understand the molecular events of photoinhibition we seek a fundamental understanding of the basis of photochemistry itself. Although some of the cofactors involved in primary photochemistry have been identified in the reaction centre of photosynthetic bacteria, we do not as yet have a clear picture of the events that lead to the reduction of the primary acceptor or the pathway which the electron follows from the primary Bchl donor pair to that acceptor. The structure of PS II is even less well understood, yet many of the components of the PS II reaction centre have been identified by analogy with those of the bacterial system.

Various optical and e.p.r. signals have been attributed to the primary donor and initial and stable electron acceptors in both photosynthetic bacteria and PS II (Okamura *et al.* 1982). In each case the characteristics of the e.p.r. signals are almost identical although the optical properties vary, as would be expected for different types of Chls and quinones. The similarity between the e.p.r. signals suggests that some analogies concerning the molecular organisation can be drawn between the bacterial reaction centre and PS II. (See Critchley 1988, for a more detailed discussion of the structural analogies between photosynthetic bacteria and PS II.) Firstly the arrangement of the molecules within the reaction centre is extremely precise. The crystal structures of the reaction centres of *Rhodopseudomonas viridis* (Deisenhofer *et al.* 1984) and *R. sphaeroides* (Chang *et al.* 1986) are very similar except that the former contains Bchl *b* and the latter Bchl *a*. This further implies that the geometry of the individual components is also significant. Devault (1986) suggested that the orientation of the prosthetic groups to form the letter T was important in electron transfer. The interactions of the phytyl tails may also be influential. The third piece of useful information is the presence of accessory Bchls and possibly also of a carotenoid molecule. Whether these pigments are involved directly in electron transfer is currently being debated (Martin *et al.* 1986; Norris *et al.* 1987). Each photosynthetic bacterial reaction centre contains 4 Bchl, 2 Bpheo, a quinone, a non-heme iron and 4 cyt heme groups (Deisenhofer *et al.* 1984). The presence of similar molecules in PS II is also suggested by the isolation of a putative reaction centre complex which contained 5 Chl *a*, 2 Pheo, 1 β-carotene and two cyt b_{559} hemes (Nanba and Satoh 1987).

In the absence of a crystal structure for PS II the three maxims described above, i.e. precision of structure, geometry of both the porphyrin and the phytyl moiety, and the presence of accessory Chls, can be used in considering the possible configuration of

PS II, the nature of the events of primary photochemistry within the reaction centre and the possible molecular events of photoinhibition. Pheo and Q_A show marked similarity to the Bpheo and quinone of photosynthetic bacteria. In contrast P680 is strikingly different from the bacterial primary donor (P_{bac}). There is evidence that P680 may be

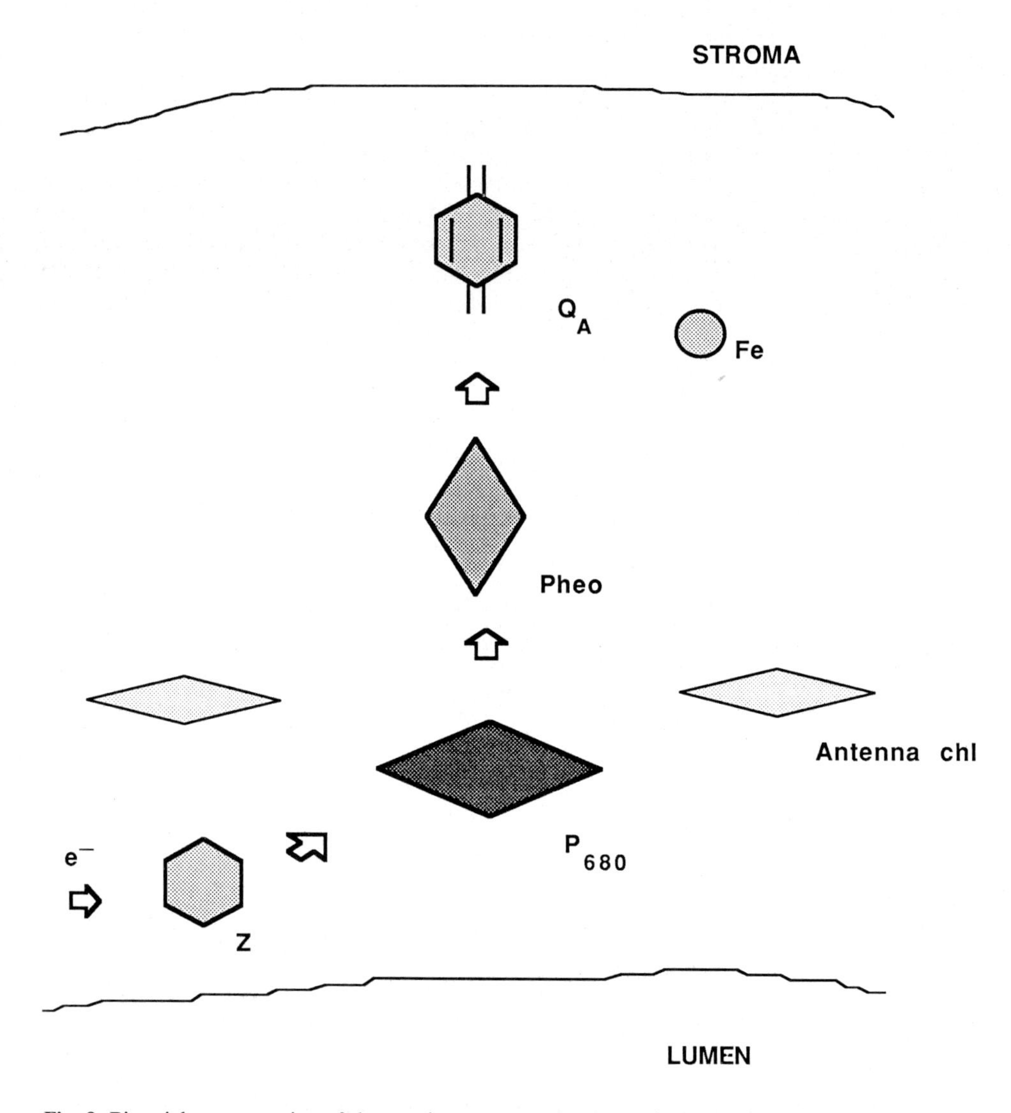

Fig. 8. Pictorial representation of the reaction centre components of PS II, their approximate locations and orientations in the membrane (not to scale). The arrows represent electron flow. *Z*, immediate electron donor to P680; P680, reaction centre Chl of PS II; Pheo, intermediate electron acceptor of PS II; Q_A, stable electron acceptor of PS II; Fe, non-heme iron which interacts both with Q_A and Q_B (Q_B, secondary electron acceptor of PS II).

a monomer (Davis *et al.* 1979) and it has a midpoint potential about 500 mV more oxidising than P_{bac} (Rutherford *et al.* 1981). The alignment of these pigments has been measured, although the orientation of the components of PS II has been determined only with respect to the plane of the membrane (Rutherford 1985) and not towards each other. The most significant difference between the bacterial reaction centre and PS II in this case is that P_{bac} lies perpendicular whereas P680 appears to lie parallel (Rutherford 1985) to the plane of the membrane. Pheo and Q_A appear to have similar orientations to Bpheo and Q in the bacterial centre, the heme plane of cyt b_{559} is oriented perpendicular to the plane of the membrane.

The crystallisation of the bacterial reaction centre did not characterise the position or possible role of lipids or water in photosynthesis. That such molecules are directly involved in photochemistry is unlikely, but they may have an important function in stabilising the prosthetic groups undergoing charge separation or in the pigment–protein interactions that contribute to the structure of the reaction centre. Particular lipids have been shown to be associated with Chl–protein complexes (Horváth *et al.* 1987) and have been shown to affect rates of electron transport (Siegenthaler *et al.* 1987). A diagrammatic representation of the PS II reaction centre is presented in Fig. 8.

The X-ray crystallography studies of *R. viridis* have not yet addressed the interactions between the reaction centre and the antenna. Obviously there must be an intimate connection between P_{bac} and the immediate Bchls of the antenna such that excitons move rapidly and efficiently onto P_{bac}. What this connection means in terms of protein–protein interactions between the reaction centre and the antenna polypeptides has not yet been elucidated. A similar state of ignorance prevails concerning LHC-II and the PS II reaction centre proteins.

Light captured by an antenna Chl reaches a reaction centre by the process of resonance energy transfer (Förster mechanism). The rate of exciton migration is much faster than photochemistry and, at low irradiance, light itself is the limiting factor. Under such conditions the dark reactions are able to keep pace with light absorption and one O_2 molecule is evolved per 8–9 quanta captured. At high light, quantum efficiency drops as electrons are not transported as rapidly as light is absorbed. Nobel (1983) calculated that, at 2000 μmol quanta $m^{-2} s^{-1}$ (full sunlight) and assuming about 300 Chl per reaction centre, only 1 photon of every 15 absorbed could be used for photochemistry. The energy of the others must be lost by alternative routes of deexcitation, through non-radiative decay or fluorescence.

This is clearly a situation in which photoinhibition could occur. Although the plant has dissipative mechanisms and can transfer excitation energy to another centre, somehow communicating that it is in the closed state, that excess energy in some cases may reach P680 and cause inhibition. We must now ask the question: what is the nature of the inhibition? Is the event of photoinhibition a result of the natural function of the reaction centre or does it arise from alterations in the light capture–charge separation process? Does the damage involve a separate chemical species such as an oxygen radical or is it an alteration in geometry or orientation?

Photoinhibitory Damage to the PS II Reaction Centre

The effect of high light treatment on fluorescence emission and excitation spectra presented above suggests that alterations in the light-harvesting complex are not the cause of photoinhibition. Such injuries probably occur later as secondary events and result in photooxidation of the bulk pigments. Possible remaining candidates for the site of a photoinhibitory lesion are P680, Pheo and Q_A. At this point Q_A seems an unlikely candidate. It is relatively removed from the actual conversion of an exciton into electron transfer, and a decline in C550, an absorbance change attributed to Pheo (Van Gorkom 1974), has been observed after incubation of isolated thylakoids in strong light (Cleland *et al.* 1986). In addition Demeter *et al.* (1987) have monitored Pheo reduction directly and found it to be inhibited by high light treatment. The role of P680 as a catalyst for the conversion of light energy into chemical work makes it a prime candidate for light-induced inhibition. Its close association with Pheo makes it likely that any adverse effect on one will affect the other. However as the flow of energy is from P680 to Pheo it is likely that P680 is the molecular component of the photosynthetic apparatus that is initially damaged by light stress.

The phenomenon of photoinhibition is likely to be caused by a specific event which involves the accumulation of energy at the reaction centre. During normal photosynthesis at saturating light the photosynthetic apparatus can probably control energy dissipation. This is evident from the observation that Q_A is not completely reduced even under saturating light conditions (Dietz *et al.* 1985). The complete reduction of Q_A may in fact correspond with the onset of photoinhibition when electron transport and non-radiative dissipation no longer provide an adequate sink for excitation energy, but experiments to show this have yet to be carried out. Obviously the rate at which energy enters the system will also influence the rate of damage and depend on its rate of capture. The substantial inhibition of PS II_α units compared to PS II_β units (Cleland *et al.* 1986) and the decreased susceptibility of the chlorina f2 barley mutant to photoinhibition (Cleland and Melis 1987) support this contention. These results provide molecular evidence that overexcitation of the photosynthetic apparatus causes photoinhibition.

Photoinhibition occurs when the amount of energy entering the system can no longer be dissipated in an orderly manner. At the molecular level this implies that energy is arriving at a component before that component can get rid of the energy it has already received. This could mean either that the molecule is still in its excited state and therefore still possesses the first energy package, or that it is in a state intermediate to the ground state. If P680 is considered to be the molecule involved, the energy arriving will be that of a second quantum. Therefore the event of photoinhibition could be caused by the arrival of this second quantum at P680. Applying the above postulates to this condition, P680 may be in either the excited state P680* or in the oxidised state $P680^+$ prior to accepting an electron from the electron donor Z when a second quantum arrives.

The fact that a second quantum may be able to reach P680 is in itself surprising. Evidence from fluorescence studies suggest that the light-harvesting antenna system has some means of diverting energy from a closed reaction centre to an open one. This is manifest as connectivity, and by this means energy can be transferred to a centre in which it can be used to do photochemistry rather than being wastefully dissipated. However when conditions approach those in which photoinhibition occurs many centres will be closed preventing dissipation by energy overflow and forcing a second quantum into the reaction centre.

Of the two conditions in which P680 may be susceptible to a second quantum of light, $P680^+$ seems the most probable. Although P680* is a highly reactive molecule, its interaction with a second quantum would be unlikely unless that quantum was of a very high energy. This is virtually impossible since the overlapping absorption and fluorescence properties of the antenna Chl mean that a small amount of energy is lost at each transfer, so that energy arriving at the reaction centre will always be slightly less than that which is originally trapped. The effect of NH_2OH treatment on the sensitivity of thylakoids to photoinhibition also supports the role of $P680^+$ as the interacting species (Cleland *et al.* 1987). NH_2OH-treated thylakoids were very much more sensitive to preillumination than untreated thylakoids. In the treated thylakoids normal electron donation to P680 is prevented, and under high light conditions P680 would be in an oxidised state for a far greater proportion of time than in normal thylakoids. This suggests that photoinhibition may involve an interaction with a quantum of light when P680 is in the oxidised state, $P680^+$.

Possible Mechanisms of Photoinhibition

The involvement of primary photochemistry in light-dependent damage is not surprising considering that it is the reaction centre components which convert that light energy into electron transfer. That only specific Chls are able to lose an electron suggests that they must have certain characteristics that distinguish them from antenna Chls. These may include a slightly different chemical composition (P700 has been suggested to be a chlorinated Chl by Dörnemann and Senger 1986), specific orientation which confers special electrochemical properties, or an interaction with charged groups of a protein. The event of photoinhibition may alter any of these parameters.

The evidence presented above suggested that the inhibition of photochemistry proceeded from an initial damaging event that under certain circumstances could ramify through the system inactivating other components including the oxygen evolving system and D1. Differential events of secondary injury could in part account for some of the discrepancies observed concerning these components. There are two possibilities concerning the nature of the primary event: (1) it is caused by an aberrant chemical species such as an oxygen radical; or (2) there is some change in the geometry of the reaction centre that results in inhibition. The result of either of these processes must be a molecule that is able to quench excitation energy or induce conditions that result in quenching and that is inactive in charge separation. That the formation of this molecule does not affect such photochemical signals as Chl fluorescence, ΔA_{320} and C550 other than to quench them suggests that its primary function is to act as an energy dissipative mechanism. One would therefore expect a large increase in heat output from photoinhibited centres.

An oxygen radical could be generated from the triplet state of Chl. Two scenarios are possible. In the first a second quantum of light arrives from the antenna at an oxidised P680 which cannot accept that energy. A Chl molecule in the immediate vicinity is then converted to the triplet form and interacts with an oxygen molecule to form a radical. This radical subsequently attacks P680, chemically inactivating it. Alternatively, P680 itself forms a triplet which subsequently interacts with oxygen forming a radical which attacks the closest molecule, P680. The triplet state is formed more often when Q is reduced (Moore *et al.* 1984). The β-keto-ester group in ring V of Chl *a* is particularly susceptible to oxidative reagents (Wasielewski 1982). Other forms of Chl degradation include splitting the porphyrin ring and removal of the Mg, and it is possible that these could be induced by an oxygen radical. Whether such modification would be detectable in the fluorescence emission spectra remains to be determined.

The second hypothesis also concerns a direct alteration to P680 itself. A second quantum is absorbed by an oxidised P680. Instead of being used to transfer an electron the energy is dissipated through a change in the orientation of the molecule. The intimate association of P680 with accessory Chls and Pheo, suggested by analogy with the reaction centre of the photosynthetic bacteria, is disrupted and charge separation cannot occur. However the connection to the antenna remains and energy continues to be dissipated through the now unconfined P680 resulting in a centre that quenches excitation energy without participating in photochemistry.

Any of these modifications of P680 would affect its electrical properties. Such alterations could affect either the forward or the back reaction of charge separation. In the former case the energy required for the forward reaction to proceed may be increased. This would prevent efficient charge separation by decreasing the rate constant of the reaction, and alternative means of energy dissipation would have to occur. If the energy required for the back reaction decreased then the rate of charge separation would also decrease as the probability that the electron would fall back to the ground state would be greater than that of its proceeding to the electron acceptor. This may be qualitatively similar to an increase in K_d suggested earlier to contribute to fluorescence quenching, and would result in an increase in non-radiative decay from the reaction centre. This also fits well with the proposed two events of photoinhibition, causing the increase and the subsequent decrease in F_0. The primary event is damage to the molecule decreasing K_P which increases F_0, the result of the event is an increase in the probability of the back reaction which dissipates any further captured energy and effectively quenches both F_0 and F_M.

The lack of precise knowledge concerning the reaction centre structure of PS II means that possible hypotheses concerning the event of photoinhibitory damage are limited only by the imagination. On the basis of the maxims taken from the crystal structure of the *R. viridis* reaction centre, any alteration in the characteristics required for electron transfer, including prosthetic group orientation and geometry, accessory Chls, unique chemical composition of the reaction centre molecule, or its interaction with the lipid and protein environment, could result in an inhibition of reaction centre function. The challenge is to elucidate a mechanism of damage that will explain the various effects of light stress on the measurable photosynthetic parameters, and to define the event which on the one hand prevents energy dissipation through charge separation but on the other quenches excitation energy.

Photoinhibition is an inevitable consequence of overexcitation of those cofactors involved in photochemistry. By measuring the individual components of PS II involved in the initial events of primary photochemistry it can indeed be shown that photoinhibition damages a cofactor necessary for the normal functioning of the centre, probably P680. The inhibition of C550 which shows similar decay kinetics to that of ΔA_{320} (Cleland *et al.* 1986), and the observation made by Demeter *et al.* (1987) that a signal directly attributable to Pheo is inhibited by high light treatment, support this conclusion. Further supportive evidence comes from the overall suppression of the Q_A^- absorption change. The inhibition of charge separation observed in all these measurements of primary photochemistry suggests a fundamental inactivation of reaction centre activity by high light stress. Although the mechanism of damage is not yet understood, the most likely candidate for the site of damage is P680.

Acknowledgments

This work was supported by a Commonwealth Post-graduate Award. The low-temperature emission and excitation spectra were measured at Division of Plant Industry, CSIRO, Canberra. The measurements of ΔA_{320} were performed in the laboratory of Professor A. Melis, University of California, Berkeley. The labelling experiment was carried out at the Carnegie Institution of Washington, Stanford, in association with Dr O. Björkman. My thanks to Dr Christa Critchley and Mr Simon Brown for helpful discussions during the course of these investigations.

References

Barényi, B., and Krause, G. H. (1985). Inhibition of photosynthetic reactions by light. A study with isolated spinach chloroplasts. *Planta* **163**, 218–26.

Björkman, O. (1987). Low-temperature chlorophyll fluorescence in leaves and its relationship to photon yield of photosynthesis in photoinhibition. In 'Photoinhibition'. (Eds D. J. Kyle, C. B. Osmond and C. J. Arntzen.) Topics in Photosynthesis, Vol. 9, pp. 123–44. (Elsevier: Amsterdam.)

Bradbury, M., and Baker, N. R. (1986). The kinetics of photoinhibition of the photosynthetic apparatus in pea chloroplasts. *Plant Cell Environ.* **9**, 289–97.

Breton, J. (1982). The 695 nm fluorescence (F_{695}) of chloroplasts at low temperature is emitted from the primary acceptor of photosystem II. *FEBS Lett.* **147**, 16–20.

Breton, J. (1983). The emission of chlorophyll *in vivo*. Antenna fluorescence or ultrafast luminescence from the reaction centre pigments. *FEBS Lett.* **159**, 1–5.

Chang, C.-H., Tiede, D., Tang, J., Smith, U., Norris, J., and Schiffer, M. (1986). Structure of *Rhodopseudomonas sphaeroides* R-26 reaction center. *FEBS Lett.* **205**, 82–6.

Cleland, R. E., and Critchley, C. (1985). Studies on the mechanism of photoinhibition in higher plants. II. Inactivation by high light of photosystem II reaction centre function in isolated spinach thylakoids and O_2 evolving particles. *Photobiochem. Photobiophys.* **10**, 83–9.

Cleland, R. E., Critchley, C., and Melis, A. (1987). Alteration of electron flow around P_{680}: the effect of photoinhibition. In 'Progress in Photosynthesis Research'. (Ed. J. Biggins.) Vol. 4, pp. 27–30. (Martinus Nijhoff: Dordrecht.)

Cleland, R. E., and Melis, A. (1987). Probing the events of photoinhibition by altering electron transport activity and light harvesting capacity in chloroplast thylakoids. *Plant Cell Environ.* **10**, in press.

Cleland, R. E., Melis, A., and Neale, P. J. (1986). Mechanism of photoinhibition: photochemical reaction center inactivation in system II of chloroplasts. *Photosynth. Res.* **9**, 79–88.

Critchley, C. (1981). Studies on the mechanism of photoinhibition in higher plants. *Plant Physiol.* **67**, 1161–5.

Critchley, C. (1988). The molecular mechanism of photoinhibition — facts and fiction. *Aust. J. Plant Physiol.* **15**, 27–41.

Critchley, C., and Smillie, R. M. (1981). Leaf chlorophyll fluorescence as an indicator of high light stress (photoinhibition) in *Cucumis sativus*. L. *Aust. J. Plant Physiol.* **8**, 133–41.

Davis, M. S., Forman, A., and Fajer, J. (1979). Ligated chlorophyll cation radicals: their function in photosystem II of plant photosynthesis. *Proc. Natl Acad. Sci. U.S.A.* **76**, 4170–4.

Deisenhofer, J., Epp, O., Miki, K., Huber, R., and Michel, H. (1984). X-ray structure analysis of a membrane protein complex. Electron density map at 3 Å resolution and a model of the chromophores of the photosynthetic reaction centre from *Rhodopseudomonas viridis*. *J. Mol. Biol.* **180**, 385–98.

Demeter, S., Neale, P. J., and Melis, A. (1987). Photoinhibition: impairment of the primary charge separation between P680 and pheophytin in photosystem II of chloroplasts. *FEBS Lett.* **214**, 370–4.

Demmig, B., and Björkman, O. (1987). Comparison of the effects of excessive light on chlorophyll fluorescence (77K) and photon yield of O_2 evolution in leaves of higher plants. *Planta* **171**, 171–84.

Demmig, B., Winter, K., Krüger, A., and Czygan, F.-C. (1987). Photoinhibition and zeaxanthin formation in intact leaves. A possible role for the xanthophyll cycle in the dissipation of excess light energy. *Plant Physiol.* **84**, 218–24.

Devault, D. (1986). Vibrionic coupling to electron transfer and the structure of the *R. viridis* reaction center. *Photosynth. Res.* **10**, 125–37.

Dietz, K.-J., Schreiber, U., and Heber, U. (1985) The relationship between the redox state of Q_A and photosynthesis in leaves at various carbon-dioxide, oxygen and light regimes. *Planta* **166**, 219–26.

Dörnemann, D., and Senger, H. (1986). The structure of chlorophyll RC 1, a chromophore of the reaction center of photosystem I. *Photochem. Photobiol.* **43**, 573–81.

Duysens, L. M. N., and Sweers, H. E. (1963). Mechanism of two photochemical reactions in algae as studied by means of fluorescence. In 'Microalgae and Photosynthetic Bacteria'. (Ed. J. Ashida.) pp. 353–72. (University of Tokyo Press: Tokyo.)

Greer, D. H., Berry, J. A., and Björkman, O. (1986). Photoinhibition of photosynthesis in intact bean leaves: role of light and temperature, and requirement for chloroplast-protein synthesis during recovery. *Planta* **168**, 253–60.

Horváth, G., Melis, A., Hideg, E., Droppa, M., and Vigh, L. (1987). Role of lipids in the organization and function of photosystem II studied by homogeneous catalytic hydrogenation of thylakoid membranes *in situ*. *Biochim. Biophys. Acta* **891**, 68–74.

Kitajima, M., and Butler, W. L. (1975). Quenching of chlorophyll fluorescence and primary photochemistry in chloroplasts by dibromothymoquinone. *Biochim. Biophys. Acta* **376**, 105–15.

Kok, B. (1956). On the inhibition of photosynthesis by intense light. *Biochim. Biophys. Acta* **21**, 45–67.

Kok, B., Gassner, E. B., and Rurainski, H. J. (1965). Photoinhibition of chloroplast reactions. *Photochem. Photobiol.* **4**, 215–27.

Krause, G. H., and Behrend, U. (1986). ΔpH-dependent chlorophyll fluorescence quenching indicating a mechanism of protection against photoinhibition of chloroplasts. *FEBS Lett.* **200**, 298–302.

Kyle, D. J., Ohad, I., and Arntzen, C. J. (1984). Membrane protein damage and repair: selective loss of a quinone-protein function in chloroplast membranes. *Proc. Natl Acad. Sci. U.S.A.* **81**, 4070–4.

Malkin, S., and Jones, L. W. (1968). Photoinhibition and excitation quenching in photosystem II of photosynthesis, from fluorescence induction measurements. *Biochim. Biophys. Acta* **162**, 297–9.

Martin, J.-L., Breton, J., Hoff, A. J., Migus, A., and Antonetti, A. (1986). Femtosecond spectroscopy of electron transfer in the reaction center of the photosynthetic bacterium *Rhodopseudomonas sphaeroides* R-26: direct electron transfer from the dimeric bacteriochlorophyll primary donor to the bacteriopheophytin acceptor with a time constant of $2{\cdot}8 \pm 0{\cdot}2$ psec. *Proc. Natl Acad. Sci. U.S.A.* **83**, 957–61.

McCauley, S. W., and Melis, A. (1986). Quantitation of photosystem II in spinach chloroplasts. *Biochim. Biophys. Acta* **849**, 175–82.

Melis, A., and Brown, J. S. (1980). Stoichiometry of system I and system II reaction centers and of plastoquinone in different photosynthetic membranes. *Proc. Natl Acad. Sci. U.S.A.* **77**, 4712–16.

Moore, T. A., Gust, D., Mathis, P., Mialocq, J.-L., Chachaty, C., Bensasson, R. V., Land, E. J., Doizi, D., Liddell, P. A., Lehman, W. R., Nemeth, G. A., and Moore, A. L. (1984). Photodriven charge separation in a carotenoporphyrin–quinone triad. *Nature* **307**, 630–2.

Myers, J., and Burr, G. O. (1940). Studies on photosynthesis. Some effects of light of high intensity on *Chlorella. J. Gen. Physiol.* **24**, 45–67.

Nanba, O., and Satoh, K. (1987). Isolation of a photosystem II reaction center consisting of D-1 and D-2 polypeptides and cytochrome *b*-559. *Proc. Natl Acad. Sci. U.S.A.* **84**, 109–12.

Nobel, P. S. (1983). 'Biophysical Plant Ecology.' (Freeman and Co.: San Francisco.)

Norris, J. R., Lin, C. P., and Budil, D. E. (1987). Magnetic resonance of ultrafast chemical reactions. *J. Chem. Soc., Faraday Trans.* **83**, 13–27.

Ohad, I., Kyle, D. J., and Arntzen, C. J. (1984). Membrane protein damage and repair: removal and replacement of inactivated 32-kilodalton polypeptides in chloroplast membranes. *J. Cell Biol.* **99**, 481–5.

Okamura, M. Y., Feher, G., and Nelson, N. (1982). Reaction centres. In 'Photosynthesis. Energy Conversion in Plants and Bacteria'. (Ed. Govindjee.) Vol. 1, pp. 195–272. (Academic Press: New York.)

Powles, S. B. (1984). Photoinhibition of photosynthesis induced by visible light. *Annu. Rev. Plant Physiol.* **35**, 15–44.

Powles, S. B., and Björkman, O. (1982). Photoinhibition of photosynthesis: effect on chlorophyll fluorescence at 77K in intact leaves and in chloroplast membranes of *Nerium oleander. Planta* **156**, 96–107.

Powles, S. B., Osmond, C. B., and Thorne, S. W. (1979). Photoinhibition of intact attached leaves of C_3 plants illuminated in the absence of both carbon dioxide and of photorespiration. *Plant Physiol.* **64**, 982–5.

Rutherford, A. W. (1985). Orientation of EPR signals arising from components in photosystem II membranes. *Biochim. Biophys. Acta* **807**, 189–201.

Rutherford, A. W., Paterson, D. A., and Mullet, J. E. (1981). A light-induced spin-polarized triplet detected by EPR in photosystem II reaction centers. *Biochim. Biophys. Acta* **635**, 205–14.

Satoh, K. (1970). Mechanism of photoinactivation in photosynthetic systems. II. The occurrence and properties of two different types of photoinactivation. *Plant Cell Physiol.* **11**, 29–38.

Siegenthaler, P.-A., Smutny, J., and Rawyler, A. (1987). Involvement of distinct populations of phosphatidylglycerol and phosphatidylcholine molecules in photosynthetic electron-flow activities. *Biochim. Biophys. Acta* **891**, 85–93.

Van Gorkom, H. J. (1974). Identification of the reduced primary electron acceptor of photosystem II as a bound semiquinone anion. *Biochim. Biophys. Acta* **347**, 439–42.

Wasielewski, M. R. (1982). Synthetic approaches to photoreaction center structure and function. In 'Light Reaction Path of Photosynthesis'. (Ed. F. K. Fong.) Molecular Biology, Biochemistry and Biophysics. Vol. 35, pp. 234–76. (Springer-Verlag: Berlin.)

Light Response of CO_2 Assimilation, Reduction State of Q, and Radiationless Energy Dissipation in Intact Leaves

Barbara Demmig and Klaus Winter

Lehrstuhl für Botanik II, Universität Würzburg, Mittlerer
Dallenbergweg 64, 8700 Würzburg, Federal Republic of Germany.

Abstract

Response curves of the rate of net CO_2 uptake and of fluorescence quenching (photochemical, q_P, and non-photochemical, q_E) to photon irradiance were obtained for leaves of eight species. Their photosynthetic capacity varied greatly due to genetic (sun/shade species) and environmental factors (growth under different light and nutrient regimes). The two fluorescence parameters were studied in response to the proportion of excess light at each irradiance, which is given by $1-A_a/A_p$, where A_a is the actual and A_p the potential rate of photosynthesis (assuming that the high photon yield of CO_2 uptake seen at low light occurred at all irradiances). All leaves exhibited a very similar response of either $1-q_P$ or q_E to $1-A_a/A_p$ indicating that the differences in $1-q_P$ or q_E among leaves at a given irradiance can be explained exclusively by differences in the balance between absorbed light and electron transport. In leaves of all species q_E, which is primarily a reflection of radiationless dissipation of excitation energy, responded linearly to $1-A_a/A_p$, except for a component already present under conditions in which light is limiting (i.e. when $A_a=A_p$). By contrast, the reduction state of the primary electron acceptor of photosystem II, $1-q_P$, when expressed as a function of $1-A_a/A_p$, increased in all leaves much less than expected if it were solely a reflection of the balance between light absorption and photochemistry. Radiationless energy dissipation is a process which has the potential to maintain Q in a low reduction state.

Introduction

Light is the driving force for photosynthesis. The light level at which photosynthesis saturates differs widely among plants (Björkman 1981) and is below that of full sunlight in most C_3 species at the presently prevailing atmospheric CO_2 partial pressure. Absorbed light energy which is not used in photochemistry may be dissipated as heat. Measurements of photochemical and non-photochemical fluorescence quenching combined with measurements of leaf gas exchange can be used to discriminate between the allocation of excitation energy to photochemistry (typically electron transport sustaining CO_2 reduction) and to radiationless dissipation processes (Bradbury and Baker 1984; Dietz *et al.* 1985; Schreiber and Bilger 1987; Weis *et al.* 1987; Winter and Demmig 1987).

This paper presents a study on the partitioning of excitation energy between these processes during steady-state photosynthesis at various irradiances. Leaves of eight C_3 species were examined. They exhibited a large range of photosynthetic capacities due to genetic factors or due to growth of plants at different irradiances and nutrient levels. The data show that non-photochemical quenching is well correlated with changes in electron transport rate in response to light and allows one to roughly estimate the rate

0310-7841/88/010151$03.00

of photosynthesis at a given irradiance. By contrast, the reduction state of Q* is maintained at a lower level than expected if it were solely a reflection of the balance between the rates of the absorption of photons and photochemistry.

Materials and Methods

Plant Material

Populus balsamifera L., *Arbutus unedo* L. and *Helianthus annuus* L. (sunflower) were grown in 10–25-litre plastic pots filled with garden soil and were kept outdoors in the Würzburg Botanic Garden during the summer of 1986, when these experiments were performed. Plants were watered daily and, once per week, received Hewitt's type nutrient solution (Wong 1979) containing 12 mM and 0·6 mM NO_3^- (*P. balsamifera*), 0·6 mM NO_3^- (*A. unedo*), or 12 mM NO_3^- (*H. annuus*). Leaves of *Hedera helix* L. were obtained from a natural population growing at a shaded site in the Würzburg Botanic Garden. *Monstera deliciosa* Liebm. was kept in a glasshouse at 20–30 μmol photons $m^{-2} s^{-1}$. *Spinacia oleracea* L. (spinach) and *Glycine max* (L.) Merrill (soybean) were maintained in growth cabinets at either 60 and 640 μmol photons $m^{-2} s^{-1}$ or 50 and 500 μmol photons $m^{-2} s^{-1}$, respectively, under a 12 h photoperiod. *Welwitschia mirabilis* Hook. fil. was grown as described previously (Winter and Schramm 1986).

Gas Exchange and Fluorescence Techniques

CO_2 assimilation rate and Chl *a* fluorescence were measured simultaneously in intact leaves (for methodological details see Winter and Demmig 1987). Leaves were either enclosed in a 600-ml leaf chamber made of glass and nickel-plated brass (Winter and Schramm 1986) or clamped within a double-sided glass and aluminium chamber (Winter and Demmig 1987). Leaf temperature was kept at 22·5°C and leaf to air vapour pressure difference at about 10 mbar bar^{-1}. Chl *a* fluorescence was measured using a pulse amplitude fluorometer (Schreiber *et al.* 1986). Following a 12-h night in complete darkness, fluorescence was excited with a measuring beam of weak light from a pulsed light-emitting diode to obtain F_O, which designates the fluorescence level when all reaction centres of PS II are open. Maximum fluorescence, F_M, was obtained during a 1 s pulse of saturating light, which transiently closes all reaction centres and completely reduces Q. Variable fluorescence, F_V, equals F_M minus F_O. To obtain light response curves, leaves were illuminated with an Osram Power Star HQI-R 250 W / NDL lamp. After 30–60 min (unless stated otherwise) at each irradiance, the modulated measuring beam of weak light was turned on to measure steady-state fluorescence emission which was composed of a variable component, F_V', plus the actual minimum fluorescence, F_O. Then a saturation pulse was applied, during which fluorescence was usually increased to a new level, composed of a variable component, F_V'', and of F_O. F_O was determined by darkening the leaf for 2–5 min. This protocol was followed at each irradiance. The term $1-q_P$ $(=F_V'/F_V'')$ was used as a measure of the reduction state of the acceptor Q of PS II. For the sake of simplicity, we have not considered a possible non-linearity between $1-q_P$ and the reduction level of Q caused by energy transfer between PS II units (Joliot and Joliot 1964). q_E is $1 - F_V''/F_V$, where F_V was determined at the end of a 12 h dark period as stated above.

During periods of pronounced quenching of fluorescence, e.g. at high irradiances, estimation of the level of fluorescence at open traps (F_O) was complicated. Upon darkening of leaves, fluorescence decreased rapidly, and increased immediately thereafter. Values of $1-q_P$ were based either on the lowest measured value of F_O during 5 min darkness or on a level of F_O obtained by extrapolation to time zero as shown previously (Winter and Demmig 1987). This extrapolation resulted in values up to 10% lower than the lowest measured ones. Two additional methods for calculating $1-q_P$ are described in the caption of Fig. 6.

*Abbreviations used: A, net CO_2 assimilation rate; A_a, actual rate of net CO_2 assimilation at a given photon irradiance; A_p, potential rate of net CO_2 assimilation at a given photon irradiance, assuming that the high photon yield at low light occurred at all irradiances; Chl, chlorophyll; F_O, instantaneous fluorescence emission; F_M, maximum fluorescence emission; F_V, variable fluorescence emission; LHC-2, light harvesting complex of photosystem II (PS II); Q, primary electron acceptor of PS II; q_P, photochemical component of fluorescence quenching; q_E, non-photochemical component of fluorescence quenching.

Results

Light Response Curves of CO_2 Uptake and of Photochemical and Non-photochemical Fluorescence Quenching in Sun Leaves of Populus *Grown with Different Nutrient Supply*

Populus balsamifera was irrigated with nutrient solutions containing either 0·6 or 12 mM NO_3^-. Leaves from the two treatments displayed different light-saturated rates of net CO_2 uptake (Fig. 1*a*). Photochemical fluorescence quenching, q_P, expressed as $1-q_P$ which is a measure of the reduction state of the acceptor *Q* of PS II, is shown in Fig. 1*b*. Fig. 1*c* shows non-photochemical fluorescence quenching, q_E, which is primarily a measure of the proportion of absorbed photons that is lost as heat instead of

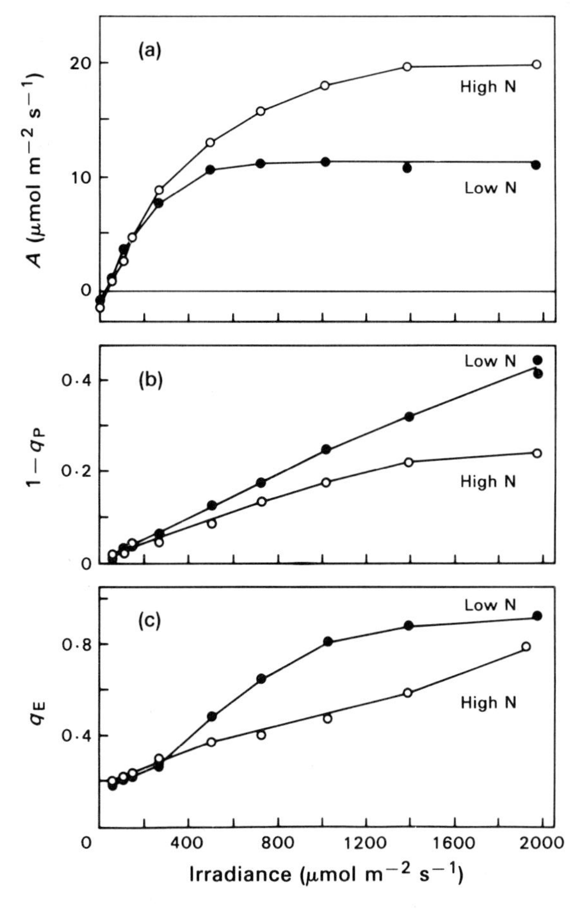

Fig. 1. Light response curves of (*a*) net CO_2 assimilation rate, *A*, (*b*) the reduction state of *Q*, $1-q_P$, and (*c*) non-photochemical quenching, q_E, for sun leaves of *Populus balsamifera* grown under conditions of high and low NO_3^- supply in the soil.

being used to drive photochemistry. It also includes a type of quenching related to an altered distribution of energy between PS II and PS I. In the limiting range of irradiances where rates of net CO_2 uptake were similar for the two leaves (Fig. 1*a*), both the reduction state of *Q*, $1-q_P$, (Fig. 1*b*), and the degree of non-photochemical quenching, q_E (Fig. 1*c*), were similar. At irradiances above 300 μmol m^{-2} s^{-1}, where the high N leaf exhibited higher rates of net CO_2 uptake, more photochemical quenching was observed in the high N leaf, i.e. $1-q_P$ was smaller, whereas non-photochemical quenching, q_E, was less in the high N leaf than in the low N leaf. In the low N leaf, q_E markedly increased at irradiances above 400 μmol m^{-2} s^{-1}, above which no further increase in net CO_2 uptake was observed. Non-photochemical quenching reached a maximum value of 0·92 in the low N leaf whereas it did not exceed 0·78 at even 2000 μmol photons $m^{-2} s^{-1}$ in the high N leaf. The reduction state of Q, $1-q_P$, showed much less pronounced changes than q_E in both types of leaves. Up to an irradiance corresponding to full sunlight, the reduction state of Q did not exceed approximately 45% in the low N leaf and 25% in the high N leaf, respectively. These values are remarkably low, considering that at these high irradiances net CO_2 uptake was light saturated in both types of leaves, and that there was much more light energy absorbed than could be dissipated through photochemistry.

Since at each irradiance the amount of excitation energy which could be utilised via photochemistry differed between the two leaves, the reduction state of Q, $1-q_P$, and q_E were expressed as a function of $1-A_a/A_p$ (Fig. 2) (see also Schreiber and Bilger 1987). The term A_a represents the actual assimilation rate, i.e. the observed rate of net

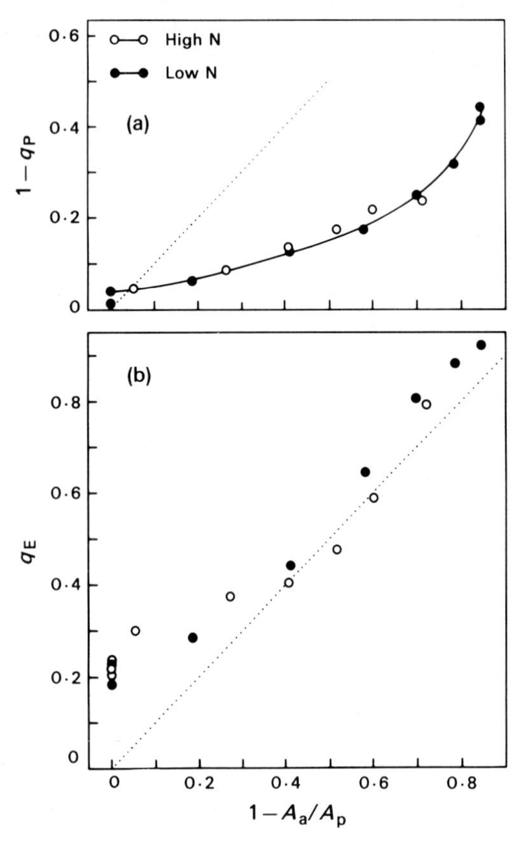

Fig. 2. Response of (*a*) the reduction state of Q, $1-q_P$, and (*b*) q_E to light in excess of that used in photosynthesis $(1-A_a/A_p)$ in *Populus balsamifera* leaves grown with high and low nitrogen supply. Data are from Fig. 1. The dotted lines indicate the 1 : 1 relationships.

CO_2 uptake at a given irradiance. The corresponding potential rate, A_p, for the same irradiance was taken from the initial slope of the A-irradiance curve assuming a constant photon yield at all levels of irradiance. As A_a falls behind the potential rate of CO_2 uptake, A_p, with successive increases in irradiance, an increasing portion of the absorbed excitation energy becomes excessive. Thus, $1-A_a/A_p$ is a measure of the proportion of photons not utilised in photosynthesis. When expressed in this way, data obtained for the low and high N leaves described the same relationship with regard to the reduction state of Q, $1-q_P$ (Fig. 2*a*), as well as to non-photochemical quenching (Fig. 2*b*). Total non-photochemical quenching increased almost linearly with increasing proportions of excess excitation at values of $1-A_a/A_p$ between 0·2 and 0·85. At low irradiance the relationship deviated from linearity and there was a component of q_E (of approx 0·2) already present at limiting irradiance, i.e. when $A_a = A_p$. By contrast, the reduction state of Q, $1 - q_P$, increased only marginally at first, and then exponentially at very high degrees of excess excitation.

Light Response of CO_2 Uptake and of Photochemical and Non-photochemical Fluorescence Quenching in Sun and Shade Species and in Spinacia oleracea *Grown under Two Contrasting Light Regimes*

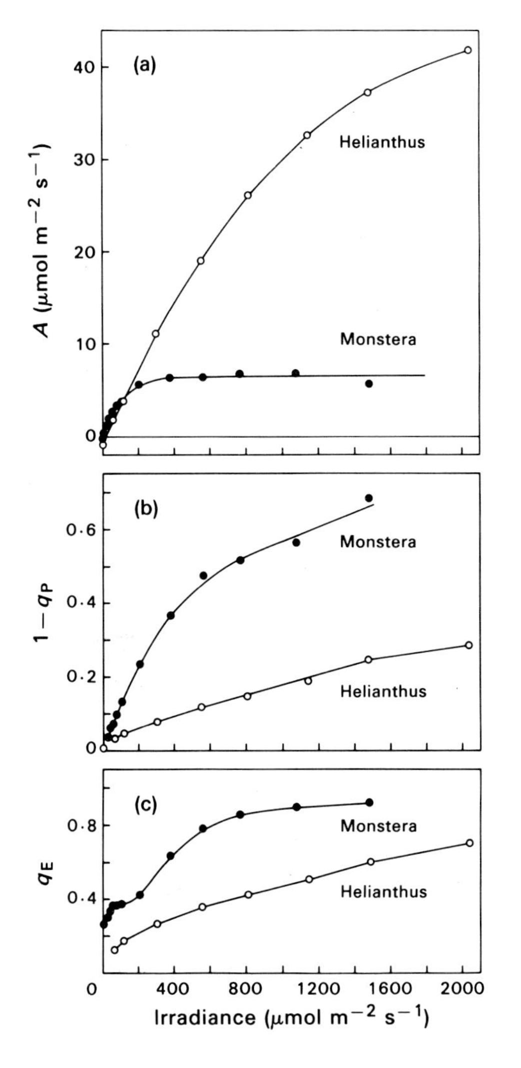

Fig. 3. Light response curves of (*a*) net CO_2 assimilation rate, A, (*b*) the reduction state of Q, $1 - q_P$, and (*c*) non-photochemical quenching, q_E, for a shade leaf of *Monstera deliciosa* and a sun leaf of *Helianthus annuus*.

The response of photosynthetically highly active sun leaves of *Helianthus annuus* was compared with the response of leaves of the shade-tolerant rain forest species *Monstera deliciosa* which had developed in deep shade. CO_2 uptake exhibited light saturation at 10% (200 μmol photons $m^{-2} s^{-1}$) of full sunlight in *M. deliciosa* (Fig. 3*a*), whereas the sunflower leaf had not reached saturation at an irradiance corresponding to full sunlight. At all irradiances, Q was much more reduced and there was therefore much less photochemical quenching in *M. deliciosa* than in *H. annuus* (Fig. 3*b*). At approximately 1500 μmol photons $m^{-2} s^{-1}$, the reduction state of Q was almost 70% in *M. deliciosa* and only 25% in sunflower. Non-photochemical quenching, q_E, was much greater in the *M. deliciosa* leaf than in the sunflower leaf (Fig. 3*c*).

In addition to the above comparison between leaves of a sun and a shade species, we compared low light and high light grown leaves of the same species, *Spinacia oleracea*. These leaves had developed under either 60 or 640 μmol photons $m^{-2} s^{-1}$, and exhibited rates of net CO_2 uptake which differed by a factor of more than two at light saturation (16 and 36 μmol CO_2 $m^{-2} s^{-1}$) (Fig. 4*a*). Again, in the leaf which displayed lower rates of CO_2 uptake, the reduction state of *Q* was higher (Fig. 4*b*) and there was more non-photochemical quenching at a given irradiance (Fig. 4*c*).

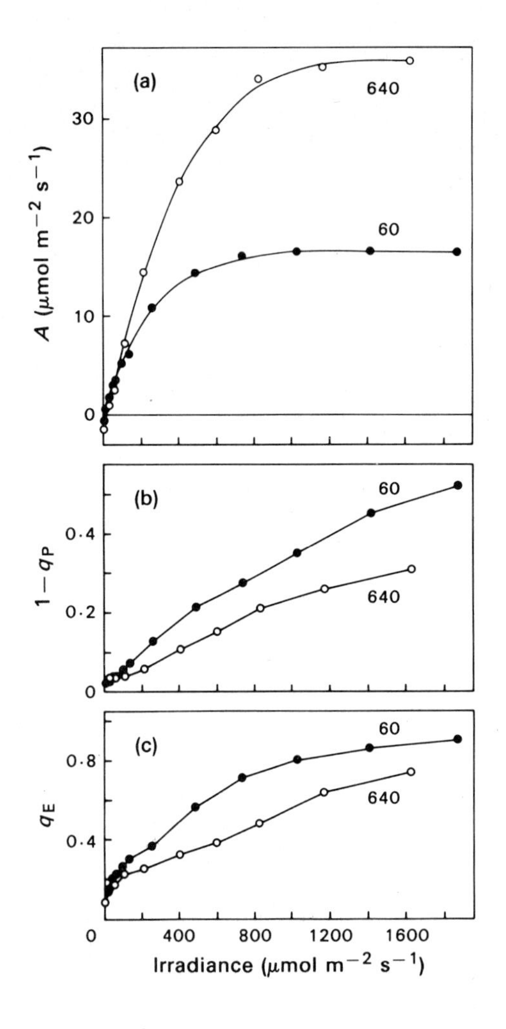

Fig. 4. Light response curves of (*a*) net CO_2 assimilation rate, *A*, (*b*) the reduction state of *Q*, $1-q_P$, and (*c*) non-photochemical quenching, q_E in spinach leaves grown at two levels of irradiance, i.e. 60 and 640 μmol photons $m^{-2} s^{-1}$.

Comparison of the Response of the Reduction State of Q *and of Radiationless Energy Dissipation to Excess Light in Species from Different Habitats Grown under a Variety of Light and Nutrient Conditions*

Fig. 5 summarises values of $1-q_P$ and q_E as a function of excess excitation ($1-A_a/A_p$) for *Populus balsamifera, Spinacia oleracea, Helianthus annuus, Monstera deliciosa,* and four additional species (*Welwitschia mirabilis, Arbutus unedo, Hedera helix, Glycine max*). In all cases, the reduction state of *Q*, $1-q_P$, did not increase linearly with $1-A_a/A_p$ (Fig. 5*a*); it increased only marginally at first, and then increased exponentially at very high values of $1-A_a/A_p$. In all species q_E also responded similarly, at least at values of $1-A_a/A_p$ between 0·2 and 0·9 (Fig. 5*b*). The magnitude of the component of q_E already present at $A_a=A_p$ varied between 0·13 and 0·3 among species.

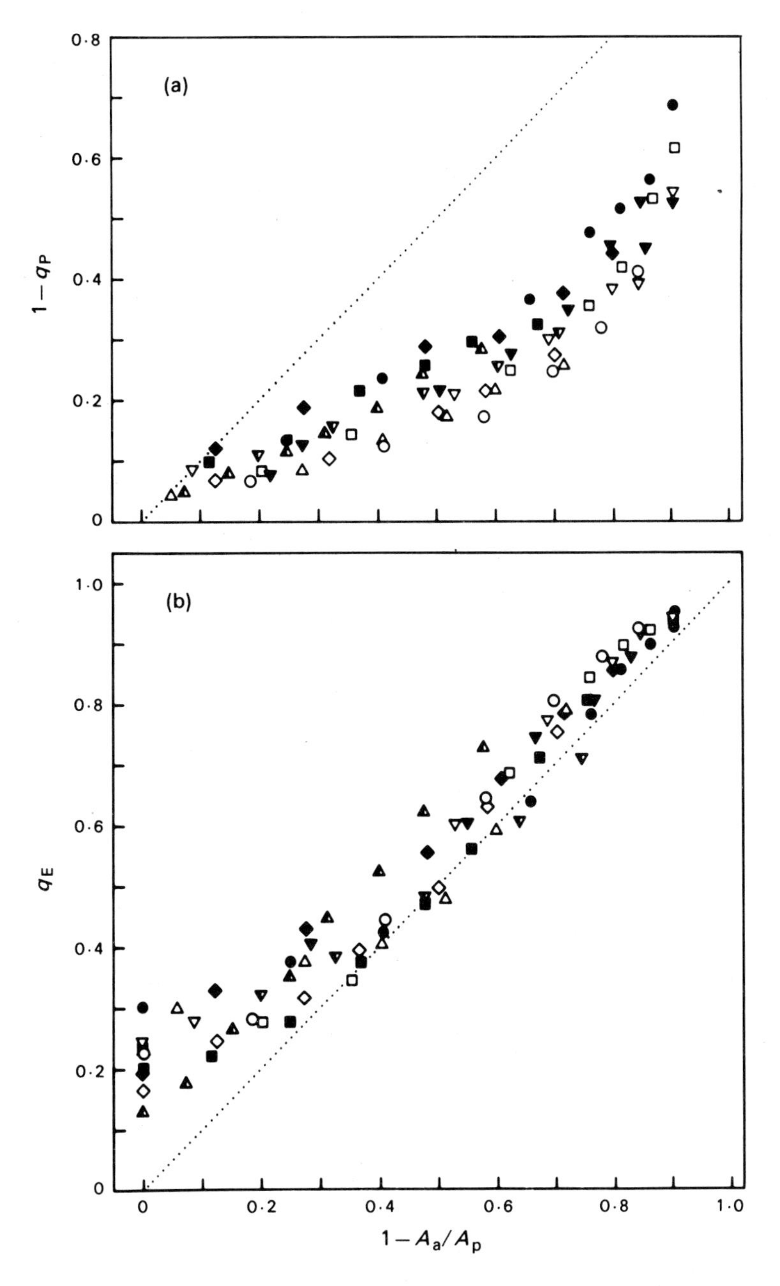

Fig. 5. Response of (*a*) the reduction state of *Q*, $1-q_P$, and of (*b*) non-photochemical quenching, q_E, to excess excitation $(1-A_a/A_p)$ in eight species. *Populus balsamifera* grown at 0·6 (○) and 12 (△) mM NO_3^-, *Hedera helix* (□), *Welwitschia mirabilis* (◇), *Arbutus unedo* (▽), *Monstera deliciosa* (●), *Helianthus annuus* (▲), *Spinacia oleracea* grown at 60 (▽) and 640 (▼) μmol photons $m^{-2} s^{-1}$, and *Glycine max* grown at 50 (♦) and 500 (■) μmol photons $m^{-2} s^{-1}$.

To investigate the possibility that the non-linearity between $1-q_P$ and $1-A_a/A_p$ was due to a varying contribution of photorespiration and/or Mehler reaction in sustaining electron transport, in addition to CO_2 reduction, we compared the light response of $1-q_P$ in air and in 20 mbar O_2, 350 μbar CO_2 in *H. annuus*. The Mehler reaction was shown to have a K_m value of 80 mbar O_2 in algae and in protoplasts (Furbank *et al.* 1982). Although rates of net CO_2 uptake were greatly stimulated in 20 mbar O_2 at all irradiances, photochemical quenching was similar in 20 mbar O_2 and in air (data not shown). This is the expected result if electron transport rates at a given irradiance are similar, irrespective of whether O_2 or CO_2 is reduced. Nevertheless, the relationship between $1-q_P$ and $1-A_a/A_p$ was the same in 20 mbar O_2, 350 μbar CO_2 and in air.

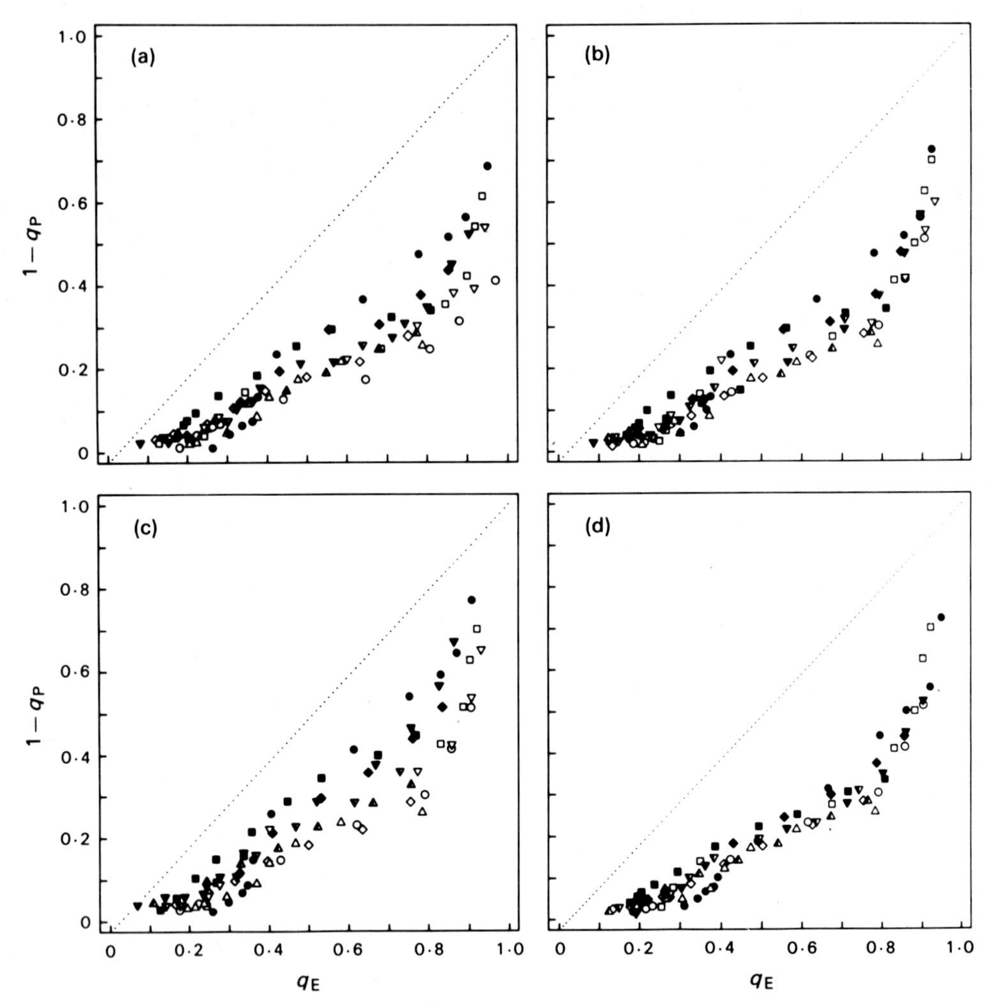

Fig. 6. Reduction state of Q, $1-q_P$, as a function of non-photochemical quenching, q_E. Calculations were based on (*a*) the lowest measured level of F_O during a 5-min dark period following each irradiance; on (*b*) values of F_O extrapolated to time zero when quenching in the light was pronounced; on (*c*) the level of F_O in long-term (12-h) dark-adapted leaves when the level of fluorescence during dark interruptions was higher than the long-term dark level; and (*d*) on the level of fluorescence observed immediately upon darkening when the level of fluorescence during dark interruptions was higher than in long-term dark-adapted leaves. Symbols as in Fig. 5.

This was due to the fact that A_p was taken from the *actual* slope of each light response curve, with the slope being higher at 20 mbar O_2. This encompasses any differences in electron transport sustained by electron acceptors other than CO_2 as well. Results similar to the response of $1-q_p$ and q_E to excessive light in normal air were obtained at high CO_2 (1000 μbar) in 210 mbar O_2. The nature of the relationship of $1-q_p$ and q_E with $1-A_a/A_p$ remained unchanged when based on estimated electron transport rates instead of rates of net CO_2 uptake. Rates of electron transport were calculated after Farquhar and von Caemmerer (1982) using a compensation point in the absence of day respiration of 33 μbar and a day respiration rate of 0·5 μmol CO_2 m^{-2} s^{-1} for spinach grown at an irradiance of 350 μmol m^{-2} s^{-1}. The ratio of electron transport rate and net CO_2 uptake was found to be quasi-constant (at 6·63–6·71) for irradiances between 211 and 1627 μmol m^{-2} s^{-1}, where intercellular CO_2 partial pressures were around 200–240 μbar in all species.

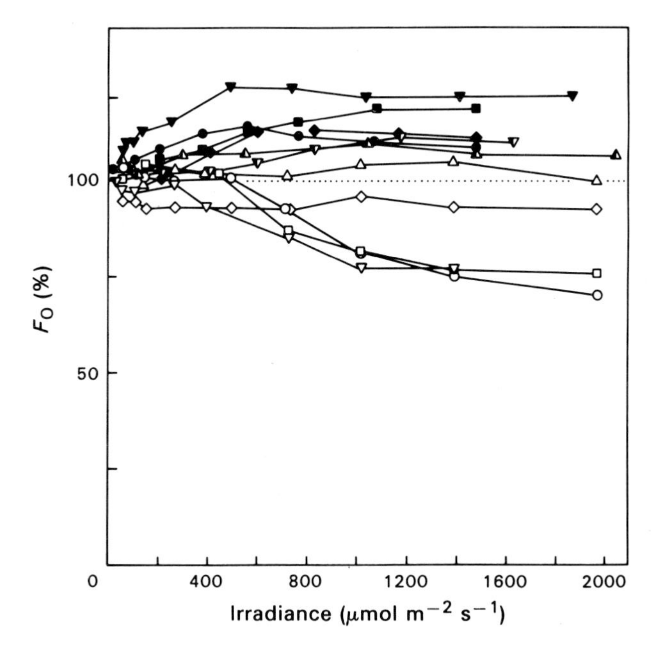

Fig. 7. Changes in F_O caused by exposure of leaves from eight species to various irradiances. The respective values of F_O represent the lowest measured value of fluorescence during 5 min dark interruptions. The level of F_O determined after 12 h of darkness was set to 100%. Symbols as in Fig. 5.

Fig. 6 shows plots of $1-q_p$ versus q_E for the eight species. At low and intermediate values of q_E, increases in q_E were accompanied by less pronounced increases in $1-q_p$. Only when q_E approached its maximum value was there a marked increase in $1-q_p$. The value of $1-q_p$ is particularly sensitive to changes in the level of F_O. Calculation of data in Figs 5*a* and 6*a* was based on the lowest measured level of F_O during a 2–5 min dark period following steady-state photosynthesis at each irradiance. After exposure to low irradiance, F_O could be determined easily since fluorescence decreased rapidly to a constant level in the dark. Estimation of F_O was complicated, however, at higher irradiances, where fluorescence upon darkening first decreased and then increased. This was taken into account in Fig. 6*b* which shows $1-q_p$ and q_E calculated from extrapolated values of F_O (see Winter and Demmig 1987). These F_O values were at most 10% lower than the lowest measured value of F_O. This correction resulted in $1-q_p$ being higher at high values of q_E, especially in *P. balsamifera* and *A. unedo*. As a result, the response of the different types of leaves became more uniform.

In addition to these rapid changes in F_O upon darkening under conditions of pronounced quenching, there were also changes in F_O with slower kinetics. Depending upon species, the lowest level of fluorescence observed during dark interruptions was either increased, unchanged or decreased (Fig. 7) compared to the F_O level of fully dark-adapted leaves determined at the end of a 12-h dark period. Illumination with weak far-red background light to promote oxidation of Q, in some typical cases, did not yield any further decreases in the level of fluorescence during the dark interruptions. When these measured values were ignored and the level of F_O determined after 12 h darkness was used to calculate $1-q_P$ and q_E, values of $1-q_P$ increased more than q_E in those cases in which the level of fluorescence during dark interruptions had increased (Fig. 6*c*). As a consequence, changes in the reduction state of Q deviated less from changes in q_E in species such as *M. deliciosa*, which exhibited an apparent increase in F_O (Fig. 7). On the other hand, the assumption could be made that a true increase in F_O had occurred during the light treatment which became reversed upon darkening. In this case the best approximation of F_O would be the value measured immediately upon darkening. When calculations of $1-q_P$ and q_E were based on the level of fluorescence immediately upon darkening (Fig. 6*d*), instead of the lowest fluorescence level measured during the subsequent 2–5 min dark period, species like *M. deliciosa* (Fig. 6*c*) exhibited the same pronounced deviation from linearity in the relationship of $1-q_P$ versus q_E as was observed in all the other species.

Discussion

A high photosynthetic capacity resulting from acclimation or adaptation to high light and high nutrient supply is accompanied by strong photochemical and little non-photochemical fluorescence quenching at a given irradiance, i.e. allocation of light to photochemistry is favoured compared to heat dissipation. This is clearly seen in the comparison between high and low nitrogen status leaves of *Populus balsamifera*, sun and shade leaves of spinach, and between leaves of sunflower and leaves of the shade-tolerant plant *Monstera deliciosa*. At a given irradiance, however, one leaf may exhibit a higher reduction state of Q than another in spite of a higher rate of CO_2 assimilation. For example, at 1600 μmol photons m^{-2} s^{-1} (near full sunlight), the reduction state of Q in the high N status sun leaf of *P. balsamifera* was 22% compared to 31% in the high-light grown leaf of spinach, although the spinach leaf showed a much higher light-saturated rate of photosynthesis (36 versus 20 μmol CO_2 m^{-2} s^{-1}, respectively).

Irrespective of the large differences in the A-irradiance curves among the great variety of leaves used in this study, there was no intrinsic difference with respect to excessive light. All leaves exhibited a remarkably uniform behaviour in terms of fluorescence quenching, when both the reduction state of Q and the non-photochemical component, q_E, were related to the proportion of excessive excitation at each irradiance expressed as $1-A_a/A_p$. Thus the observed differences in $1-q_P$ or q_E among leaves at a given irradiance can be attributed exclusively to a different balance between absorbed light and electron transport rate. Non-photochemical fluorescence quenching reflects mainly radiationless decay, but also includes a component related to alterations in energy distribution between PS II and PS I. The component of q_E, which was already present at limiting levels of irradiance when $A_a = A_p$, should not represent heat dissipation as this would be inconsistent with the finding of high photon yields of photosynthesis in non-photoinhibited leaves (Björkman and Demmig 1987). This type of quenching is possibly related to phosphorylation of LHC-2, which was found to be phosphorylated at low irradiances and dephosphorylated at high irradiances (Demmig *et al.* 1987*a*). This component of q_E, which was already present at limiting levels of irradiance, was much smaller or absent in the studies by Weis *et al.* (1987) as well as in studies by us using spinach leaves grown at extremely high light levels (1500 μmol photons $m^{-2} s^{-1}$) (unpublished data). In these leaves q_E increased linearly with $1 - A_a/A_p$. If, for the spinach leaf grown at 1500 μmol photons $m^{-2} s^{-1}$, a correction is introduced to account for the fact that only open reaction centres can perform photochemistry (Weis *et al.* 1987) and q_E is plotted versus $1 - A_a/A_p \times q_P$, then a curvilinear relationship is obtained, which is consistent with the relationship between variable fluorescence (and hence q_E) and the rate of radiationless dissipation in the antenna chlorophyll as predicted by the simple model of Kitajima and Butler (1975).

Irrespective of these considerations, the reduction state of Q responded less strongly to excess light than radiationless energy dissipation, indicating that the reduction state of Q is under some form of control. If this variable were solely a reflection of the balance between the rate of absorption of photons and the rate of photochemistry, it would respond linearly to excess light. If, however, absorbed light energy were allocated to alternative processes, the reduction state of Q would be maintained at a low level. The above findings suggest that the partitioning of excitation energy between photochemistry and radiationless dissipation is altered as the rate of absorption of photons exceeds the rate of utilisation of excitation energy. Increased radiationless energy dissipation is the most likely process which competes with the reduction of Q. The similarity of the q_E versus $1 - A_a/A_p$ curves suggests that all leaves had the same capacity for increased allocation of energy to radiationless dissipation. The observation that the reduction state of Q began to respond strongly to increasing excitation energy as soon as q_E approached its maximum possible value is consistent with the interpretation that radiationless dissipation competes with photochemistry. These results are in line with studies on the light response of *Phaseolus vulgaris* (Weis *et al.* 1987) and on the CO_2 response in spinach and in the CAM plant *Kalanchoë daigremontiana* (Winter and Demmig 1987). In these studies the reduction state of Q was a very poor indicator of photosynthetic rates. A good correlation between the rate of photosynthesis and the reduction state of Q was seen in water-stressed tissues under conditions of maximal q_E (Schreiber and Bilger 1987) and transiently upon abrupt large increases in irradiance which caused O_2 evolution and $1 - q_P$ to oscillate (M. Stitt and U. Schreiber, personal communication).

In some leaves, values of F_O were increased following photosynthesis at high levels of irradiance compared to the F_O level determined previously after prolonged (12 h) darkness. If these increased fluorescence levels were overestimates of F_O, the reduction state of Q would respond to excess light in a fashion more similar to q_E (Fig. 6*c*). However, experiments with background far-red light make this possibility unlikely. Both increases and decreases in F_O, depending on species and growth conditions, have been reported previously as a result of high light treatments (Demmig and Björkman 1987; Demmig *et al.* 1987*b*). According to Kitajima and Butler (1975), a net decrease in F_O results from increased radiationless dissipation in the antenna chlorophyll, and a net increase in F_O is caused by a decreased transfer of excitation energy to PS II reaction centres. We therefore favour the use of $1 - q_P$ based on increased or decreased values of F_O (Fig. 6*d*).

Calculations of excess light were based on the initial slope of the dependence of net CO_2 assimilation rates on incident irradiance for each leaf. Since both A_a and A_p were based on incident light, this method renders comparisons among different leaves independent of differences in light absorption. Varying ratios of CO_2 reduction rate to oxygenation rate among species and their effect on photon yield would also be accounted for by this method. Intercellular CO_2 partial pressure was between 200 and 240 μbar at irradiances beyond the photon yield region in all species. Furthermore, the response of q_E and $1 - q_P$ to excess light ($= 1 - A_a/A_p$) was similar for sunflower in normal air and in the absence of photorespiration (20 mbar O_2, 350 μbar CO_2, or 210 mbar O_2, 1000 μbar CO_2), provided that the respective initial slopes of the light response curves were considered. If the actual photon yield, based on incident light, is known for a leaf at a given temperature, A_p can be obtained for a particular irradiance. Once the above are ascertained, the actual CO_2 assimilation rate can be roughly estimated from measurements of leaf fluorescence alone, using the relationship between q_E and $1 - A_a/A_p$ shown in Fig. 5.

Acknowledgments

This work was supported by the Deutsche Forschungsgemeinschaft. We are grateful to W. W. Adams III, W. Bilger, E. Weis and U. Heber for critical comments on the paper.

References

Björkman, O. (1981). Responses to different quantum flux densities. In 'Physiological Plant Ecology I. Responses to the Physical Environment'. (Eds O. L. Lange, P. S. Nobel, C. B. Osmond and H. Ziegler.) Encycl. Plant Physiol. New Ser., Vol 12A, pp. 57–107. (Springer-Verlag: Berlin.)

Björkman, O., and Demmig, B. (1987). Photon yield of O_2 evolution and chlorophyll fluorescence characteristics at 77K among vascular plants of diverse origins. *Planta* **170**, 489–504.

Bradbury, M., and Baker, N. R. (1984). A quantitative determination of photochemical and nonphotochemical quenching during the slow phase of the chlorophyll fluorescence induction curves in bean leaves. *Biochim. Biophys. Acta* **765**, 275–81.

Demmig, B., and Björkman, O. (1987). Comparison of the effect of excessive light on chlorophyll fluorescence (77K) and photon yield of O_2 evolution in leaves of higher plants. *Planta* **171**, 171–84.

Demmig, B., Cleland, R. E., and Björkman, O. (1987*a*). Photoinhibition, 77K chlorophyll fluorescence quenching and phosphorylation of LHC-II. *Planta* **172**, 378–85.

Demmig, B., Winter, K., Krüger, A., and Czygan, F.-C. (1987*b*). Photoinhibition and zeaxanthin formation in intact leaves. A possible role of the xanthophyll cycle in the dissipation of excess light. *Plant Physiol.* **84**, 218–24.

Dietz, K.-J., Schreiber, U., and Heber, U. (1985). The relationship between the redox state of Q_A and photosynthesis in leaves at various carbon dioxide, oxygen and light regimes. *Planta* **166**, 219–26.

Farquhar, G. D., and von Caemmerer, S. (1982). Modelling of photosynthetic response to environmental conditions. In 'Physiological Plant Ecology II. Water Relations and Carbon Assimilation'. (Eds O. L. Lange, P. S. Nobel, C. B. Osmond and H. Ziegler.) Encycl. Plant Physiol. New Ser., Vol. 12B, pp. 549–88. (Springer-Verlag: Berlin.)

Furbank, R. T., Badger, M. R., and Osmond, C. B. (1982). Photosynthetic oxygen exchange in isolated cells and chloroplasts of C_3 plants. *Plant Physiol.* **70**, 927–31.

Joliot, A., and Joliot, M. P. (1964). Étude cinétique de la réaction photochimique libérant l'oxygène au cours de la photosynthèse. *C.R. Acad. Sci. Ser. D* **258**, 4622–5.

Kitajima, M., and Butler, W. (1975). Quenching of chlorophyll fluorescence and primary photochemistry in chloroplasts by dibromothymoquinone. *Biochim. Biophys. Acta* **376**, 105–15.

Schreiber, U., Schliwa, U., and Bilger, W. (1986). Continuous recording of photochemical and nonphotochemical chlorophyll fluorescence quenching with a new type of modulation fluorometer. *Photosynth. Res.* **10**, 51–62.

Schreiber, U., and Bilger, W. (1987). Rapid assessment of stress effects on plant leaves by chlorophyll fluorescence measurements. In 'Plant Response to Stress—Functional Analysis in Mediterranean Ecosystems'. (Eds J. D. Tenhunen, F. M. Catarino, O. L. Lange and W. C. Oechel.) pp. 27–53. (Springer-Verlag: Berlin.)

Weis, E., Ball, J. T., and Berry, J. (1987). Photosynthetic control of electron transport in leaves of *Phaseolus vulgaris*: evidence for regulation of photosystem 2 by the proton gradient. In 'Progress in Photosynthesis Research'. (Ed. J. Biggins) Vol. 2, pp. 553–6. (Martinus Nijhoff, Dordrecht.)

Winter, K., and Schramm, M. J. (1986). Analysis of stomatal and nonstomatal components in the environmental control of CO_2 exchange in leaves of *Welwitschia mirabilis*. *Plant Physiol.* **82**, 173–8.

Winter, K., and Demmig, B. (1987). Reduction state of Q and nonradiative energy dissipation during photosynthesis in leaves of a crassulacean acid metabolism plant, *Kalanchoë daigremontiana* Hamet et Perr. *Plant Physiol.* **85**, 1000–7.

Wong, S. C. (1979). Elevated atmospheric partial pressures of CO_2 and plant growth. *Oecologia* **44**, 68–74.

Characterisation of Three Components of Non-photochemical Fluorescence Quenching and Their Response to Photoinhibition

Barbara Demmig and Klaus Winter

Lehrstuhl für Botanik II, Universität Würzburg, Mittlerer Dallenbergweg 64, 8700 Würzburg, Federal Republic of Germany.

Abstract

Three components of non-photochemical fluorescence quenching were distinguished according to their response to irradiance and to their relaxation kinetics upon darkening. Two components of quenching were restricted to excessive irradiance and were interpreted to reflect radiationless dissipation. One relaxed rapidly upon darkening, and increased sharply when irradiance became excessive, i.e. as soon as net CO_2 assimilation rate was no longer linearly related to irradiance, and attained a maximum value with only small further increases in irradiance. The second component relaxed slowly, increased markedly when the rapidly relaxing component had reached its maximum, and continued to increase linearly with increasing irradiance. The third component was already present at low irradiances, relaxed very slowly, and may be related to an altered distribution of excitation energy between PS II and PS I.

Following exposure to weak illumination under conditions preventing photosynthetic electron transport (20 mbar O_2, zero CO_2), the reduction state of Q was initially high and decreased as non-photochemical fluorescence quenching indicative of radiationless dissipation developed.

Subsequent to photoinhibitory treatments in high light and 20 mbar O_2, zero CO_2, an increased reduction state of Q as well as increased non-photochemical quenching of the two types indicative of increased heat dissipation was observed. In sunflower a lasting increase in the reduction state of Q was observed and fluorescence characteristics reflected photoinhibitory damage. In *Nerium oleander*, increased radiationless dissipation of the slowly relaxing type was the predominant response and the reduction state of Q was increased only transiently.

Introduction

Measurements of chlorophyll fluorescence in combination with simultaneous studies of leaf gas exchange provide information about the partitioning of excitation energy between photochemical processes, responsible for CO_2 reduction, and non-photochemical processes, of which radiationless dissipation is the main one (Bradbury and Baker 1984; Dietz *et al.* 1985; Schreiber and Bilger 1987; Weis *et al.* 1987; Winter and Demmig 1987; Demmig and Winter 1988). In the preceding paper, we have examined the response of the sum of all non-photochemical quenching processes to excessive light, using a pulse amplitude modulated fluorometer at room temperature (Demmig and Winter 1988). In the present study, various components of non-photochemical quenching were determined. We have been able to distinguish between one component which is independent of excess light and two components which do respond to excess light and therefore most likely reflect radiationless dissipation. These two components are shown to differ in their response to light and in their relaxation kinetics upon darkening. Furthermore, evidence is presented that increased radiationless dissipation, as indicated

0310-7841 / 88 / 010163$03.00

by fluorescence quenching, can serve to maintain a low reduction state of Q.* Moreover, the effect of photoinhibition on the reduction state of Q and on radiationless energy dissipation is examined.

Materials and Methods

Plant material and growth conditions as well as measurements of net CO_2 assimilation rate with an open gas exchange system and simultaneous measurements of fluorescence with a pulse amplitude fluorometer were as described earlier unless stated otherwise (Demmig and Winter 1988). During measurements leaf temperature was kept at 22·5°C and leaf to air vapour pressure difference at about 10 mbar bar^{-1}. Fluorescence signals and their terminology are shown in Fig. 1. To obtain $q_{E,fast}$ and $q_{E,slow}$, we have intentionally avoided the treatment $q_{E,fast} = q_{E,total} - q_{E,slow}$ because this would force the values of $q_{E,fast}$ to decrease as soon as $q_{E,total}$ and thus $q_{E,slow}$ become large. In Fig. 4, for example, fluorescence quenching is strong at 1410 μmol photons $m^{-2} s^{-1}$, with values of variable fluorescence in a saturating pulse being $F_{V,light} = 1{\cdot}29$, $F_{V,\, 2\ min\ dark} = 5{\cdot}35$, and $F_{V,\, 12\ h\ dark} = 16{\cdot}07$ for the leaf of *Hedera helix*. Thus $q_{E,total} = 0{\cdot}92$ and $q_{E,slow} = 0{\cdot}666$. The treatment $q_{E,fast} = q_{E,total} - q_{E,slow}$ gives 0·254, a very small value which does not account for the fact that the rapidly relaxing quenching process ($q_{E,fast}$) still has the capacity to strongly reduce F_V from 5·36 to 1·29, i.e. to 24% which must correspond to a quenching coefficient of 0·76. The treatment used here yields $q_{E,fast} = 0{\cdot}76$.

Results

Three Components of Non-photochemical Fluorescence Quenching

Spinach was grown under four different light regimes. Fig. 2*a* shows light response curves of net CO_2 assimilation rate for these four treatments. Fig. 2*b* shows the results from simultaneous measurements of total non-photochemical fluorescence quenching. Upon darkening of leaves, the relaxation of fluorescence quenching showed biphasic kinetics. One portion of total non-photochemical quenching, designated $q_{E,fast}$, relaxed almost instantaneously and was, therefore, calculated from the ratio of variable fluorescence in a saturation pulse in the light, and variable fluorescence after 2 min dark (Fig. 1). The remaining component, designated $q_{E,slow}$, relaxed much more slowly and was calculated from the ratio of variable fluorescence after 2 min dark and at the end of the previous night, i.e. after long-term dark adaptation.

In spinach leaves grown at different irradiances, $q_{E,fast}$ was not measurable following illumination in the linear portion of the light response curve (Fig. 2*c*). At the irradiance at which the transition between light-limited and light-saturated photosynthesis occurred (Fig. 2*a*), i.e. at the point at which the rate of utilisation of excitation energy by photosynthesis began to fall behind the rate of absorption of photons, $q_{E,fast}$ increased strongly. In the leaf grown at 70 μmol photons $m^{-2} s^{-1}$, the light response curve of net CO_2 assimilation started to deviate from linearity above 200 μmol photons $m^{-2} s^{-1}$ and, at this irradiance, $q_{E,fast}$ began to increase. In the leaf grown at 640 μmol photons $m^{-2} s^{-1}$, the light response curve of CO_2 assimilation started to deviate markedly from linearity at irradiances above 600 μmol photons $m^{-2} s^{-1}$. This was again the irradiance at which $q_{E,fast}$ began to increase. Thus, in the leaf which had developed at 70 μmol photons $m^{-2} s^{-1}$, the fast component of q_E reached a maximum at an irradiance where

*Abbreviations used: A, net CO_2 assimilation rate; A_a, actual rate of net CO_2 assimilation at a given photon irradiance; A_p, potential rate of net CO_2 assimilation at a given photon irradiance assuming that the high photon yield at low light occurred at all irradiances; F_M, maximum fluorescence emission; F_O, instantaneous fluorescence emission; F_V, variable fluorescence emission; LHC-2, light harvesting complex of photosystem II (PS II); Q or Q_A, primary electron acceptor of PS II; Q_B, tightly bound (to the Q_B protein) plastoquinone which accepts electrons from Q_A; q_E, total non-photochemical fluorescence quenching; $q_{E,fast}$, rapidly relaxing component of non-photochemical fluorescence quenching; $q_{E,slow\ 1}$, slowly relaxing component of non-photochemical fluorescence quenching present at low photon irradiances; $q_{E,slow\ 2}$, slowly relaxing component of non-photochemical fluorescence quenching present at high photon irradiances; q_p, photochemical component of fluorescence quenching.

it was still close to zero in the leaf which had developed at 640 μmol photons $m^{-2} s^{-1}$. The intimate relationship between the shape of the curves for net CO_2 assimilation and $q_{E,fast}$ as a function of irradiance shows that the rapidly relaxing component of q_E is a sensitive indicator of the point at which light started to become excessive. This became

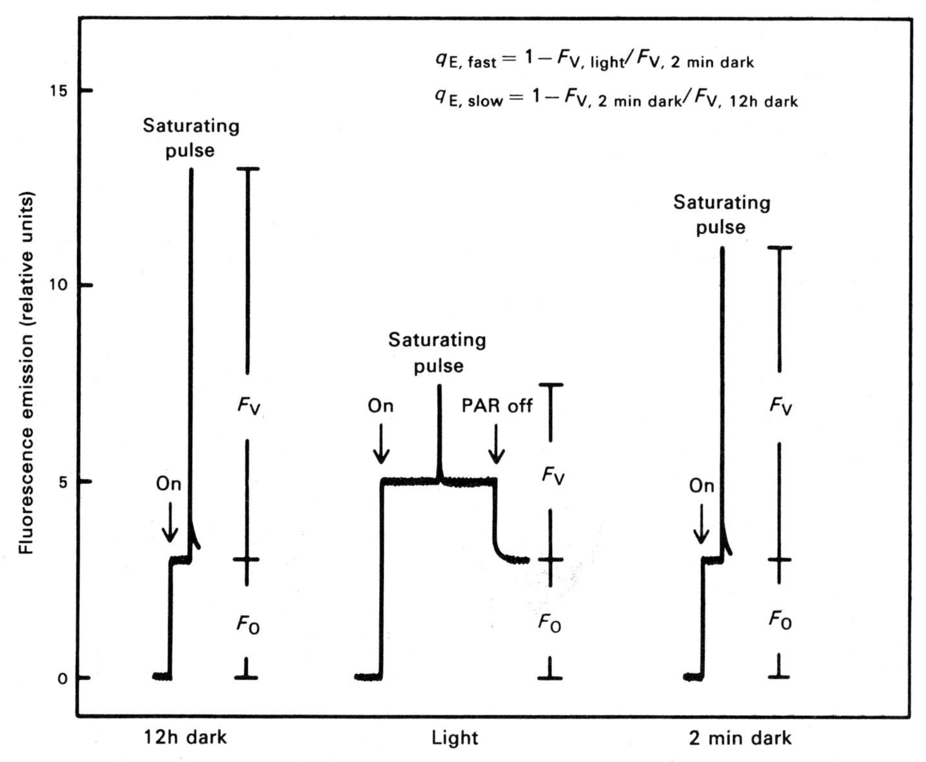

Fig. 1. Fluorescence signals and their terminology. The components of non-photochemical quenching, $q_{E,fast}$ and $q_{E,slow}$, were calculated from the variable fluorescence yields at closed PS II traps. Saturating pulses were given at the end of the 12 h dark period as well as at each irradiance in the light and following darkening of the leaves for 2 min. 'On' refers to the turning on of the weak pulsed measuring light. PAR, photosynthetically active radiation.

even more obvious when the rapidly relaxing component of q_E was expressed as a function of 1-A_a/A_p, which is used as a measure of excess excitation energy (see Demmig and Winter 1988). When plotted this way, curves of $q_{E,fast}$ from all four light treatments fell very close together (Fig. 3). The rapidly relaxing component of q_E did not exceed values of 0·45 in leaves grown at 70 and 140 μmol photons $m^{-2} s^{-1}$ (Fig. 2*c*). In leaves grown at 350 and 640 μmol photons $m^{-2} s^{-1}$, $q_{E,fast}$ was approximately 0·5 at 1670 μmol $m^{-2} s^{-1}$ and appeared to have not yet reached its maximum.

The slowly relaxing component of q_E (Fig. 2*d*) showed a complex response to increasing irradiance, indicating that $q_{E,slow}$ was composed of more than one quenching process. Firstly, there was a component which appeared to saturate in the photon yield region of the light response curve. Slightly above the point at which the rapidly relaxing component of q_E started to increase in each leaf (compare Fig. 2*c*), there was a further increase observed in the slow component of q_E (Fig. 2*d*). This second component of $q_{E,slow}$ appeared to increase almost linearly with increasing irradiances beyond the light level at which CO_2 uptake was light-saturated, with steeper increases in low light versus high light grown plants. The first component of $q_{E,slow}$, which saturated in low light, had a similar magnitude in all leaves grown at irradiances between 70 and 640 μmol $m^{-2} s^{-1}$.

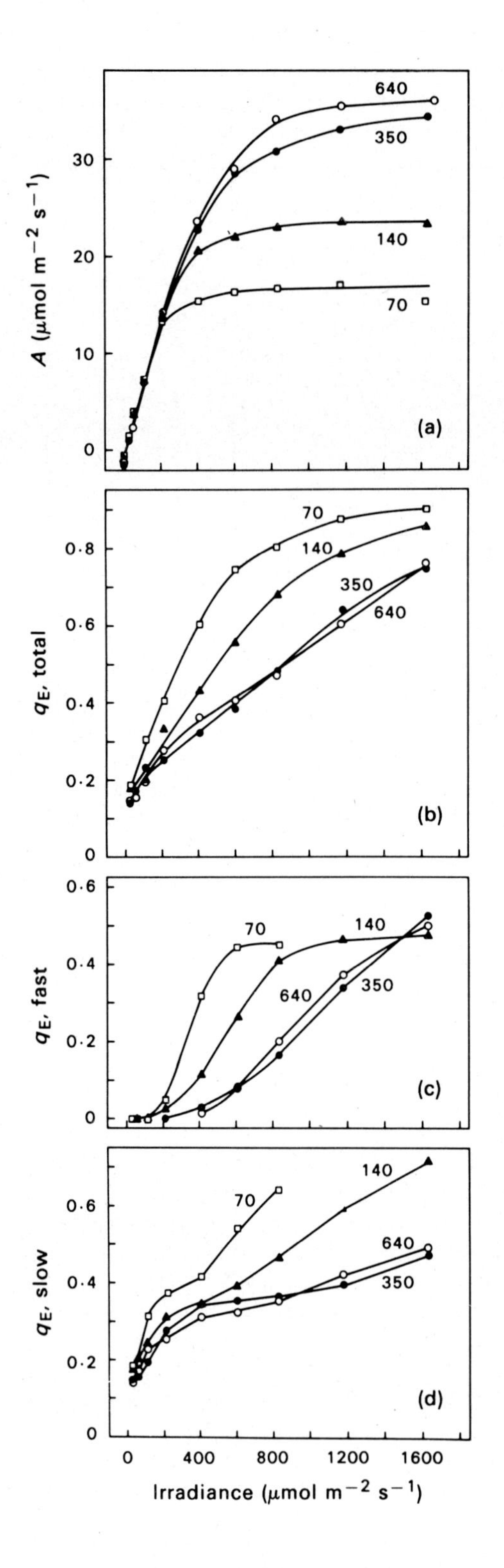

Fig. 2. Light response curves for spinach leaves of (*a*) net CO_2 assimilation rate, A, (*b*) total non-photochemical quenching, $q_{E,total}$, and of two components of q_E, one of which relaxed rapidly, $q_{E,fast}$ (*c*), and the other slowly, $q_{E,slow}$ (*d*), upon darkening. Leaves were grown at four different levels of irradiance, 70 (□), 140 (▲), 350 (●), and 640 (○) μmol m^{-2} s^{-1}. $q_{E,fast}$ was calculated from the ratio of F_V (closed traps) in the light and after 2 min darkness. $q_{E,slow}$ was calculated from the ratio of F_V (closed traps) after 2 min darkness and after 12 h darkness.

Under conditions of pronounced non-photochemical quenching, a decrease in F_O below the level of fluorescence at the end of the previous night had been observed in some species (see Demmig and Winter 1988). Two species, *Hedera helix* and *Populus balsamifera*, which displayed a strong net decrease in F_O at high irradiance, were used to examine which components of q_E were associated with a decrease in F_O. Both species were characterised by relatively low photosynthetic capacities (Fig. 4*a*). Total q_E and various components of q_E as a function of irradiance are shown in Figs 4*b* and 5*a*, respectively. The general pattern was similar to that observed with spinach. Total q_E was already 0·2–0·25 at the lowest irradiance used. There was a steep increase in total

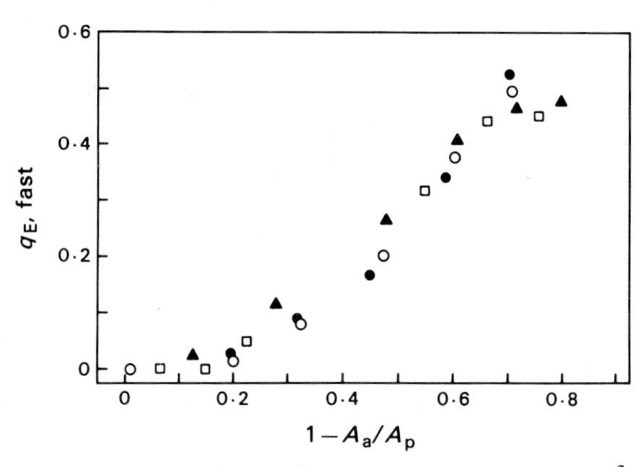

Fig. 3. Response of the rapidly relaxing component, $q_{E,fast}$, of non-photochemical quenching to excess light ($1-A_a/A_p$) in spinach leaves grown at four levels of irradiance, 70 (□), 140 (▲), 350 (●), and 640 (○) μmol m^{-2} s^{-1}. Data from Fig. 1.

q_E at irradiances between 200 and 800 μmol m^{-2} s^{-1} in *H. helix* and between 200 and 1200 μmol m^{-2} s^{-1} in *P. balsamifera*. In both types of leaves, total q_E attained maximum values of approx. 0·9. The rapidly relaxing component of q_E was zero at limiting irradiances and started to increase at irradiances above 200 μmol m^{-2} s^{-1} (Fig. 5*a*). It increased more strongly in *H. helix* than in *P. balsamifera*. Thus, the general shape of the two curves of total q_E (Fig. 4*b*) and the difference in light response between the two leaves were very much a reflection of the behaviour of the rapidly relaxing component of q_E (Fig. 5*a*). The slowly relaxing component of q_E was split into two further components. We subtracted the component which was already present at low irradiance (termed $q_{E,slow\ 1}$) from all values of the total slowly relaxing component. The remaining component was termed $q_{E,slow\ 2}$. As a consequence of the subtraction, $q_{E,slow\ 2}$ was zero at low irradiances and began to increase after $q_{E,fast}$ had already reached high values. $q_{E,slow\ 2}$ corresponds to the linear increase in $q_{E,slow}$ demonstrated for spinach (see Fig. 2*d*).

As stated above, in leaves of both *P. balsamifera* and *H. helix* there was pronounced quenching of F_O (Fig. 5*b*). At low irradiances, this quenching of F_O upon darkening relaxed within 2 min. At higher irradiances, F_O increased rapidly during the first 2 min of darkness and then continued to increase more slowly. The rapidly relaxing portion of the quenching of F_O was calculated from the ratio of the level of fluorescence observed immediately upon darkening, which was extrapolated back to time 0, and the (higher) level of fluorescence after 2 min in darkness. The slowly relaxing portion was calculated from the ratio of the level of fluorescence after 2 min darkness and the level of F_O determined at the end of the previous night. The light response of the rapidly relaxing portion of the quenching of F_O and of $q_{E,fast}$ matched exactly, as did the light response of the slowly relaxing portion of the quenching of F_O and $q_{E,slow\ 2}$. In other words, in *P. balsamifera* and *H. helix, both* types of quenching which were restricted to excessive irradiance cause a decrease not only of F_M, but also of F_O. In contrast, the slowly relaxing component already present at limiting irradiance, $q_{E,slow\ 1}$, was not associated with a decrease in F_O

Hysteresis in the Reduction State of Q *and in Non-photochemical Fluorescence Quenching*

Both the reduction state of *Q* and radiationless energy dissipation were subject to hysteresis processes. In a shade leaf of *Monstera deliciosa*, irradiance was increased in increments up to 1480 μmol $m^{-2} s^{-1}$, maintained at this level for 1 h, and then decreased again (Fig. 6). The descending light response curve showed higher values of total non-photochemical quenching than the ascending one, and the reverse was true for the reduction state of *Q* (Fig. 6*a*). The increase in total q_E was due exclusively to a strong increase in the magnitude of the slowly relaxing component of q_E (Fig. 6*b*). In contrast, the rapidly relaxing component had already decreased during the ascending curve with the increase in irradiance from 1012 to 1480 μmol $m^{-2} s^{-1}$. It decreased further when the leaf was kept for 1 h at 1480 μmol photons $m^{-2} s^{-1}$, to assume even

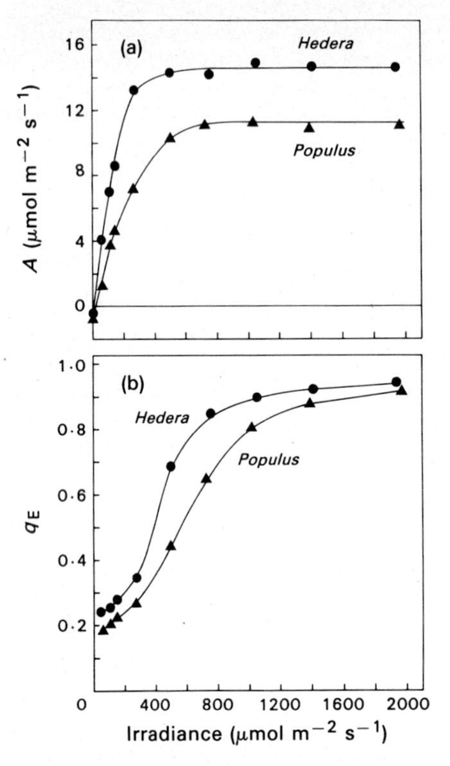

Fig. 4. Light response curves of (*a*) net CO_2 assimilation rate, *A*, and (*b*) total non-photochemical quenching, $q_{E,total}$, in leaves of *Hedera helix* and *Populus balsamifera*. *P. balsamifera* was grown in natural sunlight and received nutrient solution containing 0·6 mM NO_3^- weekly. Leaves of *H. helix* were obtained from a natural population growing at a shaded site.

lower values during the descending curve. Thus, at greatly excessive irradiances, the slowly relaxing component of non-photochemical quenching seemed to be increased *at the expense* of the fast component.

Hysteresis was also observed in sun leaves. The magnitude of $1-q_p$ and q_E appeared to vary depending on the succession of irradiances and on the dwell time at each irradiance (data not shown). With increasing length of time at a given irradiance, q_E increased and $1-q_p$ decreased to steady-state levels. In order to investigate whether the reduction state of *Q* is maintained at low levels only when there is enough time allowed for radiationless dissipation to fully develop, q_E and $1-q_p$ were determined at different times after the onset of illumination. The experiment was performed with leaves of *Glycine max* at an irradiance of 100 μmol $m^{-2} s^{-1}$ in 20 mbar O_2, zero CO_2 in order to largely prevent photosynthetic electron transport (Table 1). Under these conditions, light induces a strong increase in radiationless dissipation but not in electron transport rate. Upon the onset of illumination, the reduction state of *Q* was very high at first, then decreased to approximately 30% where it remained between 30 and 180 min. Non-photochemical quenching was initially very low and then increased for at least 30 min. The rapidly relaxing component of q_E had already attained relatively high values after 10 min and increased only slightly between 10 and 30 min. The slowly relaxing component of q_E was still relatively small after 10 min and continued to increase throughout the treatment.

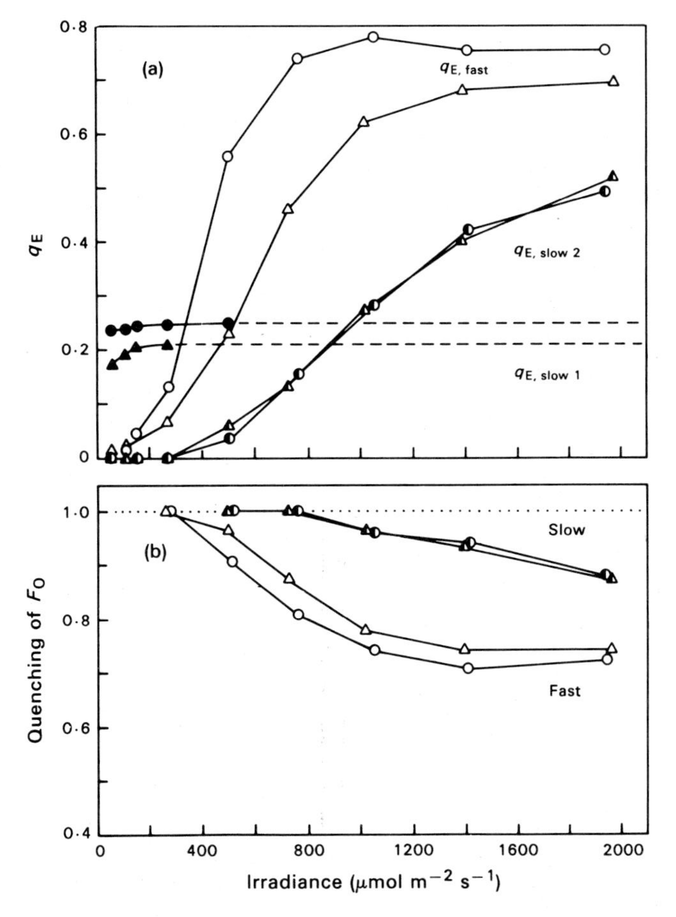

Fig. 5. Light response curves of components of non-photochemical quenching and of F_O quenching in *H. helix* and *P. balsamifera*. The rapidly (within 0–2 min) and slowly (within 2 min to 12 h) relaxing components of non-photochemical quenching were distinguished as described in the caption to Fig. 1. Values for $q_{E,slow\ 2}$ were obtained by subtracting the component present at low irradiance, termed $q_{E,slow\ 1}$, from all values of $q_{E,slow}$. The rapidly relaxing portion of the quenching of $F_O = F_O$ (extrapolated to time zero) / F_O (2 min dark); the slowly relaxing portion of the quenching of $F_O = F_O$ (2 min dark) / F_O (12 h dark). Circles, *H. helix*. Triangles, *P. balsamifera*.

Effect of Photoinhibition on Reduction State of Q *and Radiationless Dissipation*

We studied the changes in q_E and $1-q_P$ which were caused by a 3 h treatment at 2000 μmol photons $m^{-2} s^{-1}$ under conditions preventing both photosynthesis and photorespiration (20 mbar O_2, zero CO_2). Fig. 7 shows the effect of this treatment on various fluorescence characteristics in a sun leaf of *Helianthus annuus*. The ratio F_V/

F_M, determined after 5 min dark at various times throughout the experiment (Fig. 7*a*), was used to indicate changes in photochemical efficiency. Changes in F_V/F_M (5 min dark) induced by photoinhibitory high light treatments have been shown to correlate linearly with photon yield of photosynthetic O_2 evolution (Björkman and Demmig 1987; Demmig and Björkman 1987; Demmig *et al.* 1987). The level of fluorescence excited by the weak measuring beam (F_O) after 5 min darkness is also shown.

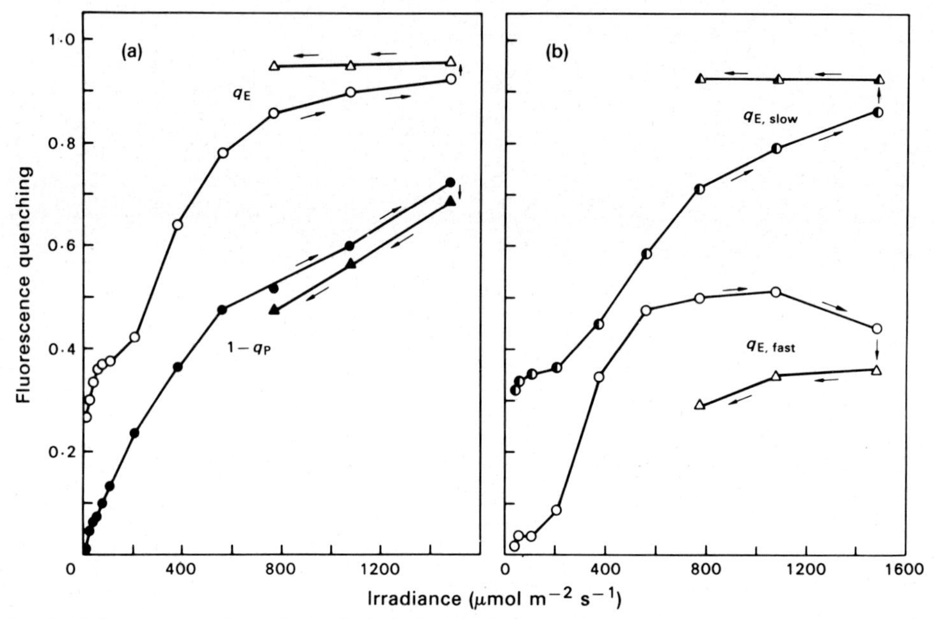

Fig. 6. Light response curves for a shade leaf of *Monstera deliciosa* (*a*) of total non-photochemical quenching, q_E, reduction state of Q, $1-q_P$, and (*b*) of the rapidly, $q_{E,fast}$, and slowly, $q_{E,slow}$, relaxing component of q_E. Following the increase in irradiance to 1480 μmol m^{-2} s^{-1} (circles) (dwell time at each irradiance 30–60 min), irradiance was maintained at 1480 μmol m^{-2} s^{-1} for 1 h and then decreased again (triangles).

Table 1. Changes in fluorescence quenching upon illumination (100 μmol photons m^{-2} s^{-1}) in 20 mbar O_2, zero CO_2

Presented are the reduction state of Q, $1-q_P$, non-photochemical quenching, q_E, as well as components of q_E, one which relaxes rapidly, $q_{E,fast}$, and one which relaxes slowly, $q_{E,slow}$, upon darkening. The experiment was with *Glycine max* grown at 500 μmol photons m^{-2} s^{-1} for 12 h per day in a growth cabinet

Time (min)	$1-q_P$	q_E	$q_{E,fast}$	$q_{E,slow}$
2	0·850	0·075	0·075	0
10	0·710	0·699	0·657	0·189
30	0·322	0·895	0·794	0·492
60	0·298	0·910	0·769	0·609
90	0·311	0·909	0·743	0·646
180	0·340	0·929	0·738	0·729

Prior to the treatment, the leaf was kept at a low irradiance (150 μmol m^{-2} s^{-1}) where F_V/F_M was close to the high value (Fig. 7*a*) typical of healthy non-photoinhibited leaves (Björkman and Demmig 1987), and where both the reduction state of Q (0·03) and total q_E (0·17) were low (Fig. 7*b*). Fig. 7*c* shows slowly and rapidly relaxing components of q_E, calculated as described above. During the high light treatment of sunflower in 20 mbar O_2 and zero CO_2, F_V/F_M declined rapidly and F_O (5 min dark) increased indicating photoinhibitory damage to PS II (Demmig and Björkman 1987).

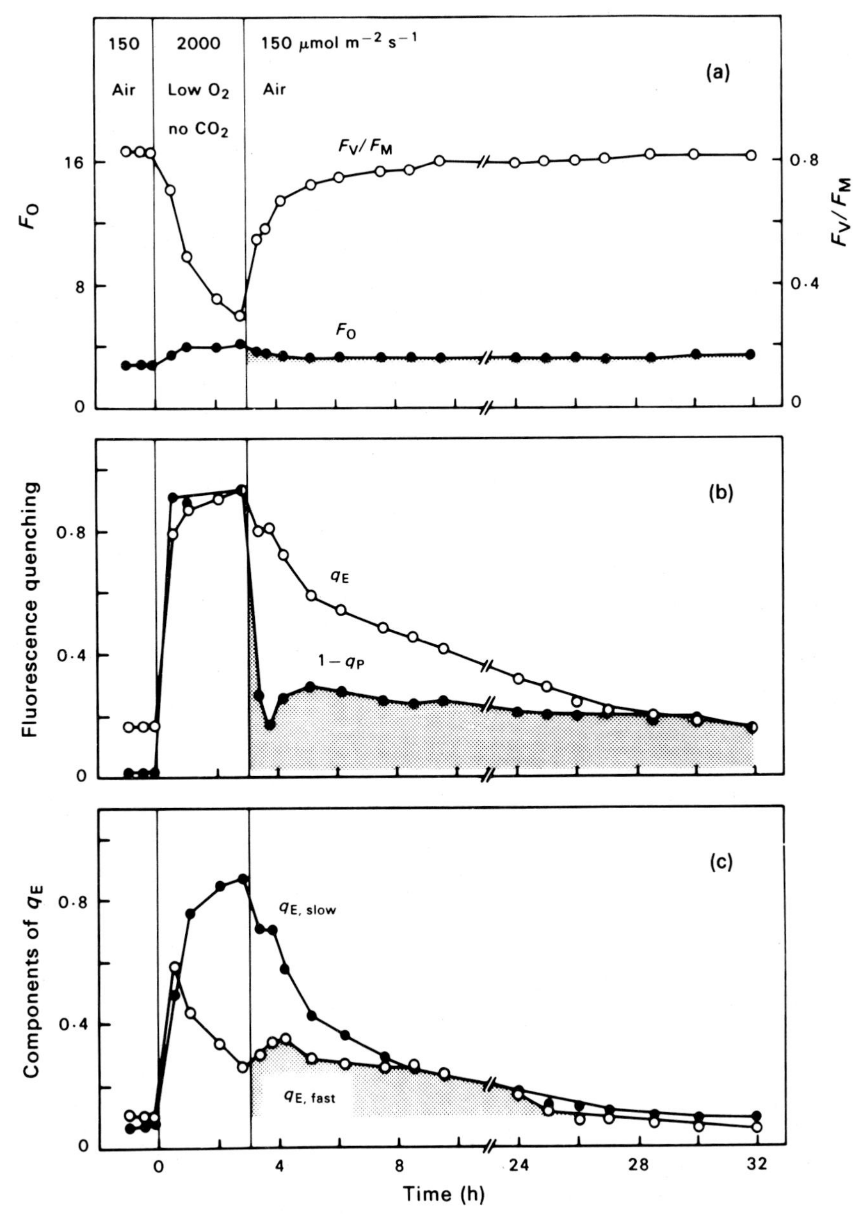

Fig. 7. Time course of changes in F_O, F_V/F_M and fluorescence quenching ($q_{E,total}$, $q_{E,fast}$, $q_{E,slow}$, and $1-q_P$) induced by an exposure to an irradiance of 2000 μmol m^{-2} s^{-1} in 20 mbar O_2, zero CO_2 for a sun leaf of *Helianthus annuus*. Levels of F_O and F_M used to calculate F_V/F_M were determined after 5 min darkness.

The reduction state of Q became extremely high since $1-q_P$ approached unity (Fig. 7*b*). Radiationless energy dissipation was also high. Total q_E was 0·8 after 30 min in 20 mbar O_2, zero CO_2 and increased further to 0·94. This increase in total q_E was initially due to an increase in *both* $q_{E,slow}$ and $q_{E,fast}$ (Fig. 7*c*). After 30 min, $q_{E,fast}$ decreased whereas $q_{E,slow}$ continued to increase.

Upon return to air and 150 μmol photons $m^{-2} s^{-1}$, F_V/F_M increased relatively rapidly from 0·3 to 0·75 within 2–3 h. The reduction state of Q transiently decreased from 90 to 17% within 40 min and then increased again to approximately 30%. Thereafter $1-q_p$ exhibited a very gradual decrease. Total q_E declined very slowly during the period of recovery. The increase in q_E was due to an increase in both $q_{E,slow}$ and $q_{E,fast}$. $q_{E,fast}$ responded in a strikingly similar fashion to the reduction state of Q. After 30 h at 150 μmol photons $m^{-2} s^{-1}$, F_V/F_M had recovered to 95% of the ratio observed at the onset of the experiment. The rate of net CO_2 uptake had also recovered to a similar extent (data not shown). However, the reduction state of Q was still much higher than prior to the treatment. This was due to an increased level of steady-state fluorescence and not due to a reduced level of fluorescence in saturation pulses.

A leaf of *N. oleander* that had been grown in natural sunlight but exhibited a much lower photosynthetic capacity than sunflower was subjected to the same photoinhibitory treatment. Prior to the treatment in 20 mbar O_2 and zero CO_2, the *N. oleander* leaf exhibited characteristics of photochemical and non-photochemical fluorescence quenching similar to the sunflower leaf (Fig. 8): values of F_V/F_M were high (Fig. 8*a*), although not quite as high as in sunflower. The reduction state of Q was low (0·05) and total q_E was small (0·2) (Fig. 8*b*). In contrast to sunflower, $q_{E,fast}$ was zero in *N. oleander* at the onset of the experiment (Fig. 8*c*). During the high light treatment, the reduction state of Q was approximately 95%. As opposed to the sunflower leaf, total q_E had already reached its maximum value when the first measurement during the photoinhibitory treatment was performed after 20 min, i.e. non-photochemical fluorescence quenching set in very strongly and rapidly in *N. oleander*. F_M dropped from 13·4 to 5·4 during the initial 20 min in *N. oleander*. In the sunflower leaf an equally low value of F_M was observed only at the end of the 3-h treatment; after 30 min F_M had dropped from 14·4 to only 11·4. The difference in magnitude of the initial response to high light between sunflower and *N. oleander* was even more pronounced when only the slowly relaxing component of q_E was considered (Figs 7*c* and 8*c*). F_V/F_M also decreased more precipitously in *N. oleander* than in sunflower (Fig. 8*a*). In *N. oleander*, F_V/F_M had decreased to 0·48 after 20 min but only to 0·71 after 30 min in sunflower. In contrast to sunflower, there was no increase in F_O in the leaf of *N. oleander* during the photoinhibitory treatment.

Upon return to air and 150 μmol photons $m^{-2} s^{-1}$, the reduction state of Q, $1-q_p$, attained the low level observed prior to the treatment within approximately 3 h, which is in sharp contrast to the response of the reduction state of Q in sunflower which remained considerably above the control level for more than 30 h. Following the photoinhibitory treatment in *N. oleander*, total q_E remained elevated for extended periods of time which is similar to what was observed in sunflower. This increase in q_E was again due predominantly to an increase in the slowly relaxing component of q_E (Fig. 8*c*). Compared to the control value at the beginning of the experiment, $q_{E,fast}$ remained elevated upon return to air and low irradiance and decreased to zero within 2–3 h. Similar to sunflower, changes in the reduction state of Q, $1-q_p$ and in the fast component of non-photochemical fluorescence quenching closely resembled each other in *N. oleander*. This was so in spite of the marked difference in the kinetics of recovery for $q_{E,fast}$ and $1-q_p$ between these two species with values of $q_{E,fast}$ and $1-q_p$ returning to control levels much faster in *N. oleander* than in sunflower. On the other hand, $q_{E,slow}$ decreased more slowly during the first 3 h of recovery in *N. oleander* than in sunflower.

Discussion

Quenching Mechanisms

Three components of non-photochemical chlorophyll fluorescence quenching were distinguished on the basis of their relaxation kinetics upon darkening. Two com-

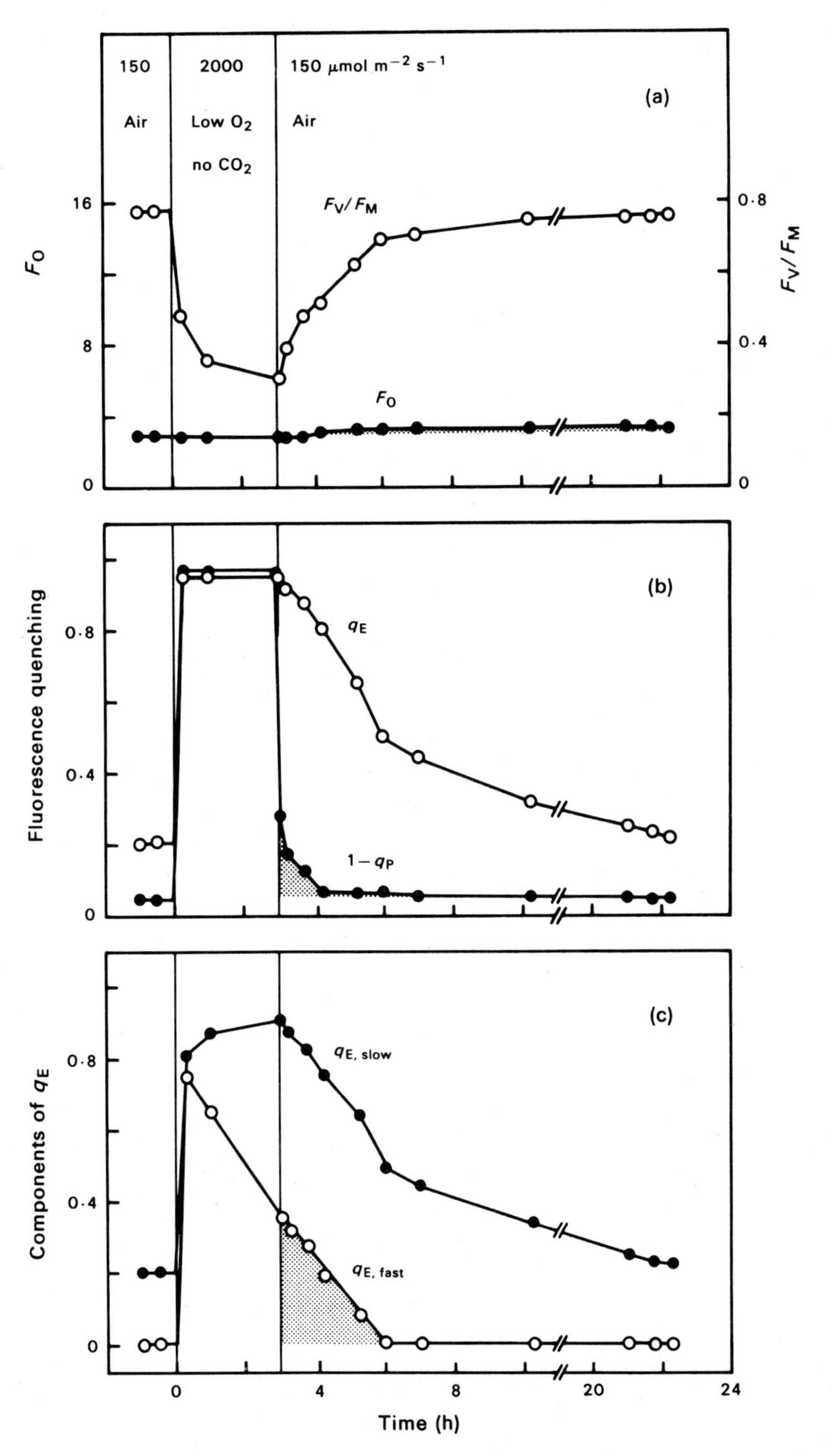

Fig. 8. Time course of changes in F_O, F_V/F_M, and fluorescence quenching ($q_{E,total}$, $q_{E,fast}$, $q_{E,slow}$, and $1-q_P$) induced by exposure to an irradiance of 2000 μmol m^{-2} s^{-1} in 20 mbar O_2, zero CO_2 for a sun leaf of *Nerium oleander*, as in Fig. 7.

ponents, $q_{E,fast}$ and $q_{E,slow\ 2}$, occurred exclusively at excessive irradiance and are considered to reflect major radiationless dissipation processes. These two components also have the same fluorescence characteristics in that they both quench F_O. $q_{E,fast}$, which relaxes almost instantaneously, corresponds to the so-called high energy state quenching (Krause 1973; Briantais *et al.* 1979; Krause *et al.* 1982; Krause and Weis 1984). $q_{E,slow\ 2}$, the more slowly relaxing component, corresponds to the component characterised previously by Demmig and Björkman (1987) as a radiationless energy dissipation process, presumably occurring in the antenna chlorophyll under potentially photoinhibitory conditions. $q_{E,slow\ 1}$ is already present under conditions of limiting light. Various processes are known which cause quenching of fluorescence in low light (Horton and Lee 1983; Staehelin and Arntzen 1983).

(*i*) $q_{E,fast}$, the component of q_E which relaxed almost instantaneously upon darkening, was zero in the photon yield region. It started to increase as soon as the light response curve of CO_2 assimilation deviated from linearity, and reached a maximum when net CO_2 uptake was light-saturated. This rapidly relaxing component is probably the major component of the quenching process characterised by Krause and coworkers, which was shown to relax within 10–30 s (Krause 1973; Krause *et al.* 1982; Krause and Weis 1984), and was linearly related to the acidity of the thylakoid lumen (Briantais *et al.* 1979). Above irradiances at which the maximum capacity for carbon metabolism was reached, both transthylakoid pH (Weis *et al.* 1987) and $q_{E,fast}$ (compare Figs 2*a,c* and 4*a,*5*a*) increased strongly.

(*ii*) $q_{E,slow\ 2}$: At low irradiances, $q_{E,slow\ 2}$ was zero and, beyond the irradiance at which net CO_2 uptake was light saturated, it increased almost linearly with increasing irradiance (Fig. 6*a*). Provided that the component of q_E already present at $A_a = A_p$ ($q_{E,slow\ 1}$) was indeed absent at high irradiances, $q_{E,slow\ 2}$ would continue to increase linearly and not level off as it appears from Fig. 5*a*. A slowly relaxing fluorescence quenching was previously shown to have the characteristics of a heat dissipation process (Demmig and Björkman 1987). This type of fluorescence quenching was suggested to be related to the operation of the xanthophyll cycle in thylakoid membranes (Hager 1969) with zeaxanthin acting as a fluorescence quencher and mediating radiationless dissipation (Demmig *et al.* 1987). In *Lemna gibba*, the zeaxanthin content was shown to be zero or close to zero at limiting irradiances and started to increase when photosynthesis became light-saturated (Siefermann 1972). These findings have been confirmed with a number of species in our laboratory (unpublished data).

There appears to be a close correspondence in the control of the rapidly ($q_{E,fast}$) and slowly relaxing ($q_{E,slow\ 2}$) types of quenching. There is evidence that ΔpH *and* the redox state of 'some' electron carrier exert synergistic control on the formation of zeaxanthin (Hager 1969; Siefermann and Yamamoto 1975), and therefore possibly on $q_{E,slow\ 2}$, as well as on the high-energy state quenching (Oxborough and Horton 1986), the latter corresponding to $q_{E,fast}$. The two quenching processes are also similar in that they both quench F_O fluorescence (Fig. 5*b*). In view of these similarities a common nature of the quenching seems possible. It is very unlikely, however, that reconversion of zeaxanthin to its precursors can account for the rapidly relaxing quenching component since the bulk zeaxanthin content in soybean did not decrease between 5 s and 5 min upon darkening of the leaves (unpublished data). The possibility that movements of zeaxanthin within the membrane may be related to the rapidly relaxing component of nonphotochemical quenching cannot be ruled out. Membrane energisation could affect the distance between the zeaxanthin and chlorophyll molecules.

The range of irradiances over which the two components of fluorescence quenching indicative of heat dissipation, $q_{E,fast}$ and $q_{E,slow\ 2}$, were predominant was not the same: $q_{E,slow\ 2}$ increased only *after* $q_{E,fast}$ had reached its maximum (Figs 2 and 5). Furthermore, $q_{E,fast}$ attained relatively small values. For example in spinach leaves grown at 70 and 140 μmol photons $m^{-2}\ s^{-1}$, $q_{E,fast}$ did not exceed 0·45. All further increases in total q_E were attributable to the slow type of quenching. During prolonged periods at high irradiances, $q_{E,slow\ 2}$ appeared to increase *at the expense* of $q_{E,fast}$ (Fig. 6, see also Figs 7*c* and 8*c*). This behaviour suggests that $q_{E,fast}$ is the major process operating in the daily routine of a plant under conditions where light is not permanently excessive, and that $q_{E,slow\ 2}$ is a process which operates predominantly when leaves are exposed to potentially photoinhibitory conditions for prolonged periods of time (compare also Winter and Demmig 1987 and Demmig and Björkman 1987).

(*iii*) $q_{E,slow\ 1}$ was already present at low irradiances (A_a/A_p). Two processes causing fluorescence quenching have been shown to saturate at low irradiance. Phosphorylation of LHC-2 was observed predominantly at low irradiance (Demmig *et al.* 1987). It results in a decrease in excitation transfer to PS II and, therefore, fluorescence quenching. Furthermore, a type of fluorescence quenching related to an increased rate of electron transport out of PS II, via a dissipative cycle of electrons around PS II, was shown to occur under limiting light and to saturate there (Horton and Lee 1983). $q_{E,slow\ 1}$ may, therefore, be related to a combination of the two types of quenching above.

Hysteresis in Photochemical and Non-photochemical Fluorescence Quenching

It has been suggested that radiationless dissipation processes may compete with the reduction of *Q*, i.e. that partitioning of trapped quanta between photochemistry and radiationless dissipation may be altered in favour of heat dissipation at excessive light levels (Weis *et al.* 1987; Demmig and Winter 1988). However, it would be expected that the reduction state of *Q* would be low only if (1) enough time was allowed for radiationless dissipation processes (indicated by q_E) to develop fully and if (2) the capacity for heat dissipation had not yet been exceeded. Such phenomena may explain why *either* values of $1-q_p$ *or* values of q_E may not be the same at a given irradiance, depending on the succession of irradiances and on the dwell time at each irradiance. The treatment shown in Table 1 clearly demonstrates that the reduction state of *Q* was very high immediately after the onset of illumination as long as q_E had not yet reached its maximum steady-state value, and that the subsequent increase in q_E, particularly of the slowly relaxing component, was accompanied by a decrease in the reduction state of *Q*. Calculation of the rate of radiationless dissipation according to Kitajima and Butler (1975) shows that the initial increase in q_E overestimates the true rate of heat dissipation whereas increases of q_E close to its maximum possible value underestimate this rate. This point will be dealt with in detail in a forthcoming paper.

The above findings clearly suggest that a low reduction state of *Q* is maintained by means of increased radiationless dissipation, since at 20 mbar O_2, zero CO_2 the observed decrease in $1-q_p$ cannot be caused by a strong increase in electron transport rate (as would be the case during dark/light transitions in air). The treatment in 20 mbar O_2, zero CO_2 was performed at low irradiance (100 μmol m^{-2} s^{-1}) because under these conditions fluorescence quenching is promoted without photoinhibitory damage (compare Demmig and Björkman 1987). Consistent with the finding shown in Table 1 are observations that, during sudden large increases in irradiance, $1-q_p$ was well correlated with changes in O_2 evolution rate (M. Stitt and U. Schreiber, personal communication) whereas in the steady state this was not the case (Demmig and Winter 1988). The reduction state of *Q* was also found to be a good reflection of assimilation rate in water-stressed tissues, under conditions where q_E was maximal, i.e. where the capacity of radiationless dissipation was exceeded (Schreiber and Bilger 1987).

Photoinhibition

In contrast to the above phenomena, photoinhibitory high light treatments resulted in *both* an increased reduction state of *Q* and a pronounced increase in q_E which, depending on species, persisted for various lengths of time during a subsequent period in non-photoinhibitory low light (Figs 7 and 8). This increase in $1-q_P$ was not fully reversible in sunflower within 30 h and is probably related to photoinhibitory damage. The increase in the reduction state of Q_A may be explained by a reduction in the number of electron acceptor sites of Q_B (Ohad *et al.* 1984). If this is so, one may speculate that $q_{E,fast}$, which changes in a fashion very similar to $1-q_P$ subsequent to the photoinhibitory high light treatment, is also sensitive to the Q_A/Q_B ratio. Alterations in connectivity among PS II centres could be another reason for the increase in $1-q_P$ (Joliot and Joliot 1964).

There were clear differences between the photoinhibitory responses of sunflower and *N. oleander*. In *N. oleander* radiationless energy dissipation, as indicated by q_E, set in more strongly than in sunflower (Figs 7*b* and 8*b*); this was particularly noticeable in the slowly relaxing type of q_E. It is interesting that these high-light grown *N. oleander* leaves contained a background level of zeaxanthin whereas sunflower did not contain any zeaxanthin at low light levels (unpublished data). Furthermore *N. oleander*, unlike sunflower, showed a capacity for rapid formation of additional amounts of zeaxanthin at the expense of β-carotene. F_V/F_M also fell much more precipitously in *N. oleander* without, however, being accompanied by a net increase in F_O which would be indicative of photoinhibitory damage to PS II reaction centres (compare Demmig and Björkman 1987). These findings indicate that the drop in F_V/F_M in *N. oleander* was due predominantly to a rapid increase in radiationless energy dissipation, and that there was less photoinhibitory damage in *N. oleander* than in sunflower. In *N. oleander*, unlike in sunflower, the reduction state of Q and $q_{E,fast}$ returned to control levels within 2–3 h after termination of the photoinhibitory high light treatment. The same conclusion, that most of the pronounced reduction in photochemical efficiency in *N. oleander* can be attributed to a strong increase in the rate of radiationless dissipation and not to damage of PS II reaction centres, was drawn from experiments in which photoinhibition in *N. oleander* was induced by prolonged exposure to high light and water stress (Björkman 1987). The nature of the more gradually developing further decrease of F_M during the photoinhibitory treatment in sunflower (Fig. 7*c*) is unclear. The possibility that it represents radiationless dissipation, as in *N. oleander*, cannot be excluded. It is more likely, however, that the gradual decrease of F_M accompanied by an increase in F_O is related to slowly developing photoinhibitory damage which may occur when the capacity for zeaxanthin formation is exceeded (Demmig *et al.* 1987).

Acknowledgments

This work was supported by the Deutsche Forschungsgemeinschaft. We are grateful to W. W. Adams III, W. Bilger, E. Weis, and U. Heber for critical comments on the paper.

References

Björkman, O. (1987). High-irradiance stress in higher plants and interaction with other stress factors. In 'Progress in Photosynthesis Research'. (Ed. J. Biggins.) Vol. 4, pp. 11–18. (Martinus Nijhoff: Dordrecht.)

Björkman, O., and Demmig, B. (1987). Photon yield of O_2 evolution and chlorophyll fluorescence characteristics at 77K among vascular plants of diverse origins. *Planta* **170**, 489–504.

Bradbury, M., and Baker, N. R. (1984). A quantitative determination of photochemical and nonphotochemical quenching during the slow phase of the chlorophyll fluorescence induction curve in bean leaves. *Biochim. Biophys. Acta* **765**, 275–81.

Briantais, J. M., Vernotte, C., Picaud, M., and Krause, G. H. (1979). A quantitative study of the slow decline of chlorophyll *a* fluorescence in isolated chloroplasts. *Biochim. Biophys. Acta* **548**, 128–38.

Demmig, B., and Björkman, O. (1987). Comparison of the effect of excessive light on chlorophyll fluorescence (77K) and photon yield of O_2 evolution in leaves of higher plants. *Planta* **171**, 171–84.

Demmig, B., Cleland, R. E., and Björkman, O. (1987). Photoinhibition, 77K chlorophyll fluorescence quenching and phosphorylation of LHC-II. *Planta* **172**, 378–85.

Demmig, B., and Winter, K. (1988). Light response of CO_2 assimilation, reduction state of *Q*, and radiationless energy dissipation in intact leaves. *Aust. J. Plant Physiol.* **15**, 151–62.

Demmig, B., Winter, K., Krüger, A., and Czygan, F.-C. (1987). Photoinhibition and zeaxanthin formation in intact leaves. A possible role of the xanthophyll cycle in the dissipation of excess light energy. *Plant Physiol.* **84**, 218–24.

Dietz, K.-J., Schreiber, U., and Heber, U. (1985). The relationship between the redox state of Q_A and photosynthesis in leaves at various carbon dioxide, oxygen and light regimes. *Planta* **166**, 219–26.

Hager, A. (1969). Lichtbedingte pH-Erniedrigung in einem Chloroplasten-Kompartiment als Ursache der enzymatischen Violaxantin/Zeaxanthin-Umwandlung; Beziehungen zur Photophosphorylierung. *Planta* **89**, 224–43.

Horton, P., and Lee, P. (1983). Stimulation of a cyclic electron-transfer pathway around photosystem II by phosphorylation of chloroplast thylakoid proteins. *FEBS Lett.* **162**, 81–4.

Joliot, A., and Joliot, M. P. (1964). Étude cinétique de la réaction photochimique libérant l'oxygène au cours de la photosynthèse. *C.R. Acad. Sci. Ser. D* **258**, 4622–5.

Kitajima, M., and Butler, W. L. (1975). Quenching of chlorophyll fluorescence and primary photochemistry in chloroplasts by dibromothymoquinone. *Biochim. Biophys. Acta* **376**, 105–15.

Krause, G. H. (1973). The high-energy state of the thylakoid system as indicated by chlorophyll fluorescence and chloroplast shrinkage. *Biochim. Biophys. Acta* **333**, 301–13.

Krause, G. H., and Weis, E. (1984). Chlorophyll fluorescence as a tool in plant physiology. II. Interpretation of fluorescence signals. *Photosynth. Res* **5**, 139–57.

Krause, G. H., Vernotte, C., and Briantais, J. M. (1982). Photoinduced quenching of chlorophyll fluorescence in intact chloroplasts and algae. Resolution into two components. *Biochim. Biophys. Acta* **679**, 116–24.

Ohad, I., Kyle, D. J., and Arntzen, C. J. (1984). Membrane protein damage and repair: removal and replacement of inactivated 32-kilodalton polypeptides in chloroplast membranes. *J. Cell Biol.* **99**, 481–5.

Oxborough, K., and Horton, P. (1986). Characterization of the effects of antimycin A upon the yield of chlorophyll fluorescence in spinach chloroplasts. *Research Institute for Photosynthesis (Sheffield) Annual Report* pp. 10–12.

Schreiber, U., and Bilger, W. (1987). Rapid assessment of stress effects on plant leaves by chlorophyll fluorescence measurements. In 'Plant Response to Stress — Functional Analysis in Mediterranean Ecosystems'. (Eds J. D. Tenhunen, F. M. Catarino, O. L. Lange and W. C. Oechel.) pp. 27–53. (Springer-Verlag: Berlin.)

Siefermann, D. (1972). Kinetic studies on the xanthophyll cycle of *Lemna gibba* L. Influence of photosynthetic oxygen and supplied reductor. In 'IInd International Congress on Photosynthesis'. Stresa 1971. (Eds G. Forti, M. Avron and A. Melandri.) Vol. 1, pp. 629–35. (Dr Junk: The Hague.)

Siefermann, D., and Yamamoto, H. Y. (1975). Light-induced de-epoxidation of violaxanthin in lettuce chloroplasts. IV. The effect of electron transport conditions on violaxanthin availability. *Biochim. Biophys. Acta* **387**, 149–58.

Staehelin, L. A., and Arntzen, C. J. (1983). Regulation of chloroplast membrane function: protein phosphorylation changes the spatial organization of membrane components. *J. Cell Biol.* **97**, 1327–37.

Weis, E., Ball, J. T., and Berry, J. (1987). Photosynthetic control of electron transport in leaves of *Phaseolus vulgaris*: evidence for regulation of photosystem 2 by the proton gradient. In 'Progress in Photosynthesis Research'. (Ed. J. Biggins.) Vol. 2, pp. 553–6. (Martinus Nijhoff: Dordrecht.)

Winter, K., and Demmig, B. (1987). Reduction state of *Q* and nonradiative energy dissipation during photosynthesis in leaves of a crassulacean acid metabolism plant, *Kalanchoë daigremontiana* Hamet et Perr. *Plant Physiol.* **85**, 1000–7.

Effects of Photoinhibition Induced by Water Stress on Growth and Yield of Grain Sorghum

M. M. Ludlow[A] *and S. B. Powles*[B]

[A]Division of Tropical Crops and Pastures, CSIRO, 306 Carmody Road, St Lucia, Qld 4067, Australia.
[B]Department of Agronomy, Waite Agricultural Research Institute, University of Adelaide, Glen Osmond, S.A. 5064, Australia.

Abstract

The significance of photoinhibition induced by water stress for growth and grain yield of sorghum was investigated by shading water-stressed plants for a 7-day period during panicle development. Thereafter, shading was removed, stressed plants were rewatered, and all plants were kept well-watered until physiological maturity.

Shading water-stressed plants for a 7-day period increased grain number and grain yield of primary heads by 10- and 13-fold, respectively. Even though plants were severely stressed, photoinhibition was only 20%. Moreover, reduction of photoinhibition by shading did not improve carbon supply to the developing panicle. Therefore, the response of grain yield to shading was attributed to a 15% higher average leaf water potential in shaded plants, which arose despite attempts to keep water status similar in shaded and unshaded plants.

The effect of water stress on the primary head was so severe that many plants produced secondary heads. The grain yield from these secondary heads compensated for the low yield of primary heads so that the total dry matter production and total grain yield of water-stressed plants were similar to those of unstressed plants. In view of the ability of this relatively determinate species to compensate for potential yield losses, as well as the mechanisms to avoid water stress, high temperatures and excessive absorption of light, it seems unlikely that photoinhibition induced by water stress will be of agronomic significance in grain sorghum.

Introduction

Water stress, like many other environmental and edaphic stresses, can predispose the primary photosynthetic reactions of chlorophyll-bearing tissues to damage by excess light (Björkman and Powles 1984; Ludlow and Björkman 1984; Powles 1984). This damage is called photoinhibition and it is similar to that incurred if shade plants are suddenly exposed to bright light or if leaves of sun plants are exposed to bright light when carbon metabolism is impaired. Photoinhibition reduces the quantum yield of photosynthesis and, to a lesser extent, the rate of light-saturated photosynthesis (Powles and Björkman 1982; Björkman and Demmig 1987; Björkman 1987; Demmig and Björkman 1987). Moreover, because photoinhibition alters the chlorophyll fluorescence characteristics of leaves (Björkman 1987), these characteristics can be used as a sensitive probe of the effect of photoinhibition on photochemical activity.

Whilst photoinhibition associated with water stress has been recorded in several vascular plants in the field (Björkman and Powles 1984; Ludlow and Björkman 1984; Adams *et al.* 1987; Björkman and Demmig 1987), there has been no attempt to assess the ecological or agronomic significance for growth, survival or yield (Ludlow 1987). Because plants have evolved with the dual stresses of water shortage and high light, it is likely that they have evolved characteristics to resist these stresses. In fact, plants both

0310-7841/88/010179$03.00

avoid and tolerate water stress (Hsiao 1973; Levitt 1980; Turner 1986) as well as avoiding or tolerating photoinhibition (Powles 1984; Ludlow 1987). Furthermore, before photoinhibition becomes appreciable, growth and photosynthesis are already severely inhibited by water stress (Björkman and Powles 1984; Ludlow and Björkman 1984). Consequently, any additional reduction in growth and photosynthesis due to photoinhibition may be relatively minor, unless it causes leaf death or impairs the photosynthetic capacity of leaves after water stress is relieved.

This paper describes a first attempt to assess the *agronomic* significance of photoinhibition induced by water stress for grain yield in sorghum grown outdoors during the summer in a subtropical environment. Grain sorghum was chosen for two main reasons. Firstly, it is grown extensively under dryland conditions in the semiarid tropics of the world, where it frequently experiences high light, high temperature and water stress (Doggett 1970; Seetharama *et al.* 1982). High temperature exacerbates photoinhibition induced by water stress (Ludlow and Björkman 1984; Ludlow 1987). Secondly, grain sorghum is relatively determinate. This minimises the possibility of subsequent flowering and grain production occurring after stress is relieved, thus compensating for damage caused by the stress event. Water stress was imposed in a period 2 weeks prior to anthesis when the panicle is developing and when the carbon supply has the greatest effect on grain yield by reducing grain number (Inuyama *et al.* 1976). We argued that, if grain yield could not be enhanced by reducing photoinhibition and improving the carbon supply to the developing heads at this critical stage of panicle growth, it was less likely to be important at other stages of development.

Photoinhibition was induced by withholding water from grain sorghum plants in full sunlight, and photoinhibition was reduced for a 7-day period by shading half of the water-stressed plants. The reduction of growth of shaded plants associated with lowered irradiance, independent of the relief of photoinhibition, was assessed by comparing well-watered plants in full sunlight with those that were also shaded for the 7-day period. Shading can improve the water status and lower the leaf temperature of water-stressed plants. Therefore, leaf temperature was monitored in all four treatments. In addition, leaf water potential was measured twice a day and small amounts of water were added to water-stressed plants in full sunlight in an attempt to keep their leaf water status similar to the water-stressed plants in the shade.

Materials and Methods

Plant Culture, Treatments and Chronology of Experiment

Seeds of grain sorghum (*Sorghum bicolor* cv. Texas 610) were sown on 27 December in pots, 0·25 m diameter and 1 m deep, containing 50 kg of air-dried soil (an Arguistoll; Soil Survey Staff 1975) at Brisbane, Queensland, Australia (27°S., 152°E.). Fertiliser was added to overcome soil deficiencies, and the pots were watered daily and allowed to drain. In addition, all plants were fertilised at a rate of 50 kg ha^{-1} of N on day 22 after sowing. The pots were maintained close together to give a density equivalent to 90 000 plants per hectare.

Twenty-four days after sowing, water was withheld from half the plants, and pots were sealed to prevent rainfall from entering the soil (Table 1). The remainder of the plants were kept well-watered. Fourteen days later when unwatered plants were moderately stressed, half of these plants and half of the unstressed, control plants were shaded with Sarlon shade cloth with a 40% transmission of photon irradiance. At the same time, five unstressed and five stressed plants were harvested, separated into leaves, stems, heads and roots, and leaf area and dry weight were determined. For the next 7 days, the following measurements were made: daily photon irradiance (LiCor photon sensor, model no. 190S and LiCor printing integrator, model LI 550); leaf temperature at 1400 h (LiCor null balance porometer, model 1600, and Heimann infrared thermometer, model KT24); leaf diffusive conductance of the abaxial surface at 1400 h (LiCor, null balance porometer, model 1600); leaf water potential at 0900 h and 1400 h using an automatic scanning dew point hygrometer (Wescor, model HR3000) and 16 sample chambers (Wescor, model CR52); chlorophyll fluorescence at 77K was measured on samples taken at noon and held in the dark for at least 20 min, using a technique similar to that described by Powles and Björkman (1982); and photon yield based on incident photon flux using a leaf disc oxygen electrode (Hansatech, model No. LD2). Pots were weighed every second day, and based on these results and leaf water potentials measured at 0900 and 1400 h, small amounts of water (*c*. 100 ml) were added to water-stressed plants in full sunlight in an attempt to keep their leaf water status similar to shaded, stressed plants.

Table 1. Chronology of experiment, crop phenology and timing of treatments and harvests for dry matter production and grain yield of sorghum

Plants were sown on 27 December

Days after sowing	Crop phenology	Treatments	Harvest No.
17	Floral initiation[A]		
24		Water withheld from half plants	
39		Half control and half water-stressed plants shaded	1
46	Beginning of head emergence	Shades removed and water-stressed plants rewatered	2
52	50% anthesis of primary heads		
73	50% anthesis of secondary heads		
96	Physiological maturity of primary heads		3
116	Physiological maturity of secondary heads		4

[A]Estimated from unpublished data of M.A. Foale.

On day 46, when heads were beginning to emerge and unwatered plants were severely stressed, shading was removed from both unstressed and stressed plants, and stressed plants were rewatered (Table 1). All plants were kept well-watered thereafter until physiological maturity. Physiological measurements continued for 4 days after rewatering to characterise the ability to recover from severe water stress. On days 3 and 9 after rewatering, the recovery of photosynthetic capacity of leaves was assessed by measuring $^{14}CO_2$ uptake (McWilliam *et al.* 1973). Six days after rewatering, heads on main stems of plants (primary heads) were at 50% anthesis (Table 1). However, because the water stress treatment had been severe, some primary stems on both full sun and shaded plants from the water stress treatment either failed to produce a head or produced very small heads. Consequently, secondary heads were produced by branching of the main stem. These heads reached 50% anthesis on day 73 after sowing. Physiological maturity occurred on days 96 and 116 after sowing, respectively, for primary and secondary heads.

All primary heads were harvested on day 96, but only those plants without secondary heads were harvested. The remainder were not harvested until the secondary heads reached physiological maturity. At both harvests, plants were separated as before, and grain number and grain weight (oven dry) were determined. The delay in determining biomass in plants with secondary heads meant that there was an unequal number of replicates at harvests 3 and 4. Consequently, analyses of variance could not be performed, and biomass of water-stressed plants in full sunlight at harvest 3 had to be estimated by taking the biomass at harvest 4 and adjusting it for the relationship between the shaded water-stressed plants at harvests 3 and 4. Where analysis of variance could not be performed, the means (± standard error) are given. Initially, there were five replicates of each water stress treatment (unstressed and water-stressed) and of each shading treatment (full sun and shade) for each harvest.

Results and Discussion

Photon Irradiance and Leaf Temperature

Four of the 7 days over which plants were either shaded or in full sunlight had very high photon irradiances, whereas the other three had about half as much (Fig. 1). Shading did not lower leaf temperature measured at 1400 h with a thermocouple in the null balance porometer compared with plants in full sunlight (Fig. 1), except on day 2. The temperature of water-stressed leaves in sun and shade treatments never differed by more than 2°C and, on average (±s.e.), were not significantly different for the 7-day period (32·4 ± 0·6°C and 32·6 ± 0·6°C for sun and shade, respectively). Leaf temperature of control plants was also not influenced by shading, but was on average 1°C cooler than water-stressed plants (31·6 ± 0·6°C and 31·5 ± 0·5°C for sun and shade, respectively). On days 1 and 5, leaf temperature (°C) was also measured with an infrared thermometer and the data (mean ± standard errors) of four determinations are given below:

	Sun		Shade	
	Unstressed	Water-stressed	Unstressed	Water-stressed
Day 1	31·9 ± 0·2	33·5 ± 0·3	31·5 ± 0·1	33·2 ± 0·1
Day 5	30·9 ± 0·1	33·7 ± 0·8	30·9 ± 0·3	33·0 ± 0·2

Again, there were no significant differences between sun and shade leaves of either unstressed or water-stressed plants, but unstressed leaves were *c.* 2°C cooler than water-stressed leaves. Shading, therefore, did not result in lower leaf temperature compared to leaves in full sunlight. Moreover, although we did not measure head temperature, it seems unlikely that it would have been differentially affected by shading as well. None of the values of leaf temperature recorded should have been harmful to grain sorghum, which has an optimum temperature for leaf photosynthesis of 38°C (Ludlow and Wilson 1971) and a lethal leaf temperature of 52–55°C (M. Paje, M. M. Ludlow and J. M. Peacock, unpublished data).

Water Use and Leaf Water Potential

Pots containing shaded plants were about 400 g lighter than those in full sunlight at the beginning of the 7-day period, and this difference was maintained such that final weights were 53·9 ± 0·5 (s.e.) and 54·3 ± 0·06 (s.e.) kg at the end of the period, respectively, for shade and sun treatments. Sun plants received four additions of 100 ml H_2O and used an average 141 ml per day, whereas the shade plants used 120 ml per day; sun plants, therefore, used about 20% more water. Both groups used water at a constant rate during the period. Consequently, neither treatment had exhausted the available soil water when plants were rewatered.

Plants were moderately stressed by 39 days after sowing, when the shading treatments began. Leaves were rolled and leaf water potential measured at 1400 h was −2·4 MPa. Values for unstressed plants were −0·6 to −0·9 MPa for the duration of the experiment. When water-stressed plants were shaded on day 0, leaf water potential recovered to a level 1 MPa higher than that in full sunlight at both 0900 (data not shown) and 1400 h (Fig. 1). Small amounts of water were added to water-stressed plants in full sunlight so that their leaf water potentials (±s.e.) were similar at 0900 h on day 1 (−1·92±0·12 and −1·82±0·42 MPa for sun and shade, respectively) and day 2 (−1·57±0·16 and −1·63±0·30 MPa for sun and shade, respectively). However 1400 h values were lower for sun than for shade leaves (Fig. 1). Fluctuating environmental conditions (Fig. 1) on days 0–2 made matching leaf water potentials very

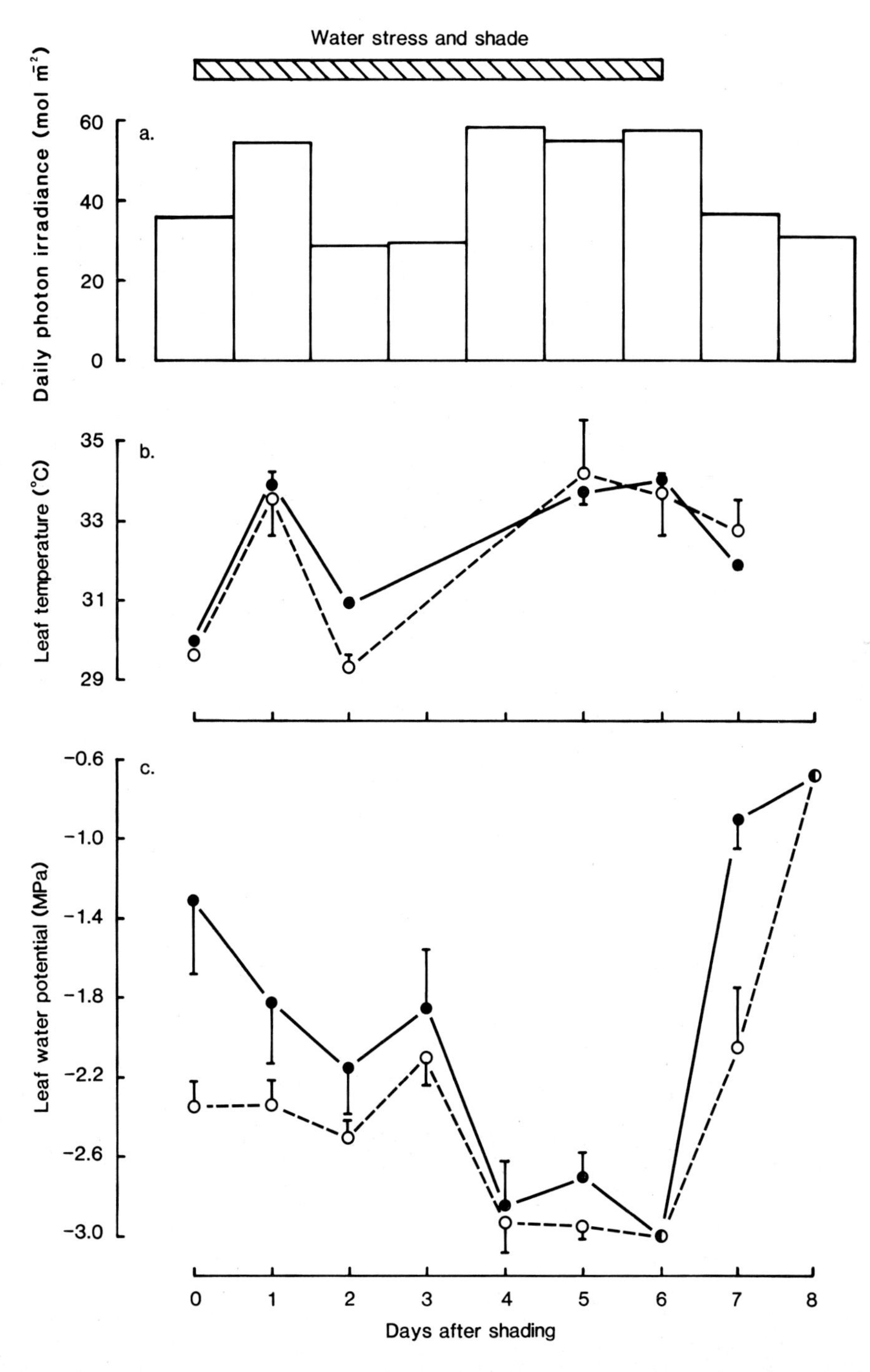

Fig. 1. Trends with time after water-stressed plants were shaded to 40% of sunlight (●) or left in full sunlight (○) of (*a*) daily photon irradiance, (*b*) leaf temperature and (*c*) leaf water potential measured at 1400 h. On day 6, shades were removed and water-stressed plants were rewatered. Thereafter, all plants were kept well-watered. Leaf temperatures and leaf water potential of unstressed plants that were in full sunlight or shaded are not presented, but are quoted in the text. Standard errors, where they exceed the size of the symbols, of five replicates are shown.

difficult. Although leaves in full sunlight were always lower in leaf water potential than those in shade, there was no significant difference for the last 4 days of the shading treatment. The average of the 0900 and 1400 h leaf water potentials in full sunlight over the 7 days (−2·44 ± 0·13 MPa) was 15% lower than that of shaded leaves (−2·12 ± 0·17 MPa), suggesting leaves in the sun had a higher level of stress. During the last 3 days of water stress, leaf water potentials were −2·7 to −3·0 MPa, which approached the lethal leaf water potential (−3·1 MPa) of the youngest fully expanded leaf of this cultivar (Santamaria 1986). Moreover, 25% of the leaves of water-stressed plants were dead on day 6 in both sun (3·6 ± 0·2 leaves) and shade (3·8 ± 0·2), indicating the severity of water stress. In contrast, unstressed plants had no dead leaves. Consequently, plants were severely stressed before they were rewatered. When shades were removed and all plants rewatered, the leaf water potential of previously shaded plants recovered more rapidly than in those that had been in full sunlight. However, by the second day, all plants had recovered to the level of unstressed controls (Fig. 1).

In general, leaf water potentials of plants in full sunlight and in the shade were similar over the 7-day shading treatment. However, leaves in full sunlight had a lower potential for the first 3 days and were, on average, 15% more stressed than those in the shade over the 7 days.

Stomatal Conductance and $^{14}CO_2$ *Uptake*

We could not measure leaf net photosynthetic rate. Consequently, we have used stomatal conductance as an estimate of photosynthetic activity, based on the known close relationship between them (Wong *et al.* 1979), especially when water stress develops slowly.

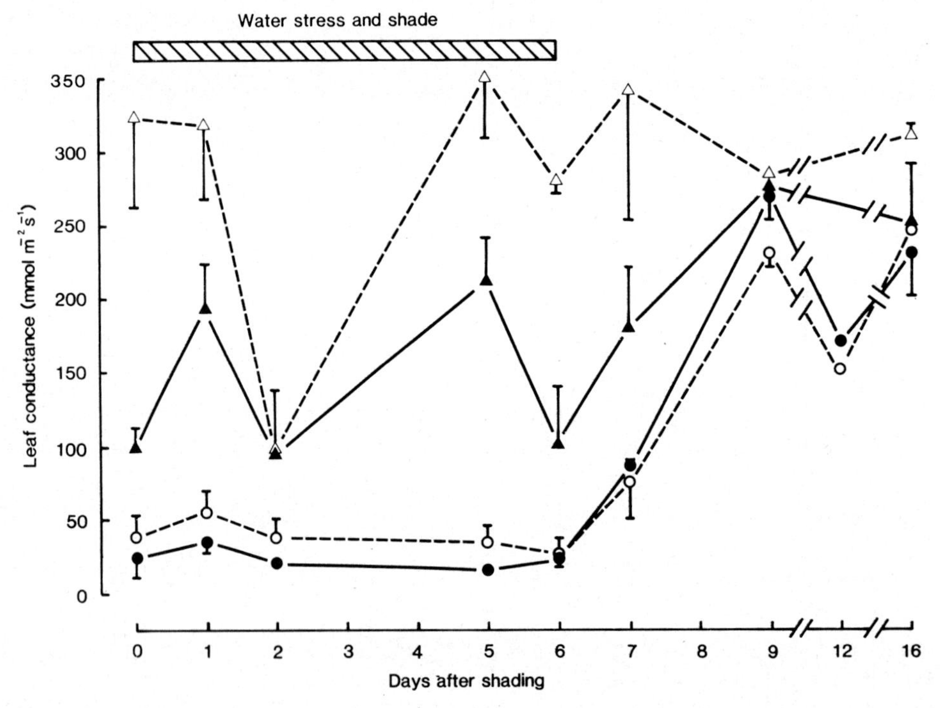

Fig. 2. Conductance of the abaxial leaf surface from well-watered plants (triangles) and water-stressed plants (circles) either shaded (closed symbols) or in full sunlight (open symbols) for a 7-day period, after which shades were removed and unstressed plants were rewatered. Bars represent ± standard error of the mean of five replicates.

Shading reduced the stomatal conductance (g_s) of unstressed plants by about 50% (Fig. 2). Day to day fluctuations in photon irradiance were responsible for the temporal variations in g_s of both shaded and unshaded plants. Stomatal conductance of water-stressed plants in both sun and shade, which were similar, was about 10% of the value of controls in full sunlight, and was constant during the whole 7 days. During the first 3 days when leaf water potential was 0·4–0·9 MPa lower in unshaded plants, g_s was higher (not lower) than in shaded plants, suggesting that shading had a bigger effect than leaf water potential on g_s, even on these water-stressed plants. When water-stressed plants were rewatered and shades removed, g_s of both water-stressed treatments recovered rapidly, and in 3 days were equivalent to unstressed values.

Shading plants to reduce photoinhibition did not result in higher stomatal conductance, and hence by analogy higher leaf photosynthetic rate, either during the period of water stress or after stressed plants were rewatered. $^{14}CO_2$ uptake rate of leaves held horizontal was measured at noon on clear days, 3 and 9 days after water-stressed plants were rewatered and shades were removed. Values ($\pm$s.e.; $n = 8$ for control and 20 for stressed; mg CO_2 m^{-2} s^{-1}) averaged for the 2 days are given below:

Sun		Shade	
Control	Water Stress	Control	Water Stress
1·92 ± 0·28	1·61 ± 0·10	2·32 ± 0·14	2·18 ± 0·13

Rates of $^{14}CO_2$ uptake for water-stressed leaves previously shaded had recovered to control values, but leaves water-stressed in the sun achieved only 84% of the rate of leaves from unstressed plants. Leaves from plants water-stressed in full sunlight had significantly ($P<1\%$ level) lower $^{14}CO_2$ uptake rates (74%) than comparable leaves that had been shaded for the 7-day period. The lower rates for unshaded plants may be due to greater photoinhibition (presented later) or to the greater degree of water stress (Fig. 1) during this period. These previously unshaded leaves had a nitrogen concentration 90% that of previously shaded leaves, which may account for part of the lower rates.

Leaf Rolling

One of the most obvious responses of grain sorghum leaves to water stress prior to anthesis was rolling (Fig. 3). Expanding leaves cupped, whereas fully expanded leaves exhibited varying degrees of rolling. These movements resulted in the adaxial leaf sur-

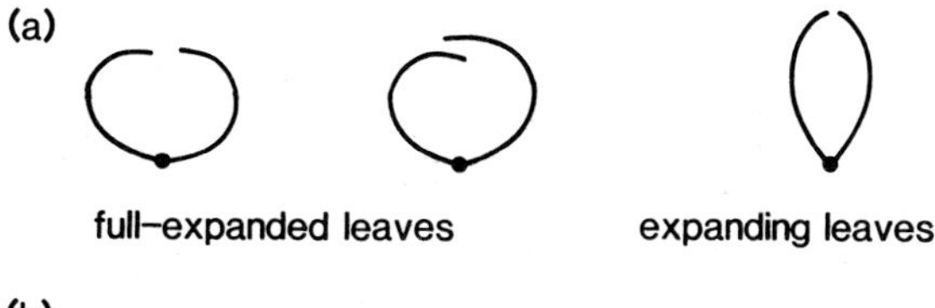

Fig. 3. Examples of (*a*) rolled fully expanded and expanding leaves, and (*b*) the unrolling of water-stressed leaves after rewatering.

faces being fully or partially protected and the normally shaded abaxial surface being exposed. After water-stressed plants were rewatered, the exposed abaxial surfaces were again shaded as leaves unrolled and the protected adaxial surfaces were exposed to full sunlight. In addition to undergoing recovery from water deficit, the need for cells of the adaxial surface to reacclimate to the higher light may have slowed recovery of leaf photosynthetic rate after rewatering.

As well as reducing the amount of radiation intercepted by individual leaves, rolling resulted in protection of the adaxial surface, which was subsequently exposed to full sunlight when water-stressed plants were rewatered and leaves unrolled.

Photoinhibition

Photoinhibition was characterised mostly using chlorophyll fluorescence measured at 77K and, to a lesser extent, using the quantum yield of O_2 evolution measured with a leaf disc oxygen electrode. Björkman (1987) and Björkman and Demmig (1987) have shown that the quantum yield of O_2 evolution is linearly and positively related to the ratio of variable to maximum fluorescence (F_V/F_M) of photosystem II. This ratio is remarkably constant for both surfaces of uninhibited leaves among higher plants. However, to remove any direct effects (as opposed to indirect effects due to photoinhibition induced by water stress) of water deficits (Adams *et al.* 1987) or leaf nitrogen content (M. M. Ludlow, P. S. M. Sirisena and J. R. Wilson, unpublished data) on F_V/F_M, we have calculated the ratio of the F_V/F_M for upper (exposed abaxial) compared to lower (protected adaxial) leaf surfaces (Fig. 4). The degree of photoinhibition is seen by the

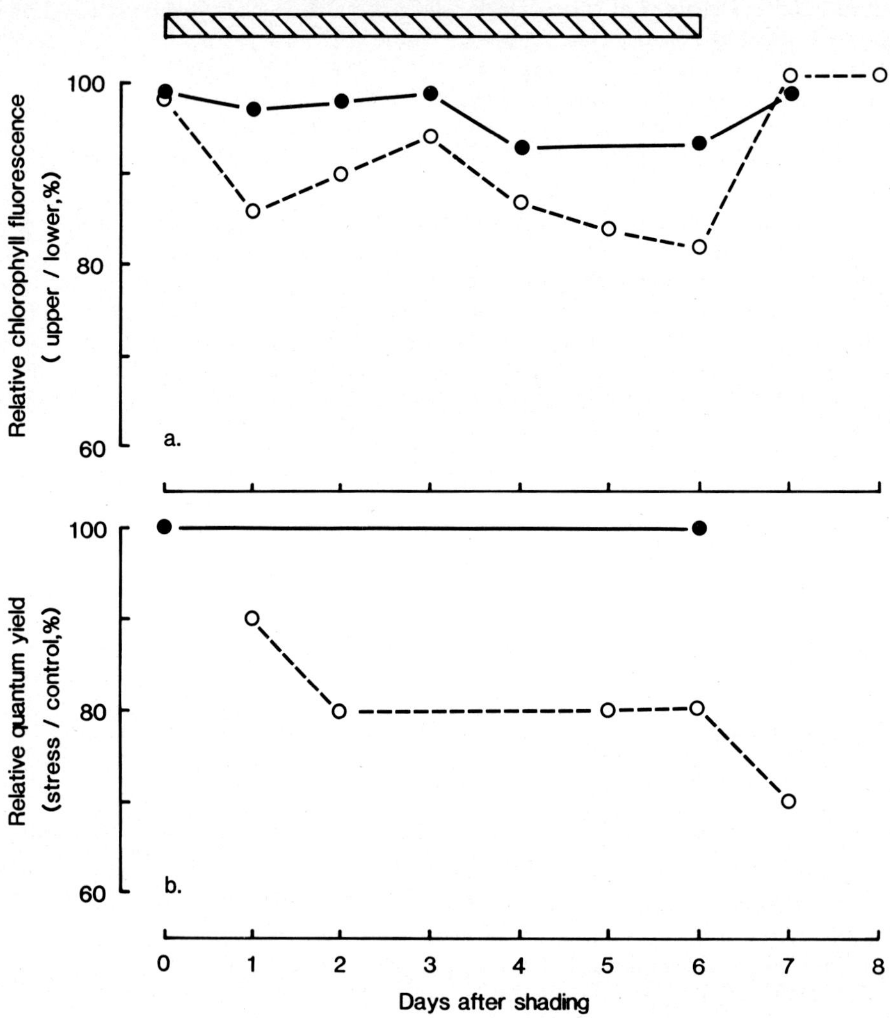

Fig. 4. Trends with time after water-stressed plants were shaded (●) or left in full sunlight (○) of (*a*) relative chlorophyll fluorescence and (*b*) relative quantum yield of oxygen evolution. Relative chlorophyll fluorescence is the ratio of F_V/F_M of the upper and lower surfaces of the same leaf, where F_V and F_M are the variable and maximum fluorescence at 692 nm measured at 77K. Relative quantum yield is the values of stressed leaves expressed as a percentage of the unstressed control.

extent that values fall below 100%. Values of this ratio were on average reduced by 10% for water-stressed leaves at the top of the canopy in full sunlight, with a maximum of 20% reduction. Shading almost eliminated photoinhibition on all but the brightest days.

The relationship between photon irradiance received from dawn until noon and the relative F_V/F_M of leaves collected at noon from plants in full sunlight shows a linear increase in photoinhibition above a threshold of 12 mol m^{-2} (Fig. 5). Photoinhibition of 20% may be a maximum value for water-stressed sorghum in this environment, because of the severe water stress and the high photon irradiances experienced in this experiment. The value of 34 mol m^{-2} is equivalent to the maximum expected in this

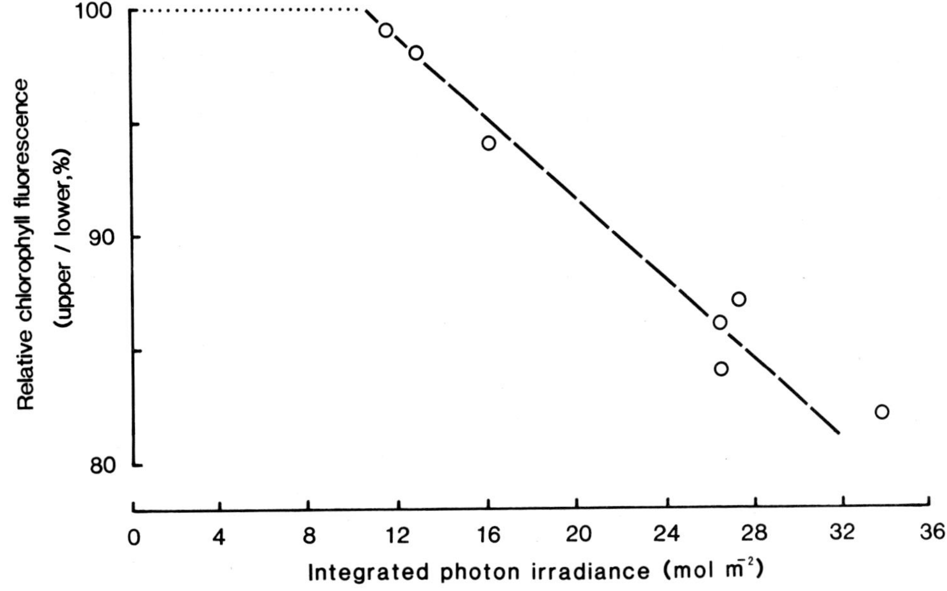

Fig. 5. Relationship between integrated photon irradiance (400–700 nm) from dawn until noon and relative chlorophyll fluorescence measured at noon of leaves from water-stressed plants in full sunlight. Relative chlorophyll fluorescence is the ratio of F_V/F_M on the upper and lower leaf surfaces, where F_V and F_M are, respectively, variable and maximum fluorescence at 692 nm measured at 77K.

environment based on calculated solar irradiance (Spencer 1979) and the general relationship between daily photon flux and daily solar irradiance (Meek *et al.* 1984). The relationship in Fig. 5 suggests that photoinhibition is cumulative and is related to the amount of light received above a threshold on each day until noon. Moreover, the damage is reversible overnight. These relationships may not hold after noon because the rate of repair (Greer *et al.* 1986) will increase relative to rate of injury as photon irradiance falls to zero at nightfall.

We had some difficulty in determining the quantum yield of this C_4 grass; values for the unstressed controls changed with time. To overcome this problem, we have expressed data for stressed plants as a percentage of unstressed controls. The relative quantum yield of leaves from plants stressed in full sunlight fell with time and reached a maximum inhibition of about 20% (Fig. 4*b*), which corresponded with the maximum reduction of relative F_V/F_M (Fig. 4*a*). There was no inhibition evident in plants in the shade.

Despite the extreme level of water stress experienced by these plants and the high photon irradiances received, the maximum photoinhibition recorded by both techniques was 20%. Furthermore, this damage occurred on the exposed abaxial surface, which recovered within 1 day of rewatering as it became the protected surface again when

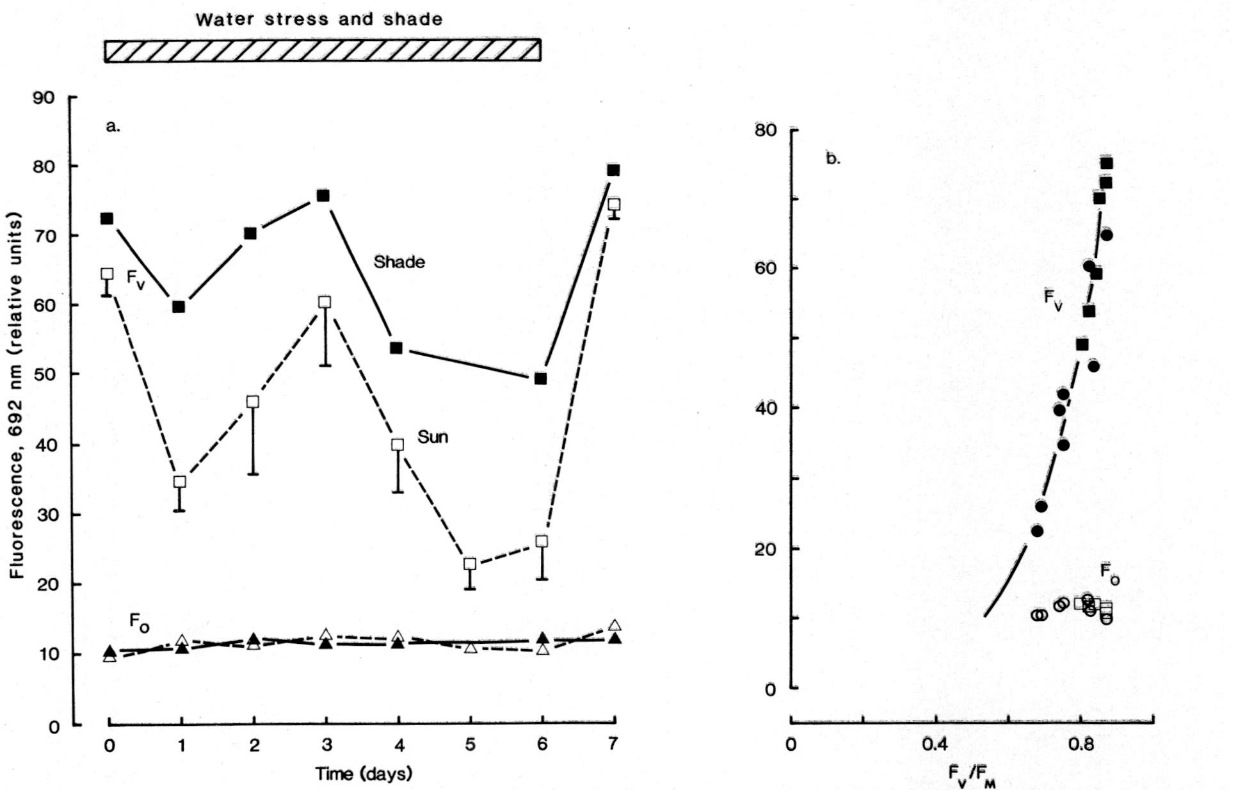

Fig. 6. (*a*) Time trends after water-stressed plants were shaded (closed symbols) or left in full sunlight (open symbols) of chlorophyll fluorescence (F_o = instantaneous and F_M = maximum fluorescence) measured at 692 nm and 77K of exposed lower surfaces of grain sorghum leaves. Standard errors of the mean of four values are shown.
(*b*) Relationship between components of chlorophyll fluorescence (F_o = instantaneous, F_M = maximum and F_V = variable fluorescence) of water-stressed leaves in shade (squares) and in full sunlight (circles) over 7 days during panicle development. F_o and F_V are open and closed symbols, respectively.

leaves unrolled. The adaxial surface that is protected from sunlight by leaf rolling during the stress period had an average F_V/F_M (0·85 ± 0·006, s.e.) similar to that of unstressed controls (0·87 ± 0·007, s.e.), and showed no damage when again exposed after leaves unrolled. On day 5 a value of F_V/F_M of 0·80 was recorded in one replicate that was most severely stressed. Therefore, in contrast to Adams *et al.* (1987), F_V/F_M of the protected leaf surface, which was only exposed to dim light, did not decline appreciably with leaf water status.

The instantaneous fluorescence (F_o) varied little over the 7 day period in both sun and shaded plants (Fig. 6*a*). Consequently, fluctuations in variable fluorescence (F_V) reflect changes in maximum fluorescence (F_M). Björkman (1987) has shown how changes in the various components of chlorophyll fluorescence (F_o and F_V) can be used to infer whether the changes in F_V/F_M are due to damage to the photochemistry (photoinhibition) or to increases in non-radiative decay. Although the latter reduces current photochemical efficiency, and as such could be considered as a form of injury, it is believed to be a protective response to dissipate excess energy and so reduce more severe and longer-term damage to the photosystems. We have assumed quantum yield used by Björkman is linearly related to F_V/F_M measured in this experiment (Demmig and Björkman 1987) and plotted F_o and F_V against F_V/F_M (Fig. 6*b*). The fact that F_o was little affected, whereas F_V fell sharply as F_V/F_M decreased, is consistent with the decline in F_V/F_M in our experiment being due to both an increase in non-radiative decay as well as damage to the photochemistry (Björkman 1987). However, the nature of our measurements does not allow us to quantify the relative importance of these two processes.

Dry Matter Production and Grain Yield

Two weeks of water stress reduced, though not significantly, leaf area and dry weight of plants and of heads (Table 2), while a further week of severe water stress reduced leaf area at harvest 2 in both the sun and shade. In contrast, both unstressed treatments remained unaltered. Total dry matter yield in unstressed plants was lowered 19% by shading, 23% by water stress, and 29% when both factors were combined. The reductions due to shading and water stress both separately and in combination occurred mostly in leaf and stem; roots were relatively unaffected (data not shown). While the effect of shading on head weight (26%) was slightly greater than its effect on dry matter, the effect of water stress on head weight was considerably greater (75%). Shading of water-stressed sorghum plants had no significant effects on either plant growth or upon the proportion of total dry matter production partitioned to the developing head.

Total dry weight at harvest 3 was unaffected by shading, but it was reduced 23% by water stress and 58% by these two combined. Moreover, the estimated total dry weight of plants water-stressed in the sun was greater than those in the shade. Grain yield however, responded in the opposite way, being 13 times greater in plants water-stressed in the shade. The reduction in grain yield by both shade and water stress was mostly due to reduced grain number. A larger single-grain weight of the three other treatments, compared with sun/unstressed, partially compensated for their lower grain number (Table 2). The proportionately lower grain yield compared with total dry weight is seen in low harvest indices in water-stressed plants, particularly those in the sun. In contrast, shading of unstressed plants had no effect on grain yield and harvest index, as well as total dry matter production. A harvest index of 0·48 for unstressed plants is high for grain sorghum and demonstrates that plants were well-grown.

The five replicates of sun/water-stressed and three replicates of shade/water-stressed plants grew for a further 20 days before the secondary heads reached physiological maturity. Several of the sun/water-stressed plants produced three secondary heads and the grain yield was higher than those water-stressed in the shade (Table 2). Again the

Table 2. Leaf area, head dry weight, total dry weight, grain yield, number and weight of grains, and harvest index recorded at four occasions during the growth of grain sorghum plants, which were either water-stressed or unstressed and either in full sunlight or shade (40% of full sunlight) for a 7-day period between harvests 1 and 2

Total grain yield and total biomass yield from harvests 3 and 4 are also given. All unstressed plants were harvested at harvest 3, whereas all five water-stressed / sun plants and three water-stressed / shade plants were allowed to continue until harvest 4 so that grains on secondary heads could reach physiological maturity. Grain yields are on an oven dry weight basis

	Sun		Shade		l.s.d. (1%)	
Plant component	Unstressed	Water-stressed	Unstressed	Water-stressed	Water stress	Shade
(a) Harvest 1 (39 days after sowing)						
Leaf area (m^2)	0·29	0·25	—	—	NS	—
Head dry wt (g)	1·94	0·74			NS	
Total dry wt (g)	67	64			NS	—
(b) Harvest 2 (46 days after sowing)						
Leaf area (m^2)	0·30	0·19	0·29	0·21	0·04	0·04
Head dry wt (g)	10·5	2·7	7·8	1·6	5·4	4·3
Total dry wt (g)	100	77·2	81·1	70·9	13·1	15·3
(c) Harvest 3 (96 days after sowing)						
Total dry wt (g)	170 ± 15	(131[A])	165 ± 11	108[B]	—	—
Grain yield (g)	82·1	2·4	79·6	32·4	25·9	25·5
Grain number	3130	114	2392	1062	777	919
Grain weight (mg)	26	35	33	33	6·1	6·7
Harvest index	0·48	0·0015	0·48	0·30	—	—
(d) Harvest 4 (116 days after sowing)						
Total dry wt (g)	—	169 ± 6	—	139 ± 16[C]	—	—
Grain yield (g)	—	73 ± 3	—	47 ± 9		
Grain number	—	2472 ± 135	—	1350 ± 27		
Grain weight (mg)	—	30 ± 2	—	35 ± 0·4		
Harvest index		0·43		0·34		
Total biomass yield (g)	170	169	165	127	58	49
Total grain yield (g)	82	75	80	60	29	25

[A]Estimated from total dry weight of stressed plants at harvest 4 and shade / stressed plants at harvest 3.
[B]Based on two replicates. [C]Based on three replicates.

higher yields were due mostly to greater grain number, despite a lower grain weight. Total dry matter yield was higher in sun than in shade plants that had experienced water stress. The relativity of the differences in grain yield and dry matter yield resulted in a higher harvest index in the sun than in the shade plants.

The total grain yield for both primary and secondary heads and total biomass production of plants water-stressed in the sun was not significantly different from unstressed plants (Table 2). Thus a relatively determinate plant like grain sorghum was able to compensate for a severe reduction in grain yield and to a lesser extent dry matter production by producing a further cycle of flowering and grain production following a return to conditions favourable to growth.

Changes during 7-day Shading Period

Grain number, which was the main determinant of grain yield in all the treatments (Table 2), is established during a period of rapid development in the head (Inuyama *et al.* 1976), which coincided with the 7-day shading treatment in this study. If the higher grain yield and the associated higher grain number of plants that were shaded during water stress was due to the absence of photoinhibition, then shading should have improved the carbon balance of the plant. This would be reflected in a higher dry weight change, but more importantly in a higher change in head weight, over the 7-day period.

Table 3. Change in (*a*) total plant dry weight and (*b*) head dry weight over a 7-day period when plants were either water-stressed or unstressed and either in full sunlight or 40% shade

Values in parentheses are relative to the unstressed/sun treatment, whereas values in brackets are predicted for the water-stressed/shade treatment based on effects of shading and water stress separately. Least significant differences at 5% for shade and water stress are given in { }

	Sun	Shade		
(*a*) Change in total dry matter (g)				
Unstressed	32·9	13·8		
	(100%)	(42%)		{13·3}
Water-stressed	13·5	7·3	[5·6]	
	(41%)	(22%)	[17%]	
	{15·1}			
(*b*) Change in head dry weight (g)				
Unstressed	9·33	6·67		
	(100%)	(71%)		
				{3·5}
Water-stressed	1·94	0·90	[1·4]	
	(21%)	(10%)	[15%]	
	{2·9}			

However, shade and water stress had a similar depressing effect on total dry matter production during the period. Plants subjected to both stresses simultaneously were reduced only to 22% of unstressed plants in the sun, which is similar to a value of 17% predicted from the combined effect of shade and water stress (Table 3*a*). Therefore, reduction of photoinhibition by shading water-stressed plants did not increase dry matter production.

In a similar analysis for change in head dry weight, water stress had a much greater effect than did shading (Table 3*b*). When both stresses were present the change in head weight (10%) was less than would be predicted from their individual effects. Similar conclusions were reached when this analysis was done on head weight at harvest 2 or on the ratio of head weight to total dry weight at harvest 2 (data not shown). Therefore, shading, designed to reduce photoinhibition induced by water stress, did not result in a better carbon supply to the developing head. This conclusion is supported by the absence of differences in stomatal conductance of the water-stressed plants over the 7-day period (Fig. 2). Consequently, the higher grain number and higher grain yield of the shade / water stress treatment compared with sun / water stress cannot be attributed to a decrease in photoinhibition that was induced by water stress.

The marked effect of shading on grain yield of water-stressed plants is probably due to the small, but important, differences in the amount of water stress experienced by the two treatments, because panicle development is very sensitive to small changes in leaf water potential. Despite our efforts to keep leaf water potentials similar in the sun and the shade, plants in the sun were on average 15% more stressed than those in the shade. This difference mostly occurred in the first 3 days when environmental conditions were highly variable (Fig. 1). Bennett (1979) found that even a 0·2 MPa reduction of leaf water potential for 12 days during panicle development was sufficient to cause a 20% reduction in photosynthesis but a 40% reduction in grain yield, due to a lower grain number.

General Discussion and Conclusions

Even though grain sorghum plants were subjected to extreme water stress when photon irradiances were high, the amount of photoinhibition was less than 20%. A similar degree of photoinhibition occurred in a subsequent experiment when sorghum plants were water-stressed during the early grain-filling stage (M. M. Ludlow, K. S. Fischer and M. Paje, unpublished data). Greater levels of photoinhibition might be expected in other situations such as the semiarid tropics of Australia, India and Africa, when irradiance and leaf temperature are higher; photoinhibition induced by water stress increases with leaf temperature (Ludlow and Björkman 1984). Because of the severity of the water stress in this experiment, it is unlikely that higher levels of water stress would cause more photoinhibition; if stress was higher, plants would die. However, grain sorghum has several avoidance and tolerance mechanisms that reduce the possibility of water stress, high leaf temperature and high incident photon irradiances on leaf surfaces. Avoidance mechanisms include: deep-rootedness that increases water uptake; reduction in leaf area, stomatal closure and leaf rolling that reduce water loss; and leaf waxiness and rolling that reduce leaf temperature. An increase in non-radiative decay is a tolerance mechanism that potentially reduces damage to the photosystems. However, the significance of non-radiative decay in grain sorghum has not been assessed, and more work is required. Thus, in view of these avoidance and tolerance mechanisms, it seems unlikely that higher levels of photoinhibition will occur in this species in the subtropics. However, it is possible that higher levels may be recorded in the semiarid tropics where temperatures are higher.

The degree of water stress in this experiment was so severe that it reduced the level of dry matter production by 23% and head weight by 74% during the 7-day period. However, this relatively determinate species produced secondary heads upon rewatering and completely compensated for the severe reduction in grain yield of primary heads and the reductions in dry matter production. Consequently, the reduction in carbon supply by photoinhibition induced by water stress would need to be even larger than the growth depression due to water stress alone in this experiment to exceed the compensatory responses.

The importance of photoinhibition induced by water stress was investigated during panicle development because carbon supply has the greatest effect on grain yield via grain number. Independent of the development of secondary heads, there is more scope for compensation when photoinhibition occurs during early vegetative growth and during grain filling than during panicle development. For example, unless substantial leaf death occurs, there is time for leaves to recover from damage during early vegetative growth. Moreover, retranslocation of preanthesis dry matter can partially or completely compensate for reduction in current photosynthesis during grain filling (Inuyama *et al.* 1976). Thus there seems less scope for photoinhibition to cause yield reduction at other stages of phenology than during panicle development. The scope for compensation is even greater in indeterminate crops, such as cowpea, mung bean, chickpea and pigeonpea, and the likelihood of photoinhibition induced by water stress causing significant yield reduction is much less than in grain sorghum.

In view of the compensatory responses and the avoidance and tolerance mechanisms, we consider it unlikely that photoinhibition induced by water stress in full sunlight is of any agronomic consequence in grain sorghum.

Acknowledgments

The skilful technical assistance of Bob Kerslake, Frank Lamaree and Maria Paje are gratefully acknowledged. Dr S. B. Powles received financial assistance from the Waite Research Committee to travel to Brisbane to participate in this experiment.

References

Adams, W. W., Smith, S. D., and Osmond, C. B. (1987). Photoinhibition of the CAM succulent *Opuntia basilaris* growing in Death Valley: evidence from 77K fluorescence and quantum yield. *Oecologia* **71**, 221–8.

Björkman, O. (1987). Low temperature chlorophyll fluorescence in leaves and its relationship to photon yield of photosynthesis in photoinhibition. In 'Photoinhibition'. (Eds D. J. Kyle, C. B. Osmond and C. J. Arntzen.) Topics in Photosynthesis, Vol. 9, pp. 123–44. (Elsevier: Amsterdam.)

Björkman, O., and Powles, S. B. (1984). Inhibition of photosynthetic reactions under water stress: interaction with light level. *Planta* **161**, 490–504.

Björkman, O., and Demmig, B. (1987). Photon yield of O_2 evolution and chlorophyll fluorescence characteristics at 77K among vascular plants of diverse origins. *Planta* **170**, 489–504.

Bennett, J. M. (1979). Responses of grain sorghum [*Sorghum bicolor* (L.) Moench] to osmotic stresses imposed at various growth stages. Ph.D. Thesis, University of Nebraska.

Demmig, B., and Björkman, O. (1987). Comparison of the effect of excessive light on chlorophyll fluorescence (77K) and photon yield of O_2 evolution in leaves of higher plants. *Planta* **171**, 171–84.

Doggett, H. (1970). 'Sorghum.' (Longmans Green: London.)

Greer, D., Berry, J. A., and Björkman, O. (1986). Photoinhibition of photosynthesis in intact bean leaves: role of light and temperature, and requirement for chloroplast-protein synthesis during recovery. *Planta* **168**, 253–60.

Hsiao, T. C. (1973). Plant responses to water stress. *Annu. Rev. Plant Physiol.* **24**, 519–70.

Inuyama, S., Musick, J. T., and Dusek, D. A. (1976). Effect of plant water deficits at various growth stages on growth, grain yield and leaf water potential of irrigated grain sorghum. *Proc. Crop Sci. Soc. Jpn* **45**, 298–307.

Levitt, J. V. (1980). 'Responses of Plants to Environmental Stresses.' Vols 1 and 2, 2nd edition. (Academic Press: New York.)

Ludlow, M. M. (1987). Light stress at high temperature. In 'Photoinhibition.' (Eds D. J. Kyle, C. B. Osmond and C. J. Arntzen.) Topics in Photosynthesis, Vol. 9, pp. 89–110. (Elsevier: Amsterdam.)

Ludlow, M. M., and Björkman, O. (1984). Paraheliotropic leaf movement in Siratro as a protective mechanism against drought-induced damage to primary photosynthetic reactions: damage by excessive light and heat. *Planta* **61**, 505–18.

Ludlow, M. M., and Wilson, G. L. (1971). Photosynthesis of tropical pasture plants. I. Illuminance, carbon dioxide concentration, leaf temperature, and leaf–air vapour pressure difference. *Aust. J. Biol. Sci.* **24**, 449–70.

McWilliam, J. R., Phillips, P. J., and Parkes, R. R. (1973). Measurement of photosynthetic rate using labelled carbon dioxide. CSIRO Aust. Div. Plant Ind. Tech. Pap. No. 31.

Meek, D. W., Hatfield, J. L., Howell, T. A., Idso, S. B., and Reginato, R. J. (1984). A generalised relationship between photosynthetically active radiation and solar radiation. *Agron. J.* **76**, 939–45.

Powles, S. B., and Björkman, O. (1982). Photoinhibition of photosynthesis: effect on chlorophyll fluorescence at 77K in intact leaves and in chloroplast membranes. *Planta* **156**, 97–107.

Powles, S. B. (1984). Photoinhibition of photosynthesis induced by visible light. *Annu. Rev. Plant Physiol.* **35**, 15–44.

Santamaria, J. (1987). Study of traits for drought resistance in *Sorghum bicolor* (L.) Moench, with emphasis on osmotic adjustment. M.Agr.Sci. Thesis, University of Queensland.

Seetharama, N., Subba Reddy B. V., Peacock, J. M., and Bidinger, F. R. (1982). Sorghum improvement for drought resistance. In 'Drought Resistance in Crops with Special Emphasis on Rice'. pp. 317–38. (International Rice Research Institute: Los Baños.)

Spencer, J. W. (1979). Brisbane solar tables. Tables of solar position and radiation for Brisbane (latitude 27·5°S.) in SI units. CSIRO, Australia, Division of Building Research, Tech. Pap. (Second Series) No. 29.

Turner, N. C. (1986). Adaptation to water deficits: a changing perspective. *Aust. J. Plant Physiol.* **13**, 175–90.

Wong, S. C., Farquhar, G. D. and Cowan, I. R. (1979). Stomatal conductance correlates with photosynthetic capacity. *Nature* **282**, 424–6.

Effect of Temperature on Photoinhibition and Recovery in *Actinidia deliciosa*

Dennis H. Greer

Plant Physiology Division, DSIR,
Private Bag, Palmerston North, New Zealand.

Abstract

Photoinhibition of photosynthesis was induced in intact leaves of kiwifruit (*Actinidia deliciosa*) grown in natural light not exceeding a photon irradiance (PI) of 300 μmol $m^{-2} s^{-1}$ by exposing them to a PI of 1500 μmol $m^{-2} s^{-1}$. The temperature was held constant during the high-light exposure between 5 and 35°C. Recovery was followed at temperatures between 10 and 35°C, after photoinhibition was induced by a 240 min exposure to high light. The kinetics of photoinhibition and recovery were followed by chlorophyll fluorescence at 692 nm and 77K.

Photoinhibition occurred at all temperatures but was greatest at low temperatures. Temperature affected the severity of photoinhibitory damage but not the kinetics of photoinhibition. Recovery was also temperature-dependent with little or no recovery occurring below about 20°C and rapid recovery at 30–35°C. The extent of photoinhibition also affected the rates of recovery which were reduced as the severity of photoinhibition increased.

An analysis of the rate constants for energy transfer within photosystem II indicated that kiwifruit leaves have some capacity to prevent photoinhibition by increasing the amount of non-radiative energy dissipation. However, the analysis also indicates that this protection mechanism was not wholly effective since the primary photochemical reactions apparently become inactivated during exposure of these leaves to high light.

Introduction

The number of studies that have investigated the effect of temperature on photoinhibition is relatively small in spite of a number purporting to have done so. Most studies have examined no more than two temperatures, which is an inadequate number to evaluate the effect of temperature on any process. To gain a coherent understanding of the effect of temperature, it is essential that a wide range of temperatures be examined.

Of those studies where temperature effects on photoinhibition have been studied, by far the greatest number of them have been on chilling temperatures (0–15°C). These include the studies of Taylor and Rowley (1971) on a range of C_3 and C_4 grasses, Long *et al.* (1983) on *Zea mays*, Ögren and Öquist (1984) on *Lemna gibba*, and Yakir *et al.* (1985) on *Lycopersicon esculentum*. Such studies uniformly show that most species are particularly susceptible to photoinhibition at low temperatures. However, Powles and Björkman (1982) and Greer *et al.* (1986) have shown for *Nerium oleander* and *Phaseolus vulgaris*, respectively, that photoinhibition can occur over a wide temperature range.

If studies examining the effects of temperature on photoinhibition are scarce then those on the effects of temperature on recovery from photoinhibition are even more so. However, the effect of temperature on recovery in *Phaseolus vulgaris* has been determined (Greer *et al.* 1986) and this process has been shown to be temperature-dependent

0310-7841/88/010195$03.00

with little or no recovery below 15°C and maximum recovery at 30°C. This study also demonstrated that recovery occurs concomitantly with photoinhibition, suggesting that net photoinhibition is reduced by the concomitant recovery process.

Björkman (1987*a*, 1987*b*) has proposed that, during the exposure of leaves to excess light, there is an increase in non-radiative energy dissipation and a decrease in photosystem II (PS II) primary photochemical activity. The former process is activated to regulate the level of excitation energy impinging on the reaction centres while the latter results from direct photoinhibitory damage to the reaction centres. Whether or not temperature affects the regulation of excess excitation energy is unknown but this would clearly be of importance in understanding the susceptibility of leaves to photoinhibition.

Kiwifruit (*Actinidia deliciosa*) is an important horticultural crop plant in New Zealand. It has been shown from controlled-environment studies that leaves of this plant suffer severe damage, photo-oxidation and leaf death when exposed to moderate light at 10°C (Morgan *et al.* 1985). It would appear that this plant is susceptible to low-temperature-induced photoinhibition since, at the same temperature and a reduced light level, no apparent symptoms of damage were evident. The severity of damage also suggests photoinhibition greatly exceeded the recovery process at this temperature.

In this paper, the effect of temperature on both the photoinhibition and recovery processes in shade-grown kiwifruit leaves is described. The changes in non-radiative energy dissipation and photochemistry that might contribute to these processes are also examined.

Materials and Methods

Plant Material

Rooted cuttings of *Actinidia deliciosa* [(A. Chev.) C. F. Liang et A. R. Ferguson] were grown in 9-litre pots as described by Greer *et al.* (1988). The plants were grown outdoors in a shade enclosure where they experienced natural fluctuations in temperature and in the photon irradiance (PI) which did not exceed 300 μmol $m^{-2} s^{-1}$. Plants were watered twice daily.

Photoinhibition and Recovery

To induce photoinhibition, each intact leaf was sealed into a gas exchange chamber described by Greer *et al.* (1988) and exposed to a PI of 1500 μmol $m^{-2} s^{-1}$ from a high pressure discharge lamp (GTE, Sylvania, Metal-arc, 1 kW) equipped with a water-screen and neutral density filters. The leaf temperature was held constant during the light exposure at between 5 and 35°C. The rest of the plant was at approximately 22°C and 100 μmol $m^{-2} s^{-1}$.

For experiments on recovery, the leaves were initially exposed for 240 min to a PI of 1500 μmol $m^{-2} s^{-1}$ at 20°C as described by Greer and Laing (1988). Recovery was subsequently followed at a PI of 20 μmol $m^{-2} s^{-1}$, provided by a 40 W tungsten filament lamp. Leaf temperature was held constant at between 10 and 35°C. In additional experiments, the duration of the high-light exposure was varied between 90 and 400 min.

Photoinhibition and Recovery Assay

At intervals throughout each photoinhibition or recovery experiment, duplicate leaf samples (10 mm diam.) were collected from the leaf. These were kept in the dark at room temperature for at least 30 min to ensure complete oxidation of the photosynthetic reaction centres. Chlorophyll fluorescence at 692 nm and 77K was then determined as described by Greer *et al.* (1986). The initial fluorescence (F_0) and the maximum fluorescence (F_M) were recorded and the variable fluorescence (F_V) and the fluorescence ratio (F_V/F_M) where $F_V = F_M - F_0$ were calculated.

Time-course data of photoinhibition and recovery were fitted to an exponential equation by non-linear regression analysis as described by Greer *et al.* (1988). The rate constants for PS II photochemistry (K_P) and for non-radiative energy dissipation (K_D) were calculated according to Björkman (1987*a*, 1987*b*) from the following equations:

$$K_D = (\epsilon - F_M)/F_M \qquad K_P = [(\epsilon - F_0) - (K_D \times F_0)]/F_0,$$

where ϵ is a correction coefficient to correct F_0 and F_M to the same units on the excitation light causing the fluorescence emission (see Appendix 1 for derivation of these equations).

Results

Effect of Temperature on Photoinhibition

When the leaves were exposed to a PI about 5-fold higher than that at which they were grown, there were marked changes in F_V. However, as shown in Fig. 1, temperature had a significant effect on these changes with a higher initial rate of decline and a greater reduction in F_V at 5°C compared to 25°C. That temperature has a significant

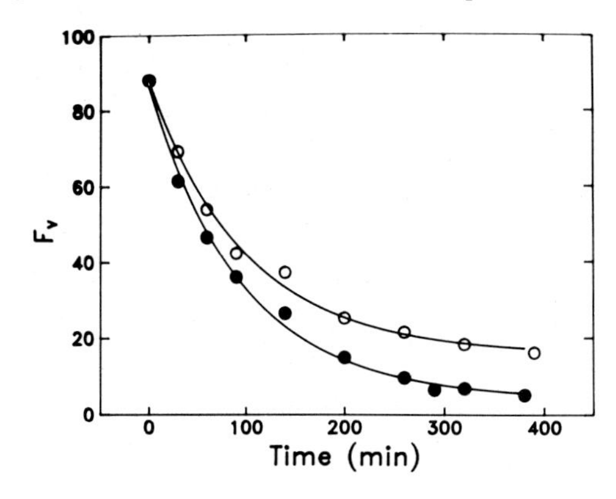

Fig. 1. The time-course of changes in F_V for intact kiwifruit leaves exposed to a photon irradiance of 1500 μmol m^{-2} s^{-1} at 25°C (○), and 5°C (●). (From Greer *et al.* 1988.)

effect on the severity of photoinhibition can be seen from Table 1, where the extent of the decline in F_V increased dramatically with decreasing temperatures. The kinetics of photoinhibition were, however, only weakly dependent on temperature since the first-order rate constant ($k(F_p)$, Table 1) varied little across a broad temperature range. These data demonstrate that kiwifruit leaves are susceptible to photoinhibition at all tempera-

Table 1. Effect of temperature on the rate constant for photoinhibition, $k(F_p)$, and the steady-state extent of photoinhibition, F_∞ as measured by F_V, for intact kiwifruit leaves exposed to a photon irradiance of 1500 μmol m^{-2} s^{-1}

F_V at the start of each treatment was 88·0±0·9. Values given are means ± s.e.

Temp. (°C)	$10^3 \times k(F_p)$ (min^{-1})	F_∞
5	11·1 ± 0·4	5·4 ± 1
10	9·0 ± 0·4	1·1 ± 2
15	9·6 ± 0·9	4·7 ± 3
20	8·5 ± 0·6	8·2 ± 2
25	10·5 ± 0·7	16·4 ± 2
30	9·1 ± 1·2	18·2 ± 4
35	10·7 ± 1·5	23·7 ± 4

tures between 5 and 35°C when exposed to photon irradiance which greatly exceeds that experienced during growth, though more so at the lower temperatures.

The relative rate constant for photochemistry (K_P) decreased during a typical high-light exposure while the rate constant for non-radiative energy dissipation (K_D) increased (Fig. 2). Temperature affected the extent of the decline in K_P, with the greatest reduction at lower temperatures (Table 2). However, the extent of the increase in K_D was not particularly temperature-dependent although the initial rate of increase in K_D (ΔK_D) was temperature-dependent with a higher rate at lower temperatures (Table 2).

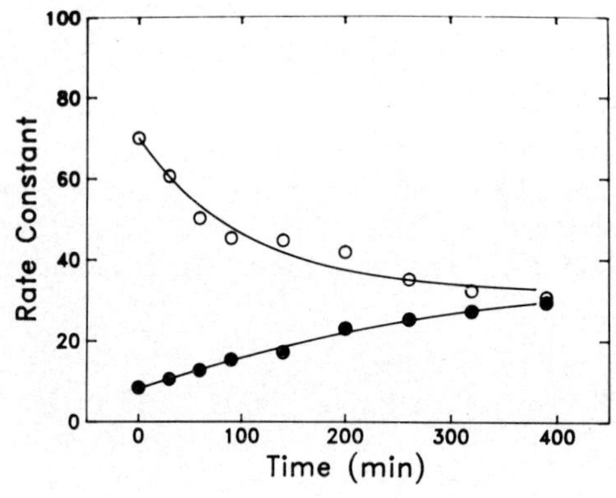

Fig. 2. The time-course of the decline in the rate constant for photochemistry, K_P (○), and of the increase in the rate constant for non-radiative energy dissipation, K_D (●), for intact kiwifruit leaves exposed to a photon irradiance of 1500 μmol m^{-2} s^{-1} at 25°C. (From Greer *et al.* 1988.)

The contribution of the decrease in K_P and the increase in K_D to the overall reduction in efficiency of PS II photochemistry (ϕ_{pc}) during exposure of kiwifruit leaves to high light at the various temperatures is shown in Table 3. There was a 55% decrease in ϕ_{pc} after 200 min at 5°C and this progressively decreased with increasing temperature to 24% at 35°C. Damage to PS II reaction centres contributed mostly to this decline at 5°C whereas at higher temperatures, notably at 25°C, the increases in non-radiative energy dissipation contributed mostly to the decline in ϕ_{pc} (see Appendix 2 for details of this analysis).

Table 2. Effect of temperature on the steady-state level of the rate constant for photochemistry, K_P and on the initial rate of increase in the rate constant for non-radiative energy dissipation, ΔK_D

K_P = 70 at t = 0. Values given are means ± s.e.

Temp. (°C)	K_P at $t = \infty$	$10^2 \times \Delta K_D$ (min^{-1})
5	0·7 ± 2·8	7·6 ± 0·5
10	3·0 ± 3·4	9·6 ± 0·4
15	16·4 ± 3·3	9·7 ± 0·7
20	30·2 ± 2·6	6·9 ± 0·4
25	33·7 ± 2·6	7·1 ± 0·3
30	32·3 ± 1·8	4·7 ± 0·4
35	38·0 ± 3·3	5·6 ± 0·8

Effect of Temperature on Recovery

The time-courses of recovery of kiwifruit leaves from photoinhibition at temperatures of 10, 20 and 30°C are shown in Fig. 3. Over 400 min no recovery was apparent at 10°C while, over the same time, only about 20% recovery had occurred at 20°C. Recovery was significantly faster at 30°C and was about 90% complete in about 400 min. These data show recovery from photoinhibition in kiwifruit is strongly temperature-dependent. However, recovery was also dependent on the extent of photoinhibition as established by varying the initial duration of exposure to high light (Fig. 4). The first-order rate constant for recovery, $k(F_r)$, declined in a linear fashion as the extent of photoinhibition increased, the rate of decrease in $k(F_r)$ being more rapid at 30 than at 20°C. This shows that the effect of the extent of photoinhibition on recovery also depended on temperature.

Taking into account the effect of the extent of photoinhibition on $k(F_r)$, the rate constant for recovery increased in a linear fashion with increasing temperature (Fig. 5). From 15°C, the lowest temperature at which recovery could be detected, to 35°C there was a 4-fold increase in $k(F_r)$, confirming that recovery is strongly temperature-dependent. However, a similar temperature-dependency was observed over a range of different extents of photoinhibition.

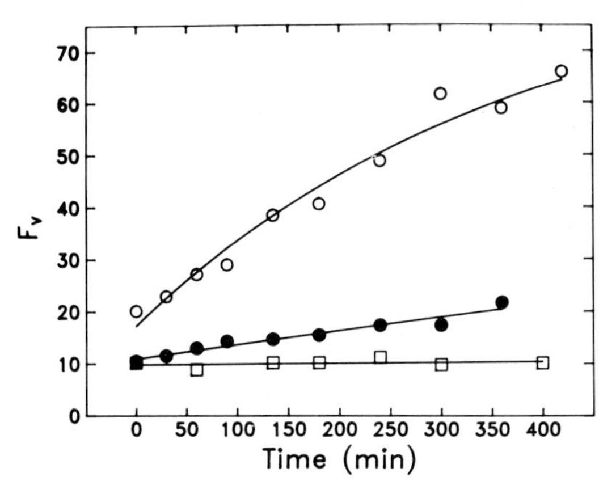

Fig. 3. The time-course of recovery of F_V for intact kiwifruit leaves at 10 (□), 20 (●), and 30°C (○) from photoinhibition following a 240 min exposure to a photon irradiance of 1500 μmol m^{-2} s^{-1} at 20°C. (From Greer and Laing 1988.)

The changes in K_P and K_D that occurred during photoinhibition were readily reversed once the high-light stress was removed (Fig. 6). Both K_P and K_D recovered, apparently with first-order kinetics, with increasing times (that is, decreasing rate constants) for

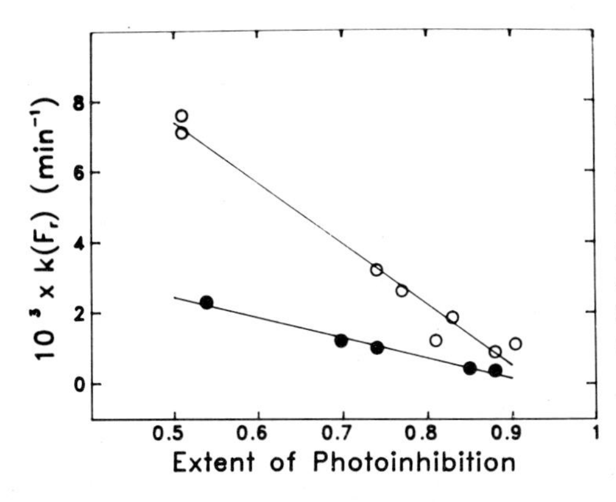

Fig. 4. The rate constant for recovery of F_V, $k(F_r)$, at temperatures of 20 (●) and 30°C (○) in intact kiwifruit leaves as a function of the extent of photoinhibition. The extent of photoinhibition was determined from the ratio of F_V at the end of the photoinhibition treatment to that of the control before photoinhibition. (From Greer and Laing 1988.)

complete recovery with decreasing temperatures. Half-times for recovery of both K_P and K_D were in excess of 100 min under the most favourable temperatures (30–35°C) and greater than 500 min at the lower temperatures where recovery could be detected (15–20°C).

The recovery in the efficiency of PS II photochemistry (ϕ_{pc}) and the contribution that K_P and K_D have in this recovery at the various temperatures is shown in Table 4. There was a progressive increase in the efficiency of photochemistry after 240 min, from 25% at 15°C to 74% at 35°C. At the lowest temperature, repair of the damage to reaction centres was contributing relatively more to this increase in efficiency whereas, at the higher temperatures, the relaxation in non-radiative energy dissipation contributed relatively more.

Discussion

When shade-grown kiwifruit leaves are exposed to high light, they exhibit typical symptoms of photoinhibition. An exposure of 140 min to a PI more than five times in excess of that in which they were grown resulted in a 55% decrease in variable fluorescence (F_V). These results are comparable with those for shade-grown *Phaseolus vulgaris* leaves exposed to high light for 200 min (Greer *et al.* 1986) and with those for shade-grown *Nerium oleander* leaves exposed for 240 min (Powles and Björkman 1982).

Table 3. Effect of temperature on the efficiency of photochemistry (ϕ_{pc}, measured as F_V/F_M) on intact kiwifruit leaves exposed for 200 min to a photon irradiance of 1500 μmol m^{-2} s^{-1} and the contribution that the decrease in photochemical activity (K_P) and increases in non-radiative energy dissipation (K_D) have on the decrease in ϕ_{pc}

ϕ_{pc} = 0·88 prior to photoinhibition

Temp. (°C)	ϕ_{pc}	% change in ϕ_{pc} due to: Rise in K_D	Fall in K_P
5	0·39	38	62
10	0·45	49	51
15	0·51	57	43
20	0·56	57	43
25	0·63	63	37
30	0·66	54	46
35	0·67	56	44

These data might suggest kiwifruit leaves are intrinsically more susceptible to photoinhibition than either of the other two species. For plants grown in comparable growth conditions, Demmig and Björkman (1987) have shown that the extent of the reduction in F_V can vary depending on the species. Thus, interspecific variation in susceptibility to photoinhibition does apparently occur, and conforms with the above conclusion.

Table 4. Effect of temperature on the efficiency of photochemistry (ϕ_{pc}) after 240 min of recovery of intact kiwifruit leaves from photoinhibition and the contribution that the increase in photochemical activity (K_P) and the decrease in non-radiative energy dissipation (K_D) make to the recovery in ϕ_{pc}

ϕ_{pc} = 0·53 at the start of recovery

Temp. (°C)	ϕ_{pc}	% change in ϕ_{pc} due to: Fall in K_D	Rise in K_P
15	0·62	43	57
20	0·63	50	50
25	0·67	54	46
30	0·78	51	49
35	0·79	56	44

The time-course of photoinhibition in kiwifruit leaves followed pseudo-first-order kinetics with an initially high rate that declined to approach a steady state. Temperature had only a small effect on the kinetics of photoinhibition but a significant effect on the steady state or the severity of damage. At temperatures below about 15°C there was an almost complete loss of variable fluorescence, indicating very severe photoinhibition.

At higher temperatures, there were also losses of variable fluorescence; for instance at 35°C, F_V declined to 27% of the initial level. This would indicate severe photoinhibition at the higher temperatures and implies that these shade-grown kiwifruit leaves have a propensity to be photoinhibited on exposure to high light, though more so at low than high temperatures.

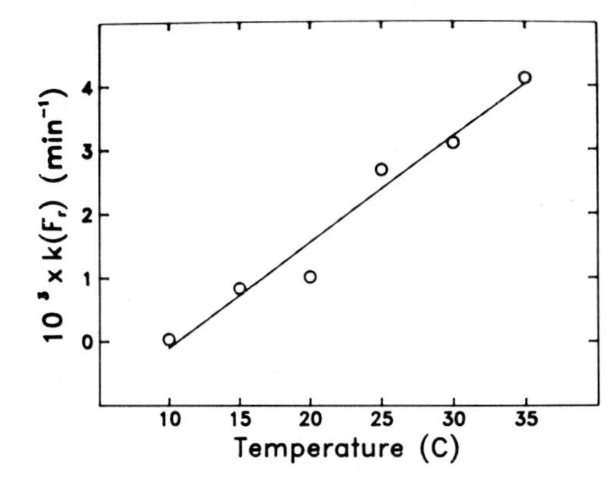

Fig. 5. The rate constant for recovery, $k(F_r)$, for intact kiwifruit leaves as a function of temperature at a constant level of 75% photoinhibition. (From Greer and Laing 1988.)

The effect of temperature on the long-term extent of photoinhibition has not apparently been reported in other species. Greer *et al.* (1986) determined the effect of temperature on the initial rate of photoinhibition in *Phaseolus vulgaris*, as measured by F_V/F_M, and noted that the rate of photoinhibition was temperature-dependent with the

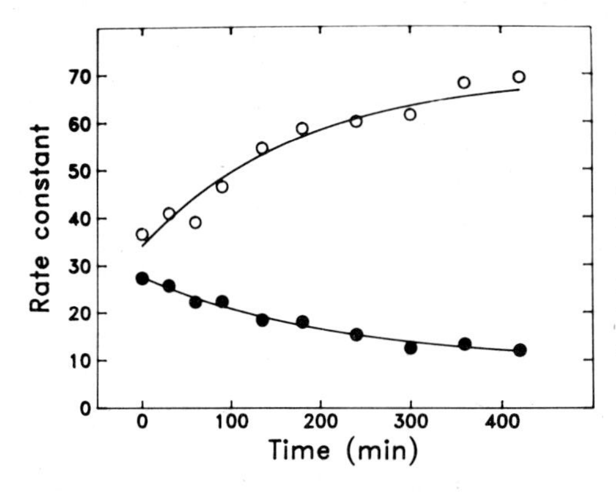

Fig. 6. The time-course of the decrease in the rate constant for non-radiative energy dissipation, K_D (●), and for the increase in the rate constant for photochemistry, K_P (○), during the recovery from photoinhibition of intact kiwifruit leaves at 30°C and a photon irradiance of 20 μmol m^{-2} s^{-1}. (From Greer and Laing 1988.)

lowest rate at 30°C and the highest rate at 10–20°C. However, changes in F_V (Greer, unpublished data) followed exponential kinetics and the estimated steady-state value for photoinhibition at 10°C was 23% of the initial F_V while, at 30°C, the correspondent value was 69%. Thus in *Phaseolus vulgaris* there too was a tendency for the severity of photoinhibition to decrease with increasing temperature. Similar apparent results were evident when shade-grown *Nerium oleander* leaves were exposed to high light at a range of temperatures, at least after a 100 min exposure (Powles and Björkman 1982). By contrast, Powles *et al.* (1983) showed that high-light-grown leaves of *Phaseolus vulgaris* did not suffer any inhibition of the apparent photon yield at temperatures above about 12°C when exposed to 2000 μmol m^{-2} s^{-1}. This could suggest the light regime during leaf growth and development affects the temperature-dependency of photoinhibition although D. H. Greer and W. A. Laing (unpublished data) found no evidence of this in kiwifruit. It seems likely, then, that in general photoinhibition is exacerbated by low temperatures. Whether this is caused by a lower photosynthetic capacity at low temperatures, a low-temperature-induced lesion within the thylakoid membrane, or a lack of concomitant recovery (see later) is unknown. However, a low-temperature-induced membrane lesion (chilling damage) seems unlikely since a low-temperature exposure in the dark apparently has no effect on the subsequent susceptibility of kiwifruit leaves to photoinhibition (Greer, unpublished data).

Once the high-light stress was removed, the photoinhibited kiwifruit leaves began to recover. However, the recovery that occurred was dependent on temperature with maximum rates of recovery at temperatures of 25–35°C, slow rates below 20°C, and no apparent recovery at 10°C. This effect of temperature is thus similar to that observed for *Phaseolus vulgaris* (Greer *et al.* 1986). However, the highest rate constant for recovery in kiwifruit ($0{\cdot}0040\ min^{-1}$) was less than 50% of the highest rate constant for recovery in *Phaseolus* ($0{\cdot}0087\ min^{-1}$). The generally lower rates of recovery in kiwifruit are consistent with this species being apparently more susceptible to photoinhibition than *Phaseolus*.

Recovery from photoinhibition in kiwifruit leaves is dependent not only on temperature but also on the extent of photoinhibition. Similar effects of the extent of photoinhibition on subsequent recovery have been reported for *Lemna gibba* (Nilsen *et al.* 1984; Ögren *et al.* 1984), *Spinacia oleracea* (Krause *et al.* 1985) and *Hedera canariensis* and *Monstera deliciosa* (Demmig and Björkman 1987). However, it is only in kiwifruit that the interactive effects of temperature and extent of photoinhibition on recovery have been defined. The dependence of the rate constant for recovery on the extent of photoinhibition increased as the temperature increased. The effect of temperature on the recovery rate constant, on the other hand, was apparently independent of the extent of photoinhibition. A possible explanation for this effect of the extent of photoinhibition on recovery is that, as photoinhibition increases, secondary damage other than in the reaction centre occurs. This could lead to a more general deterioration of the biosynthetic machinery facilitating the low rates of recovery.

When the extent of photoinhibition was taken into account, recovery of kiwifruit leaves was linearly dependent on temperature. The temperature-dependency therefore differed slightly from that in *Phaseolus* which was curvilinearly dependent (Greer *et al.* 1986). However, in both species the high rates of recovery at high temperature conform with less severe photoinhibition at these temperatures. Similarly, the susceptibility of leaves to photoinhibition at low temperatures can, in part, be attributed to the low rates of recovery at these temperatures. It still remains an open question whether the low rates of photosynthesis at low temperatures further exacerbate photoinhibition through a reduced demand for excitation energy.

Björkman (1987*a*, 1987*b*) has proposed that leaves developing in high light can regulate the amount of excess excitation energy by increasing non-radiative energy dissipation. In shade-grown leaves, the capacity of this mechanism to dissipate excess energy may not be well developed and damage to the reaction centres occurs on exposure of these leaves to high light. For cotton grown in low light, it appears that K_D increases only slowly on exposure to high light whereas, for leaves developed in high light, K_D apparently increases rapidly (Björkman 1987*b*). The kinetics of the increase in non-radiative energy dissipation appeared to differ between low-light- and high-light-grown plants. Therefore, it may be a slow activation of the mechanism to dissipate excess energy rather than a limited capacity to do so that predisposes shade-grown plants to photoinhibition.

The analysis of changes in K_D for kiwifruit leaves indicates that there is some capacity for these leaves to dissipate excess energy. K_D tended to increase about 4-fold during photoinhibition and this was largely independent of temperature. A similar increase in K_D was observed in leaves of *Monstera deliciosa* when exposed to high light for 180 min (Björkman 1987*a*). That the increase in K_D has a 'cost', however, can be seen (Table 3) by the 40–60% reduction in efficiency of photochemistry as a consequence of increasing non-radiative energy dissipation. This is further borne out by the slow relaxation in K_D

once the leaves are returned to low light. At optimal conditions, K_D required over 100 min to complete 50% recovery. Over longer periods (Table 4), the increased non-radiative dissipation was still contributing about 50% to the reduced efficiency of PS II photochemistry. The slow relaxation in K_D has also been reported in *Monstera deliciosa* (Björkman 1987*a*) and it would seem likely that the mechanism for dissipating non-radiative energy involves more than a build-up of a proton gradient (Krause and Behrend 1986). An alternative mechanism reported by Demmig *et al.* (1987), that the energy dissipation is manifest by the reversible conversion of violaxanthin to zeaxanthin, is more consistent with the kinetics and the effect of temperature on changes in K_D reported here.

Direct damage to the primary photochemical reactions of PS II was evident when the kiwifruit leaves were exposed to high light and this damage could be subsequently repaired in weak light. The rate constant for photochemistry, K_P, declined exponentially during the exposure to high light and recovered exponentially when the light stress was removed. Both of these processes were temperature-dependent with more damage occurring at low temperatures and higher rates of repair at high temperatures. These temperature responses are typical of many biosynthetic processes and are consistent with the damage to reaction centres and their subsequent repair involving protein breakdown and *de novo* chloroplast-directed protein synthesis (Greer *et al.* 1986; Kyle and Ohad 1986). Damage to reaction centres contributed most to the reduction in efficiency of photochemistry at low temperatures (Table 3). This suggests that protection by non-radiative energy dissipation is too limited, particularly at low temperatures, to prevent substantial damage to the reaction centres. Whether this is specific to kiwifruit or is more a reflection of leaf development in the shade remains uncertain.

In conclusion, it is apparent that leaves of shade-grown plants are susceptible to photoinhibition over a wide temperature range, though more so at low temperatures. This, in part, can be attributed to the temperature-dependent recovery process but also occurs possibly because of an ineffective mechanism to dissipate excess energy. There is a need to determine if more protection is inherent and whether or not the recovery process can be activated further in high-light-grown plants to reduce net photoinhibition. However, it remains that photoinhibition and recovery are temperature-dependent processes and this temperature-dependency has important ramifications for the effect of photoinhibition on crop growth.

References

Björkman, O. (1987*a*). High-irradiance stress in higher plants and interaction with other stress factors. In 'Progress in Photosynthesis Research'. (Ed. J. Biggins.) Vol. 4, pp. 11-18. (Martinus Nijhoff: Dordrecht.)

Björkman, O. (1987*b*). Low-temperature chlorophyll fluorescence in leaves and its relationship to photon yield of photosynthesis in photoinhibition. In 'Photoinhibition'. (Eds D. Kyle, C. B. Osmond and C. J. Arntzen.) Topics in Photosynthesis, Vol. 9, pp. 123-44. (Elsevier: Amsterdam.)

Demmig, B., and Björkman, O. (1987). Comparison of the effect of excessive light on chlorophyll fluorescence (77K) and photon yield of O_2 evolution in leaves of higher plants. *Planta* **171**, 171-84.

Demmig, B., Winter, K., Krüger, A., and Czygan, F.-C. (1987). Photoinhibition and zeaxanthin formation in intact leaves. A possible role of the xanthophyll cycle in the dissipation of excess light energy. *Plant Physiol.* **84**, 218-24.

Greer, D. H., Berry, J. A., and Björkman, O. (1986). Photoinhibition of photosynthesis in intact bean leaves: role of light and temperature, and requirement for chloroplast-protein synthesis during recovery. *Planta* **168**, 253-60.

Greer, D. H., and Laing, W. A. (1988). Photoinhibition of photosynthesis in intact kiwifruit (*Actinidia deliciosa*) leaves: recovery and its dependence on temperature. *Planta* (in press).

Greer, D. H., Laing, W. A., and Kipnis, T. (1988). Photoinhibition of photosynthesis in intact kiwifruit (*Actinidia deliciosa*) leaves: effect of temperature. *Planta* (in press).

Krause, G. H., and Behrend, U. (1986). ΔpH-dependent chlorophyll fluorescence quenching indicating a mechanism of protection against photoinhibition of chloroplasts. *FEBS Lett.* **200**, 298–302.
Krause, G. H., Köster, S., and Wong, S. C. (1985). Photoinhibition of photosynthesis under anaerobic conditions studied with leaves and chloroplasts of *Spinacia oleracea* L. *Planta* **165**, 430–8.
Kyle, D. J., and Ohad, I. (1986). The mechanism of photoinhibition in higher plants and green algae. In 'Photosynthesis III. Photosynthetic Membranes and Light Harvesting Systems'. (Eds L. A. Staehelin and C. J. Arntzen.) Encycl. Plant Physiol. New Ser., Vol. 19, pp. 468–75. (Springer-Verlag: Berlin.)
Long, S. P., East, T. M., and Baker, N. R. (1983). Chilling damage to photosynthesis in young *Zea mays*. I. Effects of light and temperature variation on photosynthetic CO_2 assimilation. *J. Exp. Bot.* **34**, 177–88.
Morgan, D. C., Warrington, I. J., and Halligan, E. A. (1985). Effect of temperature and photosynthetic photon flux density on vegetative growth of kiwifruit (*Actinidia chinensis*). *N.Z. J. Agric. Res.* **28**, 109–16.
Nilsen, S., Chaturvedi, R., and Dons, C. (1984). Photoinhibition of photosynthesis in *Lemna gibba:* the effect of different O_2 and CO_2 concentrations during photoinhibitory treatment. *Acta Hortic.* **162**, 129–35.
Ögren, E., and Öquist, G. (1984). Photoinhibition of photosynthesis in *Lemna gibba* as induced by the interaction between light and temperature. III. Chlorophyll fluorescence at 77K. *Physiol. Plant.* **62**, 193–200.
Ögren, E., Öquist, G., and Hallgren, J.-E. (1984). Photoinhibition of photosynthesis in *Lemna gibba* as induced by the interaction between light and temperature. I. Photosynthesis *in vivo*. *Physiol. Plant.* **62**, 181–6.
Powles, S. B., and Björkman, O. (1982). Photoinhibition of photosynthesis: effect on chlorophyll fluorescence at 77K in intact leaves and in chloroplast membranes of *Nerium oleander*. *Planta* **156**, 97–107.
Powles, S. B., Berry, J. A., and Björkman, O. (1983). Interaction between light and chilling temperature on the inhibition of photosynthesis in chilling-sensitive plants. *Plant Cell Environ.* **6**, 117–23.
Taylor, A. O., and Rowley, J. A. (1971). Plants under climatic stress. I. Low temperature, high light effects on photosynthesis. *Plant Physiol.* **47**, 713–18.
Yakir, D., Rudich, J., and Bravdo, B.-A. (1985). Photoacoustic and fluorescence measurements of the chilling response and their relationship to carbon dioxide uptake in tomato plants. *Planta* **164**, 345–53.

Appendix 1. The Relationship between Chlorophyll Fluorescence and Photochemistry

Light energy absorbed by PS II can be dissipated in four major ways (Björkman 1987*a*, 1987*b*). These are fluorescence (F), non-radiative dissipation (D), transfer to PS I (T) and photochemical activity (P). The fluorescence yield when the PS II reaction centres are open, ϕF_0, is given by

$$\phi F_0 = F_0/\epsilon = K_F/(K_F+K_D+K_T+K_P)$$

and the fluorescence yield when the reaction centres are closed, ϕF_M, is given by

$$\phi F_M = F_M/\epsilon = K_F/(K_F+K_D+K_T),$$

where K_F, K_D, K_T and K_P are the appropriate rate constants for each process and ϵ is the coefficient to correct F to the same units as the excitation light causing the fluorescence emission. The efficiency of photochemistry of PS II, (ϕ_{pc}), is given by

$$\phi_{pc} = K_P/(K_F+K_D+K_T+K_P) = F_V/F_M.$$

Following Björkman (1987*b*) we have merged the rate constants K_T and K_D into a single rate constant, K_D $(= K_D+K_T)$.

The relative rate constants were calculated using the equations above with the following assumptions.

(1) K_F is constant at all times and set to equal 1.

(2) The correction coefficient for fluorescence measurement is constant at all times.
(3) $K_P = 70\,K_F$ at $t = 0$.

Using these assumptions, the value of ϵ at $t = 0$ is calculated. Then using this value of ϵ, K_P and K_D can be calculated for all $t > 0$ where:

$$K_D = (\epsilon - F_M)/F_M \qquad K_P = [(\epsilon - F_0) - K_D \times F_0]/F_0.$$

Appendix 2. Calculation of the Relative Contribution that Changes in the Rate Constants for Photochemistry (K_P) and Non-radiative Energy Dissipation (K_D) make to Changes in the Efficiency of Photochemistry, ϕ_{pc}

By definition, $K_{P,0} = 70$ at $t = 0$, $K_F = 1$.

The efficiency of PS II photochemistry prior to photoinhibition ($t = 0$) is given by

$$(\phi_{pc})_0 = \frac{K_{P,0}}{K_F + K_{D,0} + K_{P,0}}.$$

The efficiency of PS II photochemistry at any time, t, during photoinhibition or recovery when K_D changes (i.e. K_P is held constant) is given by

$$(\phi_{pc})_t^{K_P} = \frac{K_{P,0}}{K_F + K_{D,t} + K_{P,0}}.$$

The efficiency of PS II photochemistry at any time, t, during photoinhibition or recovery when K_P changes (i.e. K_D is held constant) is given by

$$(\phi_{pc})_t^{K_D} = \frac{K_{P,t}}{K_F + K_{D,0} + K_{P,t}}.$$

The relative change in ϕ_{pc} due to changes in K_D is thus given by

$$\frac{(\phi_{pc})_0 - (\phi_{pc})_t^{K_P}}{(\phi_{pc})_0}$$

and the relative change in ϕ_{pc} due to changes in K_P is given by

$$\frac{(\phi_{pc})_0 - (\phi_{pc})_t^{K_D}}{(\phi_{pc})_0}.$$

Photoinhibition at Chilling Temperatures

Robert M. Smillie, Suzan E. Hetherington, Jie He and Robyn Nott

Division of Horticulture, CSIRO, P.O. Box 52, North Ryde, Sydney, N.S.W. 2113, Australia.

Abstract

Relative susceptibilities of chilled leaves to photoinhibition were determined for 15 species of crop annuals showing a wide range of chilling tolerance. Leaf tissue at 7°C was exposed to a moderate photon irradiance of 300 μmol m^{-2} s^{-1} and photoinhibition was measured by the decrease in chlorophyll fluorescence (F_V/F_M) measured at 77K. All chilling-resistant plants surveyed (barley, broad bean, oat, pea and wheat) were photoinhibited at chilling temperatures. The chilling-sensitive plants (bean, cucumber, lablab, maize, pearl millet, pigeon pea, sesame, sorghum and tomato) were more susceptible, the mean of values for susceptibility to photoinhibition being twice that of the chilling-resistant plants. Rice, however, showed a tolerance to photoinhibition at 7°C comparable to that of some of the chilling-resistant plants. Indica rices were more susceptible than japonica rices. Photoinhibition increased with decreasing temperature and with increasing photon irradiance in both the chilling-resistant and sensitive plants. In pea and cucumber, photoinhibition at 7°C was correlated linearly with the decrease in photosystem II activity assayed in chloroplast thylakoids isolated from similarly treated tissue.

Relative tolerances of leaves of the same 15 species to chilling injury in the dark were also measured. No linear correlation was found between susceptibility of chilled leaves to photoinhibition and susceptibility to dark chilling injury. The pattern of differences between species for photoinhibition at 7°C was largely preserved when photoinhibitory treatments were given at a non-chilling temperature (21°C) by increasing the photon irradiance to 900 μmol m^{-2} s^{-1}. We conclude that, while the chilling-sensitive plants were generally more susceptible than the chilling-resistant ones to photoinhibition at low temperatures, this arose from a greater sensitivity to the irradiance rather than from the chilling sensitivity.

Photoinhibition associated with low temperatures was also demonstrated in the tropical fruit species, banana, pawpaw and *Monstera*. Low values of F_V/F_M recorded in leaves exposed to full sunlight during the winter month of July (range 0·39–0·56 compared with 0·70–0·79 in January) indicated that photoinhibition could adversely affect some tropical perennial fruit species cultivated in semitropical or warm temperate areas experiencing recurrent cool to cold winters.

Introduction

Temperatures below about 12°C greatly slow or stop the growth and development of plants of tropical origin. Prolonging the exposure can also induce chilling injury, a time and low-temperature dependent dysfunction of cellular processes, initially reversible but ultimately irreversible, leading to cell death. A feature of chilling injury is the inhibition of photosynthetic (Öquist and Martin 1986) and other metabolic processes (Graham and Patterson 1982) and a general deterioration of cellular functions which are often exacerbated with the return to higher temperatures (Lyons 1973). Chilling stress can therefore be a constraint on yields of tropical annual crops such as maize, tomato and legumes cultivated as summer crops in temperate environments or of tropical perennial crops grown in semitropical regions. In contrast, while low chilling temperatures may slow growth and metabolism of plants of temperate origin such as pea, wheat and barley,

0310–7841/88/010207$03.00

these plants are resistant to the development of chilling injury and growth rapidly resumes when the temperature rises (Pollack *et al.* 1983).

Chilling injury is most likely to develop at night but will also occur during the day if temperatures remain low. Experimentally, however, treatments to induce chilling injury are customarily given in darkness as the presence of light can elicit another sort of low temperature injury. The combination of chilling temperatures and high or even moderate light intensities results in photoinhibition of photosynthesis in leaves of chilling-sensitive plants. These include bean (Powles *et al.* 1983; Greer *et al.* 1986), cotton (Powles *et al.* 1983), cucumber (Kislyuk and Vas'kovskii 1972; Lasley *et al.* 1979; Van Hasselt and Van Berlo 1980; Hodgson *et al.* 1987), maize (Taylor and Rowley 1971; Baker *et al.* 1983; Long *et al.* 1983), rice (He *et al.* 1987), sorghum (Taylor and Rowley 1971) and tomato (Powles *et al.* 1983; Martin and Ort 1985; Yakir *et al.* 1985, 1986; Kee *et al.* 1986). Photoinhibition in these plants is characterised by decreases in the

Table 1. List of crop annuals used in the experiments on photoinhibition at chilling temperatures

Common name	Botanical name	Common name	Botanical name
Chilling sensitive species:		Sesame	*Sesamum indicum* L.
Bean	*Phaseolus vulgaris* L. (cv. Redlands Pioneer)	Sorghum	*Sorghum bicolor* (L.) Moench (cv. F64A hybrid)
Bean, dwarf	*P. vulgaris* L. (cv. Windsor Long Pod)	Tomato	*Lycopersicon esculentum* Mill. (cv. Rouge de Marmande)
Cucumber	*Cucumis sativus* L. (cv. Palomar)	*Chilling resistant species:*	
Lablab	*Lablab purpureus* (L.) Sweet	Barley	*Hordeum vulgare* L. (cv. Weeah)
Maize	*Zea mays* L. (cv. GH390 hybrid)	Broad bean	*Vicia faba* L. (cv. Coles Early)
Pearl millet	*Pennisetum typhoides* (Burm.f.) Stapf & Hubbard	Oat	*Avena sativa* L.
Pigeon pea	*Cajanus cajun* (L.) Millsp.	Pea	*Pisum sativum* L. (cv. Greenfeast)
Rice	*Oryza sativa* L.[A] (cv. Dee-geo-woo-gen)	Wheat	*Triticum aestivum* L. (cv. Gatcher)
Rice	*O. sativa*[B] (cv. Asahi)		

[A]Indica rice. [B]Japonica rice.

light-saturated rate of photosynthesis, the quantum yield for photosynthesis, and the photosystem II (PS II) activity of chloroplast membranes isolated from the leaves. Changes in chloroplast ultrastructure also take place in photoinhibited leaves (Taylor and Craig 1971; Van Hasselt 1974; Wise *et al.* 1983; Wise and Naylor 1987) and long-term exposure of leaves to the cold under high light results in photooxidative destruction of carotenoids, chlorophyll (Van Hasselt 1972) and lipids (Wise and Naylor 1987).

In several of the papers cited above the issue is raised that chilling in the light appears to be potentially more damaging to the plant than 'dark' chilling, that is, chilling effects due to temperature alone. Whether or not this is the case will depend on a number of variables such as climate and plant adaptability, but it nonetheless raises questions such as what is the precise nature of the interaction of chilling and light, what are the critical ranges of temperature and light intensity, which plants are most vulnerable, and what is the incidence of photoinhibition in plants under cultivation or in the wild. From an agricultural point of view it will be also important to define the extent to which photoinhibition at low temperatures can affect subsequent growth and development and, ultimately, crop yields.

Nearly all of the studies on photoinhibition at chilling temperatures have been carried out on chilling-sensitive plants and in the literature there is a clearly implied connection between chilling sensitivity and susceptibility of chilled leaves to photoinhibition. However, a cool temperate species *Lemna gibba* was photoinhibited by irradiating it at a low chilling temperature, 3°C (Ögren *et al.* 1984), and two species of potato, intermediate in chilling tolerance between the chilling-sensitive plants listed above and resistant cool temperate species like pea and barley, have also been reported to show light-dependent inhibition of photosynthetic oxygen evolution at chilling temperatures (Steffen and Palta 1986). Could it be that susceptibility to photoinhibition at chilling temperatures is widespread in the plant kingdom?

We have begun to address some of these questions by using plant species known to differ markedly in chilling tolerance and measuring photoinhibition and recovery from photoinhibition as a function of the temperature and the photon irradiance. As rapid chilling of leaves of chilling-sensitive plants can impose additional stress through water loss and wilting depending on the atmospheric relative humidity (McWilliam *et al.* 1982; Mustárdy *et al.* 1984), all photoinhibitory and chilling treatments were carried out with leaf tissue kept in a water-saturated atmosphere. This paper reviews our current investigations on the interaction of chilling and light, the results of which point to light sensitivity rather than chilling sensitivity as the overriding factor responsible for the differential responses displayed by the chilling-sensitive and resistant species to photoinhibition at low temperatures.

Table 2. Experimental conditions for photoinhibition and recovery

Condition	Photoinhibition	Recovery
Temperature (°C)	7	25
Photon irradiance (μmol $m^{-2} s^{-1}$)	300	20
Time (h)	20	24

Photoinhibition Induced by Chilling

(1) Photoinhibition at 7°C Compared in Fifteen Species

In order to ascertain if photoinhibition under conditions of chilling stress is confined in the main to chilling-sensitive plants or whether it affects a much wider spectrum of plants, relative susceptibilities to photoinhibition at 7°C were determined for leaves of 15 species. The experiments were confined to annual crop plants, but included both chilling-sensitive and resistant plants (Table 1). Two cultivars of bean were used and two of rice: one an indica type, the other a japonica type considered to be more cold tolerant. Recovery from the photoinhibitory treatment was also measured. The experimental conditions used for photoinhibition and recovery are summarised in Table 2. The photon irradiance chosen for the photoinhibitory treatment at 7°C did not induce photoinhibition at 21°C (data not shown).

The photoinhibitory treatments were given as described below. Plants were raised in pots using drip nutrient irrigation during summer under shade cloth in a glasshouse (temperature range 18–28°C; maximum photon irradiance 700–800 μmol $m^{-2} s^{-1}$). With rice, permanent flood was applied at the 3-leaf stage. Penultimate fully expanded leaves were harvested and floated on water in subdued light for 1–2 h. Discs (1·3 cm diameter) or 2-cm lengths of leaves (cereals only) were cut and placed on wet filter paper laid on an aluminium plate. The leaf pieces were covered with thin polyethylene film and the plate was then placed in a controlled environment cabinet and, after temperature equilibration, exposed to light from metal halide lamps. Following the irradiation (Table 2), photoinhibition of the samples was measured. Three samples were used for each determination and each experiment was repeated at least twice. Determinations were also made on freshly harvested leaves and on samples treated as described above but kept in darkness. Recovery from photoinhibition was measured after exposing photoinhibited samples to the conditions listed in Table 2.

Photoinhibition and recovery from photoinhibition were determined from changes in F_V/F_M measured on individual leaf samples cooled to 77K (Greer *et al.* 1986). Each sample was attached to one end of a Perspex rod and, following at least 10 min in darkness, was immersed in liquid nitrogen. The other end of the rod was positioned by means of an adaptor to the terminal end of a bifurcated light pipe system. Actinic blue light (Corning 4–96 filter, photon irradiance 28 μmol m^{-2} s^{-1}) was applied via a shutter through one arm of the pipe and induced chlorophyll (Chl) fluorescence emission from the leaf sample piped through the other arm to a photomultiplier protected by a 693-nm interference filter and a 2–64 Corning red cut-off filter. Output from the

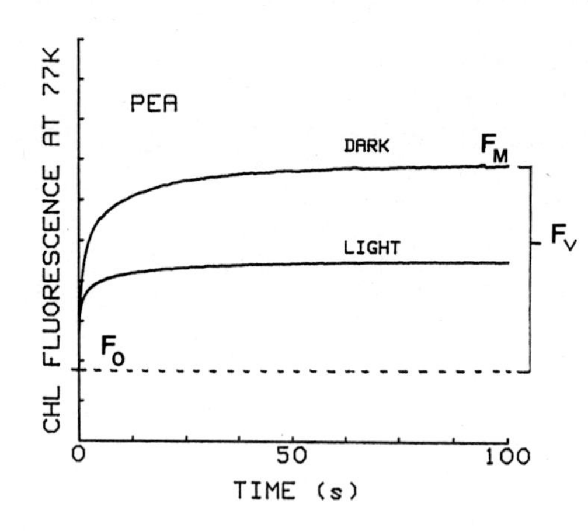

Fig. 1. Chl induction kinetics from PS II measured at 77K in pea leaf after treatment for 20 h in darkness at 7°C (dark) or at a photon irradiance of 300 μmol m^{-2} s^{-1} at 7°C (light). F_0, F_M and F_V are indicated on the graph for the dark treatment. F_0 was taken as the fluorescence emission obtained 5 ms after opening the shutter.

photomultiplier was fed via a fast A/D converter into a HP9826 computer (Hewlett Packard, Sydney) programmed to record F_0, F_V and F_M (Fig. 1). Data collection was initiated by opening the shutter. Fig. 1 shows the kinetics of induced Chl fluorescence emission at 77K from pea leaves kept either in the dark at 7°C for 20 h or under photoinhibitory conditions (Table 2). Both the maximum Chl fluorescence (F_M) and F_V, the difference between F_M and F_0, was decreased in the leaf irradiated at 7°C. Although not shown in Fig. 1, F_0 tended to show a net small increase following photoinhibitory treatments, but the kinetics were complex and in most species an initial slow increase in F_0 was followed by a decrease. At 77K, Chl fluorescence emission at 693 nm reflects the photochemistry of PS II and, as PS II is the primary site affected by photoinhibition, this accounts for the decrease in F_V. In this study the ratio F_V/F_M is used as the measure of photoinhibition as Björkman and Demmig (1987) have shown that the decrease in quantum yield for photosynthesis determined by gas exchange during photoinhibition and the decrease in F_V/F_M are linearly related. Measurements of Chl fluorescence made at 77K would not be expected to disclose any photoinhibitory stress effects on photosynthesis other than those affecting PS II photochemistry. The control values for F_V/F_M reported here are 5–10% lower than values reported by Björkman and Demmig (1987) because of a small breakthrough of actinic light to the photomultiplier.

Values for F_V/F_M at the start of the experiment and after chilling and recovery treatments in the dark are given in Table 3. The first column shows the values of F_V/F_M for freshly harvested leaf samples from each of the 17 cultivars. Values obtained varied little between cultivars. Leaves were harvested 4–5 h after dawn and slightly higher values were obtained for most species if the plants were kept in darkness from the end of the previous day. Maintaining leaves at 7°C in darkness for 20 h did not result in a significant change in F_V/F_M (column 2). Nor was there any appreciable change in F_V/F_M when these leaves were placed at 25°C in darkness for an additional 24 h (column 4) or for 24 h at 25°C under the low irradiance used for the recovery treatment (column 3). Thus any dark chilling or ageing effects on leaf samples during the 44-h treatment period had little effect on F_V/F_M.

Table 3. Control values for F_V/F_M

Values for F_V/F_M of leaf material used in photoinhibition and recovery experiments are shown in this table

Plant	Experimental condition			
	At harvest	7°C, dark 20 h	Recovery in light[A]	Recovery in dark[B]
Maize	0·70	0·67	0·68	0·66
Pearl millet	0·67	0·72	0·66	0·66
Lablab	0·76	0·77	0·70	0·72
Cucumber	0·76	0·75	0·78	0·71
Sesame	0·71	0·74	0·75	0·71
Sorghum	0·71	0·71	0·67	0·68
Pigeon pea	0·74	0·75	0·76	0·72
Bean[C]	0·77	0·74	0·74	0·74
Tomato	0·73	0·74	0·71	0·69
Bean[D]	0·77	0·76	0·75	0·75
Oat	0·73	0·75	0·70	0·69
Pea	0·73	0·74	0·72	0·68
Rice[E]	0·78	0·77	0·77	0·76
Rice[F]	0·77	0·78	0·75	0·77
Broad bean	0·74	0·71	0·69	0·67
Wheat	0·74	0·73	0·69	—
Barley	0·73	0·74	0·68	0·67
Mean ± s.e.m.	0·74 0·01	0·74 0·01	0·72 0·01	0·71 0·01

[A]Recovery: 7°C, dark for 20 h then 25°C, photon irradiance of 25 μmol m^{-2} s^{-1} for 24 h.
[B]Recovery: 7°C, dark for 20 h then 25°C, dark for 24 h.
[C]cv. Windsor Long Pod.
[D]cv. Redlands Pioneer.
[E]cv. Asahi.
[F]cv. Dee-geo-woo-gen.

The combination of light and low temperature did, however, result in a decrease in F_V/F_M in all cultivars. This is shown in Fig. 2 which also shows values of F_V/F_M determined after the recovery treatment. Also included in the figure for purposes of comparison are the control values shown in column 3 of Table 3, i.e. the treatment at 7°C in the dark for 20 h followed by recovery treatment in light. The names of the plants used are arranged in the figure in order of decreasing susceptibility to photoinhibition and, with the exception of rice, all chilling-sensitive species were more susceptible than the chilling-resistant species. The japonica rice was slightly less susceptible than the indica rice. In relation to chilling sensitivity, several important points emerged from this comparison. (1) Chilling-resistant as well as the chilling-sensitive cultivars were photoinhibited. (2) While the majority of the chilling-sensitive cultivars

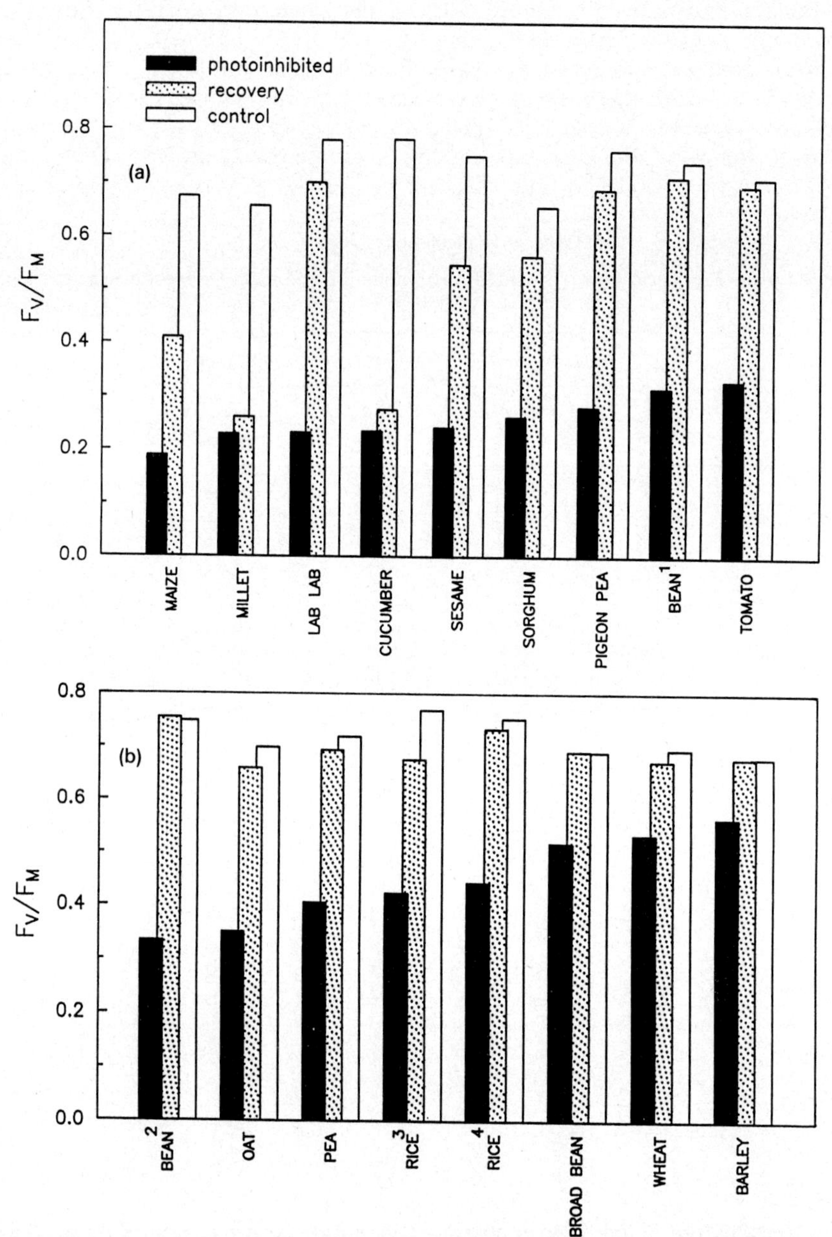

Fig. 2. Photoinhibition at 7°C and recovery from photoinhibition in leaves of chilling-sensitive and resistant plants. Experimental conditions were those shown in Table 2. After photoinhibition, F_V/F_M was determined at 77K. Comparable samples after photoinhibition were placed under recovery conditions (Table 2) and F_V/F_M subsequently determined at 77K. Control values after recovery treatment (column 3, Table 3) are included in the figure so that the extent of reversibility of photoinhibition can be compared. [1]cv. Windsor Long Pod; [2]cv. Redlands Pioneer; [3]cv. Dee-Geo-Woo-Gen; [4]cv. Asahi. (S. E. Hetherington, J. He and R. M. Smillie, unpublished data.)

were more susceptible to photoinhibition at 7°C, the differences between these and the chilling-resistant ones were no more than 2–3-fold (maize, millet, lablab and cucumber compared with broad bean, wheat and barley). (3) Some species recognised as being highly contrasting in chilling sensitivity showed almost the same susceptibility to photoinhibition at 7°C (tomato and bean versus oat and pea). (4) Amongst the chilling-sensitive species, rice behaved more like the chilling-resistant plants. The reason for the greater resistance of rice is not known, but it may be that a successful adaptation to thrive in flooded soils, an environment unsuitable for the other crop plants, necessitates an increase in generalised stress tolerance. Mean percentage decreases in F_V/F_M as the result of photoinhibition at 7°C (relative to dark control, column 2 in Table 3) for chilling-resistant plants, rice and the other chilling-sensitive plants are compared in Table 4. On average, species in the chilling-sensitive group excluding rice were twice as susceptible as those in the chilling-resistant group. Thus from the data presented in

Table 4. Relative susceptibility of chilling-resistant species, rice, and other chilling-sensitive species to photoinhibition at 7°C and photon irradiance of 300 μmol m^{-2} s^{-1} for 20 h

Parameter	Chilling-resistant species	Rice	Chilling-sensitive species
No. of cultivars	5	2	10
Mean % decrease in F_V/F_M ± s.e.m.	35·2 ± 5·9	44·1 ± 1·1	64·1 ± 1·0
Relative susceptibility	1·0	1·25	1·82

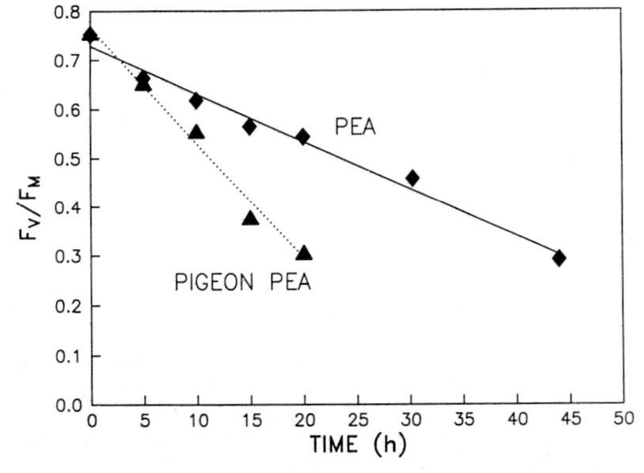

Fig. 3. Time course of changes in F_V/F_M in leaves exposed at 7°C to a photon irradiance of 300 μmol m^{-2} s^{-1}. $r^2 = 0·98$ for both sets of data.

Fig. 2 and Table 4 the observed relative susceptibilities to light plus chilling in the various species are very different from what might have been expected from the disparate differences in their chilling sensitivities. This aspect is considered in more detail in a later section of this paper.

(2) *Kinetics of Photoinhibition at 7°C*

Fig. 3 shows the change in F_V/F_M with time for the chilling-sensitive pigeon pea and the chilling-resistant pea. Similar results were obtained with cucumber versus barley. Greer *et al.* (1986) showed that F_V/F_M declined linearly in bean leaves photoinhibited at higher light (1400 μmol m^{-2} s^{-1}) and temperature (20°C). In maize (Long *et al.* 1983) and in rice (He *et al.* 1987), inhibition of photosynthetic CO_2 assimilation following irradiations at 5°C also showed an almost linear dependence on the duration of the chilling period.

(*3*) *Recovery from Photoinhibition*

Photoinhibition incurred by the chilling-resistant plants was reversible (Fig. 2*b*). Amongst the chilling-sensitive species, tomato and the legumes showed good recovery from photoinhibition. In contrast, pearl millet and cucumber recovered very little while maize and sesame recovered partially. Hence, while the capacity of a species to recover may have been related, in part, to the degree of photoinhibition incurred, other factors were involved and distinct differences existed between species in their capacity to recover from severe photoinhibition. For instance, F_V/F_M was decreased by about 80% by the photoinhibitory treatment in both lablab and cucumber, yet lablab recovered to 90% of the control value while cucumber showed almost negligible reversal of photoinhibition (Fig. 2*a*). This difference in recovery was unlikely to be related to chilling tolerance since these two species showed almost equal susceptibility to chilling injury in the dark (see Table 6 in later section).

Table 5. Photoinhibition of rice cultivars at chilling temperatures

Leaves harvested from different cultivars of rice were exposed to a photon irradiance of 275 μmol $m^{-2} s^{-1}$ at either 8°C or 0°C for 20 h and photoinhibition was measured by determining F_V/F_M at 77K. Reversibility of photoinhibition at 0°C was checked by placing the photoinhibited leaves under recovery conditions (Table 2) and later determining F_V/F_M (S. E. Hetherington and J. He, unpublished). I, indica. J, japonica

Rice cultivar	Type	F_V/F_M for photoinhibition at 8°C	0°C	0°C then recovery
Asahi	J	0·60	0·49	0·73
Calrose	J	0·61	0·44	0·67
Inga	I×J	0·61	0·42	0·65
Century Patna	I×J	0·60	0·46	0·69
IR2070-747-6-6	I	0·48	0·32	0·46
Dee-geo-woo-gen	I	0·58	0·35	0·57
IR8	I	0·46	0·36	0·42
Emata Yin	I	0·54	0·37	0·56

(*4*) *Photoinhibition of Indica and Japonica Rices at Chilling Temperatures*

While rice proved to be somewhat more resistant to photoinhibition at 7°C than the other chilling-sensitive species, indica rice was more intolerant than japonica rice. Indica rices are mostly grown in regions with warm climates such as South-East Asia and southern China, whereas japonica rices, which are adapted to cooler areas, are grown in Japan, Korea, central and northern China and south-eastern Australia (Yoshida 1981). To test the effectiveness of the Chl fluorescence assay at 77K to separate cultivar susceptibility, photoinhibition was measured in several indica and japonica rices and in their hybrids. Leaf samples from these rices were exposed during irradiation to either a severe chilling temperature of 0°C or a milder one of 8°C. Recovery following photoinhibition at 0°C was also measured. Values of F_V/F_M after any of these treatments (Table 5) served to separate the rices into two groups. The more tolerant group consisted of the japonica and hybrid indica × japonica rices, the less tolerant group the four indica rices. The similar grouping of the japonica and the hybrid rices is not surprising as the latter were hybrids selected basically for the cool tolerance traits of the japonica type and the long-grain characteristics of the indica type. These results suggest that Chl fluorescence determinations may be useful as a means of selecting cultivars with decreased susceptibility to low-temperature induced photoinhibition.

Relative Roles of Temperature and Light in Photoinhibition at Chilling Temperatures

(*1*) *Temperature*

(*i*) *Photoinhibition in response to temperature*

Photoinhibition of photosynthetic CO_2 assimilation in maize leaves following exposure to a photon irradiance of 1500 μmol m^{-2} s^{-1} at temperatures ranging from 20°C to 5°C showed a close to linear dependence on the temperature, increasing with decreasing temperature (Long *et al.* 1983) although Powles *et al.* (1983), using an even higher photon irradiance of 2000 μmol m^{-2} s^{-1}, reported that inhibition of CO_2 assimilation did not occur at temperatures above 11·5°C. The degree of photoinhibition in leaves of cucumber, measured by the decrease in F_V/F_M, increased with decreasing temperature in the range 17·3–4·1°C (Fig. 4), in agreement with the results obtained with maize by Long *et al.* (1983). Photoinhibition in leaves of pigeon pea and sesame showed similar temperature response curves in the range 21·5°C to 3°C (unpublished data). However, the incidence of photoinhibition in leaves of the chilling-resistant pea

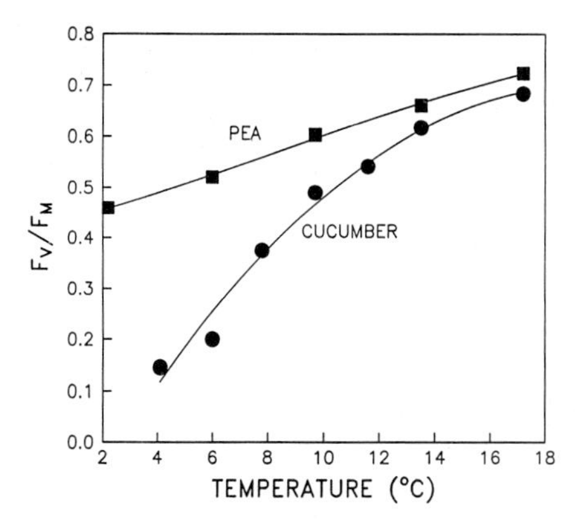

Fig. 4. Photoinhibition at different temperatures in pea and cucumber leaves. Leaf tissue was exposed to a photon irradiance of 275 μmol m^{-2} s^{-1} for 20 h at the temperatures indicated and photoinhibition measured by determining F_V/F_M at 77K.

also showed a dependence on the chill temperature, with photoinhibition increasing as the temperature decreased. Similar results were obtained using indica rice, broad bean and barley, except for a more pronounced downward deviation from linearity with decreasing temperature (unpublished data). Clearly, photoinhibition at chilling temperatures is not restricted to chilling-sensitive plants and occurs also in quite chilling-resistant plants, albeit at a slower rate. Another important conclusion from these results and those of Long *et al.* (1983) is that low-temperature enhanced photoinhibition is not confined to the generally accepted chilling-sensitive temperature range, i.e., 0°C to about 12°C (Lyons 1973), but can be demonstrated to occur at temperatures higher than this, in both chilling-sensitive and resistant plants.

(*ii*) *Relationship of F_V/F_M decrease to PS II activity*

In both early (Kok *et al.* 1965) and more recent (Cleland and Critchley 1985) studies, photoinhibition has been characterised by a decrease in PS II activity. While decreases in F_V/F_M in photoinhibited leaf tissue have been related to decreases in the quantum yield for photosynthesis (Björkman and Demmig 1987), neither have been directly correlated with PS II activity. In order to ascertain if the decrease in F_V/F_M is paralleled by a decrease in PS II activity, leaf sections adjacent to those used to obtain the results shown in Fig. 4 were homogenised in a chloroplast isolation medium and PS II activity was determined in the resulting suspension by photoreduction of ferricyanide in the presence of *p*-phenylenediamine (Hetherington and Smillie 1982). In Fig. 5 the results of these determinations are plotted against the values for F_V/F_M given in Fig. 4. For both pea and cucumber there was a linear correlation between F_V/F_M measured on frozen but otherwise intact leaf tissue and PS II activity measured on released chloroplast thylakoids. Hence in both chilling-resistant and sensitive plants, photoinhibition involves inhibition of PS II activity which, as with inhibition of induced Chl fluorescence at 77K, increases with decreasing temperature during photoinhibitory treatment.

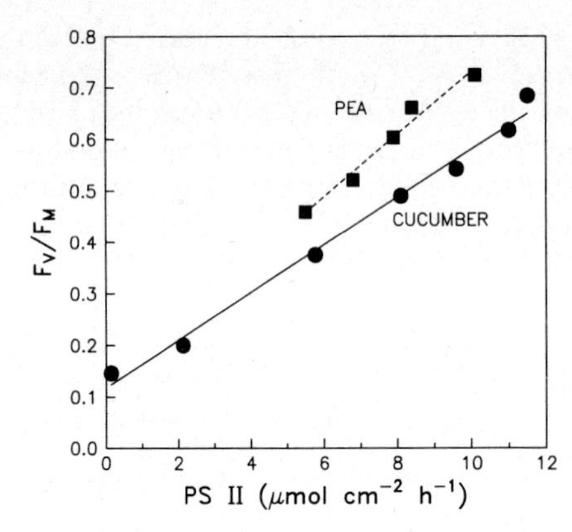

Fig. 5. Relationship between F_V/F_M measured on leaf tissue at 77K and PS II activity (μmol ferricyanide reduced expressed on a leaf area basis) measured in chloroplast thylakoids isolated from leaf tissue after photoinhibiting the tissue at 7°C and a photon irradiance of 275 μmol m^{-2} s^{-1} for 20 h. Values obtained using pea (r^2 = 0·95) and cucumber leaves (r^2 = 0·99) are shown. (R. M. Smillie and R. Nott, unpublished data.)

Table 6. Relative susceptibilities of different cultivars to chilling at 0°C in darkness

Values given are means for chilling tolerance (time in h for 50% decrease in F_R at 0°C) ± s.e.m. n = 16. (S. E. Hetherington, J. He and R. M. Smillie, unpublished data)

Plant	Chilling tolerance (h)	Plant	Chilling tolerance (h)
Pigeon pea	2·9 ± 0·3	Sorghum	37 ± 2
Lablab	3·4 ± 0·3	Rice	
Sesame	3·6 ± 0·5	cv. Asahi	40 ± 3
Cucumber	3·6 ± 0·3	cv. Dee-Geo-Woo-Gen	92 ± 7
Bean		Broad bean	214 ± 14
cv. Windsor Long Pod	3·8 ± 0·2	Wheat	259 ± 20
cv. Redlands Pioneer	4·7 ± 0·3	Oat	267 ± 21
Tomato	7·0 ± 0·6	Barley	355 ± 25
Maize	9·3 ± 0·8	Pea	490 ± 42
Pearl millet	36 ± 5		

(*iii*) *Comparison of chilling in the light with chilling in the dark*

Because differences in photoinhibition between chilling-sensitive and resistant plants amounted to no more than 3-fold or less, it was decided to measure the 'dark' chilling tolerance of each species so that a comparison could be made directly between susceptibility to the combined effects of light and chilling and susceptibility to chilling in the dark. Relative dark chilling tolerances were determined on leaf samples similar to those used to obtain the results shown in Fig. 2. A temperature of 0°C was chosen rather than

7°C because chilling injury develops very slowly at the latter temperature. The development of chilling stress at 0°C in the leaf samples was monitored by the change in induced Chl fluorescence, measured also at 0°C (Hetherington *et al.* 1983*a*, 1983*b*). The Chl fluorescence value measured was F_R, the maximal rate of rise of induced Chl fluorescence (I to P rise). As the decrease in log F_R due to the development of chilling injury in the leaf tissue is linearly related to the time of chilling exposure, relative susceptibility to dark chilling stress (chilling tolerance) was taken as the time of exposure at 0°C to give a 50% decrease in F_R (Hetherington *et al.* 1983*a*; Smillie and Hetherington 1983).

Table 6 shows values obtained for the 15 species used in the study. As might have

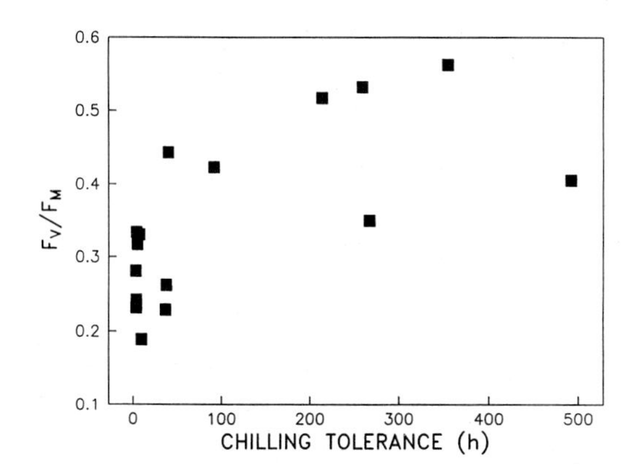

Fig. 6. Plot of values for photoinhibition at 7°C and a photon irradiance of 300 μmol m^{-2} s^{-1} (Fig. 2) versus values for chilling tolerance (0°C, dark; see Table 6). Each point represents a different cultivar.

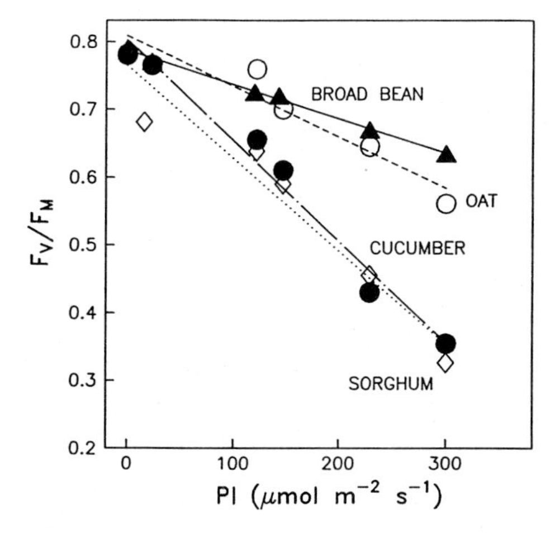

Fig. 7. Photoinhibition at 7°C as a function of photon irradiance (PI). Leaf tissues of broad bean, oat, cucumber and sorghum were maintained at 7°C while exposed for 20 h to the light intensities indicated (r^2 = 1·00, 0·95, 0·90 and 0·98 for broad bean, oat, cucumber and sorghum, respectively). (S. E. Hetherington, J. He and R. M. Smillie, unpublished data.)

been expected from the known behaviour of the different plants to chilling conditions, differences in dark chilling tolerance between the most sensitive and resistant plants were very large indeed. For instance, the dark chilling tolerance of pea was 166 times that of pigeon pea. In Fig. 6 the values for photoinhibition at 7°C are plotted against dark chilling tolerance. Obviously, no linear relationship exists between susceptibility to photoinhibition at chilling temperatures and susceptibility to dark chilling injury. Predisposition to dark chilling injury does not impart a high degree of susceptibility to photoinhibition at 7°C, while at the other end of the chilling spectrum a high degree of chilling tolerance does not guarantee protection against enhanced photoinhibition at chilling temperatures.

(2) Light

During chilling in the light, the extent of inhibition of photosynthesis in leaves of chilling-sensitive plants at a particular temperature depends on the photon irradiance (Long *et al.* 1983; Powles *et al.* 1983; Yakir *et al.* 1985). Photoinhibition measured by Chl fluorescence changes was likewise dependent on the light intensity in both chilling-sensitive and resistant plants (Fig. 7). Of particular interest in this experiment was the finding that when the irradiance was varied instead of the temperature, a difference in photoinhibition between the chilling-sensitive (cucumber and sorghum) and resistant (broad bean and oat) plants was still evident (Fig. 7). Since, as discussed above, this difference did not seem to be linked closely to susceptibility to dark chilling injury, we considered the possibility that it could result from a greater sensitivity of the chilling-sensitive plants to light. If this were so, then it might be expected that the differences in photoinhibition between the chilling-sensitive and resistant plants should be preserved at non-chilling temperatures.

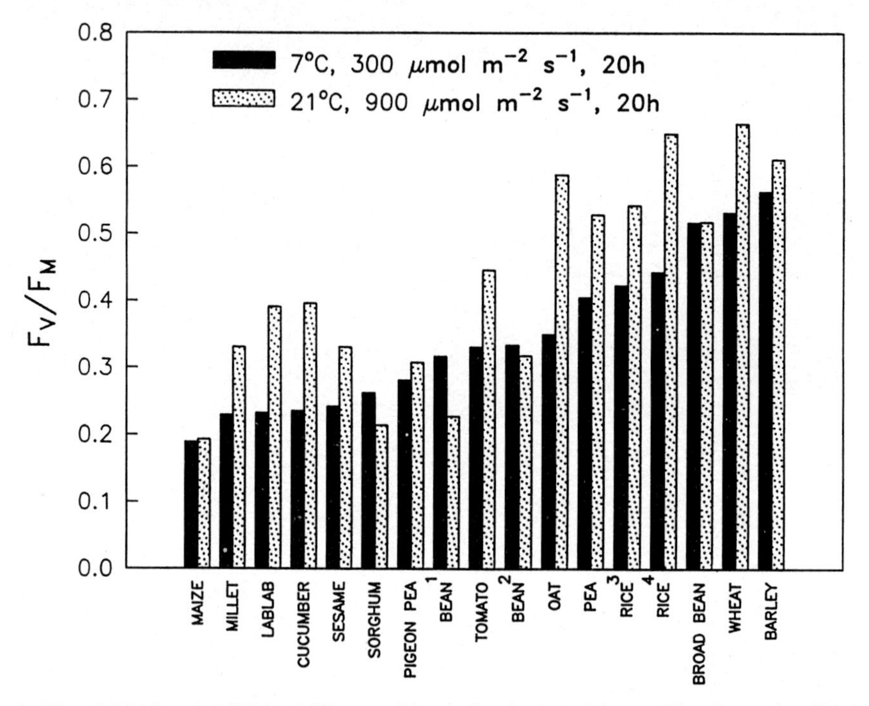

Fig. 8. Photoinhibition at 21°C in chilling-sensitive and resistant cultivars. The photon irradiance was 900 μmol m^{-2} s^{-1} for 20 h. For purposes of comparison, data for photoinhibition at 7°C (Fig. 3) are included in the figure. Footnotes are as in Fig. 3.

Accordingly, photoinhibition was determined at 21°C for each of the 15 species used previously. As photoinhibition at a photon irradiance of 300 μmol m^{-2} s^{-1} was negligible at 21°C, the irradiance was increased to 900 μmol m^{-2} s^{-1} for this experiment. The results are given in Fig. 8. For comparison, values for photoinhibition at 7°C (Fig. 2) are included. Although the experiment at 21°C was carried out on a different batch of plants to those used for the experiment at 7°C, the same pronounced trends seen at 7°C were also seen at 21°C, namely more photoinhibition occurred in chilling-sensitive plants than in chilling-resistant ones and rice. At both temperatures the indica rice was more sensitive than the japonica rice. In another experiment in which one half of the leaf was treated at 7°C and the other half at 21°C, a plot of photoinhibition at

21°C against photoinhibition at 7°C gave a linear correlation with an r^2 of 0·86 (unpublished data).

These results suggest that the enhanced susceptibility of chilling-sensitive plants to photoinhibition at chilling temperatures is the result of a reduced capacity to cope with light at any temperature rather than plant intolerance of chilling temperature. While the concept put forth here of an inherent higher light sensitivity being primarily responsible for the greater susceptibility to photoinhibition in tropical herbaceous species is in sharp contrast with the conclusions reached in previous studies that emphasised the plant chilling sensitivity (see Introduction), it does not appear to be in conflict with the results of these studies. In this connection it is interesting that, when photosynthetic parameters were found to decrease more in *Solanum tuberosum* than in the more cold-tolerant wild potato *S. acaule* after chilling at moderate light intensities, Steffen and Palta (1986) speculated that this differential response might be due, in part, to protection from light afforded by the considerably thicker palisade layer present in *S. acaule* leaves. However, anatomical differences are unlikely to account for the different responses of the chilling-sensitive and resistant species described in this paper and it may seem paradoxical that species adapted to a high-light tropical environment are more light-sensitive with respect to photoinhibition than species from temperate climes. The reason for this may lie in the recovery process. Recovery from photoinhibition in bean takes several hours, appears to require chloroplast protein synthesis and is highly temperature dependent, increasing in rate more than 3-fold between 20 and 30°C (Greer *et al.* 1986). Consequently, overnight recovery from photoinhibition will be more difficult during the growing seasons in temperate regions than in the tropical zone because of the shorter nights and lower temperatures. With less latitude for recovery, temperate summer annuals may require higher initial resistances to photoinhibition.

Table 7. Winter photoinhibition

Values given are means of determinations of F_V/F_M at 77K made directly on harvested leaves. The number of leaves sampled is given in parentheses (R. M. Smillie, unpublished data). Min/max temperatures were: July, 6/17°C; January, 18/26°C

Plant	Exposure	F_V/F_M July	F_V/F_M January
Monstera deliciosa Liebm.	Shade	0·74(2)	0·79(3)
	Sun/shade	0·66(8)	0·71(4)
	Sun	0·39(16)	0·73(7)
Banana (*Musa*, AAA group)	Sun/shade	0·67(15)	0·70(5)
	Sun	0·56(9)	0·72(4)
Pawpaw (*Carica papaya* L.)	Sun/shade	0·54(3)	—
	Sun	0·42(4)	0·72(3)

Seasonal Low-temperature Photoinhibition under Natural Conditions

The demonstration of enhanced photoinhibition at low temperature in the laboratory raises the question of whether it is possible to detect adverse effects of photoinhibition in plants growing in the field. As one approach to this problem, the question of demonstrating a cold seasonal incidence of photoinhibition was addressed by examining leaves of three perennial tropical fruit species for signs of photoinhibition (decrease in F_V/F_M at 77K) in midwinter. The plants were growing in a frost-free harbourside area of Sydney which is subject to cool to cold winters with minimum temperatures only 1–2°C lower than those of the main banana producing area located 400 km to the north. Leaves were sampled in the coldest month July and also during the summer in January.

Two categories of leaves were used, those exposed to sunlight for most of the day and those exposed for about 50% of the day. With *Monstera* it was also possible to obtain leaves which had not received any direct exposure to sunlight. The results given in Table 7 provide evidence of the occurrence of photoinhibition. Appreciable photoinhibition was found only in leaves sampled in July, leaves exposed to sunlight showing more photoinhibition than partially exposed or shaded leaves. These results suggest that photoinhibitory damage can occur in tropical fruit species grown in climates with recurrent cool winters and that the incidence of damage depends both on the temperature and the degree of exposure to sunlight.

Conclusions

Photoinhibition is a temperature- and light-dependent process which results in a decrease in the efficiency of utilisation of the energy of captured photons for photosynthesis. While net photoinhibition may be negligible at warm growth temperatures, the situation may change if the temperature drops during daylight hours as evidenced by the demonstration that photoinhibition occurred during winter in leaves of tropical fruit species growing in a warm temperate climate. In laboratory experiments with herbaceous species, it was shown that the extent of photoinhibition increases with decreasing temperature and there is near-linear dependence on light intensity and duration of chilling exposure in light. Up to a point, photoinhibition induced by low temperatures is completely reversible and might hence be thought of as a survival mechanism to protect leaves from damage from unused absorbed light energy. However, it can also lower photosynthetic rates and this effect persists for some time after the environmental stress has passed. For intensively cultivated plants, selection for the correct balance between these two aspects of photoinhibition, that is avoidance of damage from excess light versus reduced photosynthesis, may be necessary to maximise yields.

Photoinhibition at low temperatures is not specifically associated with chilling-sensitive plants of tropical origin. Both tropical chilling-sensitive cultivars and temperate chilling-resistant cultivars were photoinhibited during exposure to chilling temperatures. Appreciable photoinhibition also occurred at moderate light intensities in the range 10–20°C in both chilling-sensitive and resistant cultivars. Further, photoinhibition induced by chilling did not show a linear correlation with susceptibility to chilling in the dark and indeed some cultivars differing by more than 50-fold in rates of development of chilling injury in the dark showed similar rates of photoinhibition in the cold.

While the chilling-sensitive plants are on average more susceptible than the resistant ones to photoinhibition at chilling temperatures, this is attributed to an increased sensitivity to photon irradiance rather than to their greater chilling sensitivity.

Acknowledgment

This work was supported in part by a grant from the Australian Centre for International Agricultural Research.

References

Baker, N. R., East, T. M., and Long, S. P. (1983). Chilling damage to photosynthesis in young *Zea mays*. II. Photochemical function of thylakoids *in vivo*. *J. Exp. Bot.* **34**, 189–97.

Björkman, O., and Demmig, B. (1987). Photon yield of O_2 evolution and chlorophyll fluorescence characteristics at 77K among vascular plants of diverse origins. *Planta* **170**, 489–504.

Cleland, R. E., and Critchley, C. (1985). Studies on the mechanism of photoinhibition in higher plants. II. Inactivation by high light of photosystem II reaction center function in isolated spinach thylakoids and O_2 evolving particles. *Photobiochem. Photobiophys.* **10**, 83–92.

Graham, D., and Patterson, B. D. (1982). Responses of plants to low, nonfreezing temperatures: proteins, metabolism, and acclimation. *Annu. Rev. Plant Physiol.* **33**, 347–72.

Greer, D. H., Berry, J. A., and Björkman, O. (1986). Photoinhibition of photosynthesis in intact bean leaves. Role of light and temperature, and requirement for chloroplast-protein synthesis during recovery. *Planta* **168**, 253–60.

He, J., Wang, Y.-R., Liu, H.-X., and Guo, J.-Y. (1987). The effect of chilling temperature and light on photosynthesis in rice flag leaves in the milk stage. *Plant Physiol. Acta Sinica*, in press.

Hetherington, S. E., and Smillie, R. M. (1982). Humidity-sensitive degreening and regreening of leaves of *Borya nitida* Labill. as followed by changes in chlorophyll fluorescence. *Aust. J. Plant Physiol.* **9**, 587–99.

Hetherington, S. E., Smillie, R. M., Hardacre, A. K., and Eagles, H. A. (1983*a*). Using chlorophyll fluorescence *in vivo* to measure the chilling tolerances of different populations of maize. *Aust. J. Plant Physiol.* **10**, 247–56.

Hetherington, S. E., Smillie, R. M., Malagamba, P., and Huamán, Z. (1983*b*). Heat tolerance and cold tolerance of cultivated potatoes measured by the chlorophyll-fluorescence method. *Planta* **159**, 119–24.

Hodgson, R. A. J., Orr, G. R., and Raison, J. K. (1987). Inhibition of photosynthesis by chilling in light. *Plant Sci. Lett.* **49**, 75–9.

Kee, S. C., Martin, B., and Ort, D. (1986). The effects of chilling in the dark and in the light on photosynthesis of tomato: electron transfer reactions. *Photosynth. Res.* **8**, 41–51.

Kislyuk, I. M., and Vas'kovskii, M. D. (1972). Effect of cooling cucumber leaves on photosynthesis and photochemical reactions. *Sov. Plant Physiol.* **19**, 688–92.

Kok, B., Gassner, E. B., and Rurainski, H. J. (1965). Photoinhibition of chloroplast reactions. *Photochem. Photobiol.* **4**, 215–27.

Lasley, S. E., Garber, M. P., and Hodges, C. F. (1979). After effects of light and chilling temperatures on photosynthesis in excised cucumber cotyledons. *J. Am. Soc. Hortic. Sci.* **104**, 477–80.

Long, S. P., East, T. M., and Baker, N. R. (1983). Chilling damage to photosynthesis in young *Zea mays*. I. Effects of light and temperature variation on photosynthetic CO_2 assimilation. *J. Exp. Bot.* **34**, 177–88.

Lyons, J. M. (1973). Chilling injury in plants. *Annu. Rev. Plant Physiol.* **24**, 445–51.

Martin, B., and Ort, D. R. (1985). The recovery of photosynthesis in tomato subsequent to chilling exposure. *Photosynth. Res.* **6**, 121–32.

McWilliam, J. R., Kramer, P. J., and Musser, R. L. (1982). Temperature-induced water stress in chilling-sensitive plants. *Aust. J. Plant Physiol.* **9**, 343–52.

Mustárdy, L. A., Sz-Róssa, Z., and Faludi-Dániel, Á. (1984). Chilling syndrome in light-exposed maize leaves and its easing by low doses of DCMU. *Physiol. Plant.* **60**, 572–6.

Ögren, E., Öquist, G., and Hällgren, J-E. (1984). Photoinhibition of photosynthesis in *Lemna gibba* as induced by the interaction between light and temperature. I. Photosynthesis *in vivo*. *Physiol. Plant.* **62**, 181–6.

Öquist, G., and Martin, B. (1986). Cool climates. In 'Photosynthesis in Contrasting Environments'. (Eds N. R. Baker and S. P. Long.) pp. 237–93. (Elsevier: Amsterdam.)

Pollock, C. J., Lloyd, E. J., Stoddard, J. L., and Thomas, H. (1983). Growth, photosynthesis and assimilate partitioning in *Lolium temulentum* exposed to chilling temperatures. *Physiol. Plant.* **59**, 257–62.

Powles, S. B., Berry, J. A., and Björkman, O. (1983). Interaction between light and chilling temperature on the inhibition of photosynthesis in chilling-sensitive plants. *Plant Cell Environ.* **6**, 117–23.

Smillie, R. M., and Hetherington, S. E. (1983). Stress tolerance and stress-induced injury in crop plants measured by chlorophyll fluorescence *in vivo*. Chilling, freezing, ice cover, heat, and high light. *Plant Physiol.* **72**, 1043–50.

Steffen, K. L., and Palta, J. P. (1986). Effect of light on photosynthetic capacity during cold acclimation in a cold-sensitive and a cold-tolerant potato species. *Physiol. Plant.* **66**, 353–9.

Taylor, A. O., and Craig, A. S. (1971). Plants under climatic stress. II. Low temperature, high light effects on chloroplast ultrastructure. *Plant Physiol.* **47**, 719–25.

Taylor, A. O., and Rowley, J. A. (1971). Plants under climatic stress. I. Low temperature, high light effects on photosynthesis. *Plant Physiol.* **47**, 713–18.

Van Hasselt, P. R. (1972). Photooxidation of leaf pigments in *Cucumis* leaf discs during chilling. *Acta Bot. Neerl.* **21**, 539–48.

Van Hasselt, P. R. (1974). Photo-oxidative damage to the ultrastructure of *Cucumis* chloroplasts during chilling. *Proc. K. Ned. Akad. Wet.* **C77**, 50–6.

Van Hasselt, P. R., and Van Berlo, H. A. C. (1980). Photooxidative damage to the photosynthetic apparatus during chilling. *Physiol. Plant.* **50**, 52–6.

Wise, R. R., McWilliam, J. R., and Naylor, A. W. (1983). A comparative study of low-temperature-induced ultrastructural alterations of three species with differing chilling sensitivities. *Plant Cell Environ.* **6**, 525–35.

Wise, R. R., and Naylor, A. W. (1987). Chilling-enhanced photooxidation. I. The peroxidative destruction of lipids during chilling injury to photosynthesis and ultrastructure. *Plant Physiol.* **83**, 272–7.

Yakir, D., Rudich, J., and Bravdo, B.-A. (1985). Photoacoustic and fluorescence measurements of the chilling response and their relationship to carbon dioxide uptake in tomato plants. *Planta* **164**, 345–53.

Yakir, D., Rudich, J., Bravdo, B.-A., and Malkin, S. (1986). Prolonged chilling under moderate light: effect on photosynthetic activity measured with the photoacoustic method. *Plant Cell Environ.* **9**, 581–8.

Yoshida, S. (1981). 'Fundamentals of Rice Crop Science.' (The International Rice Research Institute: Los Baños, Philippines.)

Photosynthetic Utilisation of Lightflecks by Understory Plants

Robert W. Pearcy

Department of Botany, University of California,
Davis, CA 95616, U.S.A.

Abstract

The light environment in forest understories is highly dynamic because the weak shade light is periodically punctuated by lightflecks lasting from a second or less to tens of minutes. Although present for only a small fraction of the day, these lightflecks can contribute more than two-thirds of the photosynthetically active radiation. Several factors are of importance in determining the capacity of a leaf to utilise lightflecks. Following long low-light periods the induction state of the photosynthetic apparatus is limiting. During induction, 20–60 min may be required before maximum assimilation rates are reached due first to a light activation requirement of ribulose-1,5-bisphosphate carboxylase/oxygenase and later to the light-induced stomatal opening. Continuous light is not required and induction occurring during a series of lightflecks results in higher carbon gain for later as compared to earlier lightflecks. Post-illumination CO_2 fixation resulting from utilisation of metabolite pools built up during the lightfleck can significantly enhance carbon gain during short (5–20 s) lightflecks. The carbon gain of a leaf in response to a lightfleck is a consequence of the limitations imposed by induction state plus the enhancements due to post-illumination CO_2 fixation. In the field, this will depend on the frequency and duration of the lightflecks and the duration of the intervening low-light periods.

Introduction

Understory light environments are characterised by a very low level of diffuse light that is punctuated by intense sunflecks lasting from a second or less to 15 min or more. The importance of sunflecks has been discussed since the pioneering studies of Lundegarth (1921) on photosynthesis of understory herbs and the measurements of light by Richards (1952) and Evans (1956) in tropical forests. Indeed, on clear days, 30–60% of the daily carbon gain by understory plants in tropical forests can be attributed to utilisation of sunflecks (Björkman *et al.* 1972; Pearcy and Calkin 1983; Pearcy 1987). Moreover, the growth of tree seedlings in a Hawaiian forest understory has been shown to be highly correlated with estimates of sunfleck radiation derived from hemispherical canopy photographs (Pearcy 1983).

Studies of photosynthetic adaptation to shaded understory conditions have concentrated on steady-state responses to light in which measurements are recorded after constant conditions have been achieved. These studies have elegantly elucidated the mechanisms at the chloroplast to the whole leaf level that allow positive net photosynthesis in the very low diffuse light of the understory as well as those involved in long-term acclimation to different light environments during growth (Björkman 1973, 1981; Boardman 1977). However, responses to the rapid changes in light characterising sunflecks are inherently dynamic in nature and thus do not necessarily follow from steady-state gas exchange characteristics. Dynamic responses have largely been ignored because of methodological difficulties involved in measuring the rapid changes in CO_2 uptake. However, with recent improvements in infrared gas analysers and the avail-

0310-7841/88/010223$03.00

ability of computer-based data acquisition systems, measurement of faster responses as well as handling of the resulting large amount of data is now possible.

In this paper, I will discuss some recent studies on the dynamic responses of photosynthesis and their relationship to the utilisation of sunflecks. Much of it will focus on studies of sunfleck utilisation by *Alocasia macrorrhiza* (L.) G. Don, which is common in understories of tropical forests of north-eastern Australia and indeed has previously served as the material for studies of shade adaptation (Björkman *et al.* 1972; Boardman *et al.* 1972; Boardman 1977). Field measurements were made at the Curtain Fig State Forest, near Atherton, Qld (lat. 17°S.), where the CSIRO Division of Forest Research has an ongoing experimental program in tropical plant ecophysiology. Laboratory studies were carried out with plants growing under shadecloth in a glasshouse that gave about 1 mol photons m^{-2} day^{-1} (midday photon irradiance of 25–35 μmol photons m^{-2} s^{-1}), a value comparable to those measured in the understory. The laboratory gas exchange measurements in part examine the response of CO_2 assimilation to short pulses of light (lightflecks) designed to simulate naturally occurring sunflecks. For convenience in this paper, both naturally occurring sunflecks and their artificial counterparts will be referred to as lightflecks.

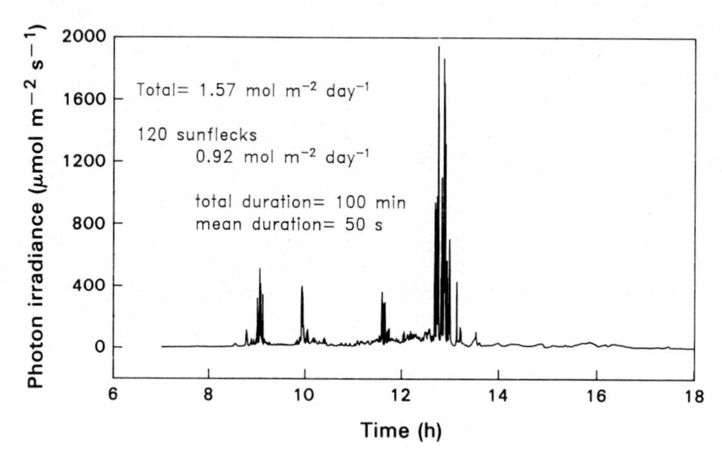

Fig. 1. Diurnal course of photon irradiance received on the surface of an *Alocasia macrorrhiza* leaf in the understory of the Curtain Fig Forest on 22 February 1987. The data were obtained with a GaAsP photodiode light sensor connected to a Campbell Scientific CR21X micrologger that recorded the signal at 1 s intervals. The light sensor was calibrated against a LI-COR model 190s quantum sensor.

Dynamics of the Light Environment

Lightflecks resulting from the penetration of direct-beam solar radiation through holes in the overstory canopy are a common feature of understory light environments. As is shown in Fig. 1, diffuse light levels range from 10 to 20 μmol photons m^{-2} s^{-1} but are punctuated by lightflecks that range in duration from a second or less to over 10 min. At this site, the daily photon irradiance was 1·57 mol photons m^{-2} day^{-1}, of which 59% was received in 120 lightflecks each exceeding 50 μmol photons m^{-2} s^{-1}. The values at other sites within a 5 m radius on the same day ranged from 0·47 to 1·5 mol m^{-2} day^{-1} (mean = 0·95 mol m^{-2} s^{-1}). The percentage contributed by lightflecks ranged from 12 to 65% (mean = 38%). These values are similar to measurements made on clear days in other tropical forests (Björkman and Ludlow 1972; Pearcy 1983, 1987; Chazdon and Fetcher 1984). The maximum photon irradiance during these lightflecks can approach full sunlight but is usually less because of penumbral effects (Miller and Norman 1971; Norman *et al.* 1971). This is particularly evident under a tall canopy such as the Curtain Fig forest where only 1% of the lightflecks exceeded 1200 μmol m^{-2} s^{-1}. Within crop canopies, shorter forests, or near the top of tall forests, lightflecks are typically brighter.

An important characteristic of understory light environments is the high degree of spatial and temporal variability caused by lightflecks. The foliage of tree canopies is clumped both vertically and horizontally, leading to regions of dense cover and others with only sparse cover (Baldocchi *et al.* 1986). Consequently the usual diurnal pattern of light in the understory is periods with relatively frequent lightflecks separated by periods with few or no lightflecks. Records for eight sensors in the Curtain Fig Forest understory showed that 70% of the lightflecks occurred within 1 min of the preceding lightfleck while only 5% were preceded by low-light periods of an hour or more. Similar results have been obtained in a Hawaiian forest (Pearcy 1983). Large spatial variations also occur, as is evident in the measurements in the Curtain Fig forest discussed above as well as in a Hawaiian forest (Pearcy 1983). Even sites separated by only 0·5 m can differ by a factor of two in numbers of lightflecks and the photon irradiance contributed by them. Daily and seasonal variations are caused by cloud cover, changes in solar angle and canopy changes.

While the dynamic nature of the light environment is perhaps most evident in understories, rapid light fluctuations are also characteristic of other microenvironments. A light sensor mounted in the middle of a soybean canopy on a windy day received over 1800 lightflecks, with most reaching full sunlight photon irradiance, but lasting on the average only 2·3 s (Pearcy, unpublished data). Moreover, cloud movement or leaf fluttering in wind can also cause rapid light changes even for fully exposed leaves. In aquatic environments, algae are exposed to highly dynamic light regimes because of lensing effects of the water and movements of the thalli due to wave action (Dera and Gordon 1968; Gerard 1984). Thus, dynamic changes in light environments on short time scales occur in a wide variety of environments and attention should be given to the potential role of the dynamic response characteristics in them.

Dynamic Responses of Photosynthesis

Photosynthetic responses to light changes such as those occurring in lightflecks are complex because several components, each with a different time constant, are involved. A particular difficulty in studying these dynamic responses is that some occur within seconds and are thus difficult to resolve with standard gas exchange techniques. For whole-leaf chambers, the time constants usually cannot easily be reduced below about 30 s even with high flow rates. However, if the chamber is well mixed then it is possible to apply a washout kinetic correction (Bartholomew *et al.* 1981) to calculate the instantaneous rate of gas exchange within the chamber independent of lags caused by the chamber volume or tubing. When these corrections are made the time required for a full response to a step change within the chamber can be reduced to as little as 4 s if high flow rates through the chamber and analyser are also used (Pearcy *et al.* 1985). An alternative approach that can be used with hypostomatous leaves, the most common type in shaded environments, is a single-sided chamber. Pearcy *et al.* (1987) used a single-sided chamber that clamped onto the abaxial leaf surface and enclosed approximately 22 cm^2 of leaf area, but had a volume of only 4·5 cm^3. With either approach, the primary limitation to the response speed of the system is the volume of the analyser cells and the time constant of analyser electronic response. The configuration of the analyser is important since it is necessary for the gas to enter and exit right at the ends of the cells in order to avoid dead volumes which slow down the response. Also, the analyser must have an inherently good signal-to-noise ratio so that electronic filtering can be greatly reduced or removed.

Induction Limitations

It has long been known that, after a long period in darkness or low light, there is an induction requirement in the photosynthetic apparatus that results in a fairly slow rise in assimilation rate when the light is increased (Osterhout and Hass 1918;

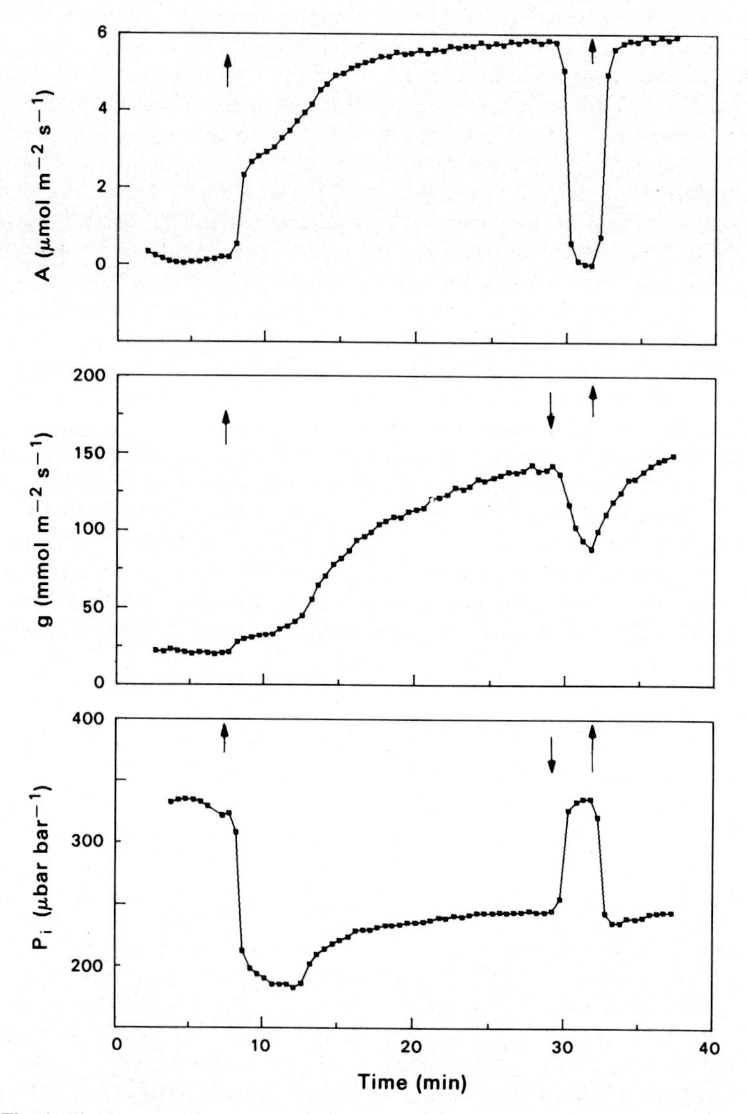

Fig. 2. Time course of CO_2 assimilation, stomatal conductance to H_2O, and intercellular partial pressure of CO_2 in response to an increase in photon irradiance from 6 to 520 μmol photons m^{-2} s^{-1} for an *Argyrodendron peralatum* leaf on a sapling in the Curtain Fig Forest understory.

Rabinowitch 1956). An example of an induction response measured on an *Argyrodendron peralatum* leaf in the understory of the Curtain Fig forest is shown in Fig. 2. Assimilation exhibited a fast initial rise in the first minute, followed by a slower sigmoidal rise up to the maximum light-saturated rates after about 20 min. When light was lowered and increased again, the second response was much faster. Paralleling the sigmoidal increase in assimilation were increases in stomatal conductance. Over this period, assimilation rates nearly doubled whereas intercellular CO_2 pressures (p_i) only increased from about 185 to 240 μbar. Similar, although usually slower, induction responses have been reported for shade-grown *Alocasia macrorrhiza* leaves as well for other C_3 tropical trees (Pearcy *et al.* 1985; Chazdon and Pearcy 1986*a*). The response in C_4 plants is also similar except for lower but still relatively constant p_i values (Usuda *et al.* 1984; Pearcy *et al.* 1985).

The similar time courses for stomatal opening and induction of assimilation are suggestive of a causative role for the former, but the situation is certainly more complex. An induction response, albeit a generally faster one, is found in algae and protoplasts, which has been used as an argument that stomata are not the primary limitation (Edwards and Walker 1983). Moreover, the calculated p_i often remains high during induction and assimilation can increase markedly while p_i shows little change. This observation has also been used as evidence that the primary limitation during induction is biochemical in nature (Usuda and Edwards 1984; Pearcy *et al.* 1985).

Kirschbaum and Pearcy (1987) have recently reconsidered the roles of stomatal versus biochemical limitations during the induction response of *A. macrorrhiza*. Since conductances in low light are often very low, calculations of p_i can be subject to considerable uncertainty during the first phase of induction. In particular, any significant cuticular conductance to water vapour can potentially cause an error leading to an overestimation of p_i. Moreover, p_i will be overestimated if the stomata do not open uniformly so that all areas of the leaf have the same stomatal conductance (Farquhar *et al.* 1987). The cuticular conductance for CO_2 should be very much less than that for water vapour since CO_2 would need to traverse the epidermal cell before reaching the mesophyll. Consequently, cuticular evaporation should be subtracted from the total evaporation rate to derive the evaporation rate through the stomatal pore and hence a correct stomatal conductance for CO_2.

Unfortunately, there is no way to measure the cuticular evaporation from a leaf without uncertainties about the contribution from stomata. Kirschbaum and Pearcy (1987) made the assumption that any biochemical induction should be the same in leaves with high and low stomatal conductances at the beginning of induction. They then chose a cuticular conductance that resulted in similar values of the carboxylation conductance (slope of the dependence of assimilation on p_i) during induction when the initial stomatal conductance was either high or low. The value of the cuticular conductance estimated in this way ranged from 1 to 3 mmol $m^{-2} s^{-1}$, which had a large effect on the calculated p_i when initial conductances were low (7–10 mmol $m^{-2} s^{-1}$) but little effect when they were high (20–40 mmol $m^{-2} s^{-1}$). Thus, even when a cuticular conductance was included there was still clear evidence of an increase in the biochemical capacity for CO_2 fixation over the first 10–15 min of induction as shown by the increase in the carboxylation efficiency (Fig. 3*a*). After 10–15 min any further increases in assimilation appeared to be due to stomatal opening. Implicit in this study was the assumption of a uniform stomatal opening. Any patchiness in the opening would increase the estimate for the role of stomata in induction. However, the effect may be relatively small in *A. macrorrhiza* since lateral diffusion of CO_2 within the leaf should be enhanced by the relatively open internal anatomy and lack of bundle sheath extensions that may isolate sections of mesophyll from other sections.

The biochemical basis of induction has been subject to considerable debate with the two major hypotheses originally put forward by Osterhout and Hass (1918), light activation of enzymes and a build-up of photosynthetic metabolites, still serving as the focus. The evidence for a role of metabolite build-up is mostly from isolated chloroplasts (Leegood and Walker 1980), where metabolites may be depleted in the

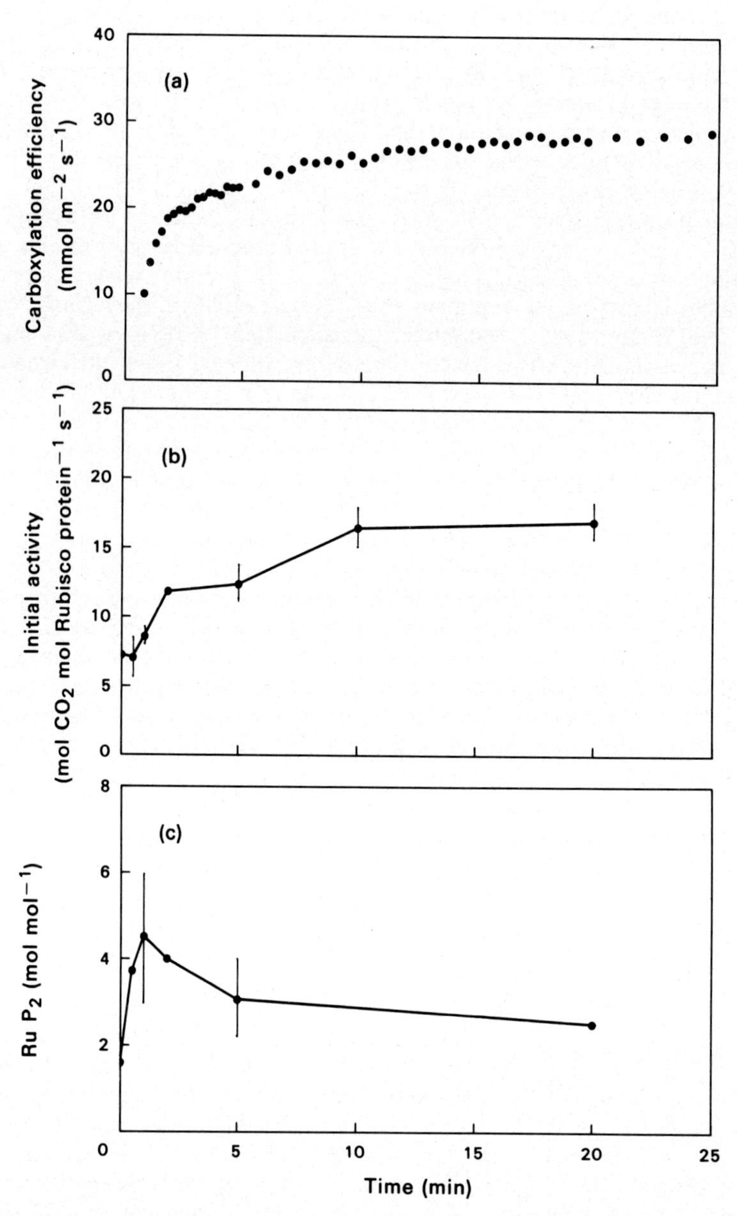

Fig. 3. Time courses of the increase in (*a*) carboxylation efficiency, (*b*) initial Rubisco activity, and (*c*) RuP_2 pool sizes during induction for *Alocasia macrorrhiza*. At time 0 the light was increased from 7–10 μmol photons m^{-2} s^{-1} to approximately 500 μmol photons m^{-2} s^{-1}. Carboxylation efficiencies are shown for 1 min onwards to avoid uncertainties due to the system response in the first minute. Carboxylation efficiencies were calculated from data of Kirschbaum and Pearcy (1987). The Rubisco activities and RuP_2 pool sizes are from Seemann, Kirschbaum, Sharkey and Pearcy (unpublished data).

isolation procedure. Its role in leaves or protoplasts has been questioned (Stitt *et al.* 1980). Five enzymes in the Calvin cycle exhibit light activation (Edwards and Walker 1983) but, with the exception of ribulose-1,5-bisphosphate (RuP_2) carboxylase/oxygenase (Rubisco), all activate within 1–2 min of the light increase, which is too fast to account for but a small part of the induction response. The general parallel between induction and light activation of Rubisco was first noted by Perchorowicz *et al.* (1981), but its potential contribution to induction has been questioned (Edwards and Walker 1983; Usuda 1985). This enzyme exhibits light regulation involving both a tight binding inhibitor, 2-carboxyarabinitol 1-phosphate (CA1P), that is present in low light and in the dark but is degraded in high light (Seemann *et al.* 1985; Berry *et al.* 1986), and a Mg^{2+}–CO_2 activation (Miziorko and Lorimer 1983). There appears to be some variation in the roles of these two mechanisms with some species lacking the inhibitor while in others the inhibitor may assume the primary role (Seemann *et al.* 1985; Vu *et al.* 1984).

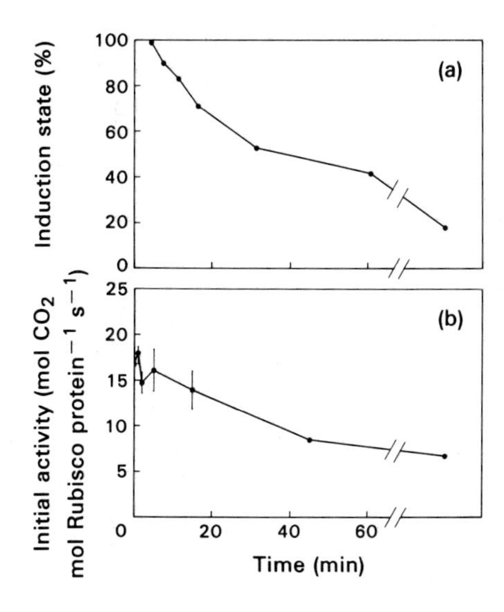

Fig. 4. The effect of time in low light on (*a*) the induction state and (*b*) the initial Rubisco activity for shade-grown *Alocasia macrorrhiza* leaves. The leaves were brought to full induction by exposure to 500 μmol photons m^{-2} s^{-1} before lowering the light to 7–10 μmol photons m^{-2} s^{-1}. Induction state was measured as the assimilation rate achieved 1 min after the light was increased again. Adapted from Chazdon and Pearcy (1986*a*) and unpublished data of Seemann, Kirschbaum, Sharkey and Pearcy.

The role of Rubisco activation in induction of *A. macrorrhiza* was evaluated by freeze-clamping leaves at various times after light increases and decreases and then measuring both metabolite pool sizes and Rubisco activity (Fig 3*b*). During induction the initial Rubisco activity, which is measured before any activation by the CO_2 and Mg^{2+} in the assay medium can occur and therefore should reflect the *in vivo* capacity of Rubisco, increased 2·3-fold with a half-time of approximately 3 min to a final steady state in about 10 min. While the time courses of the increase in carboxylation efficiency and initial Rubisco activity differ somewhat, the 10–15 min required for completion of each are in good agreement. The differences in the shapes may be due to uncertainties in the estimate of carboxylation resistance or to the fact that different leaves had to be used for each measurement of Rubisco activity. The increase in initial Rubisco activity is consistent with a joint mediation by Mg^{2+}–CO_2 activation and loss of an inhibitor. Pool sizes of RuP_2 increased to high levels within 1 min of the light increase and then gradually declined as Rubisco activity increased, which is consistent with a limitation by Rubisco that is gradually removed during induction. However, the lower p_i early in induction, which in itself reduces *in vivo* Rubisco activity, may therefore also contribute to the increased RuP_2 level. When light was decreased, the decrease in measured Rubisco activity and recovery of RuP_2 pools to the steady state levels was much slower than found for the light increase (Fig. 4). This was presumably due primarily to a slower rate of synthesis than degradation of the inhibitor. The slow loss of Rubisco activity correlates well with the slow loss of induction state for CO_2 assimilation in *A. macrorrhiza* reported by Chazdon and Pearcy (1986*a*).

The light regulation of Rubisco appears to function to maintain a homeostasis in RuP_2 concentrations despite large changes in light and consequently assimilation rate. This may serve to minimise negative feedback interactions within the Calvin cycle or as a mechanism for control of phosphate supply within the chloroplast (Seemann 1986; Sharkey *et al.* 1986). Whatever the reason, it is clear that it plays an important role in the induction response and ultimately in the capacity of a leaf in the understory to respond to lightflecks. The response of an *A. macrorrhiza* leaf to a series of 1 min lightflecks (500 μmol photons $m^{-2} s^{-1}$) separated by 2 min of low light (10 μmol photons $m^{-2} s^{-1}$) is shown in Fig. 5. The assimilation rate achieved in the first lightfleck

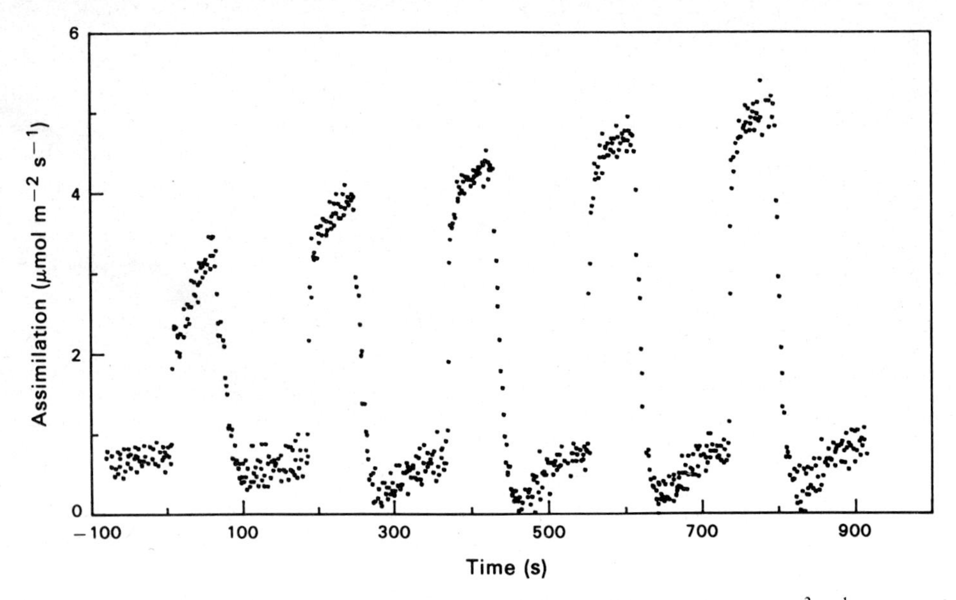

Fig. 5. Response of assimilation to a series of 1 min lightflecks (500 μmol photons $m^{-2} s^{-1}$) separated by 2 min low-light (10 μmol photons $m^{-2} s^{-1}$) periods. The data are from a shade grown *Alocasia macrorrhiza* leaf. Adapted from Chazdon and Pearcy (1986*a*).

is clearly much less than in later lightflecks in the sequence. Thus, induction occurring in response to one lightfleck can, in effect, prime the leaf so that it is better able to utilise subsequent lightflecks. It is likely that the slower deactivation than activation of Rubisco is important in allowing this response but other factors such as stomatal dynamics are clearly also involved. When a single lightfleck is given, stomatal conductance does not reach its maximum value until 20 min after the lightfleck (Kirschbaum *et al.* 1987). It then decreases very slowly back to the steady state level in low light over the next hour. The continued stomatal opening after a lightfleck also acts to enhance assimilation during subsequent lightflecks. Because lightflecks occur in clusters separated by low-light periods, perhaps the best predictor of a lightfleck is a preceding lightfleck. The priming effect may therefore help to ensure that induction carries over from one lightfleck to the next.

Responses to Light Fluctuations

There have been many measurements that show that, during flashing light, the efficiency of light utilisation by leaves for CO_2 uptake can be higher than observed under continuous light (Pollard 1970; Kriedemann *et al.* 1973; Lasko and Barnes 1978; Pearcy *et al.* 1985; Stitt 1986). Measurements of assimilation in flashing light can take advantage of a slow analyser response that averages the assimilation rate over the high and low light periods. This average is typically higher than one derived from steady-state measurements in continuous high and low light, indicating that an enhancement of assimilation is occurring in the flashing light. Enhancements of assimilation have been shown to occur at flash frequencies similar to those of natural light variations within canopies (Pollard 1970; Kriedemann *et al.* 1973). Similar observations have been made for algae (Marra 1978; Gallegos *et al.* 1980; Salvidge 1980).

The higher efficiencies during flashing light as compared to the steady state has been attributed to an averaging of light rather than photosynthesis during oscillations (Gross 1982). The enhanced assimilation is a consequence of the convex light-dependence curve. While this explanation has some predictive value, it provides no insight into the underlying mechanism. Any mechanistic interpretation must take into account the observation that light must be saturating, or at least not strictly limiting, during the flashes in order for enhanced assimilation to occur (Pollard 1970; Pearcy *et al.* 1985). In addition, there appear to be optimum durations for both the high- and low-light periods. For example, Kriedemann *et al.* (1973) found an optimum dark interval of 1·8 s was required for maximum utilisation of 0·05 s flashes. The optimum flash duration was between 0·6 and 1 s. A plausible interpretation of these results is that the enhanced assimilation is a consequence of a build-up of an intermediate during the flash that requires a finite interval in low light or dark for metabolism to be completed. Stitt (1986) has explained the enhanced O_2 evolution rates during flashing light from leaf discs in a 5% CO_2 atmosphere on the basis of a transient removal of a phosphate limitation that is present during steady-state light-saturated assimilation. While phosphate limitations to assimilation rate can be important at high CO_2 pressures or low temperatures (Sharkey 1985), they usually are not present at normal CO_2 pressures and temperatures above 20°C (Sage and Sharkey 1987).

Determination of the response of assimilation to individual lightflecks can provide insights not readily obtained from the average response to flashing light. Fig. 6 shows the response of assimilation to 20 s lightflecks when the leaf is either fully induced by a previous treatment in high light or is uninduced. When the leaf was fully induced (Fig. 6*b*), assimilation increased rapidly to a value that was greater than 90% of the steady-state light-saturated rate within 8 s. The rate of decrease of assimilation after the lightfleck was much slower than the increase, taking nearly 30 s to reach the steady-state value in low light. Uninduced leaves (Fig. 6*a*) showed only a very small increase in assimilation rate during the lightfleck but the decrease in assimilation rate was very slow, taking nearly 2 min to return to the low-light value. These responses show that assimilation continues at a relatively high rate for some time after the lightfleck, due to post-illumination CO_2 fixation. When the rate achieved during the lightfleck is low, as in an uninduced leaf, then the increased assimilation occurring in response to the lightfleck can continue for a period much longer than the lightfleck itself. Thus, post-illumination CO_2 fixation is an important component of the transient photosynthetic responses to a lightfleck.

The effect of lightfleck duration on assimilation has been investigated by Pearcy *et al.* (1985) and Chazdon and Pearcy (1986*b*). As could be expected, the total carbon gain during a lightfleck increased as the duration of the lightfleck increased. The total carbon gain, as determined by integration of the area under the response curve, can be compared to that expected from the steady-state assimilation rates determined after full induction if the assimilation responded instantaneously to a light change (i.e. a square-wave response of assimilation to a square-wave pulse of light). The integrations of assimilation rate were carried out from just before the lightfleck to 2 min after it so that losses due to a post-illumination dip in net CO_2 assimilation (see Fig. 5) were taken into account. The post-illumination dip was pronounced in fully induced leaves and lightflecks longer than 30 s.

From the ratio of the observed to the expected carbon gain, an 'efficiency of lightfleck utilisation' was determined to facilitate comparisons (Fig. 7). Efficiencies of 100% mean

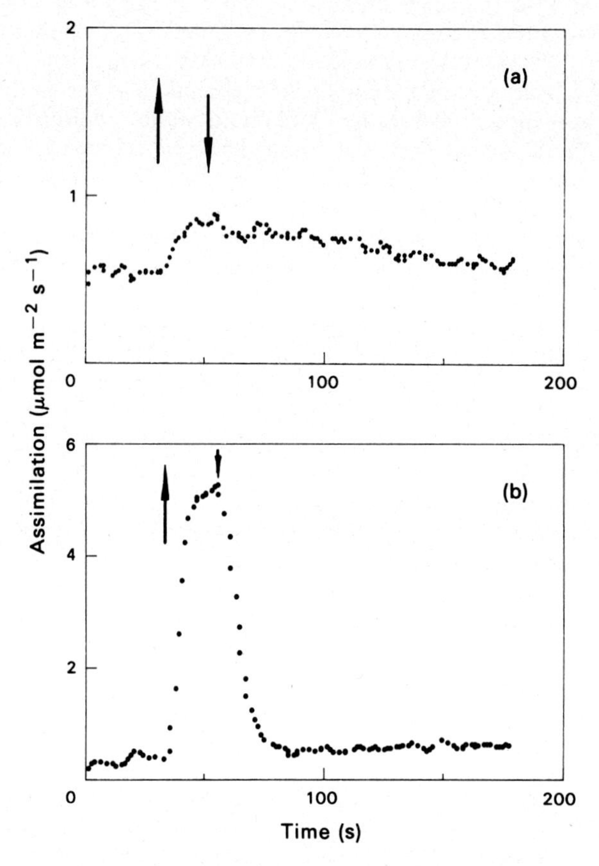

Fig. 6. Response of assimilation to 20 s lightflecks (500 μmol photons $m^{-2} s^{-1}$ for an *Alocasia macrorrhiza* leaf either (*a*) before induction or (*b*) after full induction. Redrawn from Chazdon and Pearcy (1986*b*).

therefore that the post-illumination CO_2 fixation was sufficient to compensate for the losses due to the less than instantaneous increase in assimilation and those due to the post-illumination dip. It was apparent that, for fully induced leaves, short (5–10 s) lightflecks resulted in considerably higher carbon gain than predicted from the steady-state assimilation rates because of post-illumination CO_2 fixation. For uninduced leaves, carbon gain was considerably less than for the fully induced leaves. Nevertheless, for short (5-s) lightflecks, the post-illumination CO_2 fixation was able to compensate for the very low photosynthetic rates during the lightfleck, so that the total carbon gain was equal to or slightly above the prediction from the steady-state, light-saturated assimilation rates. For longer lightflecks, however, carbon gain was much lower than predicted, illustrating the strong limitation imposed by the induction state of the leaf on utilisation of lightflecks (Chazdon and Pearcy 1986*a*).

In order for CO_2 uptake during a lightfleck to exceed the steady-state prediction, some of the light energy received during the lightfleck must be stored for utilisation in CO_2 fixation following the lightfleck. During steady-state photosynthesis, the rates of electron transport, RuP_2 regeneration and carboxylation are all constrained by the slowest step. However, Sharkey *et al.* (1986) have postulated that during a transient lightfleck this constraint can be overcome provided that a 'capacitance' exists within the system. During the lightfleck, this capacitance can be filled by the faster steps that are no longer constrained, while the slower steps can catch up during the low-light period following the lightfleck. The capacitance would be provided by the energy-rich, phosphorylated compounds, RuP_2 and its precursors, and the transthylakoid proton motive force (PMF).

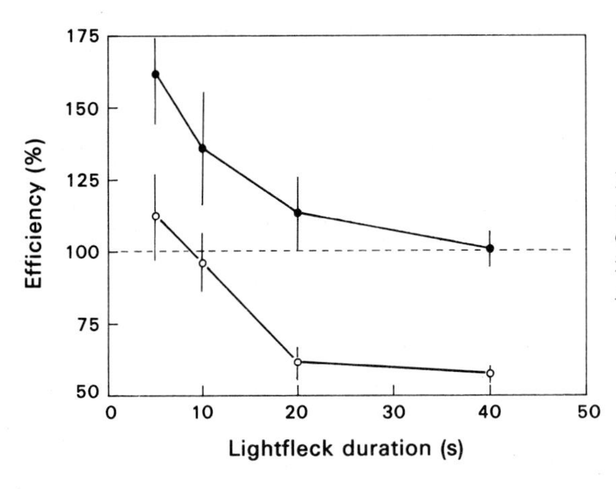

Fig. 7. Dependence of lightfleck utilisation efficiency on the lightfleck duration for fully induced (●) or non-induced (○) *Alocasia macrorrhiza* leaves. Adapted from Pearcy *et al.* (1987).

Sharkey *et al.* (1986) examined the biochemical basis of this capacitance with determination of metabolite pool sizes changes in leaves of *Alocasia macrorrhiza* during 5 s lightflecks. Both RuP_2 and triose phosphate (TP) pools increased during the lightfleck and then decreased back to the levels found in low light 1 min after the lightfleck, while the 3-phosphoglyceric acid pool behaved oppositely (Fig. 8). These changes are consistent with the utilisation of TP and RuP_2 in post-illumination CO_2 fixation. Pool size changes during and after the lightfleck were much larger for TP than for RuP_2, indicating that TP pools contributed more to supporting post-illumination CO_2 fixation. The observed pool sizes established during the lightfleck were sufficient to give a lightfleck utilisation efficiency of 190%, a value somewhat greater than those typically observed for *A. macrorrhiza*. Similar metabolite pool changes occurred in low-light grown beans (*Phaseolus vulgaris*) but high-light grown beans and *Alocasia macrorrhiza* showed small changes in TP with larger changes in RuP_2. Sharkey *et al.* (1986) postulate the build-up of TP rather than RuP_2 could favour high efficiencies by preventing the 'two kinase competition' described by Sivak and Walker (1986). The results of Sharkey *et al.* (1986) show that, for shade leaves, metabolite pools sufficient to support high efficiencies of lightfleck utilisation can be established during short lightflecks. Utilisation of TP would require a post-illumination supply of ATP. Pools of ATP in leaves are small but continued photophosphorylation supported by the PMF gradient established during the lightfleck is likely. The extensive grana stacking evident in shade-leaf chloroplasts may be important in this regard. Post-illumination ATP synthesis measured by Hangarter and Good (1982) in isolated chloroplasts is adequate to support the observed post-illumination CO_2 fixation. However, estimates for intact leaves of the proton capacitance associated with the PMF across the thylakoid membranes and the capacity for post-illumination phosphorylation are needed but difficult to obtain.

In order for efficiencies of lightfleck utilisation to exceed 100%, it is clear that electron transport must be capable of transiently greater rates than the rates of CO_2 fixation during the lightfleck. Pearcy *et al.* (1987) measured the light dependence of the 'extra assimilation' (the assimilation above that required to give a 100% lightfleck utilisation efficiency) during a 5-s lightfleck and found that it approached zero at about 50 μmol

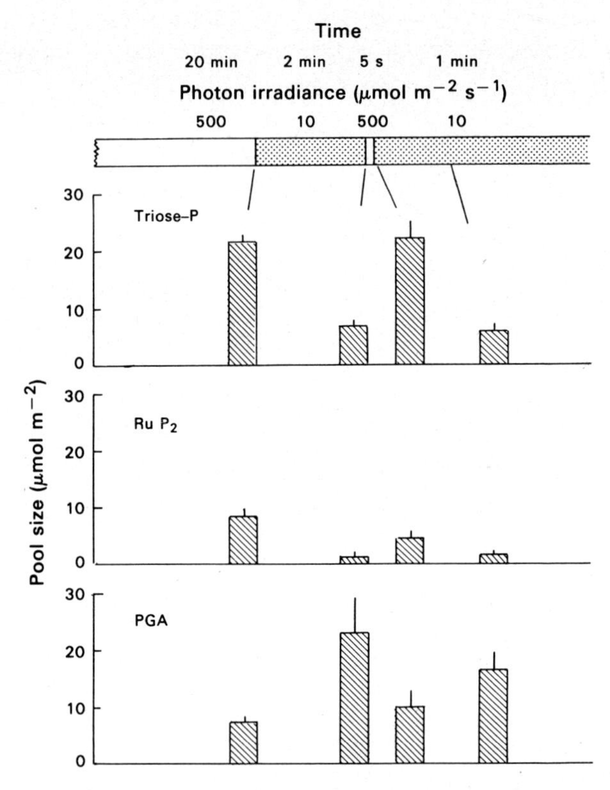

Fig. 8. Pool sizes of RuP_2, TP, and PGA in *Alocasia macrorrhiza* in response to a 5-s lightfleck. Redrawn from Sharkey *et al.* (1986).

photons $m^{-2} s^{-1}$ (Fig. 9). This was the point where the steady-state assimilation versus light curve for shade-grown *Alocasia macrorrhiza* plants began to deviate from linearity indicating that a higher photon irradiance would no longer be strictly limiting assimilation. The maximum extra assimilation did not appear to saturate before 500 μmol photons $m^{-2} s^{-1}$ were received, even though assimilation is light saturated at 150–200 μmol photons $m^{-2} s^{-1}$ of continuous light. From estimates based on assimilation rates at high CO_2 pressures using the Farquhar and von Caemmerer (1982) model, it is clear that electron transport capacity generally exceeds carboxylation rates at normal ambient CO_2 pressures. However, these may even be underestimates since considerably higher values are suggested from O_2 evolution measurements during flashing light (Stitt

1986). A similar conclusion can be reached from the transiently higher O_2 evolution rates during photosynthetic oscillations (Walker *et al.* 1983).

A high ratio of electron transport capacity to carboxylation capacity could be expected to enhance lightfleck utilisation efficiencies. Estimates by Stitt (1986) from O_2 evolution in flashing light indeed show a higher ratio in shade than sun leaves. However, for peas, the ratio was reported to be lower in shade than sun leaves (Evans 1987), based on estimates derived from A/p_i curves and the Farquhar and von Caemmerer (1982) model. Using a similar approach, no change was noted for sun and shade-grown *Alocasia macrorrhiza* (Sims and Pearcy, unpublished data). The differences may be reconciled if there is a limitation in RuP_2 regeneration capacity under steady-state conditions, and this is the limitation estimated by the model. There is evidence from modulated fluorescence techniques that, under steady-state conditions, electron transport rate is limited by the carboxylation rate (Dietz 1985; Weis *et al.* 1987).

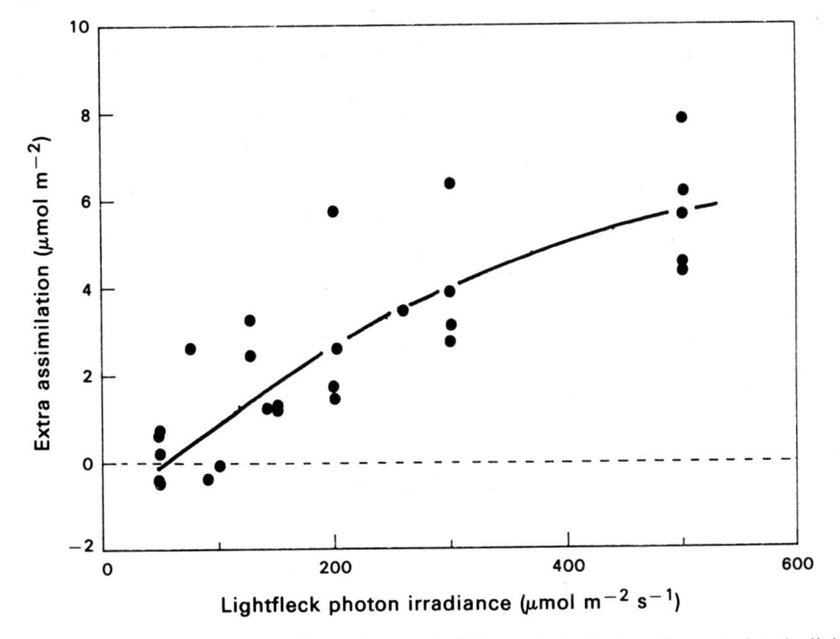

Fig. 9. Dependence of the extra assimilation due to a lightfleck on photon irradiance during the lightfleck for *Alocasia macrorrhiza* leaves. All lightflecks were 5 s duration. Redrawn from Pearcy *et al.* (1987).

The high efficiencies of lightfleck utilisation reported for *Alocasia macrorrhiza* have also been found in the C_3 tree, *Claoxylon sandwicense,* and the C_4 tree, *Euphorbia forbesii*, which are native to shaded understories in Hawaii (Pearcy *et al.* 1985). It has been shown to occur in algae (Miyachi 1979) and in a variety of C_3 crops and shrubs (Laisk *et al.* 1984). The crops and shrubs, however, exhibited considerable variation in 'assimilatory power', which Laisk *et al.* defined as the sum of energetically rich compounds available to support post-illumination CO_2 uptake. Chazdon and Pearcy (1986*b*) found lower efficiencies of lightfleck utilisation in high-light than low-light grown *Alocasia macrorrhiza*. Moreover, efficiencies of lightfleck utilisation of a high-light adapted tropical tree, *Toona australis*, were much lower than those of *Alocasia macrorrhiza*. The capacity for post-illumination CO_2 fixation may be a universal consequence of the organisation of the photosynthetic apparatus rather than a specific adaptation for utilisation of fluctuating light. However, quantitative changes in the organisation may lead to enhanced capacities that are important in variable light environments. At this time there is too little information to draw any ecological correlates concerning variation in the capacity to utilise lightflecks.

Conclusions

The limitations due to induction state and the enhancements due to post-illumination CO_2 fixation are important components of the dynamic responses of assimilation to light and hence strongly influence the capacity of a leaf to utilise lightflecks. When lightflecks are infrequent and of relatively long duration, losses in assimilation due to a low induction state are likely to be significant. However, with frequent, short-duration lightflecks, post-illumination CO_2 fixation may result in an enhancement of carbon gain over that expected from steady-state measurements. In forest understories, where there are often periods of relatively frequent, short-duration lightflecks separated by long periods of low light, daily carbon gain may be in part a function of both processes. Assimilation measurements in understories have established the importance of lightflecks but further work is clearly needed to determine the relative roles of induction and post-illumination CO_2 fixation and their importance in natural environments. However, quantifying the contributions of each under natural light regimes is difficult because of the spatial variation in light even on the scale of a leaf. Consequently, a more fruitful approach may be to examine the responses under light regimes that simulate the natural temporal but not the spatial variations due to lightflecks.

Acknowledgments

This research was supported by grants from the National Science Foundation and the USDA Competitive Grants Program. I thank Dr Miko Kirschbaum for his helpful comments on the manuscript.

References

Baldocchi, D., and Hutchison, B. A. (1986). On estimating canopy photosynthesis and stomatal conductance in a deciduous forest with clumped foliage. *Tree Physiol.* **2**, 155–68.

Bartholomew, G. A., Vleck, D., and Vleck, C. (1981). Instantaneous measurements of oxygen consumption during pre-flight cooling in spingid and saturniid moths. *J. Exp. Biol.* **90**, 17–32.

Berry, J. A., Lorimer, G. H., Pierce, J., Seemann, J. R., Meeks, J., and Freas, S. (1986). Isolation, identification and synthesis of carboxyarabinitol-1-phosphate, a diurnal regulator of ribulosebisphosphate carboxylase activity. *Proc. Natl Acad. Sci. U.S.A.* **84**, 734–8.

Björkman, O. (1973). Comparative studies of photosynthesis in higher plants. In 'Current Topics in Photobiology, Photochemistry and Photophysiology'. (Ed. A. C. Giese.) Vol. 8, pp. 1–63. (Academic Press: New York.)

Björkman, O. (1981) Responses to different quantum flux densities. In 'Physiological Plant Ecology I. Responses to the Physical Environment'. (Eds O. L. Lange, P. S. Nobel, C. B. Osmond, and H. Ziegler.) Encycl. Plant Physiol. New Ser., Vol. 12A, pp. 57–107. (Springer-Verlag: Berlin.)

Björkman, O., and Ludlow, M. (1972). Characterization of the light climate of a Queensland rainforest. *Carnegie Inst. Wash. Year Book* **71**, 85–94.

Björkman, O., Ludlow, M., and Morrow, P. (1972). Photosynthetic performance of two rainforest species in their habitat and analysis of their gas exchange. *Carnegie Inst. Wash. Year Book* **71**, 94–102.

Boardman, N. K. (1977). Comparative photosynthesis of sun and shade plants. *Annu. Rev. Plant Physiol.* **28**, 355–77.

Boardman, N. K., Anderson, J. M., Thorne, S. W., and Björkman, O. (1972). Photochemical reactions of chloroplasts and components of the photosynthetic electron transport chain in two rainforest species. *Carnegie Inst. Wash. Year Book* **71**, 107–15.

Chazdon, R. L., and Fetcher, N. (1984). Photosynthetic light environments on a lowland tropical forest in Costa Rica. *J. Ecol.* **72**, 553–64.

Chazdon, R. L., and Pearcy, R. W. (1986*a*). Photosynthetic responses to light variation in rain forest species. I. Induction under constant and fluctuating light conditions. *Oecologia* **69**, 517–23.
Chazdon, R. L., and Pearcy, R. W. (1986*b*). Photosynthetic responses to light variation in rain forest species. II. Carbon gain and light utilization during lightflecks. *Oecologia* **69**, 524–31.
Dera, J., and Gordon, H. R. (1968). Light field fluctuations in the photic zone. *Limmnol. Oceanogr.* **13**, 697–9.
Dietz, K. J., Schreiber, U., and Heber, U. (1985). The relationship between the redox state of Q_a and photosynthesis in leaves at various carbon dioxide, oxygen and light regimes. *Planta* **166**, 219–26.
Edwards, G., and Walker, D. (1983). 'C_3 and C_4: Mechanisms, and Cellular and Environmental Regulation of Photosynthesis.' (University of California Press: Berkeley.)
Evans, G. C. (1939). Ecological studies on the rainforest of southern Nigeria. II. The atmospheric environmental conditions. *J. Ecol.* **44**, 391–428.
Evans, J. (1987). The relationship between electron transport components and photosynthetic capacity in pea leaves grown at different irradiances. *Aust. J. Plant Physiol.* **14**, 157–70.
Farquhar, G. D., and Caemmerer, S. von (1982). Modelling of photosynthetic response to environmental conditions. In 'Physiological Plant Ecology II. Water Relations and Carbon Assimilation'. (Eds O. L. Lange, P. S. Nobel, C. B. Osmond, and H. Ziegler.) Encycl. Plant Physiol. New Ser., Vol. 12B, pp. 549–88. (Springer-Verlag: Berlin.)
Farquhar, G. D., Hubick, K. T., Terishima, I., Condon, A. G., and Richards, R. A. (1987). Genetic variation in the relationship between photosynthetic CO_2 assimilation rate and stomatal conductance to water loss. In 'Progress in Photosynthesis Research'. (Ed. J. Biggins.) Vol. 4, pp. 209–12. (Martinus Nijhoff: Dordrecht.)
Gallegos, C. L., Hornberger, G. M., and Kelley, M. G. (1980). Photosynthesis–light relationships of a mixed culture of phytoplankton in fluctuating light. *Limmnol. Oceanogr.* **25**, 1082–92.
Gerard, V. A. (1984). The light environment of a giant kelp forest: influence of *Macrocystis pyrifera* on spatial and temporal variability. *Marine Biol.* **84**, 189–95.
Gross, L. J. (1982). Photosynthetic dynamics in varying light environments: a model and its application to whole leaf carbon gain. *Ecology* **63**, 84–93.
Hangarter, R. P., and Good, N. E. (1982). Energy thresholds for ATP synthesis in chloroplasts. *Biochim. Biophys. Acta* **430**, 154–64.
Kirschbaum, M., Gross, L., and Pearcy, R. (1987). Observed and modelled stomatal responses to dynamic light environments in the shade plant *Alocasia macrorrhiza*. *Plant Cell Environ.*, in press.
Kirschbaum, M. U. F., and Pearcy, R. W. (1987). Gas exchange analysis of the relative importance of stomatal and biochemical factors in photosynthetic induction in *Alocasia macrorrhiza*. *Plant Physiol.*, in press.
Kriedemann, P. E., Törökfalvy, E., and Smart, R. E. (1973). Natural occurrence and photosynthetic utilization of sunflecks in grapevine leaves. *Photosynthetica* **7**, 18–27.
Laisk, A., Kirats, O., and Oja, V. (1984). Assimilatory power (postillumination CO_2 uptake) in leaves. Measurement, environmental dependences and kinetic properties. *Plant Physiol.* **76**, 723–9.
Lasko, A. H., and Barnes, J. E. (1978). Apple leaf photosynthesis in alternating light. *Hortic. Sci.* **13**, 473–4.
Leegood, R., and Walker, D. (1980). Autocatalysis and light activation of enzymes in relation to photosynthetic induction in wheat chloroplasts. *Arch. Biochem. Biophys.* **200**, 575–82.
Lundegarth, L. (1921). Ecological studies in the assimilation of certain forest plants and shore plants. *Svensk. Bot. Tidskr.* **15**, 46–95.
Marra, J. (1978). Effect of short term variations in light intensity on photosynthesis of a marine phytoplankter: a laboratory simulation study. *Marine Biol.* **46**, 191–202.
Miller, E., and Norman, J. M. (1971). A sunfleck theory for plant canopies. II. Penumbra effect: intensity distributions along sunfleck segments. *Agron. J.* **63**, 739–48.
Miyachi, S. (1979) Light-enhanced dark CO_2 fixation. In 'Photosynthesis II. Photosynthetic Carbon Metabolism and Related Processes'. (Eds M. Gibbs and E. Latzko.) Encycl. Plant Physiol. New Ser., Vol. 6, pp. 68–76. (Springer-Verlag: Berlin.)
Miziorko, H. M., and Lorimer, G. H. (1983). Ribulose 1,5-bisphosphate carboxylase–oxygenase. *Annu. Rev. Biochem.* **52**, 507–35.
Norman, J. M., Miller, E. E., and Tanner, C. B. (1971). Light intensity and sunfleck-size distributions in plant canopies. *Agron. J.* **63**, 743–8.
Osterhout, W. J., and Hass, J. R. C. (1918). On the dynamics of photosynthesis. *J. Gen. Physiol.* **1**, 1–16.

Pearcy, R. W. (1983). The light environment and growth of C_3 and C_4 tree species in the understory of a Hawaiian forest. *Oecologia* **58**, 19–25.
Pearcy, R. W. (1987). Photosynthetic gas exchange responses of Australian tropical forest trees in canopy, gap and understory microenvironments. *Funct. Ecol.,* in press.
Pearcy, R. W., and Calkin, H. (1983). Carbon dioxide exchange of C_3 and C_4 tree species in the understory of a Hawaiian forest. *Oecologia* **58**, 26–32.
Pearcy, R. W., Chazdon, R. L., and Kirschbaum, M. U. F. (1987). Photosynthetic utilization of lightflecks by tropical forest plants. In 'Progress in Photosynthesis Research'. (Ed. J. Biggins.) Vol. 4, pp. 257–60. (Martinus Nijhoff: Dordrecht.)
Pearcy, R. W., Osteryoung, K., and Calkin, H. W. (1985). Photosynthetic responses to dynamic light environments by Hawaiian trees. The time course of CO_2 uptake and carbon gain during sunflecks. *Plant Physiol.* **79**, 896–902.
Perchorowicz, J. T., Raynes, D. A., and Jensen, R. G. (1981). Light limitation of photosynthesis and activation of ribulose bisphosphate carboxylase in wheat seedlings. *Proc. Natl. Acad. Sci. U.S.A.* **78**, 2985–9.
Pollard, D. F. W. (1970). The effect of rapidly changing light on the rate of photosynthesis in bigtooth aspen (*Populus grandidentata*). *Can. J. Bot.* **48**, 823–9.
Rabinowitch, E. I. (1956). 'Photosynthesis and Related Processes.' Vol. 2, part 2. (Interscience Publishers: New York.)
Richards, P. W. (1952). 'The Tropical Rain Forest.' (Cambridge University Press: Cambridge.)
Sage, R. F., and Sharkey, T. D. (1987). The effect of temperature on the occurrence of O_2 and CO_2 insensitive photosynthesis in field grown plants. *Plant Physiol* **84**, 658–64.
Salvidge, G. (1980). Photosynthesis of marine phytoplankton in fluctuating light regimes. *Marine Biol. Lett.* **1**, 295–300.
Seemann, J. R. (1986). Mechanisms for the regulation of CO_2 fixation by ribulose-1,5-bisphosphate carboxylase. In 'Biological Control of Photosynthesis'. (Eds R. Marcelle, H. Clijesters and M. Van Poucke.) pp. 71–82. (Martinus Nijhoff: Dordrecht.)
Seemann, J. R., Berry, J. A., Freas, S. M., and Krump, M. A. (1985). Regulation of ribulose bisphosphate carboxylase activity *in vivo* by a light-modulated inhibitor of catalysis. *Proc. Natl Acad. Sci. U.S.A.* **82**, 8024–8.
Sharkey, T. D. (1985). O_2 insensitive photosynthesis in C_3 plants. Its occurrence and a possible explanation. *Plant Physiol.* **78**, 71–5.
Sharkey, T. D., Seemann, J. R., and Berry, J. A. (1986). Regulation of ribulose-1,5-bisphosphate carboxylase activity in response to changing partial pressure of O_2 and light in *Phaseolus vulgaris. Plant Physiol.* **81**, 788–91.
Sharkey, T. D., Seemann, J. R., and Pearcy, R. W. (1986). Contribution of metabolites of photosynthesis to post illumination CO_2 assimilation in response to lightflecks. *Plant Physiol.* **82**, 1063–8.
Sivak, M. M., and Walker, D. A. (1986). Summing-up: measuring photosynthesis *in vivo*. In 'Biological Control of Photosynthesis'. (Eds R. Marcelle, H. Clijesters and M. Van Poucke.) pp. 1–31. (Martinus Nijhoff: Dordrecht.)
Stitt, M. (1986). Limitation of photosynthesis by carbon metabolism. I. Evidence for excess electron transport capacity in leaves carrying out photosynthesis in saturating light and CO_2. *Plant Physiol.* **81**, 1115–22.
Stitt, M., Wirtz, W., and Heldt, H. (1980). Metabolite levels during induction in the chloroplast and extrachloroplast compartments of spinach protoplasts. *Biochim. Biophys. Acta* **593**, 85–102.
Usuda, H. (1985). The activation state of ribulose 1,5-bisphosphate carboxylase in maize leaves in dark and light. *Plant Cell Physiol.* **26**, 1455–63.
Usuda, H., Ku, M. S. B., and Edwards, G. E. (1984). Activation of NADP-malate dehydrogenase, pyruvate, P_i dikinase, and fructose-1,6-bisphosphatase in relation to photosynthetic rate in maize. *Plant Physiol.* **76**, 238–43.
Vu, C. V., Allen, L. H., and Bowes, G. (1984). Dark / light modulation of ribulose bisphosphate carboxylase activity in plants from different photosynthetic categories. *Plant Physiol.* **76**, 843–5.
Walker, D. A., Sivak, M. N., Prinsley, R. T., and Cheeseborough, J. K. (1983). Simultaneous measurement of oscillations in oxygen evolution and chlorophyll *a* fluorescence in leaf pieces. *Plant Physiol.* **73**, 542–9.
Weis, E., Ball, T., and Berry, J. (1987). Photosynthetic control of electron transport in leaves of *Phaseolus vulgaris:* evidence for regulation of photosystem 2 by the proton gradient. In 'Progress in Photosynthesis Research'. (Ed. J. Biggins.) pp. 553–6. (Martinus Nijhoff: Dordrecht.)

Regulation of Carboxylation and Photosynthetic Oscillations During Sun–Shade Acclimation in *Helianthus annuus* Measured with a Rapid-response Gas Exchange System

C. B. Osmond[A], *V. Oja*[B] *and A. Laisk*[B]

[A] Plant Environmental Biology Group, Research School of Biological Sciences, Australian National University, G.P.O. Box 475, Canberra, A.C.T. 2601, Australia.
[B] Laboratory of Biophysics, Institute for Astrophysics and Atmospheric Physics, Estonian Academy of Sciences, 202444 Toravere, Tartu, Estonia, U.S.S.R.

Abstract

The consequences of acclimation from shade to sun and vice versa for regulated photosynthetic metabolism were examined in *H. annuus*. A rapid-response gas exchange system was used to assess changes in carboxylation-related parameters (mesophyll conductance, assimilatory charge and CO_2 capacity) and to analyse oscillations in CO_2 fixation following transfer to high CO_2 concentration as a function of intercellular CO_2 concentration and light intensity. Data showed a two- to threefold change in all carboxylation-related parameters during acclimation in either direction. Dynamic regulation of carboxylation, indicated by changes in oscillatory response as a function of CO_2 concentration at light saturation, remained unchanged, consistent with concerted regulation of ribulose-1,5-bisphosphate carboxylase–oxygenase during acclimation. However, the light dependency of oscillations changed during acclimation from shade to sun, and the range of oscillation was closely tied to the maximum rate of steady-state photosynthesis at CO_2 saturation. These data imply that changes in the light-absorbing and electron transport components of the photosynthetic apparatus underlie the shift in regulatory behaviour during acclimation.

Introduction

Growth of herbaceous plants in shade or in bright light usually brings about substantial changes in photosynthetic properties, which may be characterised at morphological, anatomical, physiological and biochemical levels of organisation (Björkman 1981; Lichtenthaler *et al.* 1981; Anderson and Osmond 1987). A great deal of progress has been made recently in the biochemical characterisation of these responses (Leong and Anderson 1983; Wild *et al.* 1986; Chow and Anderson 1987*a*, 1987*b*; Evans 1987). Fully developed leaves of some plants, when grown under one light condition and provided with adequate root-zone nitrogen nutrition, are able to effect substantial adjustment in photosynthetic physiology when transferred from shade to bright light (Ferrar and Osmond 1985, 1986) or vice versa (Caemmerer and Farquhar 1984). Even within a leaf, the light response curves of photosynthesis measured from the upper (sun side) and lower (shade side) surface (Oja and Laisk 1975) differ and can be analysed in terms of the light gradient through the leaf, and the light response curves of photosynthesis of single chloroplasts, to provide a plausible description of the whole leaf response to light (Terashima and Saeki 1985).

None of these approaches provides insight into how key regulatory relationships between light-driven electron transport and photophosphorylation on the one hand, and carbon metabolism on the other, change in response to growth in sun and shade. We believe that oscillations in photosynthesis may provide some information about the capacities for regulation in response to change in activities of light reactions and dark reactions of photosynthesis.

0310-7841/88/010239$03.00

Oscillations in photosynthetic CO_2 uptake in leaves have been used as a tool for kinetic analysis of photosynthetic regulation by Laisk (1974, 1977). Subsequently, oscillations in O_2 evolution, chlorophyll fluorescence and light scattering by Walker *et al.* (1983) and Sivak *et al.* (1985) have been used to greatly expand our understanding of these processes (Giersch 1985; Laisk and Walker 1986). Although Walker and Osmond (1986) showed that the intensity of blue light required to bring about oscillations in fluorescence corresponds approximately to the intensity of white light required to saturate photosynthetic O_2 evolution in sun- and shade-grown spinach, there have been few quantitative analyses of oscillations in relation to light environment during growth.

Bearing in mind the idea that oscillations may provide a deeper insight into the processes that limit the maximum rate of CO_2- and light-saturated photosynthesis, a more systematic study was undertaken of conditions required for appearance of these oscillations. It was supposed that investigation of oscillations during readjustments of the photosynthetic machinery following transfer between sun and shade would give insight into the major regulatory processes. We therefore chose to do a transfer experiment and, as the studies were carried out with the rapid-response gas exchange system described by Oja (1983), other parameters of leaf photosynthesis, such as the mesophyll conductance, assimilatory charge (Laisk *et al.* 1984) and CO_2 capacity of the leaves (Oja *et al.* 1986) were also measured.

Our experiments demonstrate that the regulatory capacities of photosynthetic metabolism, indicated by changes in carboxylation parameters and by oscillations in CO_2 fixation, are closely coordinated with the rate of CO_2- and light-saturated photosynthesis. Different parts of the photosynthetic machinery that take part in generating the oscillations readjust to new light conditions in a closely coordinated way.

Materials and Methods

Individual plants of sunflower (*Helianthus annuus* L.) were grown in a growth chamber in 5-litre pots filled with soil–peat mixture; combined nutrients (10 g per pot) were added at the outset. Plants grown in bright light were irradiated by a metal arc lamp and plant height was adjusted regularly so that the illuminance at the level of the higher leaves was 30 klux. The shaded plant was grown under an illuminance of 3 klux. Other growth parameters were 18 h/6 h day/night cycle, air temperature 25±2°C/20±2°C (shaded plant), relative humidity 50–60%. At 36 days growth, the shaded plant had developed eight leaves 5–7 cm wide and its height was approximately 60 cm. Plants in bright light were planted later and at the beginning of the experiments were 22 days old with six leaves about 10–20 cm wide. At the start of the experiments, gas exchange characteristics of upper leaves of the plants were measured, and the illumination conditions for the plants then were exchanged. The gas exchange characteristics of the same leaf were repeatedly measured after intervals, and properties of newly formed leaves under the changed light environment were also checked. Thus, in the original shaded plant, leaf 7 was assayed four times, as well as newly formed leaf 17. In the plant originally in bright light, leaves 5 and 8 were assayed, the latter three times, as well as newly formed leaf 15.

Gas exchange measurements were carried out with the rapid-response measurement system described elsewhere (Oja 1983). In this system, two identical independent open systems for CO_2 exchange measurements (referred to as 'channels') containing two gas analysers ('Infralyt IV', DDR, 100 by 18-mm cuvettes) may be tuned to different background CO_2 concentrations. The leaf chamber (44 by 44 by 3 mm) may be rapidly switched into either channel. The flow rate of the gas through the leaf chamber was 20 $cm^3\ s^{-1}$ (0·8 mmol s^{-1}) and, by reducing the dimensions of all the elements, the response time (99·5% of the full deflection) of the system was reduced to 2·3 s. Leaf transpiration rate was measured by means of psychrometers. The CO_2 exchange rate and the intracellular CO_2 concentration were calculated as described by Laisk (1977), taking into account also CO_2 solubility in the liquid phase of the mesophyll cells. Leaf temperature (23–25°C) was estimated by the energy balance technique and maintained by close appression of the leaf to the thermostated water filter of the chamber.

After establishing steady-state photosynthesis in air at light saturation, the CO_2 curves and oscillatory transients were measured by repeated transitions from 310 $\mu l\ CO_2\ l^{-1}$ to higher concentrations (600, 900, 1400, 2000 $\mu l\ CO_2\ l^{-1}$). Between these measurements, the leaf was returned to 310 $\mu l\ CO_2\ l^{-1}$ and allowed to reach steady state again before the next transition. Light curves at 310 and 2000 μl CO_2 l^{-1} and oscillatory transients were measured by repeated transitions from 310 to 2000 μl CO_2 l^{-1} after prestabilising the leaf at 310 μl CO_2 l^{-1} at different light intensities. Throughout the measurement procedures a CO_2 concentration of 310 μl l^{-1} and a saturating irradiation density were used as a standard reference state from which transients to other values of the environmental parameters were made.

Measurements of other gas exchange parameters of the leaves during their readaption to the new light regime were included in the program. Apparent quantum yield was measured as the initial slope of the light response curve at saturating CO_2 (2000 μl l^{-1}). Light was routinely measured in energy units, then subsequently calibrated with a LiCor quantum sensor. Leaf absorptance was not measured.

Mesophyll conductance in air was measured as the slope of assimilation rate *v*. CO_2 concentration obtained after transfer from 310 to 90 and 310 to 0 μl CO_2 l^{-1}. The very first CO_2 exchange rates occurring a few seconds after the transition were used in the calculations. In these experiments, the

Table 1. Changes in photosynthetic parameters measured by rapid-kinetics gas exchange methods during sun–shade acclimation in *Helianthus annuus*

Photosynthetic parameter	Shade → sun (days after transfer)					Sun → shade (days after transfer)			
	0	1	4	7	10	0	3	6	8
Specific leaf weight (g m^{-2})	22	–	–	45	68	54	–	41	26
Maximum photosynthesis (μmol m^{-2} s^{-1})	25	24	35	36	73	78	48	44	32
Apparent quantum yield (mol CO_2 mol^{-1} photons)	0·072	0·067	0·072	0·070	0·074	0·070	0·067	0·079	0·084
10^3 × Mesophyll conductance in air (M_{air}, m s^{-1})	3·8	3·3	5·6	6·3	8·3	7·4	5·7	5·2	3·6
10^3 × Mesophyll conductance in N_2 (M_N, m s^{-1})	5·4	5·0	6·5	6·8	12·0	9·6	7·7	8·3	5·6
Assimilatory charge (μmol m^{-2})	104	145	145	118	231	235	180	200	123
RuP_2 pool size (μmol m^{-2})	99	131	130	107	215	212	166	188	116
10^3 × Specific carboxylation efficiency (m^{-3} s^{-1} $μmol^{-1}$)	0·075	0·047	0·065	0·085	0·073	0·061	0·070	0·065	0·066
10^3 × CO_2 storage capacity (m)	0·35	1·00	1·00	0·50	0·87	1·40	1·05	0·60	0·67

carboxylation rate had already reached a new value corresponding to the lower CO_2 concentration but the rate of photorespiratory CO_2 evolution still corresponded to the state at 310 μl CO_2 l^{-1}.

Assimilatory charge was measured as the post-illumination CO_2 uptake at a low CO_2 concentration in N_2 (Laisk *et al.* 1984). The background air (21% O_2, 310 μl CO_2 l^{-1}) was switched to low O_2 (1% O_2, 310 μl CO_2 l^{-1}) for 3–5 min to minimise photorespiration, then leaves were deprived of O_2 and CO_2 for 10 s in order to convert Calvin cycle intermediates to RuP_2* as completely as possible. Then the light was switched off and the post-illumination CO_2-uptake curve measured in 90 μl CO_2 l^{-1} in N_2. The curves representing the dependence of the mesophyll conductance on the decreasing pool of the assimilatory charge were calculated as described in Laisk *et al.* (1984) and Laisk (1985). These yield a second, independent estimate of mesophyll conductance. CO_2 solubility in the chloroplast stroma in the dark was obtained from the transient from 310 to 2000 μl CO_2 l^{-1}. The recorded transient was analysed using the models described by Oja *et al.* (1986).

*Abbreviations used: PCR, photosynthetic carbon reduction; PEP, phospho*enol*pyruvate; 3-PGA, 3-phosphoglycerate; RuP_2, ribulose 1,5-bisphosphate; SLW, specific leaf weight.

Results and Discussion

Analysis of Carboxylation Parameters During Sun–Shade Acclimation Using Gas Exchange

We have used kinetic, rather than steady-state, methods to analyse carboxylation parameters during sun–shade acclimation in *Helianthus annuus*. This plant has not often been used in such experiments because it is assumed to be shade-intolerant (Hiroi and Monsi 1963). However, the light regimes used in our experiments led to changes in steady-state photosynthetic parameters comparable with those studied in many other herbaceous plants during sun–shade acclimation. Thus, in Table 1 we see that SLW ranged from 22 g m^{-2} in shade-grown plants to 68 g m^{-2} in plants grown in bright light. Changes in SLW following transfer of the plant from one light regime to another were in the direction expected, but in existing fully expanded leaves the change was incomplete compared with leaves newly formed under the changed light regime. The maximum rate of photosynthesis, measured as a steady-state rate at CO_2 and light saturation, was much lower in shade-grown leaves (25–32 μmol m^{-2} s^{-1}) than in leaves developed in bright light 73–78 μmol m^{-2} s^{-1}).

Many studies have demonstrated good correlations between steady-state gas exchange parameters of leaf photosynthesis and the activity of RuP_2 carboxylase–oxygenase in

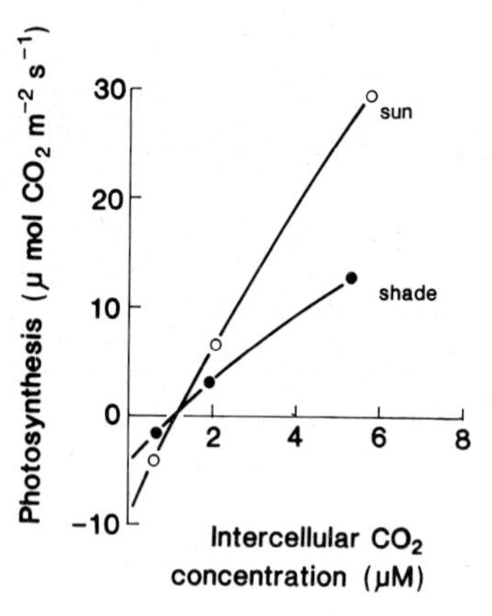

Fig. 1. Initial slope of the CO_2 response curve in leaves of *H. annuus* grown in bright light ('sun', ○) and shade (●) measured by rapid transfer from air to lower CO_2 concentrations.

extracts from leaves of C_3 plants. Björkman (1981) reviewed studies of sun–shade acclimation and found that the rate of photosynthesis in air was highly correlated with extractable RuP_2 carboxylase–oxygenase activity (r = 0·96) and soluble leaf protein (r = 0·86). As the carboxylase protein comprises a large part of total leaf soluble protein, these correlations are to be expected. SLW was less well correlated (r = 0·31), presumably because, although it is related via leaf thickness and number of chloroplasts to total leaf protein and amount of carboxylase protein, it is susceptible to wide variation due to accumulation of starch. Subsequent studies have confirmed these relationships in terms of carboxylase activity (Caemmerer and Farquhar 1981; Ferrar and Osmond 1985, 1986; Seemann *et al.* 1987), and have shown that the initial slope of the CO_2 response curve (mesophyll conductance) correlates well with the amount and activation status of RuP_2 carboxylase–oxygenase *in vivo* (Caemmerer and Edmonson 1986).

We have used two rapid-kinetic methods for estimation of mesophyll conductance in the leaves of *H. annuus* in these experiments. The first, based on rapid transfer from air to low CO_2, gives a value (M_{air}) that most closely approaches that expected on the basis of carboxylation kinetics at low CO_2 concentrations. Fig. 1 shows the relationship between photosynthesis and intercellular CO_2 concentration measured by rapid transfer from air to CO_2 concentrations in the vicinity of the CO_2 compensation point. The slopes for leaves from shade-grown plants and those grown in bright light are clearly different. Values for M_{air} obtained in these experiments are given in Table 1 and show that, following transfer from one light regime to the other, substantial adjustment in this parameter occurs in pre-existing leaves and is complete in newly formed leaves.

The second, derived from the kinetics of post-illumination CO_2 fixation, gives a value (M_N) for mesophyll conductance in N_2 at the moment of darkening. This corresponds to an *in vivo* situation in which RuP_2 carboxylase–oxygenase is functioning with maximum RuP_2 and minimum 3-PGA pools, so that active sites of the enzyme should be RuP_2-saturated. Thus, M_N is representative of the maximum carboxylase activity. When leaves are equilibrated briefly to N_2 with a low CO_2 concentration and the light is switched off, the pool of RuP_2 is sufficiently large to support post-illumination CO_2 fixation for several tens of seconds. The mesophyll conductance was calculated at any moment of the decay curve for post-illumination CO_2 fixation and plotted as a function of the remaining substrate. This pool contains more than RuP_2 itself, and includes part of the pools of ATP, triose phosphates and sugar phosphates, which can be converted to RuP_2. For these reasons, the substrates for post-illumination CO_2 fixation have been termed 'assimilatory power' (Laisk *et al.* 1984) or more dimensionally correct 'assimilatory charge' (Laisk 1985). Carboxylation kinetics are also confounded by a small additional carboxylation due to PEP carboxylase, which consumes PEP derived from 3-PGA, and the total CO_2 fixed exceeds the pool of RuP_2 by this amount.

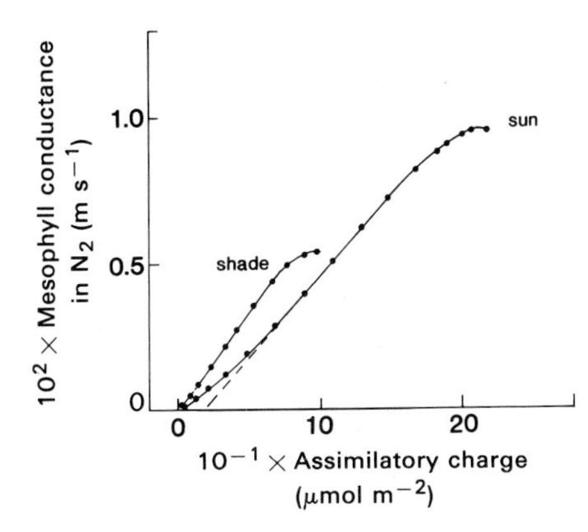

Fig. 2. Relationship between mesophyll conductance in N_2 and assimilatory charge derived from the kinetics of post-illumination CO_2 fixation (as described in the text) for leaves of *H. annuus* grown in bright light ('sun') and shade.

Fig. 2 shows the dependence of mesophyll conductance in N_2 on assimilatory charge in leaves of shade-grown *H. annuus*, and in plants grown in bright light, at the start of our experiments. M_N is taken from the maximum of the curve, at the moment of switching off the light. As shown in Table 1, estimates of M_N are greater than those of M_{air}, and are about twofold higher in leaves grown in bright light than in shade-grown leaves. On transfer from one light regime to the other, M_N changes as did M_{air}, with full transition being found in leaves newly formed under the altered light regime.

Fig. 2 also shows that there are substantial differences in assimilatory charge and estimated RuP_2 pool size in leaves of *H. annuus* in the shade and bright light. Theoretically, the left part of these curves should be linear if only CO_2 fixation by RuP_2 is responsible for carboxylation (Farquhar 1979). The sigmoidicity is thought to reflect other carboxylations, as discussed above (Laisk 1985). An accurate estimate of RuP_2 pool size can be obtained by extrapolation of the linear portion to the abscissa. The slope of RuP_2 carboxylation kinetics of the linear portion of the decay curve describes RuP_2 carboxylation kinetics (see below). Table 1 shows that assimilatory charge and RuP_2 pool size are about twofold greater in leaves from plants in bright light compared to those in shade. On transfer from one light regime to the other, there is a twofold change in these parameters.

Two other parameters can be extracted from our kinetic data. The specific efficiency of carboxylation, derived from the slope of the decay curve in Fig. 2, indicates change in mesophyll conductance (or carboxylation efficiency) with changes in RuP_2 pool size. Values *in vivo* will differ depending on the presence of 3-PGA and other sugar phosphate effectors of RuP_2 carboxylase–oxygenase and should be high when these are low. In our experiments, specific efficiency of carboxylation ranged from $0{\cdot}047 \times 10^{-3}$ to $0{\cdot}085 \times 10^{-3}$ m^{-3} s^{-1} μmol^{-1} RuP_2, but tended to a mean value of $0{\cdot}067 \pm 0{\cdot}010 \times 10^{-3}$ overall. This indicates that RuP_2 carboxylase–oxygenase activity tends to be modulated at a common level, irrespective of changes in enzyme concentration and substrate or effector concentrations. The amount of CO_2 in the leaf, the other substrate of RuP_2 carboxylase–oxygenase, can be estimated by conventional gas exchange methods as the intracellular CO_2 concentration at any particular external value. However, Oja (1985) and Oja *et al.* (1986) devised a kinetic method to estimate CO_2 solubility in the leaf, which reflects CO_2 storage capacity of the chloroplast stroma. It embraces stromal pH, the concentration of compounds with high pK (such as amino acids), and the total stromal volume for CO_2. Table 1 shows that CO_2 storage capacity, expressed as the thickness of an equivalent water layer in equilibrium with air, is low in leaves of shade-grown plants and is high in leaves from plants grown in bright light. Following transfer from one light regime to the other, there is a two- to threefold change in this parameter.

These changes in carboxylation activity and substrate parameters inferred from gas exchange kinetics match, qualitatively, direct measures of the amount and activity of RuP_2 carboxylase–oxygenase, and substrate pool sizes found in other experiments. Seemann *et al.* (1987) studied *Phaseolus vulgaris*, an herbaceous species capable of substantial sun–shade acclimation, and *Alocasia macrorrhiza*, an extremely shade-tolerant shrub from rainforests, which is capable of limited acclimation to bright light. They found that RuP_2 concentration per active site of RuP_2 carboxylase–oxygenase remained much the same ($2{\cdot}8$–$3{\cdot}9$ mol mol^{-1} at ambient CO_2 and $1{\cdot}2$–$1{\cdot}7$ mol mol^{-1} at saturating CO_2) in leaves of sun- and shade-grown plants. In our experiments (Table 1), we found a ratio of 'assimilatory charge' to mesophyll conductance (M_N) of $20{\cdot}2 \pm 5{\cdot}3$ in leaves of shade-grown plants transferred to bright light, and $21{\cdot}7 \pm 0{\cdot}9$ in leaves of plants grown in bright light and transferred to the shade. We interpret this to mean that during sun–shade acclimation, RuP_2 carboxylase–oxygenase activity and RuP_2 pool size *in vivo* tend to maintain a constant ratio, which is presumably consistent with the tendency to maintain constant intercellular CO_2 concentration in air (Ferrar and Osmond 1985).

The range of parameters measured by rapid-kinetic gas exchange methods and shown in Table 1 indicate that the two- to threefold change in RuP_2 carboxylase–oxygenase activity *in vivo* during sun–shade acclimation in *H. annuus* is accompanied by similar adjustments in RuP_2 concentration and CO_2 capacity in the chloroplast. Studies of the dynamic regulation of carboxylation, as indicated by the characteristics of oscillatory behaviour described below, indicate a similar concerted adjustment among other components of electron transport and carbon metabolism.

Oscillations in Photosynthesis as a Function of Growth Irradiance

Laisk (1977) showed that any perturbation of steady-state photosynthesis under internally limiting conditions of saturating CO_2 and light would elicit oscillations. Detailed models to account for these oscillations, based on biochemical systems such as the two-kinase system (Robinson and Walker 1979; Giersch 1985) or the orthophosphate limitation of sucrose metabolism (Walker and Robinson 1978; Laisk 1983; Laisk and Walker 1986) have been developed. We are not concerned here with the biochemical mechanisms of the oscillations *per se*. We shall simply accept that they reflect complex feed-back, and possibly other allosteric, controls, which ultimately limit the rate of RuP_2 carboxylation and O_2 evolution and which alter chlorophyll fluorescence via effects on electron transport and photochemistry. We propose, however, that comparisons of properties of oscillations *in vivo* under specified conditions might indicate the extent to which flux regulation is distributed amongst components of the photosynthetic system, and provide insight into key control processes.

Fig. 3 illustrates a typical oscillation in rate of CO_2 fixation in leaves of *H. annuus* grown in bright light. It was elicited by switching the gas phase surrounding the leaf from air (310 μl CO_2 l^{-1}) to 2000 μl CO_2 l^{-1} when the leaf was illuminated with white light at 903 μmol photons m^{-2} s^{-1}. The course of the oscillation was typically a sharp peak (P_I), due to solubilisation of CO_2 and rapid carboxylation of residual RuP_2 (Laisk 1985), followed by a quickly damped series of troughs and peaks. We found that the period of these oscillations (about 75 s between the second and the third peaks), was relatively insensitive to light conditions during growth, or time after transfer from one light regime to the other. However, the range of oscillations was markedly dependent on the light conditions and the CO_2 concentration 'jump' used in the experiment, and these requirements differed between leaves of plants grown in the shade and bright light.

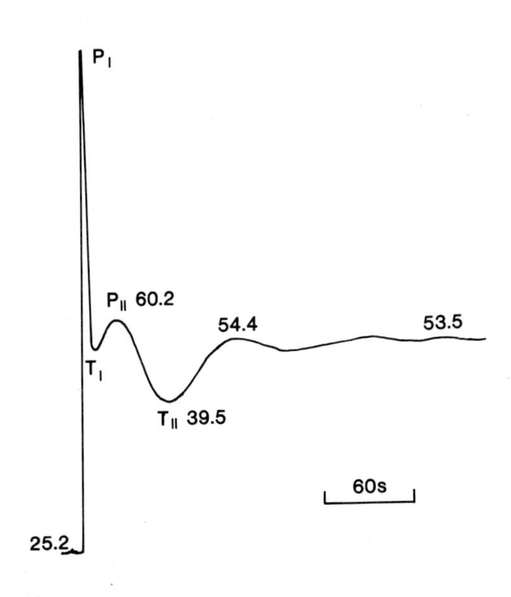

Fig. 3. Oscillation in rate of CO_2 fixation observed in a leaf of *H. annuus* grown in bright light, elicited by transfer from air to 2000 μl CO_2 l^{-1} at light saturation. Numbers refer to rate of CO_2 fixation at various points in the oscillation. The range of the oscillation P_{II}–T_{II} is presented in subsequent figures.

We chose to characterise the oscillations in terms of the range in rate of photosynthesis between the second peak (P_{II}) and the second trough (T_{II}). In terms of the orthophosphate-sucrose synthesis model of Laisk and Walker (1986), P_{II} is attained as PCR-cycle metabolites build to optimum levels, and T_{II} is reached as orthophosphate concentrations decrease. These authors assumed that the control system works with a finite delay, so that several oscillations are observed before steady state is attained. The range of the oscillation P_{II}–T_{II} can be expressed as a fraction of the final steady-state rate of photosynthesis in saturating CO_2 (P_{max}) and this indicates the range of photosynthetic rates over which these controls can be exercised at any prevailing measurement condition. Differences at any specified measuring condition can then be related to differences in the capacities for regulation arising from changes in the photosynthetic apparatus under different growth conditions.

Oscillations were observed at light saturation whenever the gas stream was switched from air (310 μl CO_2 l^{-1}) to a CO_2 concentration above the shoulder of the CO_2

response curve. Fig. 4 shows the CO_2 response curves at light saturation, and the range of oscillations is shown by the inclined bars, which are drawn between the intercellular CO_2 concentrations estimated at P_{II} and T_{II}. The trends in the CO_2 response curves in terms of mesophyll conductance (M_{air}) and maximum photosynthesis have already been described in Table 1. We were surprised that the range of the oscillation at CO_2 saturation was almost as large as the maximum rate of photosynthesis, except in newly formed leaves in which it was only about 50%. The oscillatory data are normalised and

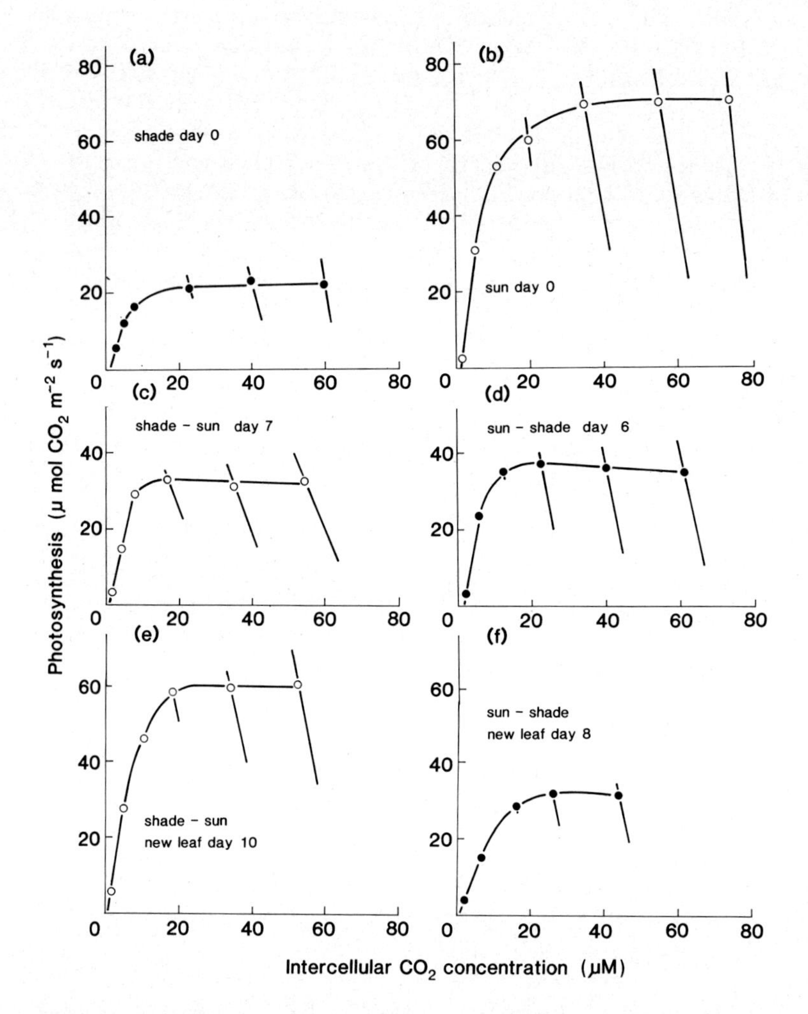

Fig. 4. CO_2 response curves of photosynthesis in leaves of *H. annuus* grown in shade (●) and bright light ('sun', ○) and then transferred to the other light regime. The range of the oscillation P_{II}–T_{II} elicited on switching from air to each higher CO_2 concentration is shown by the inclined bar above and below the final steady-state value at that CO_2 concentration.

plotted as a function of intercellular CO_2 concentration in Fig. 5. This confirms the indication in Fig. 4 that there is no difference in the CO_2 dependence of oscillations between leaves of shade-grown plants and those grown in bright light. The normalised data for measurements following transfer from one light regime to the other fall above the curve obtained for these controls at the outset. The values for newly formed leaves fall below the original curve.

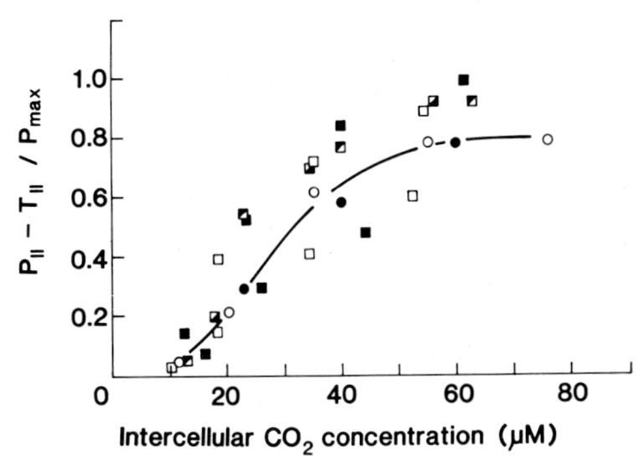

Fig. 5. Ratio of the range of oscillation (P_{II}–T_{II}) to steady-state rate of light-saturated photosynthesis (P_{max}) at each CO_2 concentration, derived from data in Fig. 4. The curve is drawn to data for leaves of *H. annuus* grown in bright light (○) and shade (●) measured at the outset. Other data refer to leaves of shade-grown plants transferred to bright light for 4 or 7 days (◪, □), or grown in bright light and transferred to shade for 3 or 6 days (□, ■). Data points below the curve refer to newly formed leaves in shade (■) or in bright light (□).

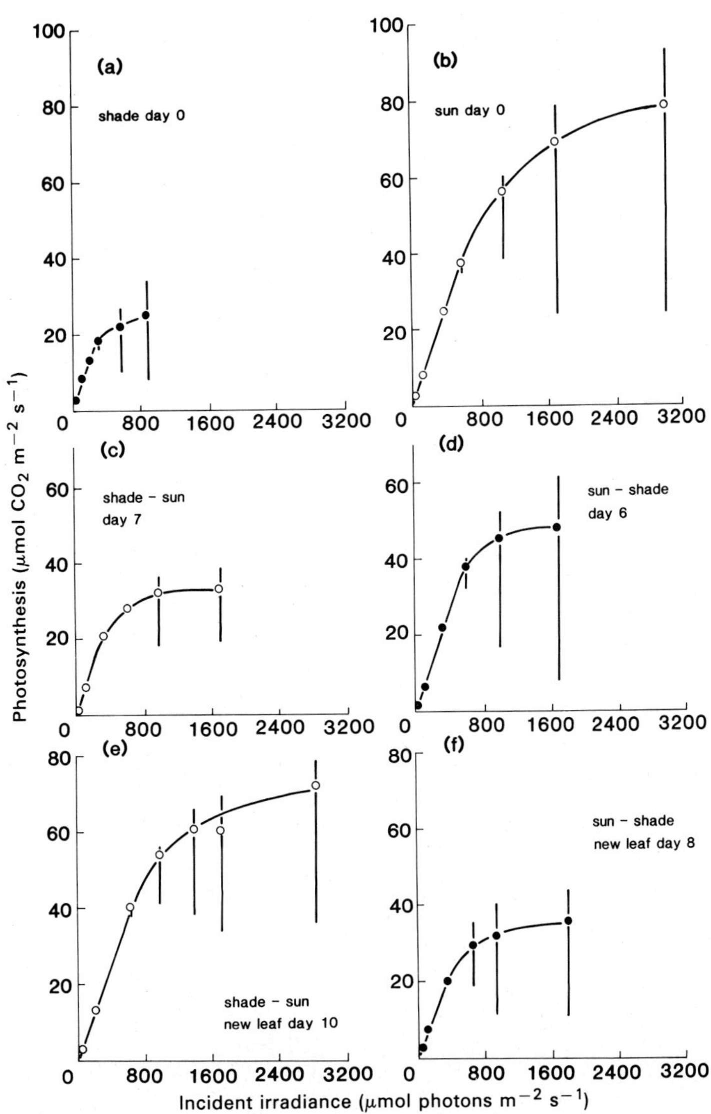

Fig. 6 Light response curves for CO_2-saturated photosynthesis in leaves of *H. annuus* grown in shade. (●) or in bright light ('sun', ○), and then transferred to the other light regime. The amplitude of the oscillation (P_{II}–T_{II}) elicited on switching from air to 2000 μl CO_2 l^{-1} is shown by the vertical bar above and below the steady-state value in 2000 μl CO_2 l^{-1} at each irradiance.

Oscillations in the rate of CO_2 fixation were observed whenever the gas stream to the leaf was switched from air (310 $\mu l\ CO_2\ l^{-1}$) to a saturating CO_2 concentration (2000 $\mu l\ l^{-1}$ in most experiments) at a photon irradiance above the shoulder of the light response curve. Fig. 6 shows the light response curves for steady-state photosynthetic CO_2 fixation in 2000 $\mu l\ CO_2\ l^{-1}$ for leaves of shade-grown *H. annuus* and for plants grown in bright light at the start of the experiment (Figs 6*a*, 6*b*), for the same leaf at an intermediate stage after transfer from one light regime to the other (Figs 6*c*, 6*d*), and for leaves formed in the new light regime (Figs 6*e*, 6*f*). The range of oscillations elicited at each point on the light response curve is indicated by the vertical bars. Leaves of shade-grown plants show small oscillations at 344 μmol photons $m^{-2}\ s^{-1}$, whereas at 860 μmol photons $m^{-2}\ s^{-1}$ the oscillation is about as large as P_{max}. Leaves of plants grown in bright light show a small oscillation at 602 μmol photons $m^{-2}\ s^{-1}$, and at 3010 μmol photons $m^{-2}\ s^{-1}$ the oscillation against is almost as large as P_{max}. Following transfer of the plants to bright light, P_{max} increased, as did the range of the oscillation

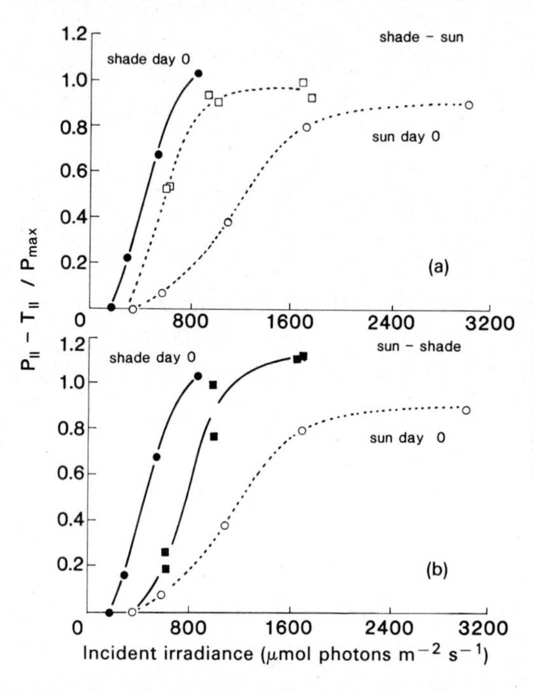

Fig. 7 Ratio of the amplitude of the oscillation P_{II}–T_{II} over the light- and CO_2-saturated rate of photosynthesis (P_{max}) at each irradiance. The relationships in leaves of *H. annuus* grown in the shade (●) and in bright light ('sun', ○) at the start of the experiment are shown in both (*a*) and (*b*). In (*a*), the intermediate position of leaves transferred from shade to bright light for 4 to 7 days (□) is shown, and in (*b*) the intermediate position of leaves transferred from bright light to shade (■) is shown.

at P_{max}. Moreover, the photon irradiance required to elicit the oscillation increased from 301 to 601 μmol photons $m^{-2}\ s^{-1}$. However, the maximum oscillation observed in the newly formed leaf of the plant transferred from shade to bright light and in the reciprocal transfer oscillation only ranged over about 50% of P_{max}. The irradiance required to elicit the oscillation did not decrease on transfer to shade, but perhaps more data points are needed to establish this point. These changes are summarised in the normalised data of Fig. 7 in which P_{II}–T_{II}/P_{max} is plotted against photon irradiance for all experiments.

These experiments indicate that the extent of regulation of carbon metabolism shown by the range of oscillations during sun–shade acclimation is closely tied to the maximum rate of steady-state photosynthesis at CO_2 saturation. It is clear that transiently higher (about 20%) rates of photosynthesis can be achieved, and that the control system can arrest the maximum rate even more markedly (by about 80%) in some circumstances (Figs 4*b*, 4*c*, 6*d*). The data indicate that, during acclimation to bright light or shade, control capacity within individual leaves changes in concert with P_{max}. However, in newly formed leaves the extent of regulation is less than that of established leaves. The CO_2 concentration response of the range of oscillations remained unchanged (Fig. 5), indicating no substantial change in the PCR-cycle relationships underlying the oscillatory behaviour. On the other hand, the light dependency of the range of oscillations changed markedly during sun-shade acclimation, in both directions (Fig. 7). This implies that alterations in the photochemical and electron transport components of the overall control system underlie sun-shade acclimation. These alterations bring about an increase in the range of high-quantum-yield photosynthesis during acclimation to bright light, and a decrease in this range during acclimation to shade as deduced from light response curves (Tooming 1977, 1984; Berry and Downton 1981).

In the accompanying paper (Woodrow and Mott 1988), these empirical indications of the interactions between carbon metabolism and electron transport activity, which are displayed by the above oscillations, are explored using a control analysis that is designed to provide a quantitative assessment of the degree to which RuP_2 carboxylase-oxygenase determines or limits the rate of photosynthesis. In broad terms, if we are able to initiate an oscillation by a sudden increase in CO_2 concentration, then the sensor in the control complex is likely to be RuP_2 carboxylase-oxygenase, even though the processes responsible for the subsequent character of an oscillation may be vested elsewhere in the complex. It follows that if the increase in CO_2 concentration is applied under conditions in which control of photosynthetic carbon flux is mainly vested in electron transport (e.g. at low light intensity), changes in activities of RuP_2 carboxylase-oxygenase and carbon reduction cycle enzymes will not produce an oscillation. If at higher light intensities the control of photosynthetic carbon flux shifts to RuP_2 carboxylase-oxygenase and the carbon reduction cycle, we would expect to observe oscillations. Thus, it is not surprising that the outcome of RuP_2 carboxylase-oxygenase control analyses yields control profiles that are graphically analogous to the normalised oscillatory data (cf. Fig. 7 and Woodrow and Mott 1988, figure 2).

Acknowledgments

These experiments were done in the Laboratory of Biophysics, Tartu, under the sponsorship of an exchange agreement between The Australian National University and The Estonian Academy of Sciences. We are grateful to many colleagues in The Institute for Astrophysics and Atmospheric Physics, and in The University of Tartu (especially to T. Laisk) for facilitating this project.

References

Anderson, J. M., and Osmond, C. B. (1987). Shade-sun responses: compromises between acclimation and photoinhibition. In 'Photoinhibition, (Eds D. J. Kyle, C. B. Osmond and C. J. Arntzen.) Topics in Photosynthesis, Vol. 9, pp. 1-38. (Elsevier: Amsterdam.)

Berry, J. A., and Downton, W. J. S. (1981). Environmental regulation of photosynthesis. In 'Photosynthesis Vol. II'. (Ed. Govindjee.) pp. 263-343. (Academic Press: New York.)

Björkman, O. (1981). Responses to different quantum flux densities. In 'Physiological Plant Ecology I. Responses to the Physical Environment'. (Eds O. L. Lange, P. S. Nobel, C. B. Osmond and H. Ziegler.) Encycl. Plant Physiol. New Ser Vol. 12A, pp. 57-107. (Springer-Verlag: Berlin.)

Caemmerer, S. von, and Edmondson, D. L. (1986). Relationship between steady-stage gas exchange, *in vivo* ribulose bisphosphate carboxylase activity and some carbon reduction in cycle intermediates in *Raphanus sativus*. *Aust J. Plant Physiol.* **13**, 669-88.

Caemmerer, S. von, and Farquhar, G. D. (1981). Some relationships between the biochemistry of photosynthesis and the gas exchange of leaves. *Planta* (*Berl.*) **153**, 367-87.

Caemmerer, S. von, and Farquhar, G. D. (1984). Effects of partial defoliation, changes in irradiance during growth, short-term water stress and growth at enhanced $p(CO_2)$ on the photosynthetic capacity of leaves of *Phaseolus vulgaris* L. *Planta* **160**, 320-9.

Chow, W. S., and Anderson, J. M. (1987*a*). Photosynthetic responses of *Pisum sativum* to an increase in irradiance during growth. I. Photosynthetic activities. *Aust. J. Plant Physiol.* **14,** 1–8.

Chow, W. S., and Anderson, J. M. (1987*b*). Photosynthetic responses of *Pisum sativum* to an increase in irradiance during growth. II. Thylakoid membrane components. *Aust. J. Plant Physiol.* **14,** 9–19.

Evans, J. R. (1987). The relationship between electron transport components and photosynthetic capacity in leaves grown at different irradiances. *Aust. J. Plant Physiol.* **14,** 157–70.

Farquhar, G. D. (1979). Models describing the kinetics of ribulose bisphosphate carboxylase–oxygenase. *Arch. Biochem. Biophys.* **193,** 456–68.

Ferrar, P. J., and Osmond, C. B. (1985). Role of nitrogen nutrition in shade to sun acclimation by *Solanum dulcamara.* In 'Kinetics of Photosynthetic Carbon Metabolism in C_3-Plants I'. (Eds J. Viil, G. Grishna and A. Laisk.) pp. 96–105. (Valgus: Tallinn.)

Ferrar, P. J., and Osmond, C. B. (1986). Nitrogen supply as a factor influencing photoinhibition and photosynthetic acclimation after transfer of shade-grown *Solanum dulcamara* to bright light. *Planta (Berl.)* **168,** 563–70.

Giersch, C. (1985). Oscillatory response of photosynthesis in leaves to environmental perturbations: a mathematical model. *Arch. Biochem. Biophys.* **245,** 263–70.

Hiroi, T., and Monsi, M. (1963). Physiological and ecological analyses of shade tolerance of plants 3. Effect of shading on growth attributes of *Helianthus annuus. Bot Mag. Tokyo* **76,** 121–9.

Laisk, A. (1977). 'Kinetics of Photosynthesis and Photorespiration in C_3 Plants.' (Nauka: Moscow.) [In Russian.]

Laisk, A. (1983). Biochemical structure and kinetic function of the photosynthetic apparatus of plants. *Fiziol. Rast.* (*Mosc.*) **30,** 837–51. (*Sov. Plant Physiol.* **30,** 621–33.)

Laisk, A. (1985). Kinetics of photosynthetic CO_2 uptake in C_3 plants. In 'Kinetics of Photosynthetic Carbon Metabolism in C_3-Plants I' (Eds J. Viil, G. Grisna, and A. Laisk.) pp. 21–34. (Valgus: Tallinn.)

Laisk, A., Kiirats, O., and Oja, V. (1984). Assimilatory power (post illumination CO_2 uptake) in leaves. Measurement, environmental dependencies and kinetic properties. *Plant Physiol.* **76,** 723–9.

Laisk, A., and Walker, D. A. (1986). Control of phosphate turnover as a rate-limiting factor and possible cause of oscillations in photosynthesis: a mathematical model. *Proc. R. Soc. Lond. B. Biol. Sci.* **227,** 281–302.

Leong, T. Y., and Anderson, J. M. (1983). Changes in composition and function of thylakoid membranes as a result of photosynthetic adaptation of chloroplasts from pea plants grown under different light conditions. *Biochim, Biophys. Acta* **723,** 391–9.

Lichtenthaler, H. K., Buschmann, C., Döll, M., Feitz, H.-J., Bach, T., Kozel, U., Meier, D., and Rahmsdorf, U. (1981). Photosynthetic activity, chloroplast ultrastructure, and leaf characteristics of high-light and low-light plants and of sun and shade leaves. *Photosynth. Res.* **2,** 115–41.

Oja, V. M. (1983). Fast gasometric device for investigation of leaf photosynthesis kinetics. *Fiziol. Rast.* (*Mosc.*) **30,** 1045–52. (*Sov. Plant Physiol.* **30,** 795–802.)

Oja, V. M. (1985). Estimation of pH and carbonic anhydrase activity in intact leaves on the basis of the kinetics of CO_2 dissolution. In 'Kinetics of Photosynthetic Carbon Metabolism in C_3-Plants II.' (Eds J. Viil, G. Grishna, and A. Laisk.) pp. 104–8. (Valgus: Tallinn.)

Oja, V. M., and Laisk, A. K. (1975). Adaptation of the photosynthetic apparatus to the light profile in the leaf. *Fiziol. Rast.* (*Mosc.*) **23,** 445–51. (*Sov. Plant Physiol,* 1976, **23,** 381–6.

Oja, V. M., Laisk, A., and Heber, U. (1986). Light induced alkalinization of the chloroplast stroma *in vivo* as estimated from the CO_2 capacity of intact sunflower leaves. *Biochim. Biophys. Acta* **849,** 355–65.

Robinson, S. P., and Walker, D. A. (1979). The control of 3-phosphoglycerate reduction in isolated chloroplasts by the concentrations of ATP, ADP and 3-phosphoglycerate. *Biochim. Biophys. Acta* **545,** 528–36.

Seemann, J. R., Sharkey, T. D., Wang, J.-L., and Osmond, C. B. (1987). Environmental effects on photosynthesis, nitrogen use efficiency, and metabolite pools in leaves of sun and shade plants. *Plant Physiol.* **84,** 796–802.

Sivak, M. N., Dietz, K.-J., Heber, U., and Walker, D. A. (1985). The relationship between light scattering and chlorophyll *a* fluorescence during oscillations in photosynthetic carbon assimilation. *Arch. Biochem. Biophys.* **237,** 513–19.

Terashima, I., and Saeki, T. (1985). A new model for leaf photosynthesis incorporating the gradients of light environment and of photosynthetic properties of chloroplasts within a leaf. *Ann. Bot.* **56,** 489–99.

Tooming, H. G. (1977). 'Solar Radiation and Yield Formation.' (Gidrometeoizdat: Leningrad.) [In Russian.]

Tooming, H. G. (1984). 'Ecological Principles of Maximum Crop Productivity.' (Gidrometeoizdat: Leningrad.) [In Russian.]

Walker, D. A., and Osmond, C. B. (1986). Measurement of photosynthesis *in vivo* with a leaf disc electrode: correlation between light dependence of steady-state photosynthetic O_2 evolution and chlorophyll *a* fluorescence transients. *Proc. R. Soc. Lond. B. Biol. Sci.* **227,** 267–80.

Walker, D. A., and Robinson, S. P. (1978). Chloroplast and cell. A contemporary view of photosynthetic carbon assimilation. *Ber. Dtsch. Bot. Ges.* **91,** 513–26.

Walker, D. A., Sivak, M. N., Prinsley, R. T., and Cheesborough, J. K. (1983). Simultaneous measurements of oscillation in oxygen evolution and chlorophyll *a* fluorescence in leaf pieces. *Plant Physiol.* **73,** 534–41.

Wild, A., Höpfner, M., Rühle, W., and Richter, M. (1986). Changes in the stoichiometry of photosystem II components as an adaptive response to high-light and low-light conditions during growth. *Z. Naturforsch. Sect. C Biosci.* **41,** 597–603.

Woodrow, I. E., and Mott, K. A. (1988). Quantitative assessment of the degree to which ribulosebisphosphate carboxylase/oxygenase determines the steady-state rate of photosynthesis during sun–shade acclimation in *Helianthus annuus* L. *Aust. J. Plant Physiol.* **15,** 252–61.

Quantitative Assessment of the Degree to which Ribulosebisphosphate Carboxylase/Oxygenase Determines the Steady-state Rate of Photosynthesis during Sun–Shade Acclimation in *Helianthus annuus* L.

Ian E. Woodrow[A] *and Keith A. Mott*[B]

[A] Division of Plant Industry, CSIRO, G.P.O. Box 1600, Canberra, A.C.T. 2601, Australia.
[B] Department of Biology, College of Science, Utah State University, Logan, Utah, 84322-5305, U.S.A.

Abstract

The degree to which ribulose-1,5-bisphosphate carboxylase/oxygenase (Rubisco) determines or limits the steady state rate of CO_2 assimilation during sun–shade acclimation in sunflower (*Helianthus annuus*) was examined and expressed as a 'flux control coefficient'. The coefficients were calculated according to the procedure developed by Woodrow *et al.* (1987) [In: Proc. Congress on Rubisco, (Ed. R. G. Jensen.) (Tucson: U.S.A.)] an outline of which is presented in this paper. The analysis shows that the Rubisco control coefficient, and therefore the degree of flux control exerted by this enzyme, increases with increasing light intensity until, at saturating light, Rubisco is the principal determinant of the photosynthetic rate. There was however a marked difference in the range of light intensities over which changes in flux control occurred between the plants at various stages of acclimation.

Introduction

Studies of plants grown in shade and bright light have revealed that differences in the amounts of the major protein components of the photosynthetic system underlie the dissimilar responses of the rates of photosynthesis of these plants to light and CO_2. Adaptation to low irradiance, for example, is generally associated with reduced levels of cytochrome *f* (Oquist *et al.* 1982; Leong and Anderson 1984*b*; Shmeleva and Ivanov 1985; Evans 1987), coupling factor (Berzborn *et al.* 1981; Leong and Anderson 1984*b*; Davies *et al.* 1986; Evans 1987), the photosystem II complex (Leong and Anderson 1984*b*; Chow and Anderson 1987; Chow and Hope 1987; Evans 1987), and enzymes of the photosynthetic carbon reduction (PCR) cycle (Medina 1971; Boardman 1977; Björkman 1981; Seeman *et al.* 1987). There is, however, little change in the photosystem I content (Leong and Anderson 1984*b*; Shmeleva and Ivanov 1985; Chow and Hope 1987; Evans 1987), and a reduced chlorophyll *a/b* ratio indicates an increase in the amount of light-harvesting chlorophyll *a/b* complexes (Leong and Anderson 1984*a*, 1984*b*; Evans 1987).

The overall effect of these changes is that the capacities for both electron transport and CO_2 fixation by the PCR cycle enzymes are reduced, but the degree of reduction does not seem to follow a fixed pattern (e.g. Caemmerer and Farquhar 1981; Evans 1987). These observations regarding protein levels and partial reaction capacities suggest that, at a given irradiance, the degree to which the major proteins limit the rate of photosyntheis may vary with the light condition to which the plant is adapted.

0310-7841/88/010253$03.00

In the current paper, we investigate this suggestion using a recently developed analysis for quantifying the degree to which ribulosebisphosphate carboxylase/oxygenase (Rubisco) limits the rate of CO_2 assimilation (Woodrow *et al.* 1987). We examine both sun-adapted and shade-adapted plants as well as a range of acclimating plants.

Materials and Methods

Sunflower (*Helianthus annuus*) plants in nursery pots containing sterile potting soil were grown in controlled environment greenhouses. Pots were watered as necessary with one-fourth strength modified Hoagland's solution. Peak irradiance for the 'shade' treatment was 250 μmol photons $m^{-2} s^{-1}$ with approximtely 7·0 mol photons m^{-2} day^{-1}. The 'sun' treatment was supplemented with low-pressure sodium lamps and received a peak irradiance of 1600 μmol photons $m^{-2} s^{-1}$ with approximately 50 mol photons m^{-2} day^{-1}. Leaves for gas exchange were selected for uniformity of age and appearance. The gas exchange characteristics of leaves from both treatments were measured and then the light treatments were reversed. Gas exchange measurements of the same leaves were then performed after 7 and 10 days and 3 and 8 days for plants acclimating to high light and low light, respectively.

The gas exchange chamber, gas mixing system, and gas analysis system are described elsewhere (Mott 1988).

To determine the response of photosynthetic rate to intercellular CO_2 partial pressure (c_i) at a particular irradiance a leaf was first brought to steady state at CO_2 and O_2 partial pressures of 290 μbar and 180 μbar, respectively. Photosynthetic rate and c_i values were recorded. Ambient CO_2 partial pressure was then changed sequentially to 235, 255, 320 and 340 μbar and the photosynthetic rate and c_i values determined at each partial pressure. The ambient CO_2 partial pressure was then returned to 290 μbar, and photosynthetic rate was measured at O_2 partial pressures of 150, 130, 200 and 230 mbar. During the O_2 experiments the ambient CO_2 level was varied to maintain c_i at the value obtained at 290 μbar CO_2 and 180 mbar O_2. After varying both c_i and ambient O_2 partial pressure the irradiance was changed, and the entire process was repeated when steady state had been achieved at 290 μbar CO_2 and 180 mbar O_2.

Definition and Derivation of Control Relationships

The degree to which Rubisco limits the photosynthetic carbon flux (rate of net CO_2 fixation) is given by a 'control' coefficient (Kacser and Burns 1973). Mathematically, this coefficient is a dimensionless quantity that relates the variation in flux (J) to an infinitesimal change in the activity at an enzyme (P):

$$C_P^J = \frac{\partial J/J}{\partial P/P},$$

where P is usually the enzyme concentration ($[E]$), the maximum catalytic velocity (V_{max}) or the catalytic rate constant (k_{cat}). In the ensuing discussion, we will write the control coefficients with respect to k_{cat} for either carboxylation or oxygenation and use an overall control coefficient for Rubisco (see Woodrow *et al.* 1987) which is defined as follows:

$$C_{\text{Rubisco}}^A = \frac{\partial A/A}{\partial k_{cat}^c/k_{cat}^c} + \frac{\partial A/A}{\partial k_{cat}^o/k_{cat}^o},$$

where A is the rate of CO_2 assimilation and k_{cat}^c and k_{cat}^o are the catalytic rate constants for carboxylation and oxygenation, respectively. In other words, C_{Rubisco}^A is the sum of the control coefficients of the oxygenase and carboxylase functions of Rubisco.

We have chosen to perturb k_{cat} theoretically to probe effects of alterations in Rubisco activity on photosynthetic flux. Such a theoretical perturbation in k_{cat} can be related to an actual change in CO_2 or O_2 concentration. For example, consider the Michaelis–Menten rate equation

$$v = \frac{k_{cat}[E][S]}{K_m+[S]}.$$

A change in v can be produced by a change in either $[S]$, $[E]$ or k_{cat}. For a given δv, equivalent $\delta[S]$, $\delta[E]$ or δk_{cat} values can be derived. Thus, to simulate a change in k_{cat} experimentally we have used $\delta[S]$, i.e. changes in CO_2 and O_2 concentration. The more complex approach, using $\delta[E]$, is developed by Woodrow *et al.* (1987). It should be noted that these theoretical analyses of the data are independent of the actual changes in $[E]$ which are well known to accompany acclimation (Björkman 1981).

Control coefficients can be assigned to each component of a system with respect to any steady state flux and can have positive or negative values: the greater the magnitude of the coefficient, the greater the role in determining the flux. In interpreting the relative importance of individual components, one should note that the sum of all of the coefficients for a given system is generally unity (for a detailed appraisal see Kacser and Burns 1973).

The equations derived by Woodrow *et al.* (1987) and outlined below express the control coefficients as functions of two other types of coefficients. The first, the flux response coefficient (Kacser and Burns 1973), reflects the influence of an external effector (Q) (i.e. an effector, the concentration of which can be treated as an independent variable) on the flux (J):

$$R_Q^J = \frac{\partial J/J}{\partial Q/Q}.$$

In our case, Q refers to either the intercellular CO_2 or O_2 partial pressure and J to the rate of CO_2 assimilation (A). R_Q^A for CO_2, for example, is the slope of the CO_2 response curve, relating photosynthetic assimilation (A) to intercellular CO_2 partial pressure (c_i) at a particular c_i, multiplied by c_i/A (see Fig. 1).

The second coefficient, a special elasticity coefficient (Kacser and Burns 1973), describes the influence of a change in external effector concentration on the catalytic velocity (v_1) of the target enzyme (E_1):

$${}^{\kappa}\epsilon_Q^1 = \frac{\partial v_1/v_1}{\partial Q/Q}.$$

In our case the enzyme is Rubisco, and the carboxylase and oxygenase functions are treated separately. It is important to note that the flux response coefficient, like the flux control coefficient, describes a property of the whole system, whereas the elasticity coefficient reflects the rate equation of the isolated enzyme.

As outlined above it is convenient to equate δk_{cat}^c and δk_{cat}^o with $\delta[CO_2]$ and $\delta[O_2]$ (Woodrow *et al.* 1987). For a small change in CO_2 ($\delta[CO_2]$),

$$^{\kappa}\epsilon^{c}_{[CO_2]} \frac{\delta[CO_2]}{[CO_2]} = \epsilon^{c}_{k^{c}_{cat}} \frac{\delta k^{c}_{cat}}{k^{c}_{cat}}, \quad (1)$$

$$^{\kappa}\epsilon^{o}_{[CO_2]} \frac{\delta[CO_2]}{[CO_2]} = \epsilon^{o}_{k^{o}_{cat}} \frac{\delta^{o} k_{cat}}{k^{o}_{cat}} \quad (2)$$

and, for small changes in O_2 ($\delta[O_2]$),

$$^{\kappa}\epsilon^{c}_{O_2} \frac{\delta[O_2]}{[O_2]} = \epsilon^{c}_{k^{c}_{cat}} \frac{\delta k^{c}_{cat}}{k^{c}_{cat}}, \quad (3)$$

$$^{\kappa}\epsilon^{o}_{O_2} \frac{\delta[O_2]}{[O_2]} = \epsilon^{o}_{k^{o}_{cat}} \frac{\delta k^{o}_{cat}}{k^{o}_{cat}}, \quad (4)$$

where

$$\epsilon^{c}_{k^{c}_{cat}} = (\partial v_c/\partial k^{c}_{cat})\,(k^{c}_{cat}/v_c) \quad (5)$$

and

$$\epsilon^{o}_{k^{o}_{cat}} = (\partial v_o/\partial k^{o}_{cat})\,(k^{o}_{cat}/v_o). \quad (6)$$

In order to derive the relationship between the response coefficients and the control coefficients, consider the situation in which a small change in $[CO_2]$ is balanced by changes in k^{c}_{cat} and k^{o}_{cat} such that there is no change in the rates of carboxylation and oxygenation (v_c and v_o) or the flux (J). Under these conditions

$$R^{J}_{CO_2} \frac{\delta[CO_2]}{[CO_2]} + C^{J}_{k^{c}_{cat}} \frac{\delta k^{c}_{cat}}{k^{c}_{cat}} + C^{J}_{k^{o}_{cat}} \frac{\delta k^{o}_{cat}}{k^{o}_{cat}} = 0 \quad (7)$$

and

$$R^{J}_{O_2} \frac{\delta[O_2]}{[O_2]} + C^{J}_{k^{o}_{cat}} \frac{\delta k^{o}_{cat}}{k^{o}_{cat}} + C^{J}_{k^{c}_{cat}} \frac{\delta k^{c}_{cat}}{k^{c}_{cat}} = 0. \quad (8)$$

Simultaneous solution of equations (1), (2) and (7) and of equations (3), (4) and (8) yields in the limit as $\delta[CO_2]$ and $\delta[O_2] \rightarrow 0$:

$$R^{J}_{CO_2} = C^{J}_{k^{c}_{cat}}\, {}^{\kappa}\epsilon^{c}_{CO_2} + C^{J}_{k^{o}_{cat}}\, {}^{\kappa}\epsilon^{o}_{CO_2}; \quad (9)$$

$$R^{J}_{O_2} = C^{J}_{k^{c}_{cat}}\, {}^{\kappa}\epsilon^{c}_{O_2} + C^{J}_{k^{o}_{cat}}\, {}^{\kappa}\epsilon^{o}_{O_2}. \quad (10)$$

Since $k^{c}_{cat} \propto v_c$ and $k^{o}_{cat} \propto v_o$, both $\epsilon^{c}_{k^{c}_{cat}}$ and $\epsilon^{o}_{k^{c}_{cat}}$ are equal to unity (see equations 5 and 6).

The solution of equations (9) and (10) gives expressions for the two control coefficients:

$$C^{J}_{k^{c}_{cat}} = \frac{{}^{\kappa}\epsilon^{o}_{O_2} R^{J}_{CO_2} - {}^{\kappa}\epsilon^{o}_{CO_2} R^{J}_{O_2}}{{}^{\kappa}\epsilon^{o}_{O_2}\, {}^{\kappa}\epsilon^{c}_{CO_2} - {}^{\kappa}\epsilon^{o}_{CO_2}\, {}^{\kappa}\epsilon^{c}_{O_2}}; \quad (11)$$

$$C^{J}_{k^{o}_{cat}} = \frac{R^{J}_{CO_2} - {}^{\kappa}\epsilon^{c}_{CO_2}\, C^{J}_{k^{c}_{cat}}}{{}^{\kappa}\epsilon^{o}_{CO_2}}. \quad (12)$$

It should be noted that, although we only examine the rate of CO_2 assimilation in the present study, equations (11) and (12) are valid for all of the fluxes of the photosynthetic system. Moreover, because the intercellular CO_2 and O_2 levels are treated as independent variables, we do not need to consider stomatal or aerodynamic boundary layer control.

Results and Discussion

In the present experiments, the response coefficients ($R^A_{CO_2}$ and $R^A_{O_2}$) were determined empirically from plots of net assimilaltion rate versus intercellular CO_2 and O_2 concentrations for sun- and shade-grown *Helianthus annuus*. An example of these calculations and the relationship between the response coefficients and the irradiance is presented in Fig. 1. The special elasticity coefficients were derived by differentiating the RuP_2-saturated rate equations for the catalytic velocities of Rubisco:

$$v_c = \frac{V_c[CO_2]K_{O_2}}{K_{CO_2}K_{O_2} + K_{O_2}[CO_2] + K_{CO_2}[O_2]} \tag{13}$$

$$v_o = \frac{V_o[O_2]K_{CO_2}}{K_{CO_2}K_{O_2} + K_{O_2}[CO_2] + K_{CO_2}[O_2]} \tag{14}$$

and are given by:

$${}^{\kappa}\epsilon^{c}_{CO_2} = \frac{K_{O_2}K_{CO_2} + [O_2]K_{CO_2}}{K_{CO_2}K_{O_2} + K_{O_2}[CO_2] + K_{CO_2}[O_2]}; \tag{15}$$

$${}^{\kappa}\epsilon^{c}_{O_2} = \frac{-[O_2]K_{CO_2}}{K_{CO_2}K_{O_2} + K_{O_2}[CO_2] + K_{CO_2}[O_2]}; \tag{16}$$

$${}^{\kappa}\epsilon^{o}_{CO_2} = {}^{\kappa}\epsilon^{c}_{CO_2} - 1; \tag{17}$$

$${}^{\kappa}\epsilon^{o}_{O_2} = {}^{\kappa}\epsilon^{c}_{O_2} + 1, \tag{18}$$

where K_{O_2} and K_{CO_2} are the Michaelis constants for O_2 and CO_2, respectively.

In the example presented in Fig. 1, the values of $[CO_2]$ and $[O_2]$ were 211 μbar and 180 mbar, respectively. Assuming values of 264 μbar for K_{CO_2} and 396 mbar for K_{O_2} (Jordan and Ogren 1984), the above equations yield the following elasticities: ${}^{\kappa}\epsilon^{c}_{CO_2} = 0{\cdot}635$; ${}^{\kappa}\epsilon^{o}_{CO_2} = -0{\cdot}365$; and ${}^{\kappa}\epsilon^{o}_{O_2} = 0{\cdot}802$. Using the values for $R^A_{CO_2}$ and $R^A_{O_2}$ shown in Fig. 1, one can calculate, using equations (11) and (12), a control coefficient for carboxylation ($C^A_{k^c_{cat}}$) of $1{\cdot}01$ and for oxygenation ($C^A_{k^o_{cat}}$) of $-0{\cdot}24$. The overall control coefficient for Rubisco ($C^A_{Rubisco}$) is therefore $0{\cdot}77$.

Two basic assumptions underlie the derivation of the equations for calculating the control coefficients. Firstly, it is assumed that changes in the activation state and kinetic properties of Rubisco in response to alterations in irradiance—and the associated changes in the levels of H^+, Mg^{2+}, RuP_2 and other enzyme effectors—do not greatly alter the shapes of the CO_2 and O_2 response curves from those described in equations 915)–(18). Studies of *H. annuus* (K. A. Mott and I. E. Woodrow, unpublished work) and other C_3 species (Perchorowicz *et al.* 1981; Perchorowcz and

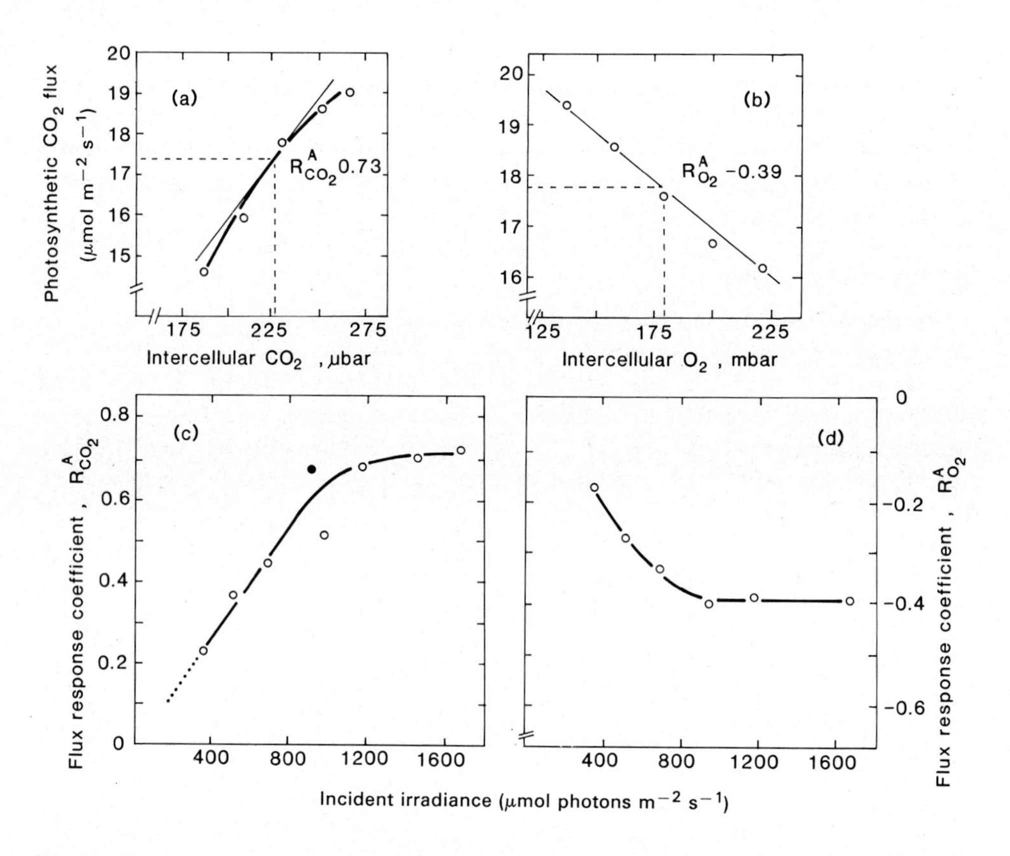

Fig. 1. Response of a leaf of *Helianthus annuus* to CO_2 and O_2. The leaf was first grown in bright light then transferred to low light (see Materials and Methods). After 3 days, the rate of photosynthesis was measured at a variety of CO_2 and O_2 partial pressures and irradiances. Examples of these curves are shown in *a* and *b* where the measurements were made at an irradiance of 1660 μmol quanta m^{-2} s^{-1}. The response coefficients for CO_2 ($R^A_{CO_2}$) and O_2 ($R^A_{O_2}$) (see text) were calculated from the slopes of the assimilation versus CO_2 (*a*) and O_2 (*b*) curves, respectively. The relationships between $R^A_{CO_2}$ and $R^A_{O_2}$ and the irradiance at which assimilation was measured are shown in (*c*) and (*d*). The $R^A_{CO_2}$ point (●) was calculated from the CO_2 response curves for acclimating sunflower presented by Osmond *et al.* (1988). These response coefficients, together with the elasticity coefficients calculated from the RuP_2-saturated rate equation for Rubisco, are used to calculate the control coefficients.

Jensen 1983; Badger *et al.* 1984; Deitz and Heber 1984, 1986; Mott *et al.* 1984; Caemmerer and Edmondson 1986) showed that the RuP_2 concentration is generally higher than the level of active sites and does not change dramatically over a large range of the irradiances at which the rate of CO_2 assimilation is changing. It is also thought that the stromal pH does not change significantly until relatively low irradiances (e.g. Heber *et al.* 1982). Therefore, unless changes in the proportion of active enzyme, which occur in response to a change in irradiance (e.g. Perchorowicz *et al.* 1981), alter the nature of the reponse of Rubisco to CO_2 and O_2, we feel that the RuP_2-saturated rate equations (equations 13 and 14) are reasonable approximations.

More complex and accurate equations could be used in place of equations (13) and (14) in order to refine the estimates of the control coefficients. This, however, will depend upon advances in our understanding of the Rubisco reaction mechanism, especially the means by which the activation state is controlled. The second assumption is that, in the vicinity of 180 mbar O_2 and the operative steady-state c_i value, changes in the concentrations of these molecules affect photosynthesis primarily by altering the catalytic properties of Rubisco. In other words, the flux changes due to alterations in the levels of CO_2 and O_2 are assumed to be brought about by Rubisco alone and any affects on the flux through, for example, the Mehler reaction are assumed to be negligible.

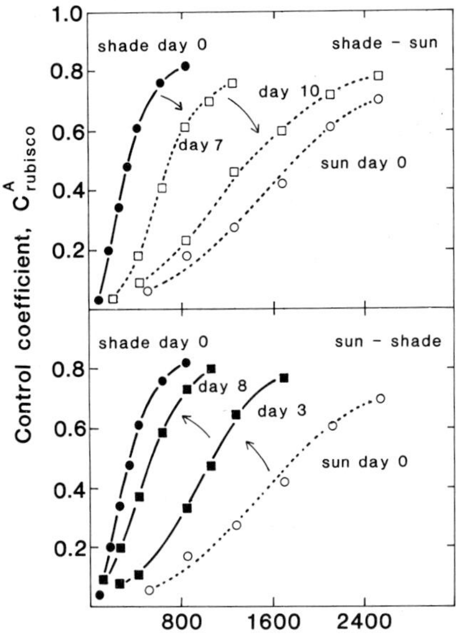

Fig. 2. Relationship between the degree to which Rubisco determines or controls the steady state rate of CO_2 assimilation (given by the control coefficient, C^A_{Rubisco} and the illumination energy density. Flux control profiles for shade- and sun-grown plants and plants acclimating to shade and bright sun conditions were determined from gas exchange measurements and the equations described in the text. In interpreting the data, it should be noted that the sum of all of the control coefficients of a system (in the present case the system is defined as all of the photosynthetic 'biochemistry', i.e. the stomata and aerodynamic boundary layer are excluded) is generally unity.

The flux control analysis shows that, in the sun- and shade-grown and acclimating plants, the degree to which Rubisco determines the steady-state rate of assimilation increases in a more-or-less sigmoidal fashion with increasing irradiance until, at saturating light, the carboxylase is the principal limitation to the rate of CO_2 assimilation (Fig. 2). This pattern is quite consistent with the analysis of quantum yield data where, as light saturation is approached, the drop in quantum yield indicates that one or more elements of the photosynthetic 'biochemistry' are becoming increasingly limiting. Our data suggest that Rubisco is probably the principal component of this limitation. Our data, and those of Woodrow *et al.* (1987), also show that, even at saturating light, there is still a significant proportion of total flux control resident in other elements of the photosynthetic system (i.e. $C^A_{\text{Rubisco}} \approx 0{\cdot}9$ at saturating light).

The data indicate that ostensible saturation of Rubisco with $\text{Ru}P_2$ (i.e. $[\text{Ru}P_2]$ > [Rubisco active sites]) does not on its own provide a reliable measure of the conditions under which the rate of photosynthesis is limited by the activity of Rubisco. In the sun-grown plants, for example, the $\text{Ru}P_2$ concentration was greater than that of active sites (data not shown) and did not vary much over the range of irradiances where the Rubisco control coefficient underwent its greatest change (Fig. 2). This observation suggests that the strength of any feedback on the electron transport system resulting from a buildup of RuP_2 is relatively low at the lower irradiances, despite the ostensibly saturating levels of this substrate. It is of course most difficult to define how closely saturation must be approached before an enzyme constitutes a 'bottleneck' without considering all of the other regulatory mechanisms affecting the activity of the enzyme (Woodrow 1986). Rubisco activity is affected not only by the three reaction substrates, but also by several intermediates in the PCR cycle (e.g. Hatch and Jensen 1980; Badger and Lorimer 1981), a specific inhibitor (Berry *et al.* 1987), and mechanisms which result in a change in the proportion of carbamilated enzyme as a function of irradiance (Perchorowicz *et al.* 1981). The relative strengths of all of these effectors must be considered before the mechanistic basis for the changes in Rubisco flux control can be properly understood.

Despite qualitative similarity in Rubisco flux control profiles of acclimating and control plants, there is a marked difference in the range of irradiances over which changes in flux control occur. In the sun-grown plant, a control coefficient for carboxylase of 0·5 occurs at about 1780 μmol quanta m^{-2} s^{-1}, whereas the same coefficient in the shade-grown control is attained at only 340 μmol quanta m^{-2} s^{-1} (Fig. 2). Acclimating plants show intermediate flux control profiles and, as one might expect, the longer the exposure to the new light regime, the closer the flux control profile resembles that of the control plant (Fig. 2). It is also evident from the control analysis that, at the irradiance to which the sun and shade plants are adapted, Rubisco only partially determines the rate of CO_2 assimilation. This is consistent with the parallel changes in electron transport capacity and Rubisco activity observed by other workers (e.g. Caemmerer and Farquhar 1981; Evans 1987).

This change in the relationship between Rubisco flux control and irradiance is also reflected by the ability of the photosynthetic system to oscillate given a rapid rise in the intercellular CO_2 concentration (Osmond *et al.* 1988). As the flux control exerted by Rubisco increases, the possibility of shared control following a step increase in CO_2 concentration grows, as does the probability of oscillations. We therefore observe a correlation between the oscillatory behaviour of the acclimating plants (Osmond *et al.* 1988) and the estimated control profiles for Rubisco (Fig. 2). The shape and movement of control profiles of Rubisco during the acclimation process also indicate that there may be a tendency in these plants to maintain the distribution of flux control in a specific configuration. This point is reinforced by the measurements described by Osmond *et al.* (1988) which show that, despite the apparent changes the concentration of Rubisco during acclimation, the ratio of maximum carboxylase activity to RuP_2 pool size, and the degree to which enzyme activity is influenced by effectors, remain quite uniform. This constancy in configuration of the Rubisco regulatory mechanisms and in the degree to which this enzyme determines net rate of photosynthesis may well reflect an optimisation of allocation of protein to various elements of the photosynthetic system.

Acknowledgments

Some of these experiments were done in the Laboratory of Biophysics, Tartu, Estonian Soviet Socialist Republic, under the sponsorship of an exchange agreement between the Australian National University and the Estonian Academy of Sciences. I.E.W. thanks Dr A. Laisk and colleagues in the Institute for Astrophysics and Atmospheric Physics for their collaboration and hospitality. I.E.W. is supported by a CSIRO Post-doctoral Award. This work was funded in part by a grant (No. DMB 8515578) to K.A.M. from the National Science Foundation.

References

Badger, M. R., and Lorimer, G. H. (1981). Interaction of sugar phosphates with the catalytic site of ribulose 1,5-bisphosphate carboxylase. *Biochemistry* **20**, 2219–25.

Badger, M. R., Sharkey, T. D., and Caemmerer, S. von (1984). The relationship between steady state gas exchange of bean leaves and the levels of carbon reduction cycle intermediates. *Planta* **160**, 305–13.

Berry, J. A., Lorimer, G. H., Pierce, J., Seemann, J. R., Meek, J., and Freas, S. (1987). Isolation, identification and synthesis of 2-carboxyarabinitol 1-phosphate, a diurnal regulator of ribulosebisphosphate carboxylase. *Proc. Natl Acad. Sci. U.S.A.* **84**, 734–8.

Berzborn, R. J., Muller, D., Roos, P., and Andersson, B. (1981). Significance of different quantitative determinations of photosynthetic ATP-synthase CF_1 for heterogenous CF_1 distribution and grana formation. In 'Photosynthesis. III. Structure and Molecular Organization of the Photosynthetic Membrane'. (Ed. G. Akoyunoglou.) pp. 107–20. (Balaban Int. Sci. Serv.: Philadelphia.)

Björkman, O. (1981). Response to different quantum flux densities. In 'Physiological Plant Ecology I. Responses to the Physical Environment'. (Eds O. L. Lange, P. S. Nobel, C. B. Osmond and H. Ziegler). Encyl. Plant Physiol. New Ser., Vol. 12A, pp. 57–107. (Springer-Verlag: Berlin.)

Boardman, N. K. (1977). Comparative photosynthesis of sun and shade plants. *Annu. Rev. Plant Physiol.* **28**, 355–77.

Caemmerer, S. von, and Farquhar, G. D. (1981). Some relationships between the biochemistry of photosynthesis and the gas exchange of leaves. *Planta* **153**, 376–87.

Caemmerer, S. von, and Edmondson, D. L. (1986). Relationship between steady state gas exchange, *in vivo* ribulose bisphosphate carboxylase activity and some carbon reduction cycle intermediate in *Raphanus sativus*. *Aust. J. Plant Physiol.* **13**, 669–88.

Chow, W. S., and Anderson, J. M. (1987). Photosynthetic responses of *Pisum sativum* to an increase in irradiance during growth. II. Thylakoid membrane components. *Aust. J. Plant Physiol.* **14**, 9–19.

Chow, W. S., and Hope, A. B. (1987). The stoichiometries of supramolecular complexes in thylakoid membranes from spinach chloroplasts. *Aust. J. Plant Physiol.* **14**, 21–8.

Davies, E. C., Chow, W. S., and Jordan, B. R. (1986). A study of factors which regulate the membrane appression of lettuce thylakoids in relation to irradiance. *Photosynth. Res.* **9**, 359–70.

Dietz, K. J., and Heber, U. (1984). Rate limiting factors in leaf photosynthesis. I. Carbon fluxes in the Calvin cycle. *Biochim. Biophys. Acta* **767**, 432–43.

Dietz, K. J., and Heber, U. (1986). Light and CO_2 limitation of photosynthesis and states of the reactions regenerating ribulose 1,5-bisphosphate or reducing 3-phosphoglycerate. *Biochim. Biophys. Acta* **848**, 392–401.

Evans, J. R. (1987). The relationship between electron transport components and photosynthetic capacity in pea leaves grown at different irradiances. *Aust. J. Plant Physiol.* **14**, 157–70.

Hatch, A. L., and Jensen, R. G. (1980). Regulation of ribulose-1,5-bisphosphate carboxylase from tobacco: changes in pH response and affinity for CO_2 and Mg^{2+} induced by chloroplast intermediates. *Arch. Biochem. Biophys.* **205**, 587–94.

Heber, U., Takahama, U., Neimanis, S., and Shimizu-Takahama, M. (1982). Transport as the basis of the Kok effect. Levels of some photosynthetic intermediates and activation of light-regulated enzymes during photosynthesis of chloroplasts and green leaf protoplasts. *Biochim. Biophys. Acta* **679**, 287–99.

Jordan, D. B., and Ogren, W. L. (1984). The CO_2/O_2 specificity of ribulose 1,5-bisphosphate carboxylase/oxygenase. The dependence on ribulose bisphosphate concentration, pH and temperature. *Planta* **161**, 308–13.

Kacser, H., and Burns, J. A. (1973). The control of flux. *Symp. Soc. Exp. Biol.* **27**, 65–104.

Leong, T.-Y., and Anderson, J. M. (1984*a*). Adaptation of the thylakoid membranes of pea chloroplasts to light intensities. I. Study on the distribution of chlorophyll–protein complexes. *Photosynth. Res.* **5**, 105–15.

Leong, T.-Y., and Anderson, J. M. (1984*b*). Adaptation of the thylakoid membranes of pea chloroplasts to light intensities. II. Regulation of electron transport capacities, electron carriers, coupling factor (CF_1) activity and rates of photosynthesis. *Photosynth. Res.* **5**, 117–28.

Medina, E. (1971). Effect of nitrogen supply and light intensity during growth on the photosynthetic capacity and carboxydismutase activity of leaves of *Atriplex patula* (ssp. *hastata*). *Carnegie Inst. Wash. Year Book* **70**, 551–9.

Mott, K. A. (1988). Do stomata respond to CO_2 concentrations other than intercellular? *Plant Physiol.* (In press.)

Mott, K. A., Jensen, R. G., O'Leary, J. W., and Berry, J. A. (1984). Photosynthesis and ribulose 1,5-bisphosphate concentrations in intact leaves of *Xanthium strumarium* L. *Plant Physiol.* **76**, 968–71.

Oquist, G., Brunes, L., and Hallgren, J. E. (1982). Photosynthetic efficiency of *Betula pendula* acclimated to different quantum flux densities. *Plant Cell Environ.* **5**, 9–15.

Osmond, C. B., Oja, V., and Laisk, A. (1988). Regulation of carboxylation and photosynthetic oscillations during sun–shade acclimation in *Helianthus annuus* measured with a rapid-response gas exchange system. *Aust. J. Plant Physiol.* **15**, 239–51.

Perchorowicz, J. T., Raynes, D. A., and Jensen, R. G. (1981). Light limitation of photosynthesis and activation of ribulose bisphosphate carboxylase in wheat seedlings. *Proc. Natl Acad. Sci. U.S.A.* **78**, 2985–9.

Perchorowicz, J. T., and Jensen, R. G. (1983). Photosynthesis and activation of ribulose bisphosphate carboxylase in wheat seedlings. Regulation by CO_2 and O_2. *Plant Physiol.* **71**, 955–60.

Seemann, J. R., Sharkey, T. D., Wang, J.-L., and Osmond, C. B. (1987). Environmental effects on photosynthesis, nitrogen use efficiency, and metabolite pools in leaves of sun and shade plants. *Plant Physiol.* (In press.)

Shmeleva, V. L., and Ivanov, B. N. (1985). Effect of irradiance during plant growth on electron transport chain in pea chloroplasts: state of cytochromes. *Photosynthetica* **19**, 402–10.

Woodrow, I. E. (1986). Control of the rate of photosynthetic carbon dioxide fixation. *Biochim. Biophys. Acta* **851**, 181–92.

Woodrow, I. E., Ball, J. T., and Berry, J. A. (1987). A quantitative assessment of the degree to which RuBP carboxylase/oxygenase determines the steady state rate of photosynthesis in C_3 plants. In 'Proc. Congress on Rubisco.' (Ed. R. G. Jensen.) (Tucson.)

Maintenance of Leaf Temperature and the Optimisation of Carbon Gain in Relation to Water Loss in a Tropical Mangrove Forest

Marilyn C. Ball[A], *I. R. Cowan*[B] *and Graham D. Farquhar*[B]

[A]North Australia Research Unit,
Department of Biogeography and Geomorphology,
Research School of Pacific Studies,
Australian National University,
G.P.O. Box 4, Canberra, A.C.T. 2601, Australia.

[B]Plant Environmental Biology Group,
Research School of Biological Sciences,
Australian National University,
G.P.O. Box 475, Canberra, A.C.T. 2601, Australia.

Abstract

Properties and display of foliage were studied in relation to gas exchange characteristics of tropical mangrove species. Rates of assimilation of CO_2 were maximal at leaf temperatures of approximately 30°C. The species operated with higher water use efficiencies than do most C_3 species, and water-use characteristics became increasingly conservative with increase in the salinity tolerance of the species. Changes in three properties of leaves, i.e. inclination, area, and succulence, contributed to maintenance of leaf temperatures near air temperatures with minimal evaporative cooling. Interspecific differences in water-use characteristics, and their relationship to the maintenance of favourable leaf temperatures, could affect the competitive abilities of mangroves and the structure of mangrove forests.

Introduction

Mangrove forests are most luxuriant and diverse along the saline coastal wetlands of the humid tropics. Despite growing in environments with an infinite supply of water, these plants characteristically are conservative in their use of water, and have water-use efficiencies which are high for C_3 plants (Ball 1986). This may be an adaptation to the saline environment, but it is not without a price. Maintenance of favourable leaf temperatures with minimal evaporative cooling is at the expense of light interception. The present study investigates the interplay between variation in the physical properties and display of leaves and the optimisation of carbon gain in relation to water loss in a tropical mangrove forest.

Materials and Methods

Leaf Properties

Studies were conducted on leaves of five mangrove species growing naturally in a tidal swamp on Hinchinbrook Island, North Queensland, Australia: *Bruguiera gymnorrhiza* (L.) Lam., *Ceriops tagal* (Perr.) C.B.Rob. var. *australis* C. White, *Rhizophora apiculata* Bl., *Rhizophora lamarckii* Montr., and *Rhizophora stylosa* Griff. A detailed description of the site is given in Boto and Wellington (1983). Leaf rosettes of each species were selected in various sites classified according to exposure as full sun, medium sun and shade. Measurements were made of the dimensions (length, width, and thickness) and the angles of orientation and inclination of each leaf in a rosette. Rosettes were harvested, and the weights of leaves were measured before and after oven drying at 80°C.

0310–7841/88/010263$03.00

Micrometeorological Studies

Leaf and air temperatures were sensed with copper–constantan thermocouples (0·0081 mm^2) referenced against a mercury and glass thermometer maintained in a water bath, the temperature of which was allowed to fluctuate naturally during the day. The voltage output was measured with a Keithley microvoltmeter. Air temperature was measured at leaf height with a thermocouple shielded from direct sunlight. Ambient vapour pressure was calculated from wet and dry bulb temperatures measured with an Assman psychrometer. A LI-COR quantum sensor model LI-1905 was used to measure the quantum flux density incident on a horizontal surface and on the surfaces of inclined leaves.

Gas Exchange Characteristics

Gas exchange characteristics were measured on intact, attached leaves of seedlings with an open flow gas exchange system essentially as described by Wong *et al.* (1978), with modifications as noted in Ball and Critchley (1982). Calculations were according to Caemmerer and Farquhar (1981). Other details are given in the text.

All measurements were made on the youngest, fully expanded leaves of seedlings grown under different conditions. Seedlings of *Avicennia marina* (Forsk.) Vierh. var. *australasica* (Walp.) Moldenke were collected from a fully exposed site on Hinchinbrook Island. Seedlings of *Bruguiera gymnorrhiza* and *Rhizophora apiculata* were grown from propagules in sand subirrigated with 50% seawater amended with 5 mol m^{-3} NH_4NO_3, 0·5 mol m^{-3} PO_4, and 2 mol m^{-3} Fe-EDTA. Solutions were maintained by daily addition of tap water and were changed weekly. The seedlings were cultivated for 6 months beneath shade cloth transmitting 30 and 70% natural sunlight in Darwin. Seedlings of *Ceriops tagal* were collected near Darwin and grown as described above under glasshouse conditions in natural sunlight and beneath shade cloth transmitting 30% sunlight in Canberra. Day/night temperatures averaged 30/20°C prior to measurement.

Results and Discussion

Leaf Gas Exchange Characteristics

Carbon cannot be gained without the expenditure of water, which in turn depends on both the leaf conductance to water vapour and the vapour gradient between the leaf and air. Diurnal variation in evaporative demand is due mainly to leaf temperature, as ambient vapour pressure changes little during the course of a day. The closer that leaf temperature stays to air temperature, the closer the evaporative demand experienced by the leaf reflects the saturation vapour deficit of the air. Leaves operating with high rates of water loss benefit from evaporative cooling which minimises increases in leaf temperature, thereby offsetting the evaporative demand to some extent. Thus, leaves operating with high evaporation rates can take advantage of the high irradiances required to maintain high rates of photosynthesis. In contrast, those operating with low evaporation rates must avoid high irradiances if they are to maintain leaf temperature within the range favourable for photosynthesis throughout the day.

In the present study, responses of the CO_2 assimilation rate, stomatal conductance and intercellular CO_2 concentration (c_i) to variation in leaf temperature and irradiance in attached leaves of three mangrove species, *Avicennia marina, Bruguiera gymnorrhiza,* and *Rhizophora apiculata*, were typical of C_3 plants (Fig. 1). The assimilation rates in field grown leaves of *A. marina* and in leaves of *B. gymnorrhiza* and *R. apiculata* grown in 70% sunlight were light saturated in approximately 1000 μmol m^{-2} s^{-1}. Assimilation rates in leaves of *B. gymnorrhiza* and *R. apiculata* grown in 30% sunlight were light saturated at approximately 500 μmol m^{-2} s^{-1}, and were lower than those in the high light plants. In all plants, stomatal conductance and assimilation rate were maximal at leaf temperatures between 25 and 30°C, a range close to the average air temperatures of 25 and 28°C at Hinchinbrook Island and Darwin (Saenger *et al.* 1977), where the plants were grown. At higher leaf temperatures, decrease in both assimilation rate and stomatal conductance was associated with a relatively constant c_i in low light leaves and increasing c_i in high light leaves. These changes in gas exchange characteristics clearly indicated a decline in the biochemical capacity of the mesophyll to fix carbon with increase in leaf temperature above 32°C. These results are consistent with other studies on photosynthetic responses to temperature in mangroves from the tropics (Andrews *et al.* 1984; Andrews and Muller 1985), and subtropics (Moore *et al.* 1972, 1973).

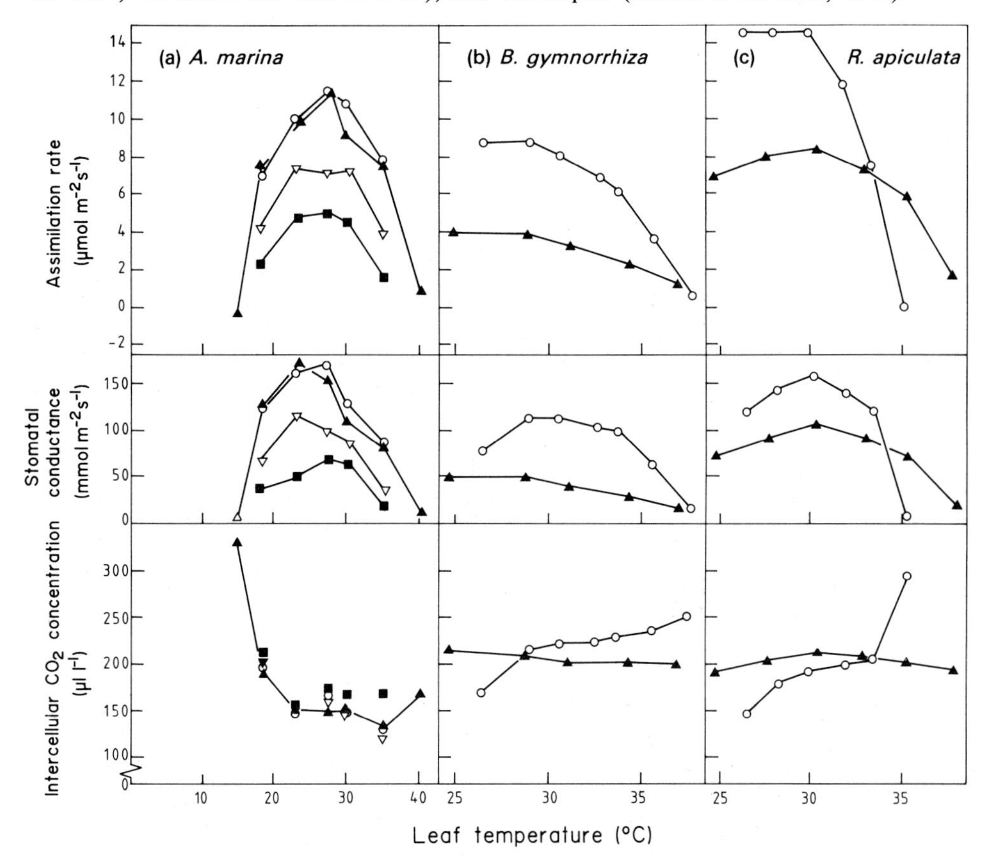

Fig. 1. Gas exchange characteristics of three tropical mangrove species in response to leaf temperature and irradiance. (*a*) *Avicennia marina* grown under fully exposed conditions at Hinchinbrook Island. Leaf temperature was varied by step changes of 5°C from 25 to 40°C followed by 25 to 15°C. Irradiance was varied at each temperature in the sequence 1000 (○), 2000 (▲), 500 (▽), 250 (■) and 1000 μmol m^{-2} s^{-1}, allowing 30 min at each treatment to permit variables relating to gas exchange to attain steady state. The vapour pressure difference between leaf and ambient air was 9 mbar at 15°C, 10 mbar at 20°C, 12 mbar at 25°C, 19 mbar at 30°C, 35 mbar at 35°C, and 53 mbar at 40°C.
(*b* and *c*). *Bruguiera gymnorrhiza* (*b*) and *Rhizophora apiculata* (*c*) grown under shade cloth transmitting 70 (○) and 30% (▲) natural sunlight in Darwin. Leaf temperature was varied by step changes from 25 to 40°C under saturating irradiances of 1000 and 500 μmol m^{-2} s^{-1} for leaves grown under high and low light conditions, respectively. Ambient vapour pressure was maintained at 12 mbar.

The gas exchange characteristics also reveal the highly conservative water use of these plants, as previously reported in field and laboratory studies of the mangroves *Aegiceras corniculatum, Avicennia marina* and *Rhizophora stylosa* (Ball 1981; Andrews *et al.* 1984; Ball and Farquhar 1984*a*, 1984*b*; Andrews and Muller 1985). Stomatal conductances are considerably lower than those typically found in well-watered C_3 species of mesic environments. Operation at low conductance restricts the influx of CO_2 and the efflux of water vapour, causing photosynthesis to operate at low c_i but with high water-use efficiency. Indeed, the water-use efficiencies (i.e. mmol CO_2 gained per mol water lost) under optimal conditions in Fig. 1 range from 3·2 in *B. gymnorrhiza* to 4·6 in *A. marina*, values which are unusually high for C_3 species under similar climatic conditions. However, the decline in stomatal conductance with increasing leaf temperature, and therefore increasing leaf-to-air vapour pressure difference (vpd), was not sufficient to reduce the evaporation rate and it increased with increasing leaf temperature. The combined effects of temperature-induced increase in evaporation rate and decrease in photosynthetic capacity caused the water cost of carbon gain to increase drastically at leaf temperatures above 32°C (Fig. 2).

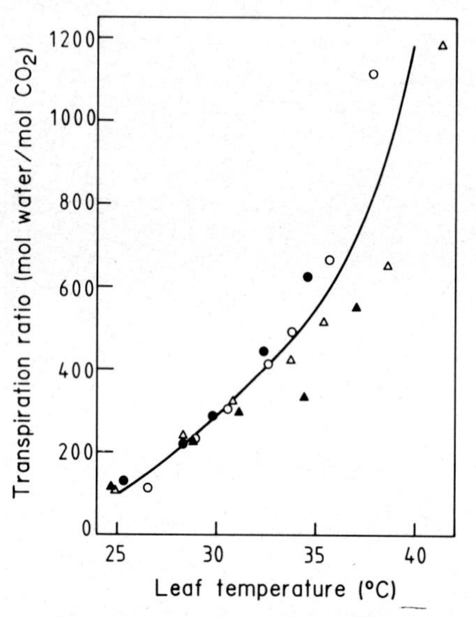

Fig. 2. Transpiration ratio (transpiration / CO_2 assimilation) as a function of leaf temperature in leaves of *Bruguiera gymnorrhiza*. Measurements were made at a range of leaf temperatures from 25 to 40°C, an ambient vapour pressure of 12 mbar, and irradiances of 250 (●), 500 (▲), 1000 (○) and 1500 (△) μmol m^{-2} s^{-1} on leaves of plants grown beneath shade cloth transmitting 70 or 30% sunlight (open and solid symbols, respectively). Line drawn by eye.

There were close relationships between assimilation rate and stomatal conductance in *B. gymnorrhiza* and *R. apiculata* which were largely independent of the source of variation, i.e. the irradiance, leaf temperature and leaf-to-air vapour pressure difference

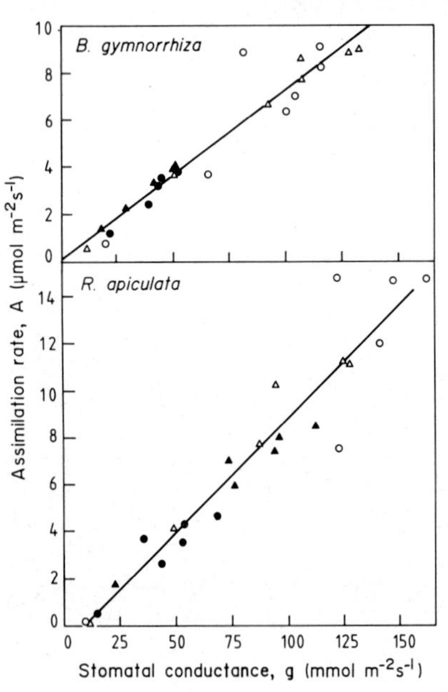

Fig. 3. Assimilation rate as a function of stomatal conductance in *Bruguiera gymnorrhiza* and *Rhizophora apiculata*. Data and symbols as in Fig. 2. Lines drawn by linear regression are $A = 0{\cdot}003 + 0{\cdot}072g$, $r^2 = 0{\cdot}96$, for *B. gymnorrhiza*, and $A = -0{\cdot}88 + 0{\cdot}096g$, $r^2 = 0{\cdot}96$, for *R. apiculata*, where A is assimilation rate (μmol m^{-2} s^{-1}) and g is stomatal conductance (mmol m^{-2} s^{-1}).

at which a measurement was made, or the irradiance at which a plant had been grown (Fig. 3). Similar results were obtained with *C. tagal*, except that the relationships differed between two groups of plants grown under high and low irradiances (Fig. 4).

In other more detailed studies of stomatal behaviour, such a constant relationship between the assimilation rate and stomatal conductance has been associated with the expenditure of a minimum amount of water consistent with maintenance of assimilation rates at or very near the capacity of the mesophyll for photosynthesis (Farquhar 1979). In this sense, the decline of stomatal conductance with increase in leaf temperature and vpd is consistent with a strategy to minimise water loss relative to carbon gain (Cowan and Farquhar 1977; Cowan 1986).

There were interspecific differences in the slopes of assimilation rate as a function of stomatal conductance, the values being 0·072 in *B. gymnorrhiza* (Fig. 3), 0·096 in *R. apiculata* (Fig. 3), 0·101 in *R. stylosa* (Andrews and Muller 1985), and 0·113 (for leaves grown at high irradiances) in *C. tagal* (Fig. 4). These differences imply differences in water-use characteristics and are apparently related to salinity tolerance, as found in previous studies on *A. marina* and *Aegiceras corniculatum* (Ball 1981; Ball and Farquhar 1984*a*, 1984*b*). Comparative growth and field studies show that salinity tolerance in the Rhizophoraceae increases in the order *B. gymnorrhiza, R. apiculata, R. stylosa*, and *C. tagal*, and is reflected in the differential distribution of these species along natural salinity gradients ranging from brackish to hypersaline, respectively, in the Northern Territory of Australia (Ball, unpublished data). Thus, the assimilation rate at a given stomatal conductance is greater the greater the salinity tolerance of the species.

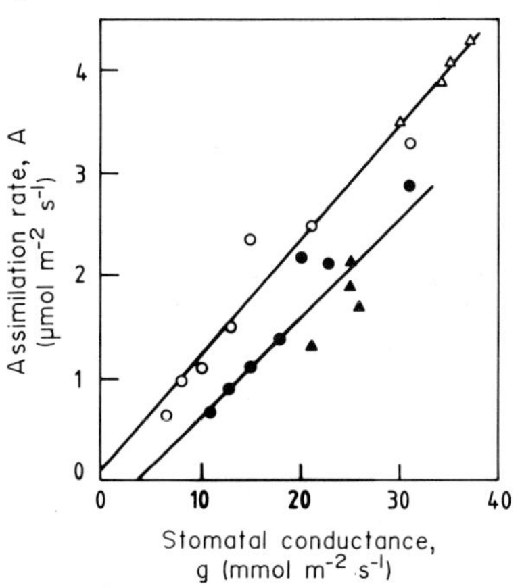

Fig. 4. Assimilation rate as a function of stomatal conductance in *Ceriops tagal*. Open and solid symbols indicate leaves of plants grown under full and 30% sunlight, respectively. Measurements were made at leaf temperatures ranging from 25 to 40°C with a constant ambient vapour pressure of 12 mbar (○, ●), and irradiances of 200, 300, 400, and 500 (▲) or 500, 750, 1000 and 1250 (△) μmol m^{-2} s^{-1} at a leaf temperature of 25°C and ambient vapour pressure of 12 mbar. Lines drawn by linear regression are $A = 0 \cdot 10 + 0 \cdot 113g$, $r^2 = 0 \cdot 99$ for leaves grown in full sunlight (○, △), and $A = -0 \cdot 33 + 0 \cdot 096g$, $r^2 = 0 \cdot 89$ for leaves grown in 30% sunlight (●, ▲), where A is assimilation rate (μmol m^{-2} s^{-1}) and g is stomatal conductance (mmol m^{-2} s^{-1}).

Finally, operation at low evaporation rates affects the maintenance of favourable leaf temperatures. In the present study, the evaporation rates (mmol m^{-2} s^{-1}) at a leaf temperature of 30°C, vpd of 20 mbar, and saturating irradiance ranged from a maximum of 3·3 in *R. apiculata* to a minimum of 0·3 in *C. tagal.* When exposed, canopy leaves of *R. apiculata* were constrained in a horizontal position, the leaf temperatures increased from 4 to 11°C above the ambient air temperature (approximately 30°C) with increase in incident irradiation from 1430 μmol m^{-2} s^{-1} at 8:25 a.m. to a maximum of 2585 μmol m^{-2} s^{-1} at 12:25 p.m., respectively (Fig. 5). Clearly, the low evaporation rates associated with low stomatal conductance were insufficient to prevent these mangrove leaves from becoming hotter than ambient air during periods of intense insolation. For comparison, exposed canopy leaves of tropical rain forest species in north Queensland operate with evaporation rates of 6–8 mmol m^{-2} s^{-1} and maintain leaf temperatures within 2°C above ambient air temperatures of 30–35°C when exposed to irradiances ranging from half to full sunlight (Doley *et al.* 1988).

In contrast, leaves of *R. apiculata* left in their natural, almost vertical orientation avoid the maximum heat load during midday when irradiance is most intense and air temperatures are greatest. During midday, these leaves received only 20% of available sunlight and were approximately 10°C cooler than they would have been if fully exposed to the sun (Fig. 5). Earlier and later in the day, the leaves received about 1000 μmol $m^{-2} s^{-1}$ and leaf temperatures were 30°C, conditions nearly optimal for photosynthesis. Leaf temperatures of a severed, non-transpiring leaf maintained in its natural orientation were only 2°C greater than those of the matching transpiring leaf (Fig. 5). Clearly, avoidance of high irradiance is a major factor contributing to the maintenance of favourable leaf temperatures with minimal evaporative cooling.

In summary, maintenance of leaf temperatures close to air temperatures has two major consequences for the carbon and water economy of the leaf. First, such vertically inclined leaves can sustain relatively constant, albeit low, assimilation rates throughout the photoperiod (Andrews and Muller 1985). This probably results in a greater net gain of carbon than would occur if the leaves were horizontal and therefore subject to inhibition of photosynthesis associated with high temperature during the middle part of the day. Second, the water cost of carbon gain can be maintained at a minimal level (Fig. 2). Indeed, Andrews and Muller (1985) found little variation in the daily total amount of water lost per carbon gained in leaves of *R. stylosa* under natural field conditions.

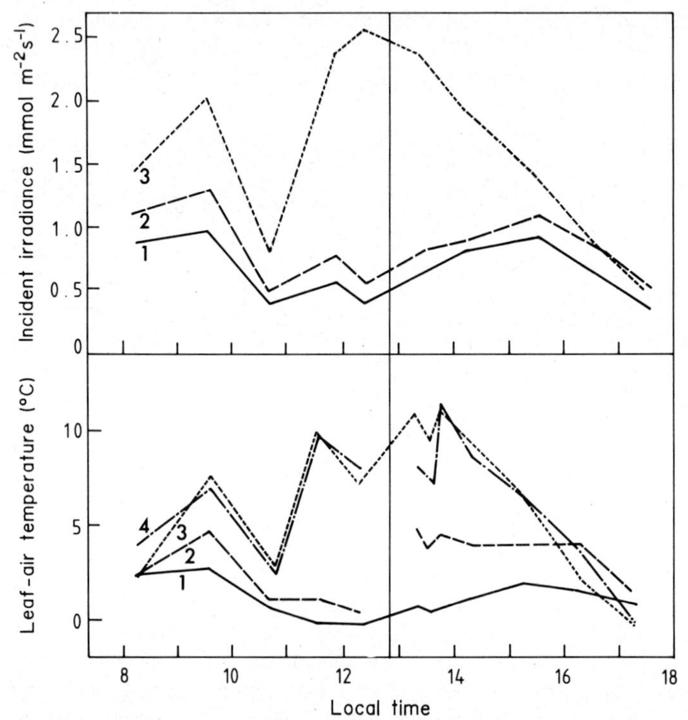

Fig. 5. Effect of leaf angle on the microclimate of transpiring and non-transpiring leaves of *R. apiculata* at Hinchinbrook Island. Measurements were made on four leaves of similar shape and size. Two leaves were left in their natural vertical inclination (curves 1, 2) and two leaves were constrained in a horizontal inclination (curves 3, 4). One leaf in each pair was detached at 12:50 (curves 2, 4). Air temperature varied from 22 to 29°C, and vapour pressure from 18 to 23 mbar during the course of the day.

Physical Properties of Leaves in Relation to Heat Balance

The present study identifies three leaf properties which contribute to maintenance of favourable temperatures in leaves of mangroves. First, increasing leaf angle (i.e. the inclination to the horizontal) decreases the intensity of radiant heat loading. Leaf angle may be regarded as a compromise between the requirements for illumination and reduction of temperature. Leaf angle was found to increase from approximately 0 in full shade to a maximum of about 75° in the most exposed sites, corresponding to a decline in the projected fraction of leaf area on a horizontal surface from about 1 to 0·3 (Table 1). The response of leaf angle to exposure varied among the five species, being greater the greater the salinity tolerance of the species. It follows from what has been said of the relationship between salinity tolerance and water use efficiency that the species which are more conservative in water use are also those that tend most to avoid intense radiation.

Table 1. Variation in leaf display with exposure in five mangrove species at Hinchinbrook Island, North Queensland

The species are listed in order of increasing salinity tolerance. Exposure categories of Sh, MS, and FS mean shade, medium sun and full sun, respectively. The projected fraction of leaf area on a horizontal surface was calculated as cos α, where α is the angle that the leaf makes with a horizontal plane. All values are mean ± s.e.m. of rosettes

Species	Exposure	Rosette area (cm^2)		Individual leaf area		Projected fraction
		Total	Projected	Total	Projected	
B. gymnorrhiza	Sh	1062 ± 32	878 ± 90	67 ± 5	55 ± 2	0·83 ± 0·04
	MS	921 ± 246	725 ± 185	70 ± 14	55 ± 10	0·79 ± 0·01
	FS	635[A]	356[A]	58[A]	32[A]	0·56[A]
R. apiculata	Sh	740 ± 98	694 ± 90	78 ± 9	73 ± 8	0·94 ± 0·01
	MS	776 ± 43	471 ± 47	75 ± 5	45 ± 3	0·60 ± 0·04
	FS	553 ± 58	196 ± 28	69 ± 7	25 ± 4	0·37 ± 0·09
R. lamarkii	Sh	1304[A]	1070[A]	109[A]	89[A]	0·82[A]
	MS	901 ± 128	726 ± 46	86 ± 8	69 ± 1	0·82 ± 0·07
	FS	548 ± 151	204 ± 62	69 ± 19	26 ± 8	0·37 ± 0·04
R. stylosa	Sh	548 ± 83	520 ± 92	64 ± 6	61 ± 7	0·95 ± 0·03
	MS	443 ± 122	247 ± 52	60 ± 20	34 ± 9	0·57 ± 0·04
	FS	419 ± 52	126 ± 14	44 ± 2	13 ± 1	0·30 ± 0·03
C. tagal	Sh	474 ± 48	441 ± 56	49 ± 1	45 ± 2	0·93 ± 0·03
	MS	315 ± 44	206 ± 39	20 ± 3	13 ± 3	0·63 ± 0·05
	FS	102 ± 15	39 ± 12	8 ± 1	3 ± 1	0·36 ± 0·08

[A]Individual samples averaged from one rosette.

A second leaf property influencing temperature is size. Convection of sensible and latent heat between a leaf and its environment depends on resistance to transfer imposed by a boundary layer, the characteristics of which are a function of wind speed and leaf geometry. Decrease in leaf size enhances boundary layer conductance, and results in the temperature of a leaf being closer to that of the ambient air.

The average areas of rosettes and of individual leaves making up the rosettes decreased with increase in exposure from shade to full sun and with increasing salinity tolerance of the species (Table 1). For example leaves of *Ceriops tagal*, the most salt-tolerant of these five species, were the smallest and most sensitive to variation in exposure. In this species, area decreased sixfold and breadth decreased from 50 to 15 mm with increase in exposure from shade to sun conditions. Thus the leaves are smallest under conditions in which, due to intense radiation and / or limitations to evaporative cooling, they sustain the greatest heat load.

Reduction in leaf temperature has another, indirect benefit in terms of photosynthesis. The cooler the leaf, the larger is the stomatal conductance corresponding to any given evaporation rate. It follows that the conductance to CO_2 transfer is also greater, as illustrated in Fig. 6. This effect on total conductance to CO_2 is more marked when reduction in temperature is associated with avoidance of radiation than when it is associated with reduction in leaf size. However, increase in the angle of inclination also reduces the light available for photosynthesis whereas decrease in leaf size does not disadvantage the leaf in this way. Therefore, selection pressure to reduce leaf size in the exposed foliage of mangroves, particularly those species with highly conservative

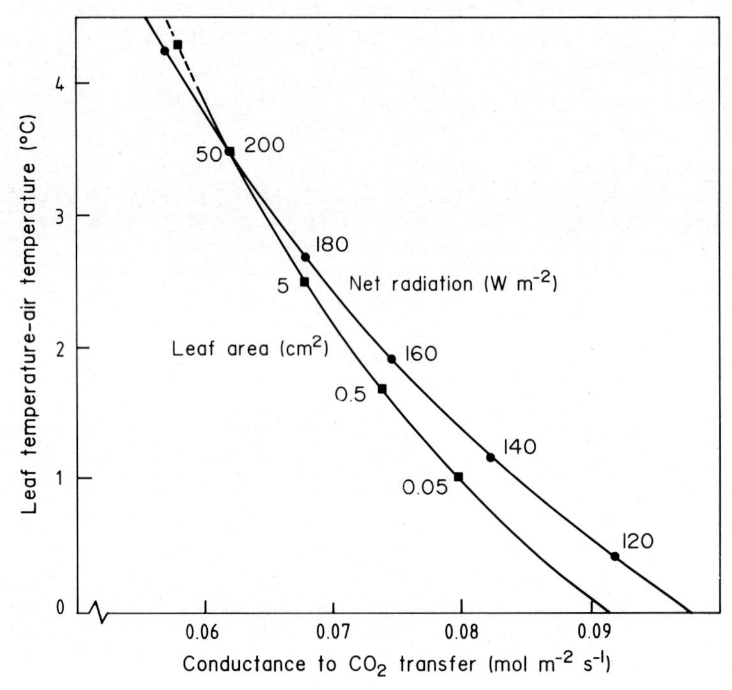

Fig. 6. Trajectories illustrating the influence of leaf size and orientation on leaf temperature and diffusive conductance of the stomata and boundary layer to CO_2 transfer in a mangrove leaf, ambient temperature being 30°C, ambient vapour pressure 25 mbar, and windspeed 0·2 m s^{-1}. The point of intersection of the two curves corresponds to a leaf area of 50 cm^2, net radiation 200 W m^{-2}, and stomatal conductance to vapour transfer 0·15 mol m^{-2} s^{-1}. The trajectories are computed according to the Appendix, with the assumption that the evaporation rate is uniform for leaves of different breadths and orientations. In interpreting the trajectory relating to net radiation, it may be assumed that 400 W m^{-2} is roughly equivalent to full sunlight.

water-use characteristics, is probably formidable. Indeed, mangrove species characteristically encountered in the hypersaline environments along arid coasts of northern Australia (i.e. *Avicennia marina, Ceriops tagal, Excoecaria agallocha, Lumnitzera racemosa* and *Osbornia octodonta*) have much smaller leaves than those dominating humid, low-salinity wetlands (e.g. *Bruguiera gymnorrhiza* and *Rhizophora apiculata*).

A third leaf property, which may be of importance in reducing leaf temperature due to transiently unfavourable microclimatic conditions, is that of heat capacity per unit area, which increases with increase in the dry weight and water content per unit area. There was a tendency for the average succulence of leaves to increase with increasing exposure in all species, and with increasing salinity tolerance between species (Table 2). The large heat capacities due to increased succulence would have the effect of damping fluctuations in leaf temperatures due to variations in irradiance, windspeed, air temperature and vapour pressure, and therefore reduce the extent of excursions into the temperature range unfavourable for photosynthesis.

The data in Table 2 were used to calculate the heat capacity per unit leaf area. Taking the specific heat of dry matter as one-third that of water, the heat capacity of the thickest leaves of *Ceriops tagal* is $2 \cdot 2 \times 10^3$ J m^{-2} °C^{-1}, and that of the thinnest leaves of *Bruguiera gymnorrhiza* is $1 \cdot 1 \times 10^3$ J m^{-2} °C^{-1}. These heat capacities can be divided by the conductance to energy transfer from a leaf to determine the time constant characterising the response of leaf temperature to a step change in environmental factors

Table 2. Variation in the average specific weight and succulence of leaves making up rosettes in five mangrove species grown under different natural light environments at Hinchinbrook Island, North Queensland

Exposure categories of Sh, MS and FS mean shade, medium sun and full sun, respectively. The species are listed in order of increasing salinity tolerance. Values are mean ± s.e.m. of rosettes. fw, fresh weight. dw, dry weight.

Species	Exposure	Specific leaf weight		Succulence
		(g fw m^{-2})	(g dw m^{-2})	(g water m^{-2})
B. gymnorrhiza	Sh	329·4 ± 7·8	92·9 ± 0·7	236·5 ± 8·0
	MS	455·4 ± 48·7	123·4 ± 13·3	332·0 ± 35·4
	FS	395·6[A]	133·1[A]	262·5[A]
R. apiculata	Sh	360·2 ± 24·1	97·8 ± 7·1	262·4 ± 21·9
	MS	410·7 ± 22·5	124·8 ± 7·9	285·9 ± 16·8
	FS	497·2 ± 51·9	148·8 ± 13·1	348·4 ± 41·1
R. lamarkii	Sh	274·0[A]	85·1[A]	188·9[A]
	MS	411·9 ± 9·2	131·8 ± 0·7	280·1 ± 6·0
	FS	496·8 ± 3·3	151·2 ± 3·4	345·6 ± 0·1
R. stylosa	Sh	365·4[A]	106·9[A]	258·5[A]
	MS	468·8 ± 28·6	147·5 ± 17·5	321·4 ± 11·2
	FS	557·3 ± 45·2	169·3 ± 11·7	387·9 ± 35·5
C. tagal	Sh	431·9 ± 6·9	121·8 ± 6·7	310·1 ± 9·2
	MS	504·3 ± 47·3	152·3 ± 13·9	351·9 ± 33·9
	FS	652·4 ± 59·7	189·2 ± 4·8	463·2 ± 55·4

[A]Individual samples averaged for one rosette.

influencing leaf temperature. The conductance to energy transfer must take into account sensible and latent heat transfer, and thermal radiation, as shown in the Appendix. For a leaf having the properties and experiencing the environmental circumstances corresponding to the intersection in Fig. 6, the conductance is 38 W m^{-2} °C^{-1}. Therefore, the time constants, τ say, for *C. tagal* and *B. gymnorrhiza* are 1·0 and 0·5 min, respectively.

The effectiveness with which leaf heat capacity dampens potential fluctuations in temperature is better expressed in Fig. 7. The period, $t_{\frac{1}{2}}$ say, of the fluctuation in environment such that the amplitude of fluctuation in leaf temperature is exactly halved is given by the relation $2\pi\tau/\sqrt{3}$. Therefore $t_{\frac{1}{2}}$ is 3·6 and 1·8 min for *C. tagal* and *B. gymnorrhiza*, respectively. Any environmental fluctuation having a period much less (as might sometimes be the case, for example, with periodic lulls in air movement and fluctuations in irradiance due to intermittent cloud) would have little effect on leaf temperature.

The discussion so far has centred on the predicament of mangrove leaves fully exposed to the sky; however, those in the shaded understory can also experience high irradiances, and leaf temperatures can become as much as 15°C greater than ambient air temperature (unpublished results, see also Fig. 8). The leaves are vulnerable to direct sunlight irradiance because of their horizontal disposition. However, most of the diffuse light on which understory foliage depends is received from above, and the amount received would be reduced if leaf angle deviated much from the horizontal. Also, reduction in leaf size may not be as efficacious in avoiding excessive heating in the

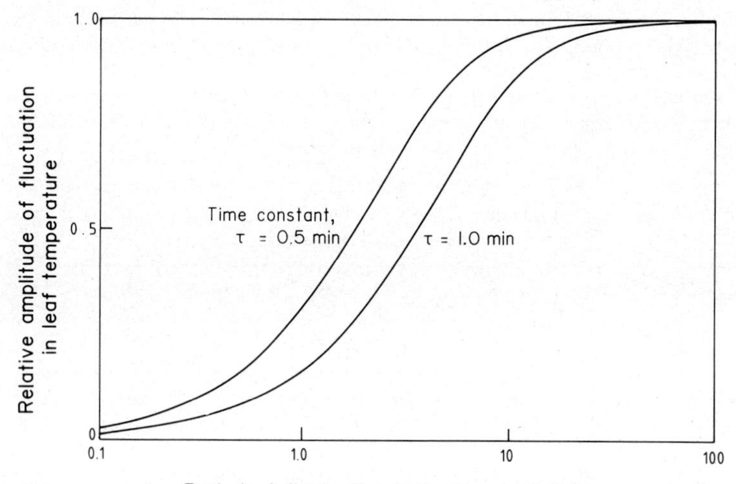

Fig. 7. Calculations of the influence of environmental fluctuations of a given amplitude and varying period on the amplitude of corresponding forced fluctuation in leaf temperature (see Appendix). Leaf properties and average environmental properties are those corresponding to the point of intersection in Fig. 6. The fluctuation in environment may be of any of the environmental factors (i.e. irradiance, windspeed, temperature or vapour pressure) that influence leaf temperature, but is assumed to be small. The amplitude of the forced fluctuation in temperature is arbitrarily taken as unity when the period is infinitely long. The two curves relate to leaves with heat capacities per unit area of $2{\cdot}2 \times 10^3$ and $1{\cdot}1 \times 10^3$ J m^{-2} °C^{-1}, i.e. the heat capacities of thick leaves of *Ceriops tagal* and thin leaves of *Bruguiera gymnorrhiza*, respectively (see Table 4).

understory as in the exposed canopy. It is unlikely that the whole of an understory leaf will become exposed to full direct irradiance and the efficiency of heat transfer by natural convection will be more nearly related to the extent of the irradiated surface than to the size of the leaf as a whole. Also, the transfer coefficient is not as sensitive to leaf size when transfer is by natural convection, as it often is at the very low windspeeds in the forest understory, rather than by the forced convection that usually predominates in exposed environments (Fig. 8).

The increase in water content may be primarily associated with dilution of accumulated salt. However, the large heat capacity per unit area may also be important to understanding the temperature regimes of understory foliage. Sunflecks in the understory are not only small, but they move quite quickly. If the height of the foliage which delineates the boundary of a sunfleck is, say, 5 m, then the speed of movement of the fleck across a horizontal surface is 2 cm min^{-1} in accordance with the angular velocity of the sun. Clearly the period of time over which any part of an understory leaf is fully exposed to solar radiation may often be much shorter than the time taken for thermal equilibration in mangrove leaves (Fig. 8).

Ecological Implications

Some of the leaf characteristics which contribute to maintenance of favourable leaf temperatures for photosynthesis may entail costs to the plant. Leaf inclinations that reduce radiant heat loading also reduce the amount of light available to support photosynthesis. The increase in dry weight per unit leaf area associated with increase in the heat capacity of leaves implies a smaller total leaf area per unit of carbon and this, too, may detract from the integrated photosynthetic capacity of the plant. Thus, maintenance of favourable leaf temperatures with minimal evaporative cooling by these means is at the expense of the assimilation of carbon, with the expense increasing as water use becomes more conservative. We are, however, unable to identify costs associated with small leaf size.

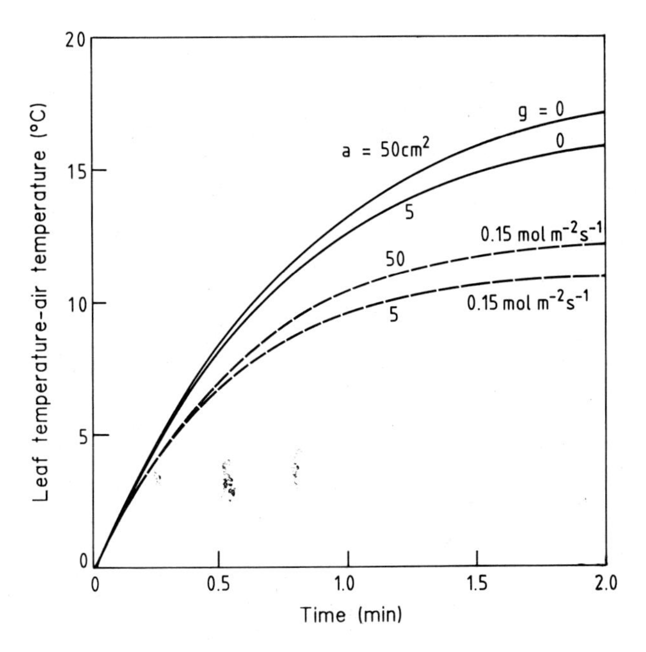

Fig. 8. Calculated increases with time of temperature in horizontal leaves receiving net radiation 400 W m^{-2} in still air, the leaves having previously been shaded and at the same temperature as the ambient air (see Appendix). Curves relate to two leaf areas, *a*, and to leaves with zero and finite stomatal conductance, *g*, to vapour transfer. Ambient temperature is 30°C and vapour pressure 25 mbar. Leaf heat capacity per unit area is $1 \cdot 1 \times 10^3$ J m^{-2} °C^{-1}.

Differences in water-use characteristics influence the organisation of mangrove forests along natural salinity gradients. Species tolerant of lower ranges of salinity, such as *Bruguiera gymnorrhiza*, operate with lower water-use efficiencies and hence can maintain larger leaves with greater projected leaf areas than those with greater salinity tolerance. Under low salinity conditions, stands of these species have dense canopies which allow little transmission of light to the forest floor. In contrast, species which are highly salt tolerant, such as *Ceriops tagal*, operate with high water-use efficiency and hence maintain small leaves with a low proportion of projected area. These slowly growing species characteristically form stands with open canopies which, under low salinity conditions, could not exclude the more rapidly growing, densely canopied species characteristic of low salinity environments. Thus, although a conservative water-use strategy may contribute to salt tolerance, it is at the expense of competitive ability under conditions of lower salinity. This may contribute to the characteristic segregation of mangrove species along salinity gradients.

Acknowledgments

The authors thank Drs John Andrews and Barry Clough for field assistance, Dr S. Chin Wong and Mr Peter Groeneveld for technical assistance with gas exchange measurements, Ms Sandra Pidsley for mangrove cultivation, and the Australian Institute of Marine Sciences for logistical support. This research was supported in part by a grant from the Rural Credits Development Fund of the Reserve Bank of Australia to M. C. Ball.

References

Andrews, T. J., Clough, B. F., and Muller, G. J. (1984). Photosynthetic gas exchange and carbon isotope ratios of some mangroves in North Queensland. In 'Physiology and Management of Mangroves'. (Ed. H. J. Teas.) Tasks for Vegetation Science, Vol. 9, pp. 15–23. (Junk: The Hague.)

Andrews, T. J., and Muller, G. J. (1985). Photosynthetic gas exchange of the mangrove, *Rhizophora stylosa* Griff., in its natural environment. *Oecologia* **65**, 449–55.

Ball, M. C. (1981). Physiology of photosynthesis in two mangrove species: responses to salinity and other environmental factors. Ph.D. Thesis, Australian National University.

Ball, M. C. (1986). Photosynthesis in mangroves. *Wetlands* (*Australia*) **6**, 12–22.

Ball, M. C., and Critchley, C. (1982). Photosynthetic responses to irradiance by the grey mangrove, *Avicennia marina*, grown under different light regimes. *Plant Physiol.* **70**, 1101–6.

Ball, M. C., and Farquhar, G. D. (1984*a*). Photosynthetic and stomatal responses of two mangrove species, *Aegiceras corniculatum* and *Avicennia marina*, to long term salinity and humidity conditions. *Plant Physiol.* **74**, 1–6.

Ball, M. C., and Farquhar, G. D. (1984*b*). Photosynthetic and stomatal responses of the grey mangrove, *Avicennia marina*, to transient salinity conditions. *Plant Physiol.* **74**, 7–11.

Boto, K. G., and Wellington, J. T. (1983). Phosphorus and nitrogen nutritional status of a northern Australian mangrove forest. *Mar. Ecol. Prog. Ser.* **11**, 63–9.

Caemmerer, S. von, and Farquhar, G. D. (1981). Some relationships between the biochemistry of photosynthesis and the gas exchange of leaves. *Planta* **153**, 376–87.

Cowan, I. R. (1972). Mass and heat transfer in laminar boundary layers with particular reference to assimilation and transpiration in leaves. *Agric. Meteorol.* **10**, 311–29.

Cowan, I. R. (1986). Economics of carbon fixation in higher plants. In 'On The Economy of Plant Form and Function'. (Ed. T. J. Givnish.) pp. 133–70. (Cambridge University Press: Cambridge.)

Cowan, I. R., and Farquhar, G. D. (1977). Stomatal function in relation to leaf metabolism and environment. In 'Integration of Activity in the Higher Plant'. (Ed. D. H. Jennings.) pp. 471–505. (Cambridge University Press: Cambridge.)

Doley, D., Unwin, G. L., and Yates, D. J. (1988). Spatial and temporal distribution of photosynthesis and transpiration by single leaves in a rainforest tree, *Argyrodendron peralatum*. *Aust. J. Plant Physiol.* **15**, 317–26.

Farquhar, G. D. (1979). Carbon assimilation in relation to transpiration and fluxes of ammonia. In 'Photosynthesis and Plant Development'. (Eds R. Marcelle, H. Clijsters and M. van Poucke.) pp. 321–8. (Junk: The Hague.)

McAdams, W. H. (1954). 'Heat Transmission.' (McGraw Hill: New York.)

Moore, R. T., Miller, P. C., Albright, D., and Tieszen, L. L. (1972). Comparative gas exchange characteristics of three mangrove species in winter. *Photosynthetica* **6**, 387–93.

Moore, R. T., Miller, P. C., Ehleringer, J., and Lawrence, W. (1973). Seasonal trends in gas exchange characteristics of three mangrove species. *Photosynthetica* **7**, 387–94.

Saenger, P., Specht, M. M., Specht, R. L., and Chapman, V. J. (1977). Mangal and coastal salt-marsh communities in Australasia. In 'Wet Coastal Ecosystems' (Ed. V. J. Chapman.) pp. 293–345. (Elsevier Scientific: Amsterdam.)

Wong, S. C., Cowan, I. R., and Farquhar, G. D. (1978). Leaf conductance in relation to assimilation in *Eucalyptus pauciflora* Sieb. ex Spreng. *Plant Physiol.* **62**, 670–4.

Appendix

The calculations underlying Figs 6, 7, and 8 are based on the following equation. The net receipt of radiant energy per unit area of leaf per unit time is

$$\Phi = \Phi_0 - 2\sigma(T_l^4 - T_a^4) = C_a \frac{dT_l}{dt} + C_p \frac{(T_l - T_a)}{r_b^*} + \frac{L[e'(T_l) - e_a]/P}{r_s + r_b}, \quad (1)$$

in which T_l and T_a are the temperatures of the leaf and ambient air, $e'(T_l)$ saturation vapour pressure at leaf temperature, e_a ambient vapour pressure, P air pressure, r_b^* boundary layer resistance to heat transfer from unit area of leaf, r_b boundary layer resistance to vapour transfer, r_s stomatal resistance to vapour transfer, σ the Stefan–Boltzmann constant ($5 \cdot 7 \times 10^{-8}$ W m^{-2} K^{-4}), C_a the heat capacity of the leaf per unit area, C_p the molar heat capacity of air ($\approx 29 \cdot 2$ J mol K^{-1}), and L the molar heat of vaporisation of water (44 kJ mol^{-1}). The term Φ_o is the net radiation that would be received if the leaf were at the temperature of the surrounding air. The last expression on the right-hand side is the rate of transpiration per unit area of leaf. For forced convection, r_b^*, has the form $c(b/u)^{\frac{1}{2}}$, where b is a characteristic linear dimension of the leaf, and u is wind speed. With the units of r_b^*, b, and u being mol m^{-2} s^{-1}, m, and m s^{-1} respectively, a theoretical value of c for an elliptically shaped leaf parallel to the wind is 3·54, with b being the downwind axis of the ellipse (Cowan 1972). Assuming there is no preferred orientation in the plane parallel to the wind, we might reasonably take, as an average estimate of b, the geometric mean of the major and minor axes. This allows us to rewrite the expression for boundary layer resistance as $r_b^* = 3 \cdot 8\, s^{\frac{1}{4}} u^{-\frac{1}{2}}$, where s is leaf surface area in m^2. Of course most leaves do not lie parallel to the wind, but we shall nevertheless adopt this expression as the best simple representation of a very complicated set of circumstances. For leaves having similar stomatal resistance at both surfaces $r_b = 0 \cdot 89 r_b^*$. The species we consider here are hypostomatous and therefore r_b is taken as $1 \cdot 78 r_b^*$.

(1) *Steady-state Calculations Underlying Fig. 6.*

dT_l/dt is set zero. We have first determined leaf temperature, T_l, and rate of transpiration, E, for assumed magnitudes of Φ_o, T_a, e_a, u, s, and r_s. We have then determined what magnitudes of r_s would maintain the same rate of transpiration, and what leaf temperature would be, at other magnitudes, firstly, of Φ_o and, secondly, of b. We have taken the resistance to CO_2 transfer across boundary layer and epidermis to be $1 \cdot 37 r_b^* + 1 \cdot 6 r_s$ (Cowan 1972).

(2) *Calculations Underlying Fig. 7.*

We use the heat balance equation to find the deviation δT_l of leaf temperature from a steady state for a small deviation $\delta\Phi_o$ in the net radiation Φ_o. We find

$$\delta\Phi_o = C_a \frac{d(\delta T_l)}{dt} + \delta T_l \left[\frac{C_p}{r_b^*} + \frac{L de'/dT_l}{P(r_s + r_b)} + 8\sigma T_l^3 \right]. \quad (2)$$

The term in square brackets is an energy transfer coefficient, K say. Consideration of the form of the equation tells us that the time constant is $C_a/K = \tau$, say. If we allow $\delta\Phi_0$ to vary periodically with amplitude a and angular frequency ω, it is easily shown that the amplitude of the fluctuation in T_l is $a/(K^2 + C_a^2\omega^2)^{\frac{1}{2}}$. Relative to the amplitude at infinitely small frequency, the amplitude is $K/(K^2 + C_a^2\omega^2)^{\frac{1}{2}}$. The angular phase lag is $\tan^{-1}(C_a\omega/K)$. These last two results would remain unchanged had we chosen to consider a periodic fluctuation in another environmental variable, T_a, e_a or u, rather than Φ_o, as the forcing function. The relative amplitude is 0·5 when $\omega = \sqrt{3}\ K/C_a$, that is to say when the period of the fluctuation is $2\pi\tau/\sqrt{3}$.

(*3*) *Calculations Underlying Fig. 8.*

Under some conditions, heat and vapour exchange take place by natural convection in response to the difference in air density, ρ, at the surface of the leaf and in the ambient atmosphere. The difference is related to difference in temperature and vapour pressure by

$$\frac{\Delta\rho}{\rho} = -\frac{\Delta T}{T} - 0{\cdot}38\frac{\Delta e}{P},$$

the numerical factor being the difference between the molecular weights of dry air and water vapour relative to the molecular weight of dry air. The boundary layer resistance to heat transfer may be written

$$r_b^* = c(b\rho|\Delta\rho|)^{\frac{1}{4}}. \qquad (3)$$

For a square, vertical plate of side b, $c = 2{\cdot}55$ (McAdams 1954). For a horizontal plate, conditions differ at the upper and lower surfaces. Taking r_b^* to be the resistance to transfer at *one* surface only, $c = 5{\cdot}4$ for the upper surface of a heated plate, or the lower surface of a cooled plate, and $c = 12{\cdot}2$ for the upper surface of a cooled plate or the lower surface of a heated plate (McAdams 1954). In applying these factors to leaves, taking b as the square root of leaf area, we have neglected what is probably a rather small influence of leaf shape.

When the difference in vapour pressure Δe is insufficiently small to be neglected, the expression for r_b^* is a little complex, for we must use the relation

$$\Delta e = \frac{r_b}{r_s + r_b}[e'(T_l) - e_a] \qquad (4)$$

with, as before, r_b being taken as $0{\cdot}89r_b^*$. The combination of equations (2), (3) and (4) yields an implicit expression for r_b^*. The curves in Fig. 8 are solutions of equation (1) with r_b^* and r_b found in this way.

Cases in which forced and natural convection occur together are exceedingly complicated. McAdams (1954) suggests that an empirical approach, usually engendering errors of no more than 25%, is to calculate magnitudes of r_b^* for forced convection and natural convection and to assume that the relevant magnitude is the least of the two. This has been adopted in computations in Fig. 8. The small, broken segment of one of the curves in Fig. 6 relates to vertical leaves which are so large that natural convection predominates on one or both sides.

Photosynthetic Characterisation of Rice Varieties in Relation to Growth Irradiance

Z. P. Tu, X. Z. Lin, Q. M. Huang, W. J. Cai, H. Y. Feng and L. Y. Ye

Rice Research Institute, Guangdong Academy of Agricultural Sciences,
Guangzhou, Guangdong, People's Republic of China.

Abstract

The growth irradiances in the primary habitats of newly popularised Chinese rice 'Gui-Chao 2' and modern U.S. rice varieties 'Bellemont' and 'Lemont' are quite different, with only 2–8 h sunshine per day in Guangzhou, China, and 8–12 h sunshine per day in Texas, U.S.A. The U.S. varieties Bellemont and Lemont adapted to higher photon irradiance, showing increase of photosynthesis and increase of leaf thickness under high photon irradiance (1000 μmol photons m^{-2} s^{-1}). Bright light reduced leaf chlorophyll content less in Bellemont than in Gui-Chao 2. The Chinese variety Gui-Chao 2 seemed adapted to low photon irradiance, showed delayed leaf senescence and retained higher photosynthetic rate after 20 days shading treatment. Thus, the yield reduction under shading of Gui-Chao 2 was less than that of Bellemont. After treatment for 9 days at 1000 μmol photons m^{-2} s^{-1}, the quantum yield and light- and CO_2-saturated photosynthesis of Gui-Chao 2 was lower than that of Lemont, suggesting possible photoinhibition in Gui-Chao 2 under high irradiance. It was proposed that these varieties represent two light-intensity ecotypes of rice. The possible application of this research to rice breeding and agricultural cultivation is discussed.

Introduction

Rice is primarily a tropical and subtropical crop but is widely distributed in 111 rice-growing countries, from 50°N. in Hie-ho, north-eastern China, to 35°S. in New South Wales, Australia (De Datta 1981). With such a range of ecological conditions, it is not surprising that many different rice varieties have been selected that are adapted to different environmental conditions.

The Chinese rice scientist Ding (Ting (1959)) classified cultivated rice into several sets of ecotypes. The first pair, hsien (indica) and keng (japonica), are geography–weather ecotypes, adapted to different latitudes and elevations. The second pair, early season (first) and late season (second), are season–weather ecotypes, but with different photoperiod responses to daylength. The third pair, wetland and dryland, are soil ecotypes adapted to different soil moisture regimes. Ding did not include light intensity among the environmental factors important in ecotypic differentiation of rice. Only recently have researchers in Japan (Tsunoda 1984) and in China (Li *et al.* 1984) recognised the importance of light intensity during growth in the ecophysiology of rice.

Huke (1976) noted that in primary habitats of rice, the growing season is characterised by high rainfall and high cloud cover. For example, in the Korat of Thailand, and in valleys of the Shan in Burma, light is incident on the crop during July for only about 110 h, or approximately 25% of the potential. In contrast to this, rice cultivation in the new world has spread to places of high incident radiation, especially in the U.S. and Australia. Introduction to China of modern rice varieties from the U.S., with a view to improving yield of local varieties, has encouraged us to compare the light response

0310-7841/88/010277$03.00

of photosynthesis of endemic and introduced varieties. Preliminary studies (Tu *et al.* 1985) showed that U.S. varieties had much higher light saturation for photosynthesis than the popular variety 'Guan-Lu-Ai 4' grown in Guangdong. In this paper, further studies to characterise the difference in light response of photosynthesis between U.S. and Chinese rice varieties, especially in relation to the irradiance received during growth, are reported.

Materials and Methods

Rice Varieties

Newly popularised, high-yielding Chinese variety 'Gui-Chao 2', released by the Rice Research Institute, Guangdong Academy of Agricultural Sciences, Guangzhou, Guangdong, China, was chosen as representative of low-light-selected local varieties. Two modern varieties of good quality from the U.S., 'Bellemont' and 'Lemont', developed at the Texas A & M Agricultural Research and Extension Center at Beaumont, by the Agricultural Research Service, USDA, in cooperation with others, were chosen as representatives of the high-light-selected rice. Both U.S. varieties have the same parents and both are photoperiod-insensitive. Their growth periods are very similar in Guangzhou: as an early-season rice the total growth period is about 130–140 days, and the late-season growth period is about 114–124 days.

Experimental Treatments

Two types of experiment were done. The rice varieties were grown in controlled environment chambers with different light intensities and nitrogen levels. They were also grown in the field under shade cloth.

In the controlled-environment studies, the plants were grown in Con-Viron chambers with two light intensities corresponding to photon irradiance of 200 and 1000 μmol photons m^{-2} s^{-1}. The former nearly equals the light intensity of a cloudy day in Guangzhou; the latter nearly equals the low range of light intensity experienced on a clear day in Guangzhou. In early experiments, a third treatment with 500 μmol photons m^{-2} s^{-1} was included. Chambers were programmed to receive 1 h of low light (50% of the maximum value) before peak illumination and 1 h of the same low light intensity before the dark period. Total daylength was 14 h, and the temperature during the light period was 28°C, and 25°C in the dark.

Two nitrogen levels were included: 2 g N per pot and 1 g N per pot. The high level nearly equals the high-yielding fertiliser level of 7·5 t ha^{-1} used in Guangzhou. The fertiliser had a ratio of $N:P_2O_5:K_2O$ of 1:1:1, and 70% of the fertiliser was mixed into the pot before transplanting. The remainder was divided into two parts and applied 7 days before inserting the plant into the chamber (before panicle differentiation) and again at the booting stage. Plants were grown in 20 cm diameter pots of 30 cm depth, using field paddy soil, with four plants per pot. Each treatment had four pots, and three pots were selected for experimental measurements.

In the field shading experiment, similar treatment times were selected as used in the growth chamber. The nitrogen applied to the already fertile experimental plots was only 150 kg ha^{-1}. Cheese-cloth was used to reduce daylight to 30% over some plots. Because the difference in light intensity was not as large as that used in the growth-chamber experiment, the duration of the experiment was increased to 20 days.

Observations on Leaf Anatomy

Rice leaf development occurs in two stages: extension followed by increasing thickness. To study the effect of light intensity on leaf thickness, the plants were transferred to the growth chamber towards the end of the extension phase. The subsequent light period lasted from 10 to 12 days. In each treatment, measurements were made on leaves from two plants in each pot. A second set of measurements was taken using flag leaves from the remaining two plants.

Leaves were fixed in FAA (formaldehyde/acetic acid/alcohol) solution and stained with hematoxylin. Leaf sections were cut with a BASE Sledge Microtome. Leaf thickness was measured under the microscope using a micrometer, and the standard section of the leaf used for comparative purposes is shown in Fig. 2.

Measurement of Net Photosynthetic Rate (P_n)

Net photosynthesis in air was measured with an ADC 225 infra-red gas analyser. Leaves to be measured were cut from the plant at about 1 cm below the node of the appropriate sheath, and immediately placed in a tube full of water. An air-sealed photosynthetic chamber was used during measurements, as described previously (Tu *et al.* 1978). Measurements were made in air with a photon irradiance of 700 μmol photons $m^{-2} s^{-1}$ at an air temperature of 28–30°C and relative humidity >95%.

Measurements of Quantum Yield and Light-saturated Photosynthesis

Light- and CO_2-saturated photosynthesis was measured using a Hansatech leaf-disc oxygen electrode, as described by Walker and Osmond (1986). Leaf segments were placed in the photosynthetic chamber, which was thermostated at 28°C and filled with CO_2 to a concentration of approximately 10%. The leaf was illuminated with a quartz iodide light source (Hansatech) giving a maximum photon irradiance of about 2000 μmol photons $m^{-2} s^{-1}$. Neutral density filters were used to reduce the light, and quantum yield was measured as the initial slope of the light response curve. Strictly, measurements are apparent quantum yields as no correction for leaf absorbance was made. Leaf chlorophyll was measured by extracting 5 cm^2 of leaf in 10 ml of *N,N*-dimethyl-formamide for 12 h (Moran and Rorath 1980). The absorption spectrum, including the peaks at 663 and 645 nm, was measured using a u.v.-visible spectrophotometer.

Comparisons were made for each of three separate experiments in years 1984–86. Most of the data in this paper were obtained in the 1985 experiment.

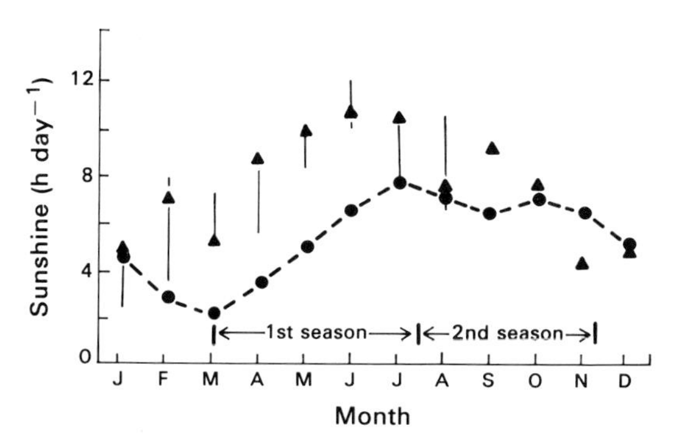

Fig. 1. Comparison of growth irradiance in primary habitats of Gui-Chao 2 and Bellemont or Lemont. Historical data for Guangzhou, China (●), compared with data for 1977–79 in Beaumont, Texas (▲ 1978).

Results

Comparison of Light Conditions During Rice-growing Seasons

Fig. 1 indicates very significant differences between the sunshine hours received during rice-growing seasons in Guangzhou and Beaumont. In Guangzhou, the primary habitat of variety Gui-Chao 2, there are only 2–8 h of sunshine during the early season. In Beaumont, the primary habitat of varieties Bellemont and Lemont, there are always 8 h and often 12 h of sunlight (April to August).

Changes in Leaf Thickness and Chlorophyll Content in Relation to Light

In our preliminary experiments in 1984, leaves of Bellemont were thicker than those of Gui-Chao 2, and the thickness of leaves increased in response to increasing light intensity in Bellemont but not in Gui-Chao 2. Fig. 2 shows photographs of leaf transsections for both varieties, and the position of the bundle measured is shown in the schematic drawing. Statistical analyses of these measurements were performed during the experiment repeated in 1985. Table 1 shows that in Bellemont there was a statistically significant 13–25% increase in leaf thickness when plants were grown in 200 and

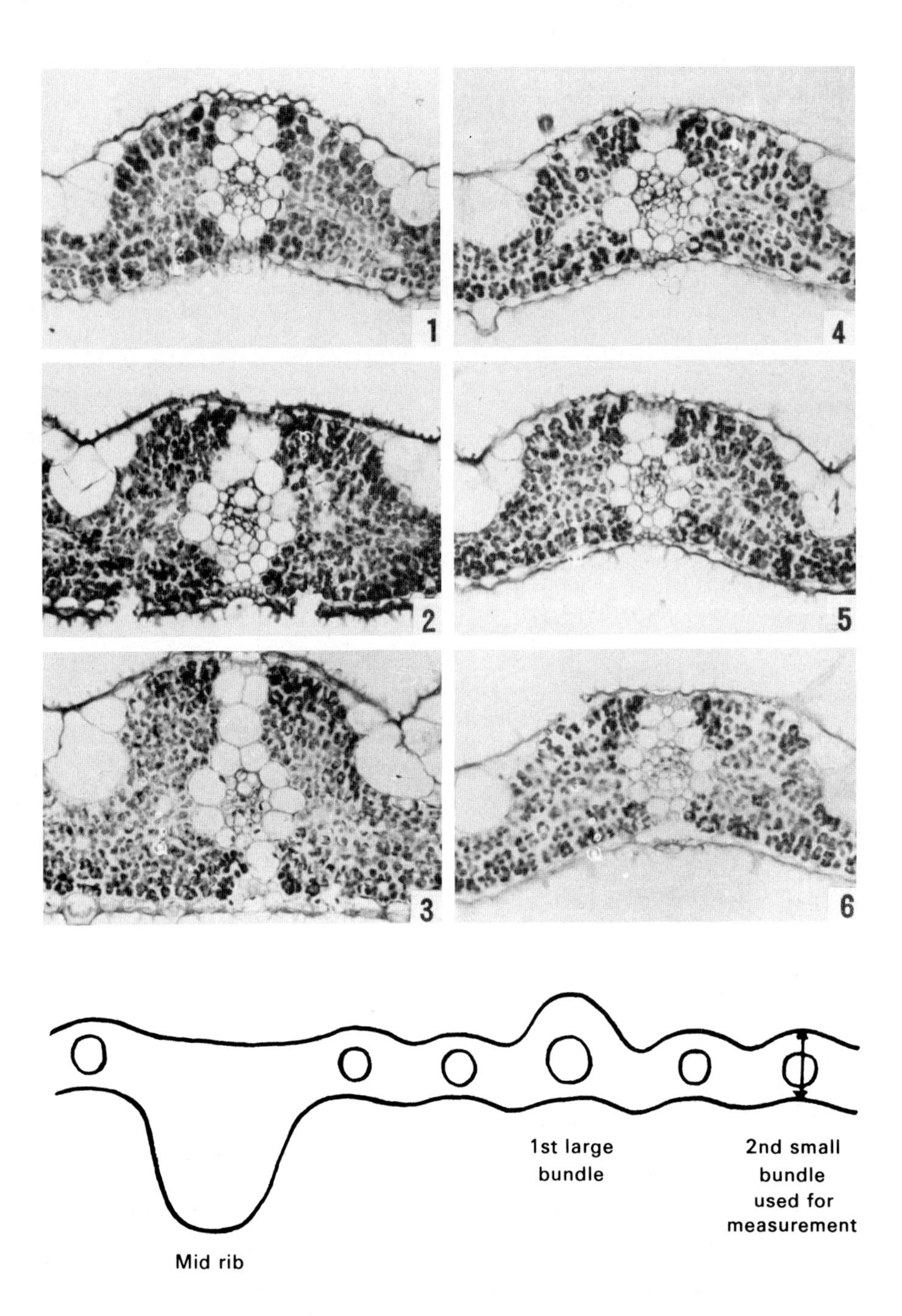

Fig. 2. Photomicrographic comparisons of leaf thickness measured at the mid line of the second small vascular bundle (see schematic drawing) at different irradiances in Bellemont (200, 500 and 1000 μmol photons m^{-2} s^{-1}: *1, 2, 3,* respectively) and in Gui-Chao 2 (200, 500 and 1000 μmol photons m^{-2} s^{-1}: *4, 5, 6,* respectively).

1000 μmol photons $m^{-2} s^{-1}$. In Gui-Chao 2, on the other hand, there was no statistical difference between leaf thickness in either treatment.

The histograms in Fig. 3 show that growth in high light caused a reduction in the chlorophyll content of leaves of both Bellemont and Gui-Chao 2, but the reduction was less in Bellemont than in Gui-Chao 2. High nitrogen levels slightly decreased the effect of high light on chlorophyll level in both varieties.

Table 1. Thickness of flag leaves in different rice varieties in response to photon irradiance and nitrogen supply

**Significant at 1% level; n.s., not significant

Variety	N level (g per pot)	Leaf thickness (μm) at irradiance (μmol photons $m^{-2} s^{-1}$) of: 200	1000	Difference in leaf thickness (μm)
Bellemont	2	116·4	145·6	+25·1%**
	1	113·8	128·7	+13·1%**
Gui-Chao 2	2	95·8	96·8	+1·0% n.s.
	1	85·8	89·7	+4·5% n.s.

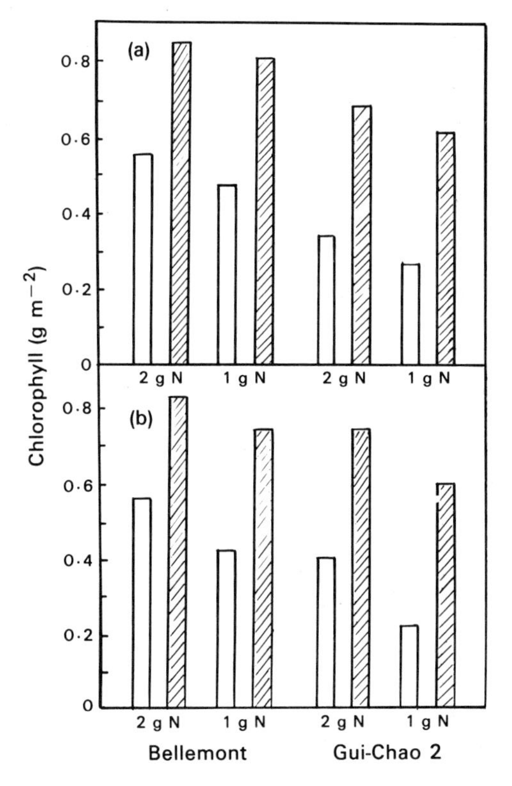

Fig. 3. Comparisons of chlorophyll content of leaves of different rice varieties that developed under 1000 μmol photons $m^{-2} s^{-1}$ (open bars) and 200 μmol photons $m^{-2} s^{-1}$ (hatched bars) and different nitrogen levels. Measurements were made at the panicle differentiation stage (*a*) and the booting stage (*b*).

Comparisons of Net Photosynthesis in Air

Preliminary studies in 1984 indicated that under natural illumination the P_n of Bellemont was higher than that of Gui-Chao 2. However, when the varieties were grown under shading treatment in the field, the P_n of Bellemont was lower or similar to that of Gui-Chao 2; shading of Gui-Chao 2 gave darker leaves than those of Bellemont. These experiments were repeated in 1985, and as shown in Fig. 4 Bellemont responded to increased illumination with an increase in photosynthetic rate. Gui-Chao 2, on the other hand, responded to increased illumination by a decrease in photosynthesis. Much the same tendencies were established at both panicle differentiation stage and booting.

A similar experiment was done under controlled environment conditions. When grown at 1000 μmol photons m^{-2} s^{-1} for 10 days, Gui-Chao 2 showed a larger reduction in P_n and a larger loss of chlorophyll than Bellemont. These results were confirmed in experiments in 1985. Gui-Chao 2 was particularly sensitive to high light during the panicle differentiation stage and Bellemont showed the greatest increase in photosynthesis in response to high light during the booting stage.

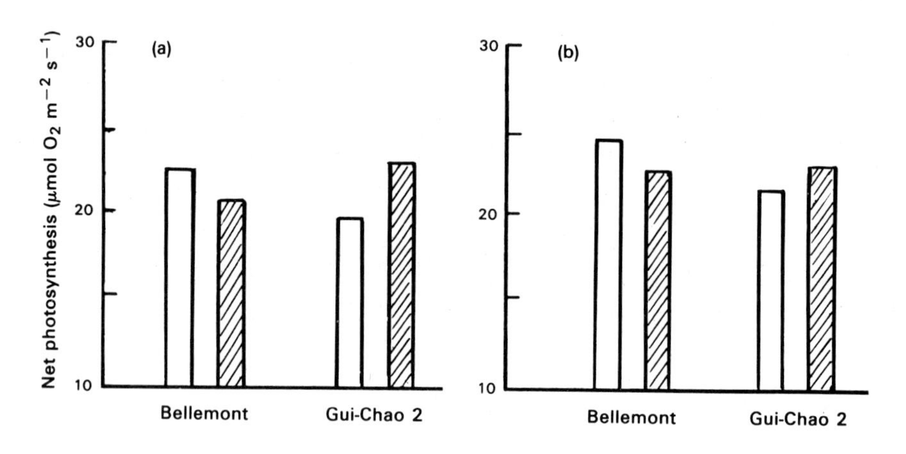

Fig. 4. Comparisons of photosynthesis in air (P_n) at natural light in leaves of rice varieties that developed in the field under natural light (open bars) and 30% daylight (hatched bars). Measurements were made at panicle differentiation (*a*) and booting stage (*b*).

Comparisons of Quantum Yield and Light-saturated Photosynthesis

The effects of high light on photosynthesis in air described above were followed up by comparisons of the light response curves for photosynthesis at CO_2 saturation. Comparisons of Lemont and Gui-Chao 2 grown for 9 days at 1000 μmol photons m^{-2} s^{-1} at panicle differentiation are shown in Fig. 5. The initial slope, indicative of apparent quantum yield (Fig. 5*b*), and the maximum rate were both lower in Gui-Chao 2 than in Lemont. The quantum yields of controls grown at 200 μmol photons m^{-2} s^{-1} were the same in both varieties (Fig. 5*c*). Comparisons of Figs 4 and 5 indicate the considerable potential of Lemont and Bellemont for light acclimation, and the limited capacity for light acclimation in Gui-Chao 2. The low quantum yield measured for Gui-Chao 2 may indicate photoinhibition (Hirata *et al.* 1983; Powles 1984; Walker and Osmond 1986), and is consistent with the reduction of photosynthesis in air at bright light compared with values at CO_2 saturation (Fig. 4).

Comparison of Yield Reduction by Shading

The plants used in the field shading experiment were further investigated as to the effect of shading on yield component changes and to estimate the yield level. The yield of Gui-Chao 2 was less depressed by shading than the yield of 'Bellemont' (Table 2).

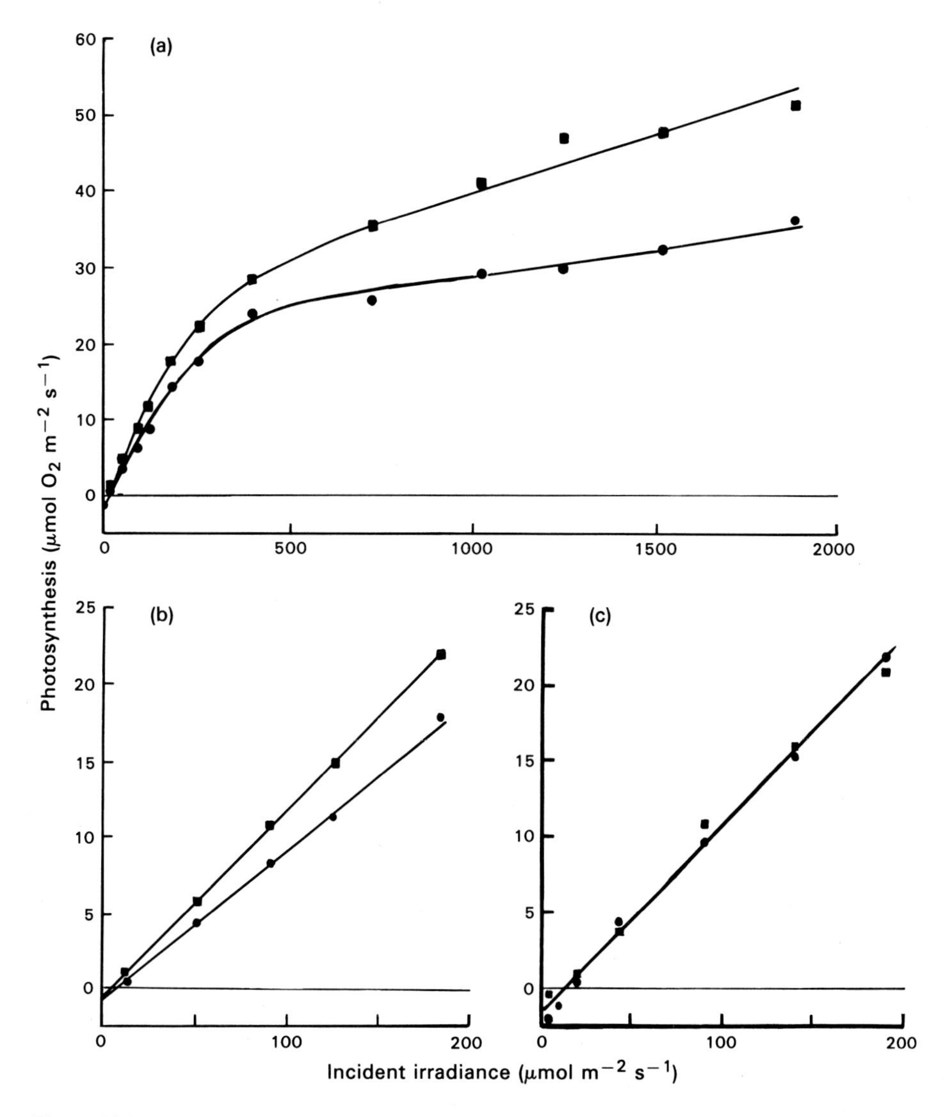

Fig. 5. Light response curve (*a*) and apparent quantum yield of photosynthesis (*b*) in leaves of rice varieties Lemont (■) and Gui-Chao 2 (●) after 9 days treatment at 1000 μmol photons m^{-2} s^{-1}. (*c*) Apparent quantum yield after treatment of Lemont (■) and Gui-Chao 2 (●) for 9 days at 200 μmol photons m^{-2} s^{-1}. Plants were treated at panicle differentiation stage; data are means of three plants.

Twenty days shading at 30% of natural light reduced yield in Bellemont at both the panicle differentiation stage and booting. Although yield was also reduced in Gui-Chao 2, the extent of this reduction was much less at both stages.

Table 2. Effect of 70% shading for 20 days on the grain yields of different rice varieties

Variety	Relative yield (%) Panicle differentiation stage	Booting stage
Bellemont	60	43
Gui-Chao 2	72	66

Discussion

Our preliminary data indicate that rice varieties can be differentiated into sun and shade ecotypes, as summarised in Table 3. Gui-Chao 2, selected under the low light conditions (few sunshine hours per day) in Guangzhou, has many characteristics of a shade plant. Leaf thickness does not increase in response to increasing illumination, photosynthesis in air is decreased in response to bright light, and in bright light, quantum yield appears to be somewhat reduced (Fig. 5*b*). On the other hand, varieties Bellemont and Lemont, from the bright light (long sunshine hours per day) habitats of Texas, have some properties of sun plants. Leaf thickness increases in response to increasing illumination during growth, photosynthesis in air increases, and quantum yields remain high. Table 3 also indicates that Bellemont and Lemont have higher carboxylase enzyme activities than Gui-Chao 2.

Our experiments suggest that selection procedures may have favoured ecotypic differentiation among rice varieties that resembles differentiation found in natural populations of other species (Björkman and Holmgren 1963). Ability to acclimate to bright light in U.S. varieties of rice correlates with increased leaf thickness and photosynthetic rate observed in other sun ecotypes (Boardman 1977; Osmond *et al.* 1980). The inability of Gui-Chao 2 to acclimate and its lower quantum yield in bright light implies sensitivity to photoinhibition normally associated with a shade plant (Powles 1984). Whether this is an intrinsic property of Gui-Chao 2, or related to some aspect of leaf nitrogen budget as demonstrated in supposed shade ecotypes of *Solanum dulcamara* (Osmond 1983; Ferrar and Osmond 1986), remains to be determined. Takano and Tsunoda (1971) showed that low nitrogen supply led to light response curves for photosynthesis in rice similar to those of shade plants.

Further studies are needed to establish these differences between characters, and such studies are necessary in relation to rice breeding and cultivation practice. The physiological characters measured here seem to be translated into yield responses to shading. It is our objective to introduce characters of high photosynthetic rate from some ecotypes, to make better use of bright light during the months of July and August. At the same time, we wish to retain yield properties under low light, and to minimise the attendant sensitivity to photoinhibition. Our studies indicate that there may well be problems associated with greater yield reduction in sun ecotypes later in the season, but introduction of these high-light-responding photosynthetic properties into local varieties such as Gui-Chao 2 might also minimise what appears to be damage by bright light in these shade varieties. Further experiments are needed to establish whether nitrogen fertilisation can mitigate the effects of bright light in shade-type rice.

We have already obtained some breeding material from the cross between high-light ecotype varieties and low-light ecotype varieties. They seem to inherit the strong points of good plant type and rapid increase in leaf area associated with local Guangdong varieties, and the thicker leaves from the U.S. varieties. Further studies are necessary to establish whether or not such changes will benefit rice production in South China.

Table 3. Comparison of photosynthetic characters of two different ecotypes of rice in relation to light intensities

	Ecotype adapted to high light intensity	Ecotype adapted to low light intensity
Representative varieties	Bellemont, Lemont	Gui-Chao 2
Growth irradiance of primary habitat	Over 8 h sunshine per day during growth season	Less than 8 h sunshine per day during growth season
Photosynthetic characters	P_n is higher under high light intensity: light saturation over 800 μmol photons $m^{-2} s^{-1}$	P_n is lower under high light intensity: light saturation about 700 μmol photons $m^{-2} s^{-1}$
Leaf thickness	Thicker	Thinner
RuP_2 carboxylase[A] activity[C]	Higher	Lower
PEP carboxylase[B] activity[C]	Higher	Lower
Leaf area index[C]	Slow development at early developmental stage	Quick development at early developmental stage
Response to high and low light intensities		
1. Photosynthesis & photoinhibition	No evidence of photoinhibition under high light intensity; P_n increases	Some evidence of photoinhibition under high light intensity; P_n decreases
2. Chlorophyll	Under long high light intensity, there is no visual change of chlorophyll content	Under long high light intensity, the decline of chlorophyll content can be seen by eye
3. Leaf thickness	Increases as the light intensity increases	No significant change as the light intensity increases
4. Grain yield	Under sunny conditions, yield potential is higher	Under conditions of cloudy and rainy days, yield loss is relatively less

[A]RuP_2 carboxylase, ribulose-1,5-bisphosphate carboxylase.
[B]PEP carboxylase, phospho*enol*pyruvate carboxylase.
[C]Unpublished data of studies by our research group, which were done in Shanghai Institute of Plant Physiology and Rice Research Institute of Guangdong Academy of Agricultural Sciences.

Acknowledgment

Part of the equipment used in this research was supplied by the Australian Centre for International Agricultural Research (ACIAR Project 8577). Barry Osmond has contributed to the guidance of this research and implementation of some of the techniques, but the authors are responsible for the opinions expressed.

References

Björkman, O., and Holmgren, P. (1963). Adaptability of the photosynthetic apparatus to light intensity in ecotypes from exposed and shaded habitats. *Physiol. Plant.* **16**, 889–914.

Boardman, N. K. (1977). Comparative photosynthesis of sun and shade plants. *Annu. Rev. Plant Physiol.* **28**, 355–77.

De Datta, S. K. (1981). 'Principles and Practices of Rice Production.' (John Wiley & Sons: New York.)

Ding (Ting), Y. (1959). Classification of rice cultivars in China. In 'Selected Rice Papers of Ding Yin'. (Eds Editor group of 'Selected Rice Papers of Ding Yin'.) pp. 74–93 (Agriculture Press: Beijing.)

Ferrar, P. J., and Osmond, C. B. (1986). Nitrogen supply as a factor influencing photoinhibition and photosynthetic acclimation after transfer of shade-grown *Solanum dulcamara* to bright light. *Planta (Berl.)* **168**, 563–70.

Hirata, M., Ishii, R., Kumura, A., and Murata, Y. (1983). Photoinhibition of photosynthesis in soybean leaves I. Effects of different intensities and durations of light irradiation on light response curves of photosynthesis. *Jpn J. Crop Sci.* **52**, 314–18.

Huke, R. (1976). Geography and climate of rice. In 'Climate and Rice'. pp. 1–13. (IRRI: Los Baños.)

Li, M. Q., -Lu, Z. R., and Deng, X. Q. (1984). A comparative study of photosynthesis and chlorophyll content of wild and cultivated rice. *Acta Phytophysiol. Sin.* **10**, 333–8.

Moran, R., and Rorath, D. (1980). Chlorophyll determination in intact tissue using *N,N*-dimethyl formamide. *Plant Physiol.* **65**, 478–9.

Osmond, C. B. (1983). Interactions between irradiance, nitrogen nutrition and water stress in the sun–shade responses of *Solanum dulcamara. Oecologia* (*Berl.*) **57**, 316–21.

Osmond, C. B., Björkman, O., and Anderson, D. J. (1980). 'Physiological Processes in Plant Ecology.' (Springer-Verlag: Berlin.)

Powles, S. B. (1984). Photoinhibition of photosynthesis induced by visible light. *Annu. Rev. Plant Physiol.* **35**, 15–44.

Takano, Y., and Tsunoda, S. (1971). Curvilinear regression of the leaf photosynthetic rate on leaf nitrogen content among strains of *Oryza* species. *Jpn J. Breed.* **21**, 69–76.

Tsunoda, S. (1984). Adjustment of photosynthetic structures in three steps of rice evolution. In 'Biology of Rice'. (Eds S. Tsunoda and N. Takahashi.) pp. 89–115. (Japan Scientific Societies Press/Elsevier: Tokyo.)

Tu, Z. P., *et al.* (1978). Varietal differences of net photosynthetic rate in rice plant and breeding for improving the efficiency of photosynthesis. *Acta Phytophysiol. Sin.* **4**, 113–20.

Tu, Z. P., Feng, H. Y., Cai, W. J., and Zhan, H. (1985). A preliminary comparative research on photosynthetic characters of several US rice varieties. *Crop Varietal Resour.* **3**, 4–6.

Walker, D. A., and Osmond, C. B. (1986). Measurement of photosynthesis in vivo with a leaf disc electrode: correlations between light dependence of steady state photosynthetic O_2 evolution and chlorophyll *a* fluorescence transients. *Proc. R. Soc. Lond.* B *Biol. Sci.* **227**, 267–80.

Compensating Effects to Growth of Changes in Dry Matter Allocation in Response to Variation in Photosynthetic Characteristics Induced by Photoperiod, Light and Nitrogen

Manfred Küppers[A,B], *George Koch*[A] *and Harold A. Mooney*[A]

[A]Department of Biological Sciences, Stanford University, Stanford, CA 94305, U.S.A.
[B]Present address: Institut für Botanik, Technische Hochschule, Schnittspahnstrasse 10, D-6100 Darmstadt, Federal Republic of Germany.

Abstract

High N-nutrition, high light and long photoperiod increased photosynthetic capacity (A_{max}) per leaf area of *Raphanus sativus* ×*raphanistrum*. The effect of A_{max} on growth was enhanced by an increased fraction of dry matter partitioned into the shoot, resulting in a larger canopy. Low N-nutrition, low light and a short photoperiod reversed these responses. At intermediate combinations, such as low light and long photoperiod or high light and short photoperiod, A_{max} and the actual net photosynthesis per leaf area (A) neither correlated with the carbon uptake rate of the canopy as a whole nor with whole plant growth. A compensatory effect to growth of changes in dry matter partitioning in response to variation in photosynthetic characteristics induced by light and N-nutrition is discussed in terms of two competitive feedbacks. For growth a long photoperiod compensated for low light or low N-nutrition. The results indicate that extrapolations from rates to gains are only valid under defined environmental conditions, especially when different photoperiods and effects on leaf ontogeny are involved.

Introduction

Many investigations relate the response of photosynthetic *activity* to environmental stresses, such as low light, water, nutrient or temperature stresses (Osmond *et al.* 1987). But very little information is available on those plant responses that compensate for effects on photosynthetic activity in order to keep the total carbon uptake of the individual as little affected as possible. For example, Mooney *et al.* (1988) have shown that photosynthetic capacity per unit leaf area declines when radishes (*Raphanus sativus* L.) are fumigated with SO_2; however, the plants partition more dry matter into leaf material and thus partially compensate. The importance of these compensations is also indicated by the observation that photosynthetic capacity declines from pioneers to late successional species of forest succession (Bazzaz 1979; Bazzaz and Pickett 1980; Küppers 1984), although later species are stronger competitors and replace the early ones. Late species have lower costs of crown formation (Küppers 1985). Therefore, photosynthetic capacity does not necessarily correlate with competitive strength. One would expect that, while these compensations are common, they are usually not evident. In contrast to compensations, enhancements are observed in situations in which different plant components respond synergistically, for example when photosynthetic capacity and growth parameters are either increased or decreased in parallel (e.g. Pearcy and Ehleringer 1984; Pearcy *et al.* 1981 discuss this for different photosynthetic pathways). Such patterns may even 'carry over' to communities (Ludlow and Wilson 1972).

Compensating effects become evident only when different components and levels of plant organisation are linked with each other. This is illustrated in Fig. 1: via a positive feedback, uptake of minerals can result in a higher carbon supply to the plant, which may enhance root growth and thus mineral uptake. However, there is a second and

0310-7841/88/010287$03.00

competitive positive feedback involving increased shoot and leaf growth—both area and number of leaves. A link between these two feedbacks, and thus a compensatory regulator, appears to be the partitioning of dry matter into shoot, root and, in the case of radish, hypocotyl. In the following the effects of photoperiod, irradiance and nitrogen nutrition on leaf and canopy photosynthesis, and on total plant growth, of wild radish will be investigated, and the results will be discussed with respect to compensating effects.

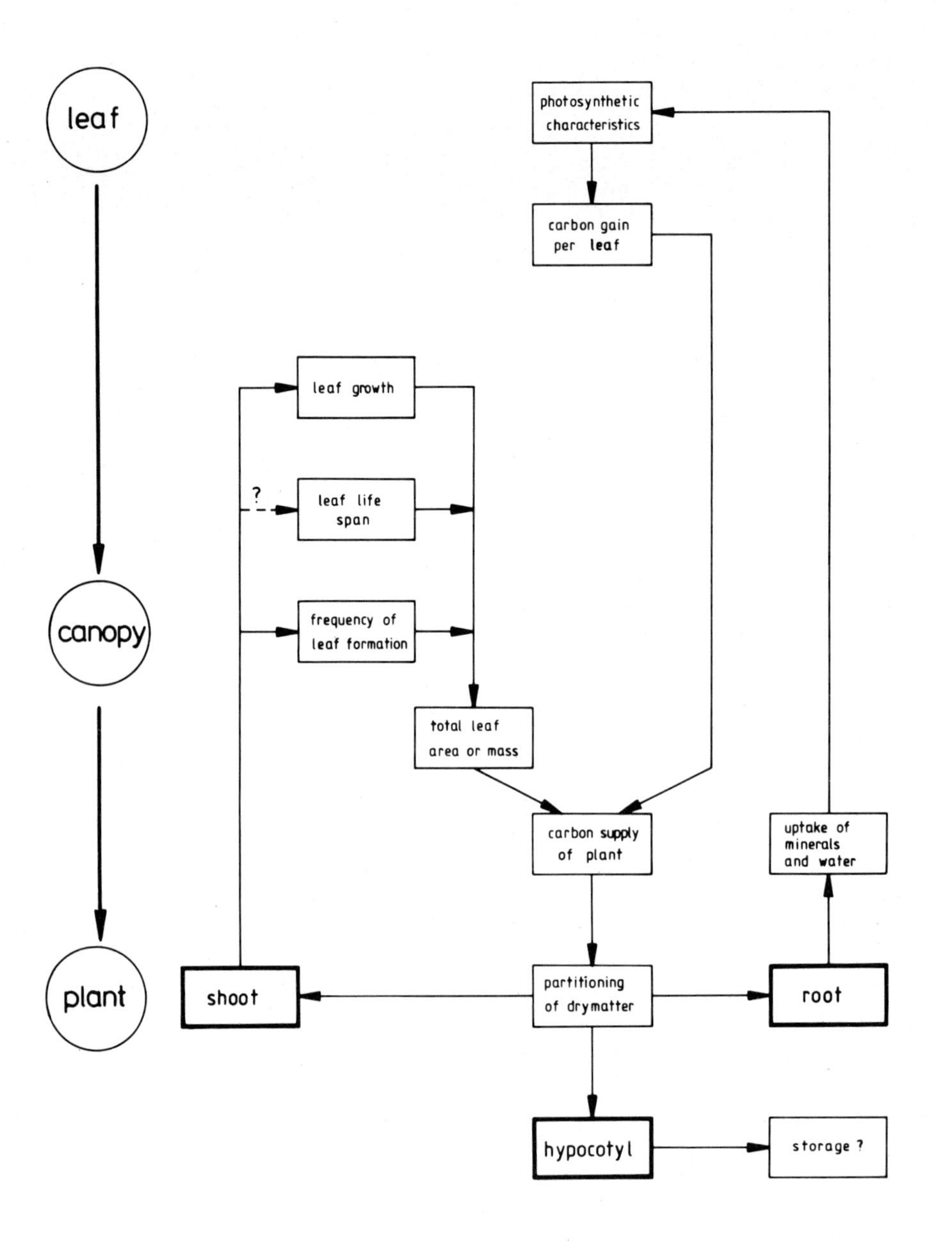

Fig. 1. Model of factors which affect the plant's carbon supply via two competitive feedback loops (thin arrows indicate a positive effect). These interactions are the result of an integration of organisation from leaf to plant (thick arrows). In the case of radish the hypocotyl is important in dry matter storage.

Table 1. Growth conditions

(a) Constant growth conditions used in all treatments

Shoot environment	
Day/night air temperatures:	25/15°C
Relative air humidity (day and night):	60%
CO_2 partial pressure in ambient air:	31–33 Pa
Root environment	
Temperature of hydroponic rooting medium:	15°C
Nutrient solution:	See Koch *et al.* (1987)

(b) Variable growth conditions used in individual treatments

Photoperiod (h)	Shoot environment: Irradiance (±15%) (μmol quanta $m^{-2} s^{-1}$)	Shoot environment: Daily light dose (mol quanta m^{-2})	Code	Root environment: Low N (10 μM)	Root environment: High N (100 μM)
14	700	35	14H	◇	□
14	280[A]	14	14L	◆	■
6	700	14	6H	△	○
6	400[A]	8	6L	▲	●

[A]The two different levels at low light result from the comparison with an intermediate photoperiod of 10 h used in an additional experiment at 14 mol m^{-2}. This experiment is omitted here since the results were similar to the more extreme treatments presented.

(c) Environmental conditions for the measurement of gas exchange

Of single leaves	
Leaf temperature	
For measurement of photosynthetic capacity (A_{max}):	25°C
For measurement of net photosynthesis (A):	25°C
Irradiance	
For measurement of A_{max}:	>1800 μmol quanta $m^{-2} s^{-1}$
For measurement of A:	According to growth conditions (Table 1*b*)
CO_2 partial pressure in ambient air:	33 Pa
Vapour concentration difference between leaf and air:	8–14 mPa Pa^{-1}
Air pressure:	101·3 kPa
Of canopies	
Air temperature:	25°C
Leaf temperature:	Maximally 1·5°C above air temperature
Irradiance	According to growth conditions (Table 1*b*)
CO_2 partial pressure in ambient air:	33 Pa
Relative air humidity:	60%
Air pressure:	101·3 kPa

Materials and Methods

Since seed size affects the growth of seedlings (Stanton 1985), wild radishes, *Raphanus sativus* × *raphanistrum*, were grown in a controlled root and shoot environment (Tables 1*a* and 1*b*) from seeds of 8·5–11·5 mg weight collected near San Francisco. The environment cabinets utilised have been described in detail by Koch *et al.* (1987). Hydroponically supplied nutrients were non-limiting with the exception of nitrate in one of the treatments (Table 1*b*). Forty plants, each transplanted to a separate growth tube 4 days after germination, were grown in each chamber. Starting 17–21 days after germination, four plants were harvested from each chamber at intervals of 5–10 days and separated into organs for analysis of dry weight, leaf area and nutrients.

For the measurement of leaf and canopy gas exchange, plants were removed from the growth chambers at periodic intervals. During measurements, roots were kept at growth conditions (Tables 1*a* and 1*b*). The gas analysis system of Winner and Mooney (1980) was partially modified after Küppers *et al.* (1987). Gas exchange of individual leaves was measured with a small ventilated leaf cuvette. In several preliminary studies performed under constant conditions in the leaf's environment (Table 1*c*), diurnal variations were not observed, either in net photosynthesis under growth conditions or in photosynthetic capacity (compare Küppers *et al.* 1986*a*, 1986*b*). For the measurement of whole canopy gas exchange the leaf cuvette was replaced by a ventilated canopy cuvette, which allowed for studies

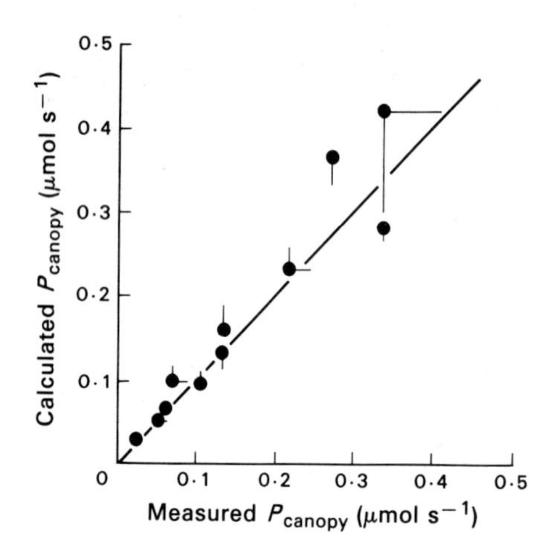

Fig. 2. Calculated versus measured canopy performance (P_{canopy}). The value for calculated P_{canopy} was obtained from measurements with a single leaf cuvette over all leaves on a plant. Bars for standard error ($n = 4$ plants) are given. Symbols irrespective of treatment.

Table 2. Nitrogen content per unit blade dry matter and photosynthetic capacity (A_{max}) per unit projected area of fully active leaves

The standard error is given ($n \geqslant 8$; N contents and A_{max} are not always for the same leaf). For abbreviations, see Table 1*b*

	Low N		High N	
	N content (mg g^{-1})	A_{max} (μmol m^{-2} s^{-1})	N content (mg g^{-1})	A_{max} (μmol m^{-2} s^{-1})
14H	27 ± 3	22 ± 1	65 ± 5	31 ± 2
14L	29 ± 2	19 ± 1	52 ± 5	22 ± 1
6H	36 ± 2	21 ± 2	44 ± 1	26 ± 4
6L	42 ± 1	18 ± 1	47 ± 2	19 ± 1

of plants up to 30 days old. Leaves of these young plants were growing almost horizontally, irrespective of treatment. A comparison of results of a preliminary investigation shows (Fig. 2) that measured canopy performance agrees well with calculated values, which have been derived by integrating the results obtained for every single leaf on a plant. In doing so, effects of *leaf* age on photosynthesis could be accounted for. At higher *plant* ages, neighbouring leaves could mechanically affect their orientation and could partially shade each other, by this deteriorating the 1 : 1 relationship given in Fig. 2.

Results

The results obtained from the different treatments are shown in the following by proceeding stepwise from the leaf level to the whole plant level.

Photosynthesis of Leaves

Higher N-availability in the rooting medium resulted in a higher photosynthetic capacity (A_{max}) per leaf area (e.g. Wong 1979; Gulmon and Chu 1981; Field and Mooney 1986). Table 2 shows for wild radish that this effect was independent of irradiance and photoperiod. High light and a long day further enhanced the effect, which is in agreement with the observations of Chabot *et al.* (1979). When A_{max} was expressed on a dry matter basis, a general relationship to the N-content of the leaf blade

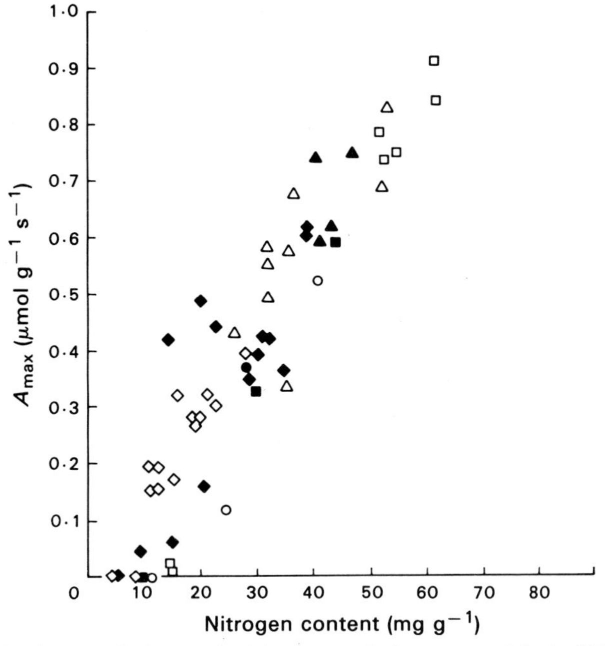

Fig. 3. Photosynthetic capacity (A_{max}) per unit dry matter of the leaf blade, versus the nitrogen content of the same blade. The decline from highest values of a certain treatment towards the lowest was obtained by simultaneously sampling older leaves. Symbols as in Table 1.

Table 3. Net photosynthesis (A) of fully active leaves and its proportion of photosynthetic capacity (A/A_{max})

The standard error is given ($n \geqslant 8$). For abbreviations see Table 1*b*

	Low N		High N	
	A (μmol $m^{-2} s^{-1}$)	A/A_{max}	A (μmol $m^{-2} s^{-1}$)	A/A_{max}
14H	16 ± 1	0·73	20 ± 1	0·65
14L	9 ± 0·5	0·47	9 ± 1	0·41
6H	16 ± 1	0·76	18 ± 1	0·69
6L	12 ± 1	0·67	11 ± 0·5	0·58

was found (Fig. 3). This pattern remained valid when, in senescing leaves, the N-content was drawn down by remobilisation (Fig. 3). Obviously, the N-content is a very important determinant of the photosynthetic potential of a leaf (Evans 1983; Evans and Terashima 1987).

However, for plant growth the actual carbon uptake rate (A) at growth conditions is of significance, and is compared to A_{max} in Table 3. At high light, high N increased

A, but at high N a lower fraction of the capacity, A/A_{max}, was utilised. Furthermore, at low photon irradiances, A_{max} had little significance for the actual uptake rate of a unit of leaf (compare Table 2 with Table 3). For the given environmental conditions this result queries the importance of A_{max} for plant growth.

Leaf Ontogeny and the Leaf Area of a Canopy

Carbon uptake of a plant as a whole is partially determined by the total leaf area or mass in the canopy. This total area is, at any given time, a result of growth of leaf blades, death rates of leaves—as a consequence of leaf aging—and the development of new leaves (Fig. 1). In order to describe leaf aging, the change of A_{max} with time has been followed (Fig. 4). A_{max} appeared to be an appropriate measure of a leaf's physiological activity, since leaves supply the plant with carbon and since A_{max} is directly related to this process.

A_{max} declined both faster and at an earlier leaf age under the long than under the

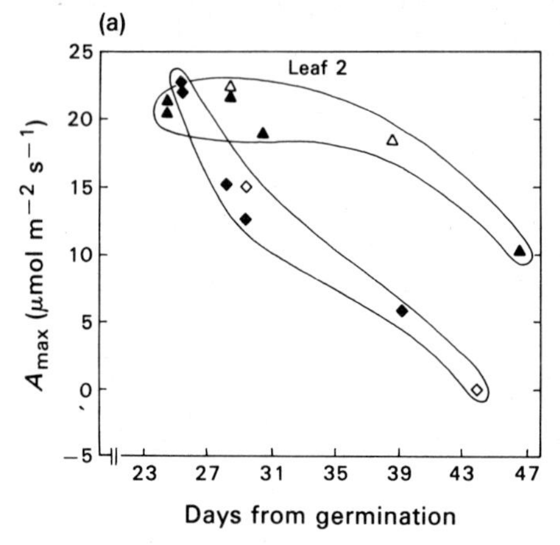

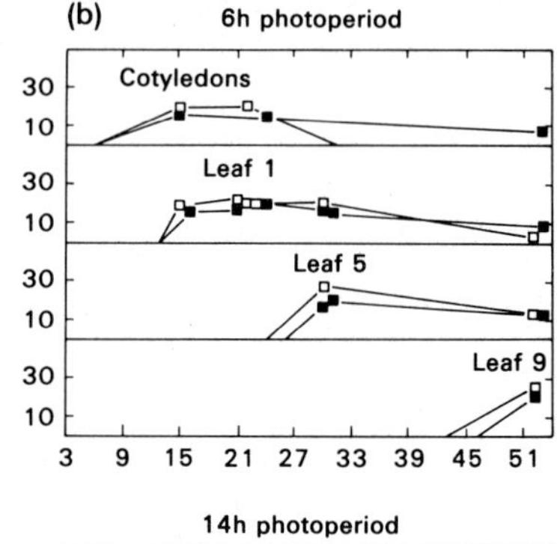

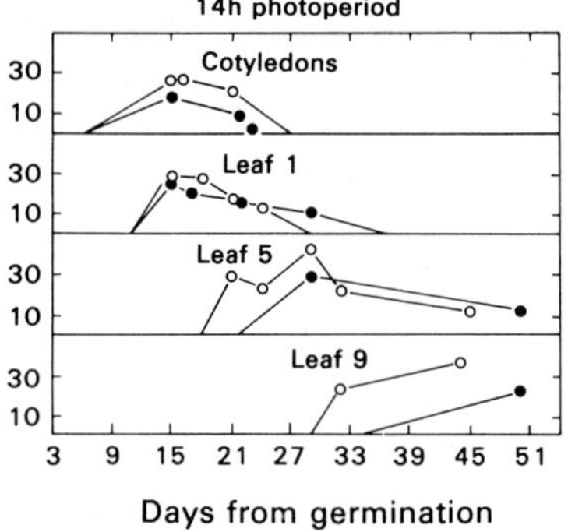

Fig. 4. The change in photosynthetic capacity (A_{max}, per projected leaf area) with increasing plant age, as affected by irradiance and photoperiod. For the same leaf, progressive plant aging is a measure of progressive leaf aging. (*a*) For low N availability. (*b*) For high N availability in the rooting medium. Symbols as in Table 1.

short day (Figs. 4*a* and 4*b*). This result was independent of N-availability, photon irradiance, daily light dose and, as a consequence of the constant light environment in an individual treatment, of the carbon balance per leaf area (not shown). Obviously, photoperiod was more important for leaf ontogeny than any other factor investigated. However, a long photoperiod and high light enhanced the frequency of leaf formation (Fig. 4*b*): 9 leaves were found on a plant after 29 days of plant growth under 14H (for abbreviations see Table 1*b*), after 35 days of plant growth under 14L, after 43 days under 6H and after 46 days under 6L. This could be a reflection of floral induction, since the plants grown in the long day started flowering earlier.

The balance of all components is shown in Fig. 5. In all cases leaf initiation and leaf area expansion compensated effectively for leaf area losses from leaf deaths, as indicated by an increasing total plant leaf area with increasing plant age. At low N-availability in the rooting medium, this pattern was clearly affected by photoperiod but not by photon irradiance or by daily light dose. However, at high N-nutrition, high light strongly enhanced the effect of day length, since it increased the relative growth rate of the leaf blade (not shown).

The effects of certain environmental parameters on two components of plant carbon supply—leaf photosynthesis and leaf area development—have been studied separately for each component. The results presented in Tables 2 and 3 in comparison with those of Fig. 5 indicate that none of A, A_{max} or its fraction being utilised in a particular environment (A/A_{max}) can be correlated with the development of total plant leaf area.

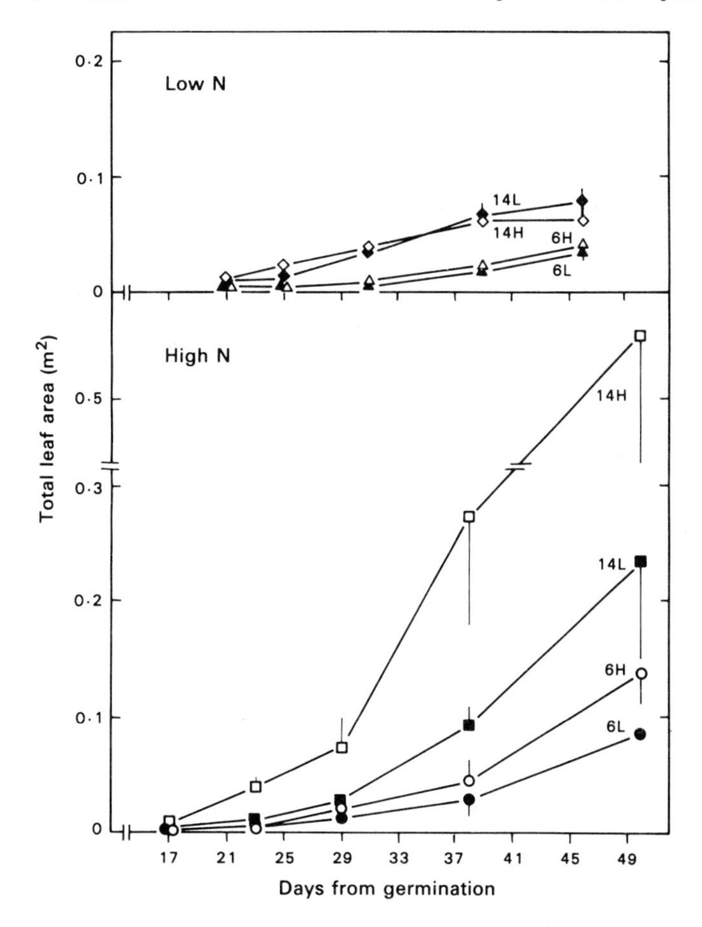

Fig. 5. The development of total leaf area with increasing plant age for the different treatments. Bars indicate standard errors ($n = 4$). Symbols as in Table 1.

Carbon Uptake by the Canopy

Carbon uptake by the canopy integrates photosynthetic activity of the individual leaf *and* total leaf mass or area of the plant (Fig. 1). Canopy photosynthesis (A_{canopy}), defined here as the carbon uptake *rate* of the canopy per *total plant leaf area*, was higher at higher irradiances (Fig. 6). This was independent of photoperiod. Over time, A_{canopy} varied only little, and N-nutrition had a small effect, apparent only at high light. The absolute CO_2 uptake *rate* of the *shoot as a whole* (P_{canopy}) showed a different pattern: it increased over time (Fig. 6), because total leaf area increased with plant age (Fig. 5). Highest P_{canopy} and fastest increments were observed under 14H, additionally enhanced by high N-availability in the rooting medium, and lowest rates and smallest increments were found under 6L and low N. This result is not surprising since these growth conditions were extremely different. However, for the intermediate conditions (6H and 14L), the pattern was not predictable although daily light doses were the same: for high N, A_{canopy} and P_{canopy} responded in parallel (Fig. 6). In contrast, for low N, P_{canopy} was higher at 14L compared to 6H, which is the reverse of the response of A_{canopy}.

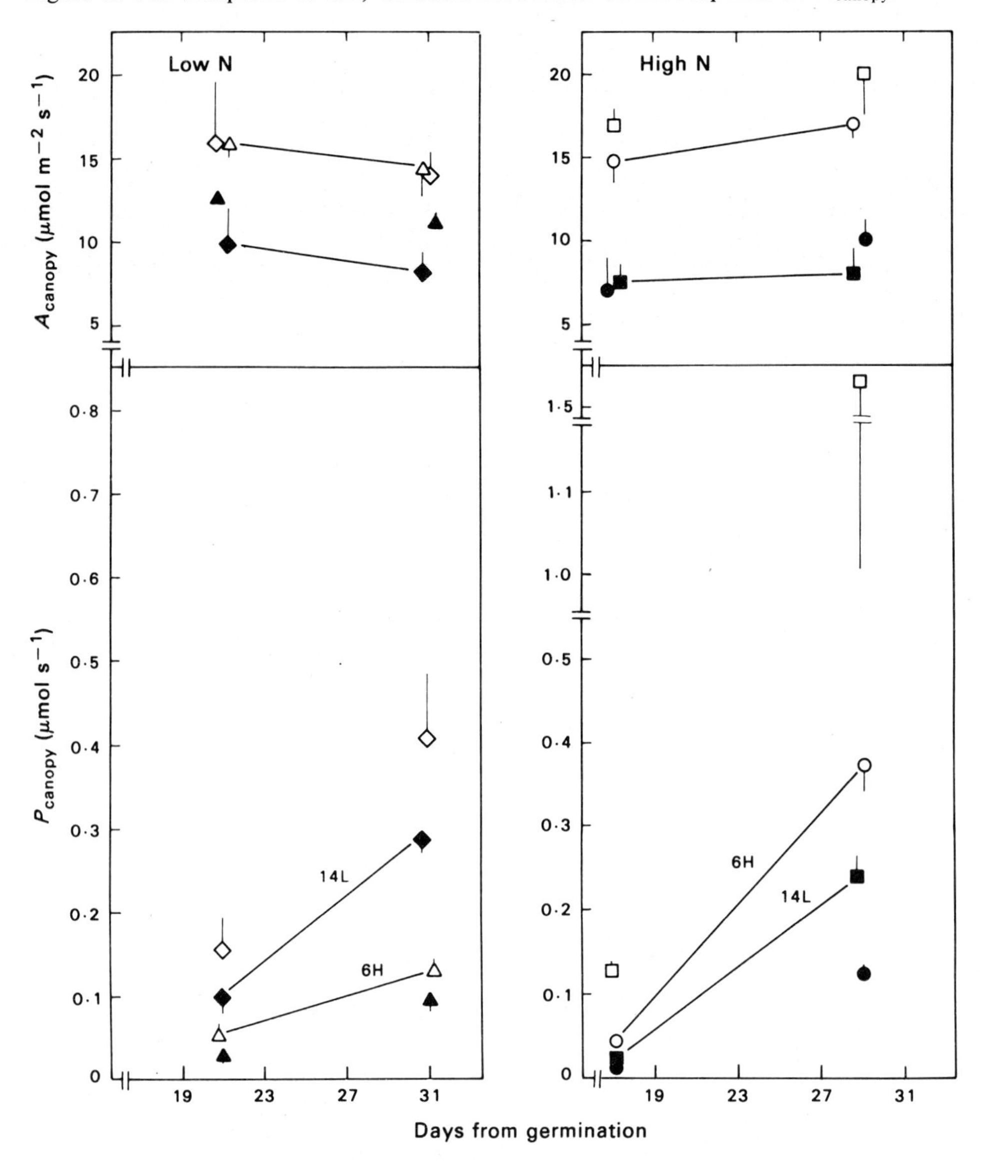

Fig. 6. Canopy photosynthesis (A_{canopy}) and canopy performance (P_{canopy}) as a function of plant age, either at low or at high N availability in the rooting medium. A_{canopy} is P_{canopy} divided by total plant leaf area. Bars indicate standard errors obtained from measurements of four plants. Symbols as in Table 1.

Obviously, a compensation to A_{canopy} (and to A (Table 3) being reflected in A_{canopy}) occurred at the canopy level, which appeared to be a consequence of the development of total leaf area. In order to understand these patterns, the partitioning of dry matter into the shoot has to be taken into consideration.

Partitioning of Dry Matter and Plant Growth

The largest fraction of dry matter was partitioned into the shoot and the lowest into the hypocotyl, which incorporated similarly under all environmental conditions (Table 4*a*). Neither photoperiod nor irradiance affected this pattern, but significantly more dry matter was partitioned into the shoot at high N-nutrition (Table 4*a*). Consequently, a plant with more N available should build, in the long term, a larger canopy, independent of whether A_{max}, A and A_{canopy} are enhanced by high N or not. This indicates the importance of the partitioning pattern for P_{canopy}.

The effect of light on final plant biomass was large at high N but not at low N (Table 4*b*). Daylength was most important for plant growth: despite the same light dose

Table 4. Partitioning of dry matter and plant growth

(*a*) *Partitioning of dry matter to shoot, root and hypocotyl*

Since the partitioning pattern remained fairly constant with increasing plant age (except for very young plants; not shown), means and standard errors for the entire growth period are given ($n=4$). For abbreviations, see Table 1*b*

	Low N			High N		
	Shoot (%)	Hypocotyl (%)	Root (%)	Shoot (%)	Hypocotyl (%)	Root (%)
14H	69 ± 2	3 ± 0·5	28 ± 2	78 ± 2	4 ± 0·5	18 ± 2
14L	71 ± 3	3 ± 0·5	26 ± 3	80 ± 1	4 ± 1	16 ± 2
6H	68 ± 2	5 ± 1	27 ± 3	79 ± 2	5 ± 1	16 ± 3
6L	74 ± 1	4 ± 0·5	22 ± 2	81 ± 1	4 ± 1	15 ± 2

(*b*) *Dry matter of whole plants at final harvest*

Standard errors are given ($n=4$). For abbreviations, see Table 1*b*

	Dry matter (g) for plant age:		
	46 days[A]	46 days[B]	50 days[A]
	Low N	High N	
14H	9·5 ± 0·5	38·7 ± 5·1	48·1 ± 5·4
14L	7·7 ± 1·3	12·4 ± 3·8	16·2 ± 5·1
6H	2·5 ± 0·4	4·6 ± 1·2	5·8 ± 1·3
6L	2·0 ± 0·3	2·8 ± 0·3	3·7 ± 0·2

[A]Mean and standard error from measured data.
[B]Interpolated for both mean and standard error to fit the same plant age in all treatments.

(under 14L and 6H), more than twice as much dry matter had been accumulated under long days (Table 4*b*). Daylength compensated for the effect of low N: under 14L more dry matter was produced than under high N and 6H. This contrasts with development of total leaf area (Fig. 5), indicating that the effects of the treatments on total leaf area did not entirely carry over to total plant growth. Several factors appeared to be responsible. At low N, leaves developed a higher specific leaf weight (not shown), thus their area was more 'expensive' in terms of carbon; additionally, less carbohydrate was partitioned into the shoot (Table 4*a*). These factors may already explain the significant difference ($P=0{\cdot}05$) between the N-treatments observed for the pattern of log plant biomass versus log total leaf area during growth (Fig. 7).

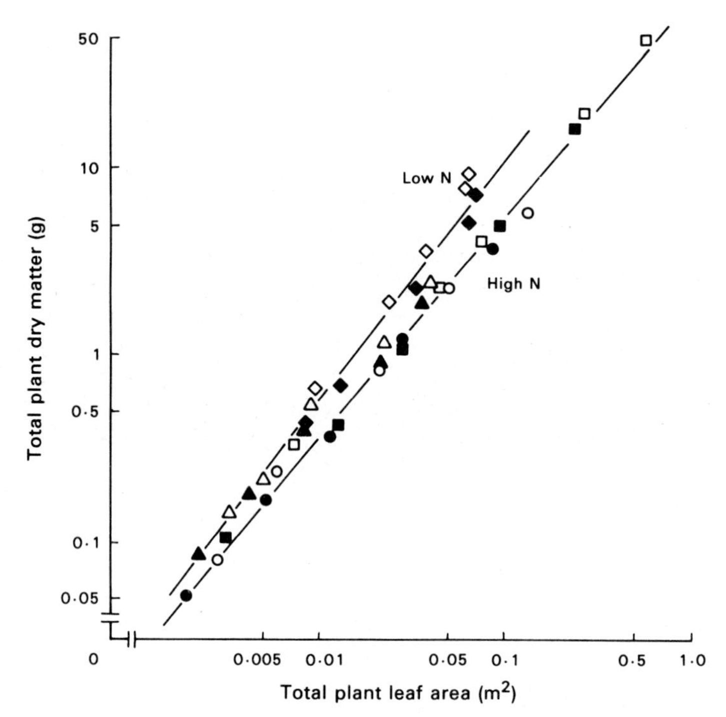

Fig. 7. Log total plant dry weight as a function of log total plant leaf area, as found during growth. Symbols as in Table 1.

Discussion

The results show that two different types of 'compensating effects' have to be distinguished—those that are under the influence of the plant, e.g. the pattern of carbon partitioning (Table 4*a*), which results in the inversion of P_{canopy} relative to A_{canopy} (Fig. 6), and those that are independent of the plant, as is daylength buffering the effect of low N and low light on plant growth (Table 4*b*). Only the first type can be discussed in relation to the diagram shown in Fig. 1.

The Significance of Interactions between Feedbacks

When the feedback loop shown in Fig. 1 is dominating, then enhancements are evident. High N clearly favours the loop via the shoot at high light, especially in the long day. It increases A and A_{max} 1·4-fold as compared to low N (Tables 2 and 3), P_{canopy} 3·8-fold (Fig. 6) and total plant growth 4·1-fold (Table 4*b*). For low N, relatively more root dry matter is grown (Table 4*a*)—presumably to minimise a limiting effect on total plant growth—and the feedback via the root is favoured. It can be shown that the effect of N-*limitation* is enhanced when proceeding from factors at the leaf level to those of the whole plant.

However, in intermediate situations, which are probably more common than extreme ones, it is difficult to predict the response of a plant since neither feedback loop dominates. For example, for low *N*, illumination with high photon irradiances enhances A_{max} 1·2-fold and A 1·8-fold, but P_{canopy} only 1·4-fold and growth only 1·3-fold. This becomes even more evident when comparing the two intermediate combinations—low N/high light and high N/low light—with each other. For the same daylength, A_{max} is not enhanced, A and P_{canopy} are decreased by a factor of 0·7 and 0·6, respectively, but growth is increased 1·3-fold. Obviously, the effects on A and P_{canopy} do not necessarily carry over to total plant growth. In general agreement, Ågren and Ingestad (1987) concluded in their study that 'photosynthetic rates do not reflect growth rates'. Similarly, Linder and Rook (1984) observed that fertilisation of conifers affected the amount of needles more than their photosynthetic rates.

The Significance of Interactions between Environmental Factors

Osmond (1983) has demonstrated the interactions between irradiance, nitrogen nutrition and water stress in the sun–shade response of *Solanum dulcamara*. Our results are in agreement: no single environmental factor but a combination of several factors determines the potential range of a plant response. For example, a high light environment can only be fully utilised at high N-availability or vice-versa (Table 3; Fig. 5; Table 4*b*). In intermediate environmental situations photosynthetic characteristics (e.g. A_{max}) may have little importance for plant growth, although carbon uptake is always essential. As a consequence, it may be difficult to predict gains from rates.

Our results, in agreement with Keating *et al.* (1985), demonstrate the importance of a long photoperiod for plant growth (Table 4*b*): a long day can effectively compensate for low light and, within a certain range, for low N availability. This may have ecological significance when comparing species growing naturally at different latitudes, especially when the temperature regime is similar. A high photosynthetic activity of a species growing in alpine areas of low latitudes (and thus in a short day) may not indicate a higher productivity as compared to a species of the boreal tundra with low photosynthetic activity in a long day. In fact, the interaction of photoperiod may explain some of the observed differences in the same species from either boreal or alpine habitats (Billings and Mooney 1968).

Conclusions

It is concluded that, at different levels of integration from leaf to plant, enhancements or compensations may occur. The physiological mechanisms that account for shifts in acclimation and allocation are not known in detail. However, the discussion shows that, especially in intermediate environmental situations, it is the relative importance of these shifts to each other that determines plant growth, although the individual mechanisms involved should be the same in all cases. Therefore, in order to predict plant growth, quantitative data on lower levels of plant organisation are necessary rather than detailed mechanistic understanding.

Acknowledgments

M. K. thanks the Deutsche Forschungsgemeinschaft (Bonn) for the fellowship 'Ausbildungsstipendium', during which this work was performed, and Drs I. R. Cowan and R. O. Slatyer for the invitation to present this paper at the conference on the ecology of photosynthesis in sun and shade at Canberra in February 1987. We thank Celia Chu, Barbara Küppers, James Gorham and Andrew Nardone (Stanford) for their help with gas exchange and biomass assessments, Mrs Schäfer and Mrs Heger (both Darmstadt) for the art work. Appreciation is also expressed to the U.S. National Science Foundation for support of this work.

References

Ågren, G. I., and Ingestad, T. (1987). Root : shoot ratio as a balance between nitrogen productivity and photosynthesis. *Plant Cell Environ.* **10**, 573–86.

Bazzaz, F. A. (1979). The physiological ecology of plant succession. *Annu. Rev. Ecol. Syst.* **10**, 361–71.

Bazzaz, F. A., and Pickett, S. T. A. (1980). Physiological ecology of tropical succession: a comparative review. *Annu. Rev. Ecol. Syst.* **11**, 287–310.

Billings, W. D., and Mooney, H. A. (1968). The ecology of arctic and alpine plants. *Biol. Rev.* **43**, 481–529.

Chabot, B. F., Jurik, T. W., and Chabot, J. F. (1979). Influence of instantaneous and integrated light-flux density on leaf anatomy and photosynthesis. *Am. J. Bot.* **66**, 940–5.

Evans, J. R. (1983). Nitrogen and photosynthesis in the flag leaf of wheat (*Triticum aestivum* L.). *Plant Physiol.* **72**, 297–302.

Evans, J. R., and Terashima, I. (1987). Effects of nitrogen nutrition on electron transport components and photosynthesis in spinach. *Aust. J. Plant Physiol.* **14**, 59–68.

Field, C., and Mooney, H. A. (1986). The photosynthesis–nitrogen relationship in wild plants. In 'On the Economy of Plant Form and Function'. (Ed. T. J. Givnish.) pp. 25–55. (Cambridge University Press: Cambridge).

Gulmon, S. L., and Chu, C. C. (1981). The effects of light and nitrogen on photosynthesis, leaf characteristics, and dry matter allocation in the chaparral shrub, *Diplacus aurantiacus. Oecologia* **49**, 207–12.

Keating, B. A., Wilson, G. L., and Evenson, J. P. (1985). Effects of photoperiod on growth and development of cassava (*Manihot esculenta* Crantz). *Aust. J. Plant Physiol.* **12**, 621–30.

Koch, G., Winner, W. E., Nardone, A., and Mooney, H. A. (1987). A system for controlling root and shoot and environment for plant growth studies. *Environ. Exp. Bot.* (in press).

Küppers, M. (1984). Carbon relations and competition between woody species in a Central European hedgerow. I. Photosynthetic characteristics. *Oecologia* **64**, 332–43.

Küppers, M. (1985). Carbon relations and competition between woody species in a Central European hedgerow. IV. Growth form and partitioning. *Oecologia* **66**, 343–52.

Küppers, M., Matyssek, R., and Schulze, E.-D. (1986*a*). Diurnal variations of light-saturated CO_2 assimilation and intercellular carbon dioxide concentration are not related to leaf water potential. *Oecologia* **69**, 477–80.

Küppers, M., Wheeler, A. M., Küppers, B. I. L., Kirschbaum, M. U. F., and Farquhar, G. D. (1986*b*). Carbon fixation in eucalypts in the field—analysis of diurnal variations in photosynthetic capacity. *Oecologia* **70**, 273–82.

Küppers, M., Swan, A. G., Tompkins, D., Gabriel W. C. L., Küppers, B. I. L., and Linder, S. (1987). A field portable system for the measurement of gas exchange of leaves under natural and controlled conditions: examples with field-grown *Eucalyptus pauciflora* Sieb. ex Spreng. ssp. *pauciflora, E. behriana* F. Muell. and *Pinus radiata* D. Don. *Plant Cell Environ.* **10**, 425–35.

Linder, S., and Rook, D. A. (1984). Effects of mineral nutrition on carbon dioxide exchange and partitioning in trees. In 'Nutrition of Plantation Forests'. (Eds G. D. Bowen and E. K. S. Nambiar.) pp. 211–36. (Academic Press: London.)

Ludlow, M. M., and Wilson, G. L. (1972). Photosynthesis of tropical pasture plants. IV. Basis and consequences of differences between grasses and legumes. *Aust. J. Biol. Sci.* **25**, 1133–45.

Mooney, H. A., Küppers, M., Koch, G., Gorham, J., Chu, C., and Winner, W. E. (1988). Compensating effects to growth of carbon partitioning changes in response to SO_2-induced photosynthetic reduction in radish. *Oecologia* (in press).

Osmond, C. B. (1983). Interactions between irradiance, nitrogen nutrition, and water stress in the sun–shade responses of *Solanum dulcamara. Oecologia* **57**, 316–21.

Osmond, C. B., Austin, M. P., Berry, J. A., Billings, W. D., Boyer, J. S., Dacey, J. W. H., Nobel, P. S., Smith, S. D., and Winner, W. E. (1987). Stress physiology and the distribution of plants. *BioScience* **37**, 38–48.

Pearcy, R. W., and Ehleringer, J. (1984). Comparative ecophysiology of C_3 and C_4 plants. *Plant Cell Environ.* **7**, 1–13.

Pearcy, R. W., Tumosa, N., and Williams, K. (1981). Relationships between growth, photosynthesis and competitive interactions for a C_3 and a C_4 plant. *Oecologia* **48**, 371–6.

Stanton, M. L. (1985). Seed size and emergence time within a stand of wild radish (*Raphanus raphanistrum* L.): the establishment of a fitness hierarchy. *Oecologia* **67**, 524–31.

Winner, W. E., and Mooney, H. A. (1980). Ecology of SO_2 resistance. I. Effects of fumigations on gas exchange of deciduous and evergreen shrubs. *Oecologia* **44**, 290–5.

Wong, S. C. (1979). Elevated atmospheric partial pressure of CO_2 and plant growth. I. Interactions of nitrogen nutrition and photosynthetic capacity in C_3 and C_4 plants. *Oecologia* **44**, 68–74.

Growth and Photosynthetic Response to Light and Nutrients of *Flindersia brayleyana* F. Muell., a Rainforest Tree with Broad Tolerance to Sun and Shade

W. A. Thompson[A], *G. C. Stocker*[B] and *P. E. Kriedemann*[A]

[A]Division of Forest Research, CSIRO,
P.O. Box 4008, Canberra, A.C.T. 2600, Australia.
[B]Tropical Forest Research Centre, CSIRO,
P.O. Box 780, Atherton, Qld 4887, Australia.

Abstract

Seed from four species of rainforest trees with widely contrasting sunlight requirements for growth and development were sown within disturbance gaps amidst mature forest on the Herberton Range in North Queensland. Observations on seedling persistence plus comparative growth of young trees of *Acacia aulacocarpa, Toona australis, Flindersia brayleyana* and *Darlingia darlingiana* (species ranked according to adaptation from full sun to deep shade) confirmed a broad tolerance of *Flindersia* to sunlight under all conditions, from wide to narrow gaps (minimum 0·6% full sun equivalent).

Photosynthetic attributes which underlie such broad tolerance were subsequently inferred from single leaf gas exchange, plus foliar analyses of nitrogen, phosphorus and chlorophyll on tree seedlings held for 180 days under two nutrient × three irradiance levels adjusted to represent natural irradiance incident upon the forest floor (low), mid-canopy (medium) and emergent crowns (high irradiance treatment). Medium irradiance plus high nutrients proved optimal for leaf expansion, chlorophyll content and photosynthesis in air.

Growth under low irradiance was characterised by thinner leaf palisade tissue, lower rates of dark respiration, increased leaf chlorophyll per unit nitrogen and lower light compensation point for photosynthesis. Such leaves retained a relatively high photosynthetic capacity despite these other shade-leaf attributes.

High irradiance plus low nutrients proved supraoptimal for leaf expansion and expression of photosynthetic activity. Chronic photoinhibition appeared to prevail because apparent quantum yield was reduced, while photosynthetic processes on a nitrogen basis were substantially impaired. Nitrogen use efficiency, as inferred from leaf chlorophyll content, light saturated CO_2 assimilation rate, electron transport rate and carboxylation rate on a nitrogen basis declined with increasing growth irradiance.

Some ecological implications for the establishment and growth of these rainforest tree species in disturbance gaps are discussed.

Introduction

Tropical rainforests on coast and hinterland of north-eastern Queensland comprise over 700 tree species (Tracey 1982). Of these, only 20 or thereabouts can rapidly colonise open situations under full sun and strong advection. The remainder regenerate in narrow disturbance gaps, are slow growing, long lived (centuries) and shade-tolerant as seedlings. Small-scale disturbance via isolated wind throw, or highly selective logging, thus preserves species richness and structural heterogeneity as typifies mature forest. By contrast, broad scale disruption after clear felling or agricultural and pastoral development favours recruitment of a relatively few species characterised by fast growth and early canopy closure (Richards 1952; Stocker 1981). Shade-tolerant tree seedlings can then begin to re-establish in a second phase succession as a substory community, but at least decades and perhaps even centuries may elapse before original floristic diversity is regained.

0310-7841/88/010299$03.00

Colonisation of disturbance gaps is thus fundamental to maintenance of species richness. Differences between taxa in response of their seedlings to strong and sustained sunlight *vis a vis* capacity for shade tolerance appears to be a key factor in determining floristic composition of that initial establishment (Whitmore 1975). Tree species in North Queensland which typically colonise wide disturbance gaps (where orientation predisposes to day-long interception of sunlight and spaces between trees are of similar order to remaining canopy height, i.e. height : width = 1 : 1) include *Acacia aulacocarpa, A. melanoxylon, Aleurites moluccana* and *Alphitonia petriei.* Such trees are sunloving, but shade-intolerant. By contrast, narrow disturbance gaps (height : width ratio from 10 : 1 up to 100 : 1) support establishment of numerous shade-tolerant tree seedlings, notably *Dysoxylon pettigrewiana, Darlingia darlingiana, Argyrodendron peralatum,* and *Syzygium wesa.* Medium gaps (height : width 4 : 1) favour recruitment of tree seedlings with intermediate shade tolerance such as *Toona australis* and *Flindersia brayleyana.*

Such ecological groupings, where tree species are ranked as wide, medium or narrow gap colonisers were arrived at empirically (Stocker, unpublished) and are consistent across a broad geographic range of submontane forest in North Queensland. Although these three categories must not be regarded as mutually exclusive, but rather as guilds along a continuum, functional differences between discernible groups of rainforest tree seedlings do raise a number of issues: (1) Are wide-gap colonisers such as *Acacia aulacocarpa* intolerant of deep shade narrow gaps as inferred from ecological observations? (2) Are shade-tolerant seedlings of narrow-gap colonisers disadvantaged by full sun? If so, does nutrient supply influence those relationships? (3) What photosynthetic attributes underlie broad tolerance of medium-gap colonisers which appear to establish over a wide range of sunlight supply?

Ecological observations on growth dynamics of mixed populations of seedling trees from each of these three categories, as a function of sunlight supply, in wide, medium and narrow gaps, provide a basis for exploring (1) and (2); a more detailed analysis of leaf photosynthesis is directed at question (3). Background concepts for irradiance $\times$ nutrient effects on leaf photosynthesis are outlined below.

Tree seedling growth within narrow disturbance gaps of upland forests on Atherton Tableland can be limited by both irradiance and nutrient supply. Successful establishment therefore necessitates full utilisation of both types of physical inputs for net carbon gain. Ideally, nutrient-based resources must be allocated to component processes of photosynthesis and respiration such that leaf-level integration is optimised for particular sets of irradiance $\times$ nutrient conditions. Relevant partial processes would include: (1) light interception and utilisation (expressed via apparent quantum yield, α); (2) photosynthetic electron transport and hence capacity for RuP_2 (ribulose 1,5-bisphosphate) regeneration (expressed as maximum rate of whole chain electron transport, J); (3) RuP_2-saturated carboxylation at high irradiance (expressed as maximum carboxylation rate, V_c; (4) metabolic cost of leaf maintenance (embodied in dark respiration, R_{dark}).

According to one hypothesis, a plant acclimates to a given irradiance and nutrient availability by physiological adjustments which serve to increase carbon gain (Mooney and Gulmon 1979; Field 1983; Hirose and Werger 1987). One consequence of this hypothesis is that in an acclimated plant light-harvesting, electron transport capacity and carboxylation capacity will be colimiting for that growing environment. Accordingly, the comparative importance of these component processes for utilisation of environmental inputs, and hence net carbon gain at leaf level, should vary in response to irradiance and nutrient supply. *Flindersia brayleyana* was chosen as a test species because foliage on seedling trees can tolerate deep shade and yet retain a latent capacity to utilise strong sun following advent of medium or wide disturbance gaps.

Acclimation from shade to sun, especially under nutrient-limiting conditions, must involve adjustments between component processes listed above. These considerations led us to the following questions regarding photosynthetic adjustments in *Flindersia*. (1) Do shade-grown leaves have increased apparent quantum yield? (2) Do sun-grown leaves have higher light-saturated rates of photosynthesis than shade-grown leaves? (3) Does nitrogen use efficiency in photosynthesis vary in response to irradiance during growth?

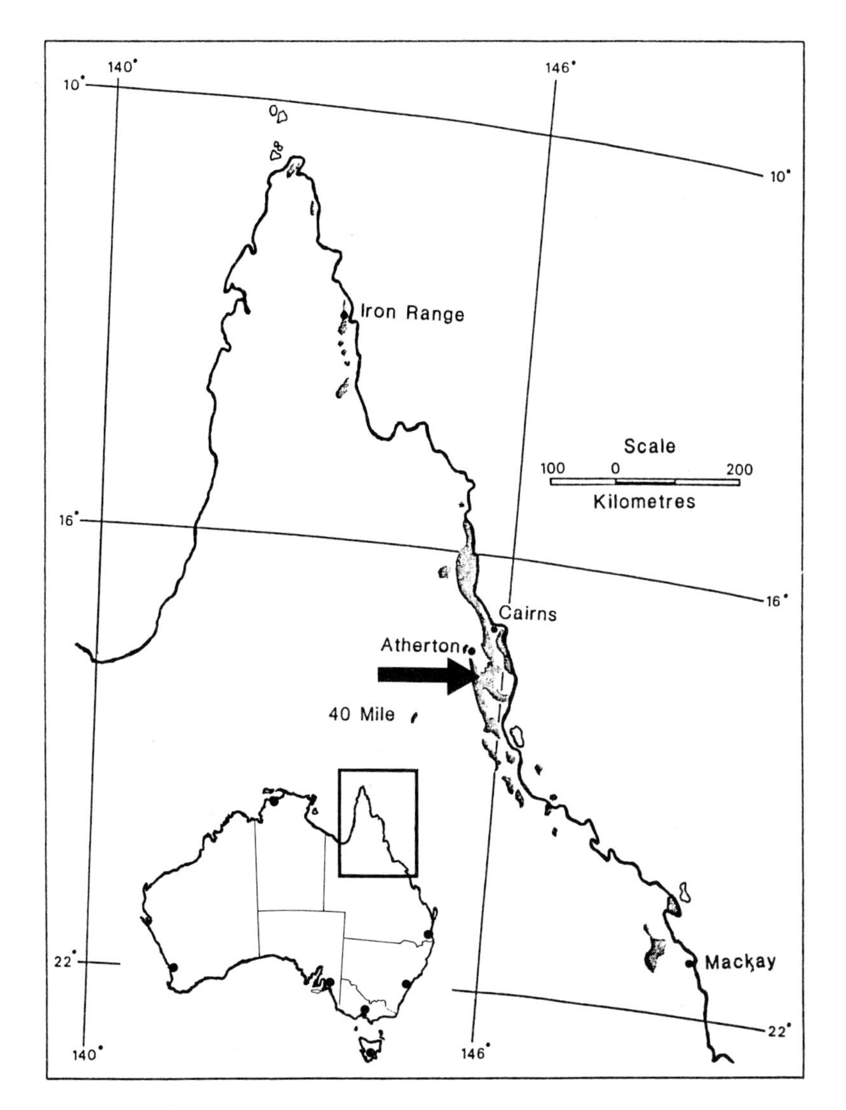

Fig. 1. Map of North Queensland with Australia inset showing site of gap regeneration experiment (arrow) and source of seed for irradiance × nutrient experiment (Atherton). Shaded areas represent existing rainforest (notional).

Materials and Methods

Forest Ecology

Field studies were concentrated in upland rainforest at 12 comparable locations on a north-facing slope along a spur of Herberton Range (lat. 17°29′ S., long. 145°30′ E.—see arrow in Fig. 1). Four locations were established in year 1 (1979). Eight additional sites were set up over two subsequent years.

Undisturbed forest in that vicinity was dominated by *Cardwellia sublimis, Flindersia pimenteliana* and *Doryphora aromatica*. Mean annual precipitation was 1425 mm year^{-1} with monthly totals ranging between 22 and 300 mm depending upon season (November–April wet season). Diurnal temperature range (annual mean max./min.) was 26/14°C with warmest months 29/17 and coldest months 22/10°C. Daily total radiation varied from *c.* 12 to 25 MJ m^{-2} day^{-1} according to season (max. values November–January).

Selective logging in Herberton Ranges during a preceding dry season had produced a variety of disturbance gaps (amidst undisturbed patches of up to 0·5 ha) that were subsequently exploited in present studies on tree seedling establishment. Sites were selected within gaps which had been ranked as wide, medium or narrow based upon cumulative annual receipt of photosynthetically active radiation at ground level, viz. 40, 9 and 0·6% of full sun for wide, medium and narrow gaps respectively. These values were derived at each site from *in situ* measurement of photon irradiance via Telefunken BPW-21 sensors, each connected to integrating 'E' cells of Plessey Ducon P/L, and housed in upturned test tubes of 20 mm internal diameter with inside bottom surfaces sandblasted to act as hemispherical diffusers. Cosine response was within 3% of ideal for angles of incidence up to 75°. Each unit was self-contained, weather-proof, and secured to a rigid stand at 50 cm above ground.

Fresh seed of high viability was obtained from *Acacia aulacocarpa, Toona australis, Flindersia brayleyana* and *Darlingia darlingiana*, representing a full range of gap-colonising types, and scattered within 1 m^2 test plots located in wide, medium and narrow gaps. A wire mesh cage (1·5 cm mesh size) was secured in place over each plot to ensure good prospects for germination free of seed predators and ground foragers for an initial 6 months. Germination was scored and seedling growth or failure was recorded after 4 months. Tree growth and general dynamics of these mixed populations were recorded over a further 7 years to March 1987.

Photosynthesis in Flindersia brayleyana

Seed from Atherton Tableland was germinated under mist in trays containing equal parts by volume of vermiculite, perlite and sand in January 1987. Seedlings were potted into 15 cm pots of sand, placed into growth cabinets, and randomly assigned to treatments when the first true leaf appeared (about 3 weeks). Six treatments, each with three replicates were set up: three irradiance regimes, low, medium and high (30, 130 and 535 μmol quanta m^{-2} s^{-1}) and two nutrient regimes, high ($\frac{1}{3}$ strength) and low ($\frac{1}{200}$ strength modified Hoagland solution) (Went 1957). Composition of full strength nutrient solution in mg l^{-1} was: $Ca(NO_3)_2.4H_2O$, 950; KNO_3, 610; $MgSO_4.7H_2O$, 490; $NH_4H_2PO_4$, 120; H_3BO_3, 0·6; $MnCl_2.4H_2O$, 0·4; $ZnSO_4.7H_2O$, 0·09; $CuSO_4.5H_2O$, 0·05; $H_2MoO_4.4H_2O$, 0·02; $Co(NO_3)_2.6H_2O$, 0·025; NaOH, 5; EDTA, 33·2; $FeSO_4.7H_2O$, 24·9. These nutrient solutions provided 71 mg l^{-1} and 1·0 mg l^{-1} of nitrogen (N) and 11 mg l^{-1} and 0·2 mg l^{-1} of phosphorus (P) respectively. Excess nutrient solution was given daily. Photoperiod was 12 h and air temperatures were 25/15°C day/night. Daytime leaf temperatures corresponding to low, medium, and high irradiance were 22, 26 and 28°C, respectively.

High irradiance treatments (23 mol quanta m^{-2} day^{-1}) approximate annual mean daily quantum flux incident upon a rainforest canopy on the Atherton Tableland, where two-thirds of summer days are overcast. Low irradiance treatments (1·3 mol quanta m^{-2} day^{-1}) correspond to the forest floor daily quantum flux. Up to 98% attenuation of sunlight between canopy top and forest floor was observed for upland rainforest on Atherton Tableland (Thompson, unpublished). Midcanopy conditions are represented by medium irradiance treatments (5·6 mol quanta m^{-2} day^{-1}).

Laboratory gas exchange of single attached leaves on *Flindersia* seedlings was measured with a double sided clamp-on cuvette (11 × 30 mm aperture) showing boundary-layer conductance of 1·0 mol m^{-2} s^{-1} and coupled to an open system. Leaf temperature (fine wire copper–constantan thermocouple in contact with lower surface) was held steady at 20·5 ± 0·2°C. Tree seedling and cuvette were housed within a 160-litre thermostatted chamber (20·5 ± 1·5°C) and illuminated by a 400-W quartz iodide lamp (Woton, West Germany). Irradiance incident upon test leaves was adjusted via lamp distance in combination with neutral density filters and measured with a LI-COR quantum sensor. CO_2 partial pressure of gas supplied to leaves was controlled by injecting via mass-flow controllers 20% CO_2 in N_2 into a flow of CO_2-scrubbed air, which was then humidified to a set dew point. Flow rate and relative humidity of gas supply were adjusted to ensure that CO_2 draw-down across photosynthesising leaves never exceeded 9 μbar and leaf–air vapour pressure difference was maintained at 8·0 ± 0·5 mbar.

H_2O and CO_2 fluxes were estimated from concurrent and independent measurements using a Cambridge dewpoint hygrometer (EG&G model 660) and ADC MK III CO_2 analyser in differential mode. Leaf gas exchange parameters were derived according to calculations outlined by von Caemmerer and Farquhar (1981).

Photosynthesis-irradiance response curves (starting at 1000 and decreasing in six steps to 0 μmol quanta $m^{-2} s^{-1}$ were measured on a fully expanded leaf of each plant (third or fourth leaf from base). Photosynthesis-CO_2 response curves were then measured at 550 μmol quanta $m^{-2} s^{-1}$ for the plant giving median irradiance response in each treatment (starting at 900 and decreasing in seven steps to 0 μbar CO_2). Leaves were excised immediately after gas exchange for specific leaf fresh weight (1 cm^2 disc). Chlorophyll content of the same tissue was subsequently extracted into *N,N*-dimethylformamide and kept in darkness at 4°C for up to 2 weeks. Clear extracts were used for absorbance measurement at 664·5 and 647·0 nm. Chlorophyll $a+b$ (Chl) was calculated according to Inskeep and Bloom (1985). Remaining leaf tissue was dried 72 h at 70°C (forced draught oven), weighed, and analysed for N and P following Kjeldahl digestion.

Leaf photosynthesis parameters and component processes were derived as follows. Assimilation-irradiance response data provided direct estimates of maximum photosynthesis (A) under saturating irradiance (1000 μmol quanta $m^{-2} s^{-1}$) at ambient CO_2 (340–350 μbar) and of dark respiration (R_{dark}). Apparent quantum yield of CO_2 to incident irradiance (α) was calculated as the slope of the irradiance response curve at low light (<50 μmol quanta $m^{-2} s^{-1}$). Light compensation point (Q_0) was calculated as

$$Q_0 = R_{dark}/\alpha. \qquad (1)$$

To estimate light saturation point (Q_l), defined as the minimum irradiance necessary to achieve light saturated assimilation rate, we fitted a truncated rectangular hyperbola to each set of assimilation-irradiance measurements using non-linear regression.

$$P = \min\{\alpha(Q-Q_0)A^*/(\alpha[Q-Q_0]+A^*), A\}, \qquad (2)$$

where P is assimilation rate, min { } means take the minimum value, Q is incident irradiance, and A^* is a curvature parameter.

Then

$$Q_l = A^*A/(\alpha[A^*-A])+Q_0. \qquad (3)$$

While this approach gave an objective estimate of light saturation, we attribute no particular significance to the parameter A^* and do not report it. The parameters A, R_{dark}, α, Q_0 and Q_l were calculated for each of three replicates per irradiance × nutrient treatment.

Maximum rates of carboxylation and electron transport were estimated with the following equations (von Caemmerer and Farquhar 1981) and parameters (Brooks and Farquhar 1985; Badger and Collatz 1977; Jensen and Bahr 1977; Jordan and Ogren 1981; Yeoh *et al.* 1981; Seemann and Berry 1982). Light saturated CO_2 assimilation rate is limited by RuP_2 carboxylation at low CO_2

$$P = V_c(p_i-\Gamma^*)/(p_i+K_c[1+O/K_o])-R_{day} \qquad (4)$$

and by RuP_2 regeneration at high CO_2

$$P = J(p_i-\Gamma^*)/(4\cdot 5p_i+10\cdot 5\Gamma^*)-R_{day}, \qquad (5)$$

where Γ^* is CO_2 photocompensation point (Laisk 1977) (35·4 μbar CO_2); p_i is intercellular CO_2 partial pressure (assumed here to be equal to CO_2 partial pressure at the site of carboxylation); R_{day} is non-photorespiratory CO_2 evolution in the light; V_c is maximum carboxylation rate; O is intercellular partial pressure of oxygen (210 mbar); K_c is Michaelis-Menten constant for RuP_2 carboxylation (252 μbar CO_2); K_o is Michaelis-Menten constant for RuP_2 oxygenation (192 mbar O_2); and J is whole-chain electron transport rate.

Maximum carboxylation rate (V_c) was estimated from the linear (low CO_2; $20<p_i<160$ μbar CO_2) end of the photosynthesis-CO_2 response curve (equation 4). Maximum electron transport rate (J) was estimated from the highest assimilation rate observed (external CO_2 *c.* 900 μbar) on the photosynthesis-CO_2 response curve (equation 5). R_{day} needed for this calculation was estimated from equation (4) as P at p_i = 35·4 μbar.

Results

Comparative Growth of Trees in Gaps

Four months after sowing, all test species were present as seedlings in wide gaps and making steady growth. *Acacia* and *Toona* were thriving despite competition from herbaceous weeds. Medium gaps were sustaining *Toona, Flindersia* and *Darlingia*, but

Acacia seedlings had already failed. Height growth after 4 years is shown in Fig. 2. *Acacia* was clearly dominant in wide gaps, having achieved an early rate of growth that far exceeded companion species. By contrast, *Acacia* failed entirely in both medium and narrow gaps where either *Toona* or *Darlingia* assumed prominence.

Fast initial growth observed in *Acacia* (Fig. 2) correlates with a relatively short lifecycle (30–40 years) in this early successional species. Height rankings after 4 years are therefore likely to change with time. After 7 years *Toona* (10 m) overtopped *Acacia* and self-sown *Alphitonia* invaded wide gaps. In medium gaps, *Darlingia* had reached 3 m and had overtopped *Toona*. *Flindersia* showed comparable capacity for height growth which eventually exceeded that for *Toona* in medium gaps. *Toona* maintained superiority in wide gaps.

Seedling persistence in narrow gaps carries significant implications for rainforest dynamics and especially for site recolonisation following minor disturbance. After 7 years in the deep shade of narrow gaps, where all other test species had failed, *Darlingia* seedlings remained. Although populations had fallen to *c*. 50% of original size due to

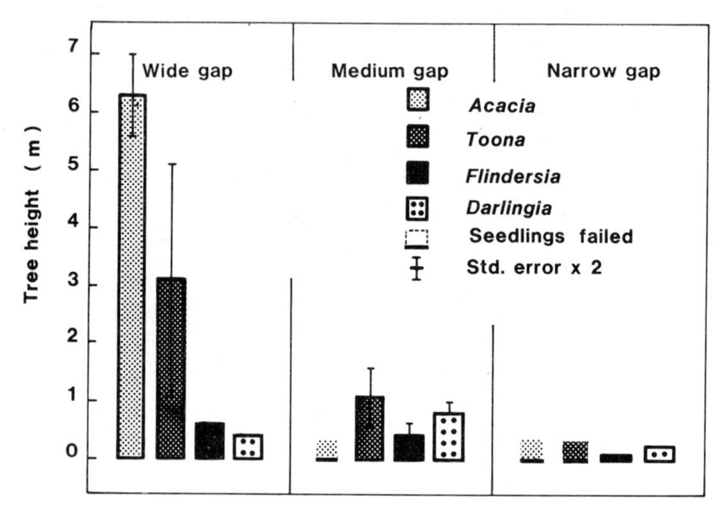

Fig. 2. Comparative height growth four years after seedling establishment of *Acacia aulacocarpa, Toona australis, Flindersia brayleyana,* and *Darlingia darlingiana* in three categories of rainforest disturbance gaps on Herberton Range in North Queensland. Solid line on zero indicates no survival.

seedling attrition, and height attained was still only 15 cm, a continuing potential for establishment of this species was evident.

As a further example of developmental plasticity during establishment phase on new sites, both *Darlingia* and *Flindersia* persisted under low sunlight regimen for at least 4 years (Fig. 2). Initial dominance in medium gaps by *Toona* over *Flindersia*, for instance, was reversed by year 7 and implies great capacity in *Flindersia* for leaf acclimation to irradiance. Since *Flindersia* does occur in mature forest as a canopy emergent, foliage must adjust from deep shade to full sun. Photosynthetic attributes which underlie such wide 'ecological amplitude' are examined below.

Growth and Photosynthesis in Flindersia brayleyana

Growth and biomass distribution

Seedling growth (whole plant biomass in Table 1) showed strong, non-additive response to treatments. For example, high irradiance in combination with high nutrient

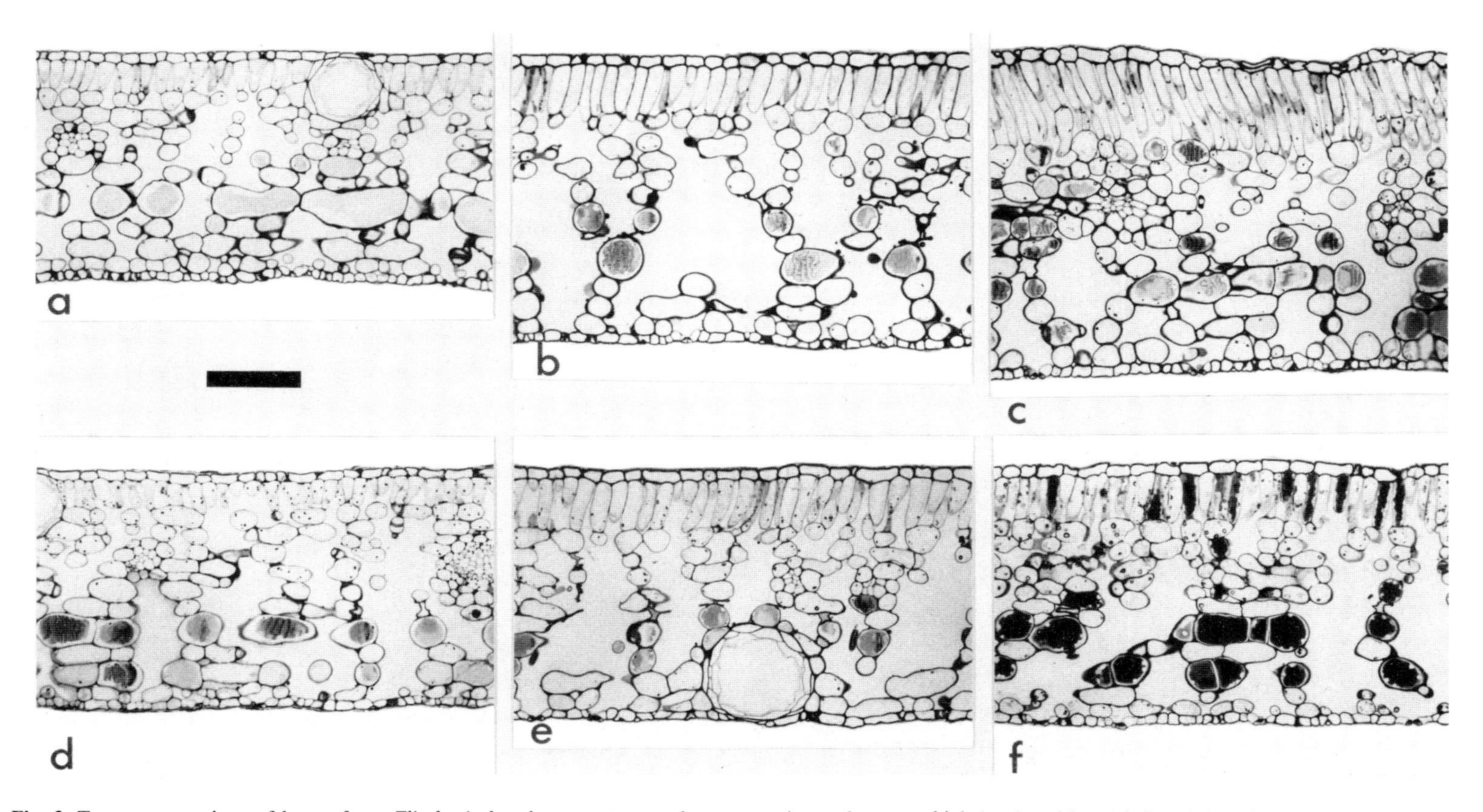

Fig. 3. Transverse sections of leaves from *Flindersia brayleyana* grown under two nutrient cultures — high (*a–c*) and low (*d–f*) and three irradiance cultures — low (*a,d*), medium (*b,e*) and high (*c,f*). For details of treatments see Table 1. Bar represents 100 μm.

elicited maximum growth, whereas high irradiance was supraoptimal for low nutrient plants. By contrast, growth of plants at low irradiance was unresponsive to nutrient supply. Presumably, seedling growth was irradiance- rather than nutrient-limited by the latter combination.

Biomass distribution also varied according to irradiance × nutrient supply. Root growth, relative to whole plant (tops : root ratio in Table 1), was enhanced under low

Table 1. Growth and biomass distribution in *Flindersia brayleyana* seedlings after 180 days in growth cabinets under low (1·3) medium (5·6) and high irradiance (23·1 mol quanta m^{-2} day^{-1}) at low (76) and high nutrient supply (5030 mmol N m^{-3})

Data shown are mean values from three replicates

	High nutrients			Low nutrients		
	Growth irradiance					
	Low	Med.	High	Low	Med.	High
Whole plant (g dry wt)	0·8	6·6	7·3	1·0	3·0	1·6
Tops : roots ratio	5·2	4·4	5·5	2·7	1·8	1·6
Leaf area (cm^2)	130	600	530	110	200	50
Specific leaf wt (mg dry wt cm^{-2})	4·2	6·4	8·1	4·7	7·5	9·9
Leaf area : whole plant biomass ratio (cm^2 g^{-1} dry wt)	163	91	73	110	67	31

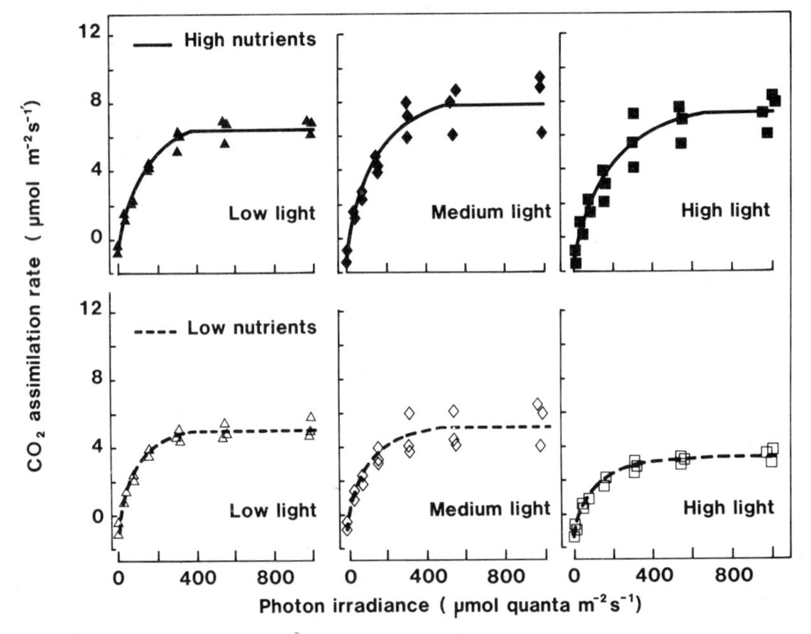

Fig. 4. Response of leaf photosynthesis in air (340–350 μbar CO_2) to photon irradiance in *Flindersia* grown at three irradiance × two nutrient treatments. Data for each of three replicate seedlings are shown, except where data points coincide. Curves were fitted by regression to equation (2). Derived parameters are shown in Table 2.

nutrient supply and especially at high irradiance. Leaf expansion was restricted by low nutrient supply, a constraint which intensified as irradiance increased. For example, ratios between leaf areas of low to high nutrient treatments were 0·85, 0·33, 0·09 for low, medium, and high irradiance respectively. Leaf expansion peaked at medium irradiance in both nutrient treatments.

While leaf expansion diminished under low nutrient supply (especially at high irradiance), laminae so formed were more dense. Specific leaf weight (Table 1) reflects both irradiance and nutrient effects, especially at high irradiance. Anatomy of leaf transverse section confirms this pattern (Fig. 3). High irradiance produced thicker leaves with a deeper palisade layer. However, greater specific leaf weight at low nutrients cannot be attributed solely to leaf thickness (see Figs 3*c* and 3*f*). Dry weight to fresh weight ratio was greater under low nutrition than under high (viz. 0·37 cf. 0·30), indicating a greater tissue density.

Leaf photosynthesis: area basis

Responses of single leaf photosynthesis to incident irradiance are given in Fig. 4. Curves were derived by regression from equation (2) of pooled data from three replicate plants. Saturating irradiance increased with irradiance during growth, but varied little in terms of nutrient supply. By contrast, both initial slope and light-saturated photosynthesis (A) showed a strong response to irradiance × nutrient treatment (Table 2).

Dark respiration (R_{dark}) and light compensation points responded to irradiance during growth, rather than to nutrient supply. Decrease in light compensation point from high to low irradiance treatment reflects respiratory adjustments which appear to be independent of nutrient supply, and which enable *Flindersia* to maintain a positive carbon balance over a wide range of irradiance and nutrient availability.

Table 2. Leaf gas exchange by mature foliage of *Flindersia brayleyana*

Data shown are mean values from three replicates; see Table 1 for treatment details

	High nutrients			Low nutrients		
	Growth irradiance					
	Low	Med.	High	Low	Med.	High
A (μmol CO_2 m^{-2} s^{-1})	6·5	8·0	7·2	5·1	5·4	3·4
R_{dark} (μmol CO_2 m^{-2} s^{-1}) (20·5°C)	0·7	1·1	1·3	0·7	0·9	1·1
α (apparent quantum yield: mol CO_2 mol^{-1} quanta incident on leaf)	0·043	0·054	0·046	0·042	0·047	0·029
Light compensation (μmol quanta m^{-2} s^{-1})	15	20	29	17	19	37
Light saturation (μmol quanta m^{-2} s^{-1})	380	440	710	340	510	650

Seedlings grown on low nutrient supply and high irradiance showed lowest values for A and apparent quantum yield (Table 2). Photosynthetic dysfunction due to high irradiance was clearly nutrient dependent. Taking α as an indicator of photosynthetic effectiveness, high nutrient enhancement [(high nutrient value – low nutrient value)/(low nutrient value),%] becomes more accentuated with increased irradiance, viz. 5, 15 and 59% for low, medium and high irradiance respectively. Comparing A between nutrient treatments on this same basis, enhancement values were 28, 55 and 118% respectively.

Photosynthesis and component processes: area basis

Photosynthesis–CO_2 response curves from single replicate seedlings which gave median response to irradiance (Fig. 4) are shown in Fig. 5. Corresponding data on component processes for those leaves are given in Table 3. Both initial slope (hence V_c)

and CO_2 saturated rates (hence J) are strongly influenced by nutrient supply. Both terms are consistently greater under high nutrients across all levels of irradiance. However, relative activity of J and V_c appeared to be highly conserved because the ratio of J/V_c remained stable (Table 3).

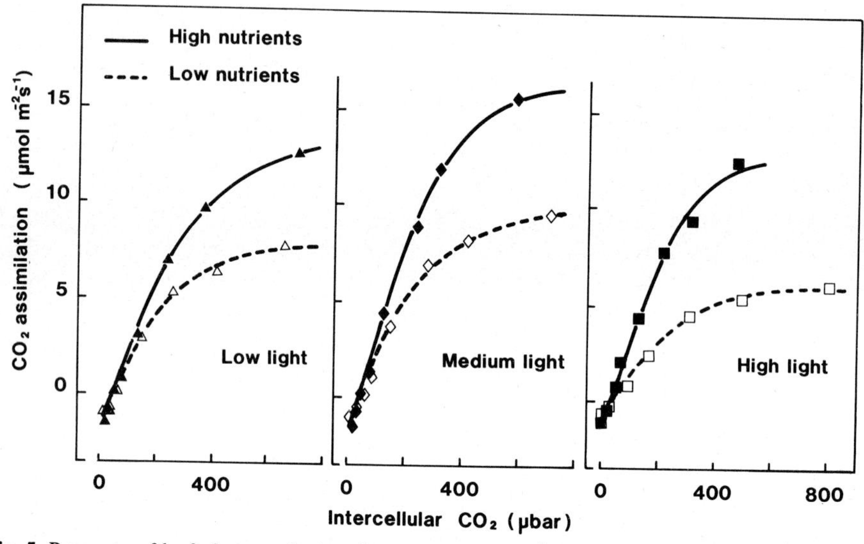

Fig. 5. Response of leaf photosynthesis to intercellular CO_2 partial pressure (irradiance 550 μmol quanta m^{-2} s^{-1}) in *Flindersia* grown at three light × two nutrient treatments. Data are from a single seedling per treatment which gave median response to irradiance (see Fig. 4). Derived parameters J and V_c are given in Table 3.

Photosynthesis and component processes: chlorophyll, nitrogen interrelations

Leaf chlorophyll content increased with higher nutrients irrespective of irradiance during growth (Table 4). Chlorophyll decreased with increasing irradiance for the low nutrient treatments. Greater chlorophyll content under high nutrient supply correlates with thicker palisade tissue (Fig. 3). Moreover, this contrast between nutrient supply treatments is strengthened when leaf chlorophyll is expressed on a tissue dry weight basis, viz. 15·9, 13·7 and 7·4 under high nutrients, compared with 12·6, 5·7 and 2·6 (μg Chl $a + b$ g^{-1} dry wt) under low nutrients for low, medium and high irradiance

Table 3. Photosynthetic component processes in mature foliage of *Flindersia brayleyana*

Gas exchange source data came from median leaves of Fig. 4; same leaves correspond to $A(p_i)$ curves in Fig. 5. Equations (4) and (5) in Methods provided the basis for present derivations of partial processes. See Table 1 captions for treatment details

	High nutrients			Low nutrients		
	Growth irradiance					
	Low	Med.	High	Low	Med.	High
Photosynthetic electron transport (J) (μmol e^- m^{-2} s^{-1})	64	86	65	39	49	34
Carboxylation capacity (V_c) (μmol CO_2 m^{-2} s^{-1})	23	32	31	19	22	13
J/V_c (mol e^- mol^{-1} CO_2)	2·8	2·7	2·1	2·1	2·2	2·6

respectively. Chlorophyll *a/b* ratio was marginally higher under high nutrients but revealed no systematic irradiance effect (ratio varied between 2·46 and 2·62 under high, cf. 2·24 and 2·37 under low nutrient supply).

Leaf N content (Table 4) was greatly influenced by nutrient supply and irradiance during growth. Ratios of leaf N under high : low nutrient treatments were 1·6, 1·8 and 2·0 for low, medium and high irradiance respectively. Leaf P varied with nutrient supply, but not with irradiance. Chlorophyll content was correlated with leaf nitrogen content, but the Chl/N ratio decreased as irradiance increased (Table 4).

Table 4. Leaf chlorophyll and nutrient levels in *Flindersia brayleyana* leaves used for detailed analysis presented in Tables 3 and 5

See Table 1 caption for treatment details

	High nutrients			Low nutrients		
	Growth irradiance					
	Low	Med.	High	Low	Med.	High
Chl $(a+b)$ (mmol m^{-2})	0·76	1·11	0·72	0·66	0·57	0·32
Total N (mmol m^{-2})	94	144	156	61	81	77
Total P (mmol m^{-2})	5·8	5·2	4·8	0·9	1·0	1·2
Chl $(a+b)$/N (mmol mol^{-1})	8·0	7·7	4·6	10·8	7·1	4·2

Photosynthetic responses to growth irradiance and nutrient supply are shown in Table 5 as a function of leaf N and Chl level. Increased irradiance lowered CO_2 assimilating activity per unit N. By contrast, leaf assimilation per unit of chlorophyll, J/Chl and $V_{c,max}$/Chl tended to increase in response to irradiance during growth.

Table 5. Photosynthetic component processors on a Chl and N basis in mature foliage of *Flindersia brayleyana*

See Table 1 for further details on treatment

	High nutrients			Low nutrients		
	Growth irradiance					
	Low	Med.	High	Low	Med.	High
A/Chl$(a+b)$ (mmol CO_2 mol^{-1} Chl s^{-1})	8·6	7·2	10·0	7·7	9·5	10·6
J/Chl[A] (mmol CO_2 mol^{-1} Chl s^{-1})	87	74	102	64	82	100
V_c/Chl[A] (mmol CO_2 mol^{-1} Chl s^{-1})	31	27	49	31	37	38
J/N[A] (mmol CO_2 mol^{-1} N s^{-1})	0·72	0·53	0·42	0·69	0·53	0·40
V_c/N[A] (mmol CO_2 mol^{-1} N s^{-1})	0·26	0·20	0·20	0·33	0·24	0·15

[A]Data from specific leaves used for $A(p_i)$ curves in Fig. 5.

A more explicit presentation of chlorophyll and light saturated photosynthesis versus leaf N is given in Fig. 6. For any given leaf nitrogen, both chlorophyll and photosynthesis are generally greater in seedlings grown under low or medium irradiance (upper lines, Figs 6*a* and 6*b*). High irradiance + low nutrient plants (open squares) have more leaf N but less chlorophyll and lower rates of CO_2 assimilation than their low irradiance counterparts (open triangles). Electron transport rate per unit N also declined with increasing growth irradiance (Table 5). A corresponding response of carboxylation capacity per unit N to irradiance treatment was not so clear (Table 5). However, the strong correlation between electron transport and carboxylation rates suggests that they vary in concert (Table 3).

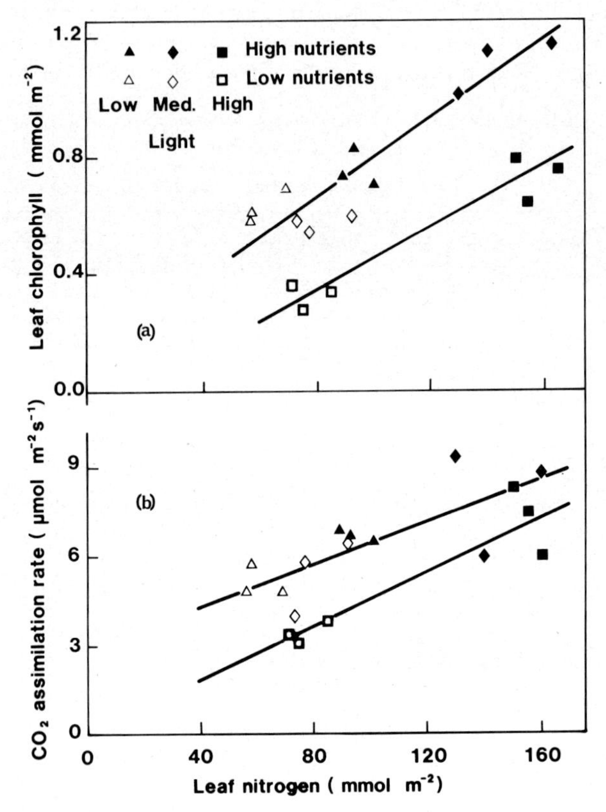

Fig. 6. Chlorophyll content (*a*) and CO_2 assimilation at saturating irradiance and ambient CO_2 (*b*), both as functions of leaf nitrogen. Regression equations are: (*a*) low and medium growth irradiance, Chl = 0.2050 + 0.0060N (r^2 = 0·80) and high growth irradiance, Chl = −0.0479 + 0.0049N (r^2 = 0·92); and (b) low and medium growth irradiance, A = 2.8020 + 0.0362N (r^2 = 0·60) and high growth irradiance, A = 0.0043 + 0.0456N (r^2 = 0·80).

Discussion

Ecology of Photosynthesis in Flindersia

Rainforest trees vary widely in capacity for sun–shade acclimation. The sun-loving species *Acacia aulacocarpa* proved incapable of acclimation to shade. It was preeminent amidst mixed populations when sown in wide gaps but failed entirely in medium or narrow gaps. By contrast later successional, shade-tolerating tree seedlings such as *Flindersia brayleyana* were able to persist under a wide range of available sunlight (down to 0·6% of full sun equivalent). Yet shade-grown *Flindersia* retains a latent capacity to acclimate from deep shade to full sun following either disturbance of closed forest or canopy emergence during later stages of succession (Unwin, personal communication, expt. No. EP14, CSIRO, TFRC, Atherton).

Species distinctions in gap colonisation on Herberton Range reported here coincide with results from a greenhouse experiment at Atherton where rainforest tree seedlings were held along a gradient in sunlight supply (Duff and Stocker, unpublished). In that experiment *Flindersia brayleyana* showed a broad tolerance to irradiance. Seedling growth rate showed a sigmoidal increase as light increased from 0·6% to 40% full sun under well nourished conditions (200 ml week^{-1} of 21·9 mol m^{-3} N) but a well defined optimum at 15% full sun when grown at one-fifth of well nourished conditions. Present data for *Flindersia* in growth cabinets (Table 1) confirm such irradiance × nutrient interactions for whole plant growth. High irradiance (535 μmol quanta m^{-2} s^{-1}) promoted maximum biomass at high nutrient supply (a *daily excess* of 5 mol m^{-3} N), whereas 130 μmol quanta m^{-2} s^{-1} proved optimum at lower nutrient supply (daily excess of 0·08 mol m^{-3} N). By extrapolation, nutrient supply on disturbance gaps in rainforest should therefore influence competitive ranking of *Flindersia vis a vis* other gap colonisers across a range of gap sizes. In particular, total daily irradiance in wide gaps (at low latitude such as Herberton Range) is likely to depress growth on depauperate sites, possibly because of photosynthetic dysfunction.

Some photosynthetic attributes of this shade-adapted species and leaf-level adjustments which underlie nutrient dependent acclimation from shade to sun are outlined below.

Shade Leaf Attributes

Light saturated photosynthesis rates in air (A) reported here in leaves of well nourished *Flindersia* are comparable to leaves of other rainforest trees (Langenheim *et al.* 1984 and citations therein; Oberbaurer and Strain 1984; Pearcy and Franceschi 1986) and evergreen perennial angiosperms, which in turn are somewhat lower than deciduous perennials and substantially less than herbs and grasses (Larcher 1980). Values for A recorded in Table 2 for cabinet-grown seedlings accord well with subsequent measurements on mature foliage of *Flindersia* in the Atherton Tropical Forest Research Centre arboretum using an ADC portable gas exchange system (Analytical Development Co., Hoddesdon, U.K.). Light-saturated rates in a 17-m tree ranged from 4·2 to 7·7 μmol CO_2 m^{-2} s^{-1} with matching conductances from 178 to 217 mmol H_2O m^{-2} s^{-1}. Present observations on cabinet-grown seedlings should therefore bear close relevance to leaf behaviour in nature. *Flindersia* seedlings held under medium irradiance and high nutrition were rich in chlorophyll, attaining a remarkably high level (1·11 mmol Chl $a+b$ m^{-2}, Table 3) which matches values subsequently found for mature foliage in the Atherton arboretum (1·15 mmol Chl m^{-2}) and is comparable to levels reported for the rainforest tree *Agathis robusta* (Langenheim *et al.* 1984) and the rainforest shrub *Alocasia macrorrhiza* (Seemann *et al.* 1987) native to the same area. Yet compared to agricultural crop plants, all of these species show meagre assimilation rates for leaves so rich in chlorophyll. In contrast, *Flindersia's* maximum electron transport rate (J) of 86 μmol e$^-$ m^{-2} s^{-1} for medium irradiance and high nutrients accords well with 85 (same units) in well nourished peas grown under a similar irradiance (Evans 1987). Increase in leaf chlorophyll per unit nitrogen under low irradiance reflects an adjustment to deep shade. Similar results have been reported for *Phaseolus* and *Alocasia* (Seemann *et al.* 1987). While apparent quantum yield was not higher under low irradiance (Table 2), reduced respiration (R_{dark}) lowered the light compensation point. Such acclimation to deep shade is a prerequisite for maintenance of a positive carbon budget.

Nutrient-dependent Leaf Features

Laminae achieved maximum expansion under medium irradiance, with high nutrient supply amplifying that optimum. Leaf cell enlargement is typically enhanced by high nutrient supply (especially N, P, K and Mg; Kriedemann 1986 and literature cited therein) and is evident in *Flindersia* as a greater depth of palisade cells (Fig. 3). High nutrient supply predisposed towards thicker laminae but lower specific leaf weight, and thus lower density leaf tissue.

Acclimation from Deep Shade to Full Sun

Low, medium and high irradiance treatments in present growth cabinet experiments roughly match total daily irradiation of deeply shaded rainforest floor, mid-canopy conditions and exposed canopy conditions for Atherton Tableland rainforest, but provide a qualitatively different radiation load. Light saturated assimilation rate (A) and electron transport capacity (J) remained high relative to other treatments in seedlings grown under low irradiance, providing a latent capacity for rapid growth in disturbance gaps. Moreover, on the rainforest floor where sun flecks and patches represent a major fraction of total daily irradiation (Björkman *et al.* 1972; Pearcy and Calkin 1983; Chazdon 1986), this latent capacity would contribute significantly to daily carbon gain. High electron transport and carboxylation capacities in leaves otherwise acclimated to deep shade would contribute to rapid photosynthetic induction necessary for effective use of sun flecks (Pearcy 1988). By contrast, seedlings in cabinets under high irradiance were permitted no respite from sustained input over 12 h. Trees in nature, though subject to about the same total daily input, experience wide fluctuations in space and time. Thus, adverse effects of strong and sustained irradiance on leaf photosynthesis, especially on depauperate sites, may be ameliorated during periods of lower irradiance.

Growth irradiance *per se* exerted a predominating effect over dark respiration, light compensation point and light saturation point. Such effects have been widely reported for annual and perennial plants (Björkman *et al.* 1972; Boardman 1977; Björkman 1981). Other photosynthetic characteristics depend upon allocation of N-based resources to component processes of photosynthesis.

Of four patterns of response of quantum yield (α) and light-saturated photosynthesis (A) to growth irradiance identified by Anderson and Osmond (1987), we anticipated that *Flindersia* would best fit their second category — reduced quantum yield but substantially increased light-saturated photosynthesis under high (compared to low) growth irradiance. However, the response in *Flindersia* depended upon nutrition. Well nourished plants fit their third category — no response of α or A to growth irradiance — while poorly nourished plants matched their fourth category — reduction in both α and A at high growth irradiance.

Leaves acclimated to deep shade require less $\mathrm{Ru}P_2$ carboxylase, hence less N, to sustain light-saturated photosynthesis in that environment compared with individuals of the same species in full sun (Björkman 1981). High irradiance generally leads to greater N contents compared with shade-grown material when nutrient availability is high. However, this greater leaf N content of high-nutrient *Flindersia* did not increase photosynthetic capacity. As demonstrated in Fig. 6, low and medium light-grown seedlings from low nutrient cultures performed as well as those from high nutrient cultures, yielding greater light-saturated photosynthesis per unit leaf N than their high irradiance counterparts. In photosynthetic terms, shade-grown leaves achieved greater N-use efficiency. Since light interception rather than energy transduction or carboxylating capacity will more commonly limit assimilation under low irradiance, acclimation to deep shade, as prevails on a rainforest floor, should involve greater partitioning of N-based resources to components associated with light harvesting. However, both maximum electron transport and carboxylation rates per unit N were highest in leaves grown under low irradiance (Table 5). This finding contrasts sharply with observations on peas (Evans 1987) and spinach (Evans and Terashima 1987).

Additional nutrients will, however, lift N status of leaves, regardless of irradiance during growth, and photosynthetic processes respond accordingly (Table 4, Fig. 5). Indeed, the scale of that high-nutrient enhancement increased with irradiance; percentage response of α to high nutrition (Table 2) was 2, 15 and 59% for low, medium, and high irradiance respectively. By analogy, relative values for A in high-nutrient compared with low-nutrient cultures (Table 2) were 1·3, 1·5, and 2·1 (same sequence). As an alternative view to this concept of high-nutrient enhancement, low-nutrient cultures could be viewed as inducing a nutrient stress which is exacerbated under strong irradiance. Thus, the higher apparent quantum yield in high-nutrient compared to low-nutrient culture at high growth irradiance is consistent with observations of nutrient-dependent protection from photooxidative damage in *Solanum* transferred from low to high light (Ferrar and Osmond 1986). Our findings for *Flindersia* that photosynthetic capacity per unit N *declines* under high growth irradiance is the opposite of results for spinach (Terashima and Evans 1988), further suggesting that *Flindersia* suffers chronic photoinhibition at sustained high irradiance (see also Field and Mooney 1986; Evans 1988).

Implications for Rainforest Ecology

Deep shade in narrow gaps limits maintenance of a positive carbon budget and, *inter alia*, diminishes resistance of seedlings to pests and diseases (Augspurger 1984). Thus survival problems are compounded, especially for those species which lack the capacity for shade acclimation (e.g. wind-dispersed, early successional, wide gap colonisers). As a related input, low nutrient supply imposes no constraint on carbon gain in *Flindersia*, as shown by increased N-use efficiency of photosynthesis in deep shade. Nutrient supply in disturbed forest, however, does assume significance for successful colonisation of wide gaps where capacity to endure full sun can dictate survival. *Flindersia* seedlings again proved highly resilient in this respect and longer term responses to present growth cabinet treatments are especially noteworthy. High-irradiance cultures (both low-nutrient and high-nutrient supply) were maintained for a further 13 weeks (phase 2) following completion of an initial 180 days (phase 1) as reported here. High irradiance, though clearly supraoptimal during phase 1, was at least sublethal! Sustained growth was maintained throughout phase 2, even on low nutrients. Whole-tree biomasses at 270 days from single replicates were 36·4 g and 5·8 g for high-nutrients and low-nutrients respectively. Their ratio of 6·3 compares with a 4·7-fold differential following phase 1. Despite putative photoinhibition (especially in low-nutrient culture), a positive carbon balance was still achieved. Tree seedlings in wide gaps on depauperate sites would match that condition and would therefore benefit substantially from nutrient flushes which allegedly follow upon disturbance in such areas (Vitousek and Melillo 1979; Ewel *et al.* 1981). Competitive ranking of *Flindersia* would be enhanced even further over shade-adapted species such as *Darlingia darlingiana*.

Given such a broad range of tolerance to irradiance × nutrient supply in *Flindersia*, heightened sensitivity to strong irradiance would be anticipated for other narrow-gap colonisers such as *Argyrodendron* spp. Their response to nutrients should be even further compounded by irradiance level, but would contrast with an early successional wide gap coloniser such as *Toona australis* (Fig. 2 and Introduction). Comparative physiology of wide, medium and narrow gap colonisers is to be reported elsewhere, and will provide a basis for (1) interpreting changes in rainforest population dynamics as a function of prior disturbance regimen, and (2) preserving species richness in regrowth forest by manipulating nutrient supply. Present analyses on *Flindersia* imply that nutrient availability will also alter competitive rankings of other rainforest seedlings along gradients in sunlight supply.

Acknowledgments

We are indebted to Paul Thomas for his able assistance with plant culture and biomass analyses, to Dean Tompkins for his participation in leaf gas exchange measurements and to Jan Murrell for her help in chlorophyll assay. Studies of leaf anatomy and photomicroscopy were facilitated by excellent technical support from Ian Craig. We thank, also, Eva Morrow and Karin Munro for typing this paper faster and more accurately than we could write it. (All staff listed are from CSIRO Division of Forest Research.)

References

Anderson, J. M., and Osmond, C. B. (1987). Shade-sun responses: compromises between acclimation and photoinhibition. In 'Photoinhibition.' (Eds D. J. Kyle, C. B. Osmond and C. J. Arntzen.) Topics in Photosynthesis Vol. 9, pp. 1-38. (Elsevier: Amsterdam.)

Augspurger, C. K. (1984). Seedling survival of tropical tree species: interactions of dispersal distance, light gaps and pathogens. *Ecology* **65**, 1705-12.

Badger, M., and Collatz, G. J. (1977). Studies on the kinetic mechanisms of ribulose-1,5-bisphospate carboxylase and oxygenase reactions, with particular reference to the effect of temperature on kinetic parameters. *Carnegie Inst. Wash. Year Book* **76**, 355-61.

Björkman, O. (1981). Responses to different quantum flux densities. In 'Physiological Plant Ecology I. Responses to the Physical Environment'. (Eds O. L. Lange, P. S. Nobel, C. B. Osmond and H. Ziegler.) Encycl. Plant Physiol. New Ser., Vol. 12A, pp. 57-107. (Springer-Verlag: Berlin.)

Björkman, O., Boardman, N. K., Anderson, J. M., Thorne, S. W., Goodchild, D. J., and Pyliotos, N. A. (1972). Effect of light intensity during growth of *Atriplex patula* on the capacity of photosynthetic reactions, chloroplast components and structure. *Carnegie Inst. Wash. Year Book* **71**, 115-35.

Boardman, N. K. (1977). Comparative photosynthesis of sun and shade plants. *Annu. Rev. Plant Physiol.* **28**, 355-77.

Brooks, A., and Farquhar, G. D. (1985). Effect of temperature on the CO_2/O_2 specificity of ribulose-1,5-bisphosphate carboxylase/oxygenase and the rate of respiration in the light. *Planta* **165**, 397-406.

Caemmerer, S. von, and Farquhar, G. D. (1981). Some relationships between the biochemistry of photosynthesis and the gas exchange of leaves. *Planta* **153**, 376-87.

Chazdon, R. L. (1986). Light variation and carbon gain in rain forest ferns. *J. Ecology* **74**, 995-1012.

Evans, J. R. (1987). The relationship between electron transport components and photosynthetic capacity in pea leaves grown at different irradiances. *Aust. J. Plant Physiol.* **14**, 157-70.

Evans, J. R. (1988). Acclimation by the thylakoid membranes to growth irradiance and the partitioning of nitrogen between soluble and thylakoid proteins. *Aust. J. Plant Physiol.* **15**, 93-106.

Evans, J. R., and Terashima, I. (1987). Effects of nitrogen nutrition on electron transport components of photosynthesis in spinach. *Aust. J. Plant Physiol.* **14**, 59-68.

Ewel, J., Berish, C., Brown, B., Price, N., and Raich, J. (1981). Slash and burn impacts on a Costa Rican wet forest site. *Ecology* **62**, 816-29.

Ferrar, P. J., and Osmond, C. B. (1986). Nitrogen supply as a factor influencing photoinhibition and photosynthetic acclimation after transfer of shade grown *Solanum dulcamara* to high light. *Planta* **168**, 563-70.

Field, C. (1983). Allocating leaf nitrogen for the maximization of carbon gain: leaf age as a control on the allocation program. *Oecologia* **56**, 341-7.

Field, C., and Mooney, H. A. (1986). The photosynthesis-nitrogen relationship in wild plants. In 'On the Economy of Form and Function'. (Ed. T. J. Givnish.) pp. 25-55. (Cambridge University Press: Cambridge.)

Hirose, T., and Werger, M. J. A. (1987). Maximizing daily canopy photosynthesis with respect to the leaf nitrogen allocation pattern in the canopy. *Oecologia* **72**, 520-6.

Inskeep, W. P., and Bloom, P. R. (1985). Extinction coefficients of chlorophyll *a* and *b* in *N,N*-dimethylformamide and 80% acetone. *Plant Physiol.* **77**, 483-5.

Jensen, R. G., and Bahr, J. T. (1977). Ribulose-1,5-bisphosphate carboxylase-oxygenase. *Annu. Rev. Plant Physiol.* **28**, 379-400.

Jordan, D. B., and Ogren, W. L. (1981). Species variation in the specificity of ribulose bisphospate carboxylase-oxygenase. *Nature* **291**, 513-15.

Kriedemann, P. E. (1986). Stomatal and photosynthetic limitations to leaf growth. *Aust. J. Plant Physiol.* **13**, 15–31.

Laisk, A. (1977). Modelling of the closed Calvin cycle. In 'Biophysikalische Analyse Pflanzlicher System'. (Ed. K. Unger.) pp. 175–82. (Fischer: Jena.)

Langenheim, J. H., Osmond, C. B., Brooks, A., and Ferrar, P. J. (1984). Photosynthetic responses to light in seedlings of selected Amazonian and Australian rainforest tree species. *Oecologia* **63**, 215–24.

Larcher, W. (1980). 'Physiological Plant Ecology.' (Springer-Verlag: Berlin.)

Mooney, H. A., and Gulmon, S. L. (1979). Environmental and evolutionary constraints on the photosynthetic characteristics of higher plants. In 'Topics in Plant Population Biology'. (Eds G. T. Sollbrig, S. Jain, G. B. Johnson and P. H. Raven.) pp. 316–37. (Columbia University Press: New York.)

Oberbauer, S. F., and Strain, B. R. (1984). Photosynthesis and successional status of Costa Rican rain forest trees. *Photosynth. Res.* **5**, 227–32.

Pearcy, R. W. (1988). Photosynthetic utilisation of lightflecks by understory plants. *Aust. J. Plant Physiol.* **15**, 223–38.

Pearcy, R. W., and Calkin, H. (1983). Carbon dioxide exchange of C_3 and C_4 tree species in the understorey of a Hawaiian forest. *Oecologia* **58**, 26–32.

Pearcy, R. W., and Francheschi, V. (1986). Photosynthetic characteristics of C_3 and C_4 tree species grown in high and low light environments. *Photosynth. Res.* **9**, 317–31.

Richards, P. W. (1952). 'The Tropical Rain Forest: An Ecological Study.' (Cambridge University Press: London.)

Seemann, J. R., and Berry, J. A. (1982). Interspecific differences in the kinetic properties of RuBP carboxylase protein. *Carnegie Inst. Wash. Year Book* **81**, 78–83.

Seemann, J. R., Sharkey, T. D., Wang, J., and Osmond, C. B. (1987). Environmental effects on photosynthesis, nitrogen-use efficiency and metabolic pools in leaves of sun and shade plants. *Plant Physiol.* **84**, 796–802.

Stocker, G. C. (1981). Regeneration of a North Queensland rain forest following felling and burning. *Biotropica* **13**, 86–92.

Terashima, I., and Evans, J. R. (1988). Effects of light and nitrogen nutrition on the organization of the photosynthetic system in spinach leaves. *Plant Cell Physiol.* (In press.)

Tracey, J. G. (1982). 'The Vegetation of the Humid Tropical Region of North Queensland.' (CSIRO: Melbourne.)

Vitousek, P. M., and Melillo, J. M. (1979). Nitrate losses from disturbed forests: patterns and mechanisms. *For. Sci.* **25**, 605–19.

Went, F. H. (1957). 'The Experimental Control of Plant Growth.' pp. 78–9. (Chemica Botanica: Waltham.)

Whitmore, T. C. (1975). 'Tropical Rain Forests of the Far East.' (Clarendon Press: Oxford.)

Yeoh, H. H., Badger, M. R., and Watson, L. (1981). Variation in kinetic properties of ribulose-1,5-bisphosphate carboxylases among plants. *Plant Physiol.* **67**, 1151–5.

and 3·4% of the total annual photosynthesis. The principal differences between the daily totals of photosynthesis in the sun and upper shade portions of the crown occurred in early spring and late autumn, with relatively lower irradiances and carbon assimilation rates at the base of the crown. However, in summer, the midday rates of photosynthesis in the upper shaded portion of the crown tended to exceed those in the lower portion of the sun crown, this being associated with much greater water vapour pressure differences between leaf and air in the upper crown position. Other studies on temperate tree species (Schulze 1970; Watts *et al.* 1976) have led to similar conclusions. In general, there are marked differences between the photosynthetic characteristics of leaves produced at the upper, middle and lower canopy levels, these differences corresponding to the light environments in which the leaves developed.

This paper describes some of the photosynthetic attributes of leaves from different positions in the crown of a mature tree growing in the upper canopy of a northern Australian rainforest under conditions of adequate water supply.

Materials and Methods

The study site was located in Curtain Fig State Forest, Yungaburra, Queensland (17°17′S., 145°34′E., altitude 730 m). A specimen of *Argyrodendron peralatum* (F. M. Bailey) H.L. Edlin ex I.H. Boas 37 m tall, with a crown about 22 m deep and located close to a canopy access tower, was studied in detail. Branches were selected at heights above the ground of 32, 31, 30 and 29 m (upper crown), 17 m (lower crown) and 6 m (understory). The first five branches were on one tree, and the branch at 6 m was from a sapling growing close by.

Physiological observations were carried out near the end of a prolonged dry season, but immediately after two rain events. A storm resulted in 22·6 mm of rain at Yungaburra recording station, 2 km from the study site, and approximately 25 mm at the site itself. Four days later, a thunderstorm resulted in 22·8 mm of rain at the site, and cloud and light rain persisted a further 5 days. Air temperature, relative humidity and irradiance were measured automatically in the open near the top of the canopy (36 m), below the level of the main crown of the study tree (22 m) and in the understory (10 m) at intervals of 10 min. Data were recorded on audio tape and transferred to digital format for processing.

Carbon dioxide and water vapour exchange were measured on 5 days following the rain events. On the first 3 days, measurements were made at six levels in the crown, and on the final 2 days at the upper two levels only. Branches were selected that were accessible from the tower, and had at least five leaves placed so that one of the leaflets could be enclosed in an assimilation chamber without alteration of its orientation. This requirement resulted in branchlets near the extremities of one side of the crown being used, but this was not considered to be a serious disadvantage as the solar track passed almost directly overhead. Leaves to be measured were selected at random at the beginning of the study, and the same leaves were used throughout except for two that became damaged. Some disturbance to leaf orientation was inevitable, particularly when wind caused branches to oscillate up to 1 m relative to the position of the tower. The illumination of leaves was not manipulated, and the illumination histories of the leaves immediately prior to measurement were not recorded. Large fluctuations in illumination often occurred during the course of measurement of photosynthesis under windy conditions.

Photosynthesis was estimated using a LI6000 portable photosynthesis system (Li-Cor Ltd, Lincoln, Nebraska) fitted with a 0·25-litre chamber. A leaf area of up to 12 cm^2 was enclosed in the chamber so that free air movement was possible both above and below the leaf. Five leaves were examined at each measuring position. The manufacturers' recommended procedures were followed to minimise instrument vibration and the entrainment of carbon dioxide exhaled from operators in the gas circulation system. Transpiration was estimated independently by a LI1600 steady state water vapour diffusion porometer (Li-Cor Ltd), using the same leaves as selected for photosynthetic measurement.

Results

Several of the physiological characteristics to be described below were associated consistently with leaf position in the crown, and consequently with leaf structure and orientation. Leaves of *A. peralatum* are trifoliate and 6–8 leaves are carried in a bunch at the end of each short ultimate branch. The laminae of young leaflets are folded upwards about the midrib, completely enclosing the astomatous adaxial surface. Leaf

Spatial and Temporal Distribution of Photosynthesis and Transpiration by Single Leaves in a Rainforest Tree, *Argyrodendron peralatum*

D. Doley[A], *G. L. Unwin*[B] *and D. J. Yates*[A]

[A]Department of Botany, University of Queensland, St Lucia, Qld 4067, Australia.
[B]CSIRO Tropical Forest Research Centre, P.O. Box 780, Atherton, Qld 4883, Australia.

Abstract

The light environments and rates of photosynthesis by leaves of *Argyrodendron peralatum* (F. M. Bailey) I.H. Boas were examined at eight positions within a natural rainforest canopy. Daily integrals of irradiance at leaf surfaces expressed as a percentage of that reaching a horizontal sensor in the open decreased from about 50% in the uppermost leaves to about 20% at a point 2 m below the top of the crown and to 10–15% in the lower crown and understory. Leaves in the upper crown had maximum instantaneous rates of apparent photosynthesis of 11·5 μmol CO_2 m^{-2} s^{-1}, whilst leaves in the lower crown of the same tree, and in understory saplings had maximum photosynthesis rates of 8·2 μmol m^{-2} s^{-1}. These rates were not significantly different, and are within the range observed in a number of rainforest species characteristic of later successional stages. Under conditions of adequate water supply, photosynthetic rate was light-saturated by 500 μmol quanta m^{-2} s^{-1}, irrespective of level in the canopy.

Introduction

Tropical rainforests are notable for their high species diversity and very low irradiances near ground level (Whitmore 1975). The structural and functional attributes of leaves differ according to the light environment in which they develop, and according to the successional stage occupied by the species (Bazzaz and Carlson 1982; Mooney *et al.* 1984; Oberbauer and Strain 1985; Kwesiga *et al.* 1986). Species occurring in the early stages of forest succession have high rates of photosynthesis (about 15 μmol CO_2 m^{-2} s^{-1}), later stage tree species have a wide variety of rates (about 4–15 μmol m^{-2} s^{-1}), and understory species have low maximum rates (about 1–3 μmol m^{-2} s^{-1}). Higher light saturation and higher light compensation points are associated with early stage species or leaves developed in the sun than with late stage or understory species. Leaves of early successional species are commonly short-lived, whilst leaves of late successional or understory species may persist for several years.

Pearcy (1988) recently demonstrated consistent differences between the photosynthetic responses of leaves of individual species that are displayed in the upper canopy or the understory. Daily integrals of photosynthesis by leaves in the upper canopy, understory, and on the edge of a large gap were 0·25, 0·024 and 0·097 mol CO_2 m^{-2} day^{-1}, respectively.

The variations in photosynthetic characteristics of leaves of a single species at different positions in the canopies of temperate forests have been described in detail, particularly for coniferous species. For example Schulze *et al.* (1977) found that, in *Picea abies*, the sunlit portion of the crown represented 63·3% of the total leaf weight, but accounted for 70·9% of the total annual photosynthesis. In contrast, the lower shaded and transition portions of the crown represented 6·4% of the total leaf weight,

0310-7841/88/010317$03.00

expansion is rapid, and at this stage the leaflets are very soft in texture and bronze in colour, and tend to droop about the petiole. Subsequent development is associated with unfolding, flattening and greening of the lamina, and its elevation towards a horizontal position. At maturity, leaves in the upper crown are generally steeply inclined relative to the horizontal, whilst those in the understory are almost horizontal. Leaf texture also varies, the upper crown leaves being stiff, leathery and markedly cupped and with a slightly revolute margin, whilst understory leaves are thin, flat and flexible. These differences in leaf orientation can be attributed to the development of low turgor potentials for longer periods of time in upper crown leaves exposed to direct sunlight than in the lower, more protected positions in the crown or in the understory. Some consequences of these differences in orientation will be discussed below.

Leaves were studied in their natural positions, so that the irradiances intercepted by leaves were not the same as those intercepted by horizontally placed quantum sensors.

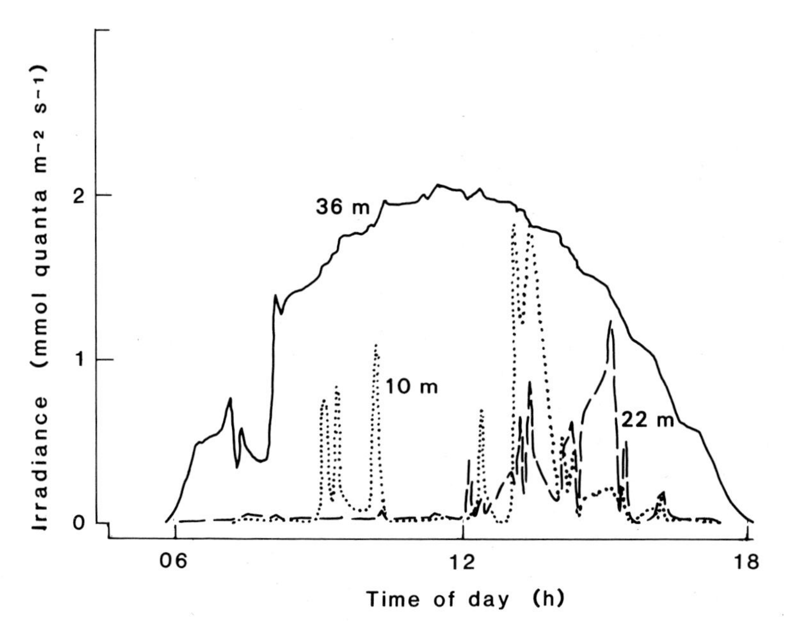

Fig. 1. Daily coures of irradiance reaching horizontal sensors above the canopy of a northern Australian rainforest at 36 m from the ground, beneath the upper canopy at 22 m and in the understory at 10 m on an almost cloudless day.

The daily courses of irradiance at the three sensors at 36, 22 and 10 m on an almost cloudless day are presented in Fig. 1. At the four upper branch positions, leaves were exposed to direct sunlight for a total of several hours per day, and no leaf was either completely exposed or completely shaded. The sensors at the lower crown (22 m) and understory (10 m) positions were exposed to direct sunlight for at least one period of 30 min duration on sunny days, and the pattern of irradiance was one of fluctuation from very low to substantially higher levels, rather than one of a uniform low level.

Fig. 2 shows the relationship between the irradiance incident on leaf surfaces in their natural orientations at the two uppermost branch positions (32 and 31 m) and at a horizontal sensor placed above the forest canopy on an almost cloudless day. The horizontal sensor data were obtained from observations made at 10-min intervals throughout the day, and the leaf data were obtained from the mean irradiance of five leaves in a cluster on each measuring occasion. The branches measured were located on the western side of the crown, so the relationship between leaf and sensor irradiances differed between morning and afternoon. Because of the orientation of leaves, they were nearly normal to the solar beam at about 1500 h but, even when they were in full sunlight in the morning, they intercepted as little as 50% of the radiation reaching a

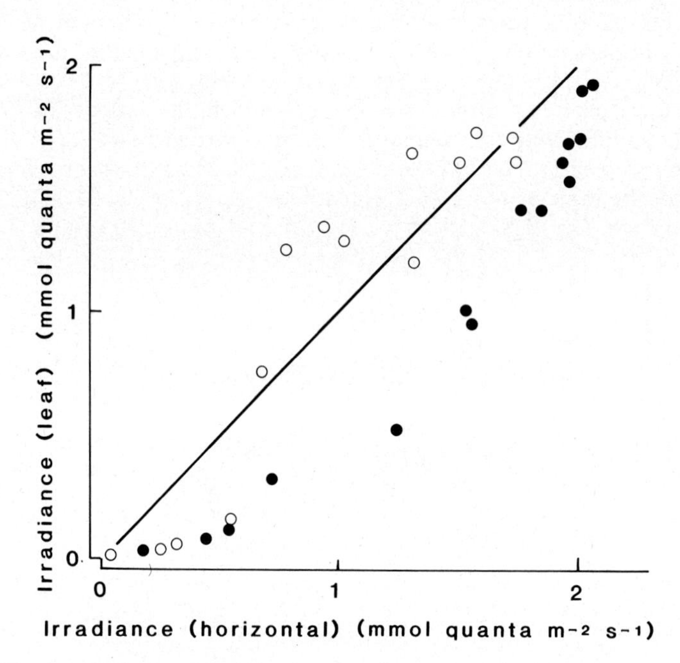

Fig. 2. Relationship between radiation reaching naturally oriented leaf surfaces and a horizontal sensor placed above the canopy. ● Values recorded before 1400 h. ○ Values recorded after 1400 h.

Table 1. Estimated daily integrals of irradiance at various positions within the crown of an overstory tree of *Argyrodendron peralatum*

I_h, irradiance received at height *h*

Day No.	Attribute	Height above ground (m)								
		36[A]	32[B]	31[B]	30[B]	29[B]	22[A]	17[B]	10[A]	6[B]
1	I_h (mol m^{-2} day^{-1})	54·6	37·5	30·0	8·7	16·7	5·6	13·1	13·0	10·3
	I_h/I_{36} (%)	100	68	55	16	30	10	24	12	18
2	I_h (mol m^{-2} day^{-1})	58·7	44·2	20·1	8·7	13·7	6·2	10·4	7·3	4·0
	I_h/I_{36} (%)	100	75	34	15	23	11	18	12	7

[A]Horizontally mounted quantum sensor. [B]Leaves measured in natural orientation.

horizontal surface. Over the course of a day, leaves would receive slightly less radiation than a horizontal surface, as indicated by the distribution of points about the 1 : 1 line in Fig. 2.

Table 1 compares daily integrals of irradiance at three sensors at different levels in the canopy with the integrals of irradiance at leaves at six positions within the

A. peralatum crown, determined by graphical interpolation of the daily courses of irradiance at each level. In the upper two measuring positions in the crown, leaves intercepted up to about 70% of the radiation reaching a horizontal sensor. Beneath these positions, light penetration to leaves was variable, but leaves in the lower crown

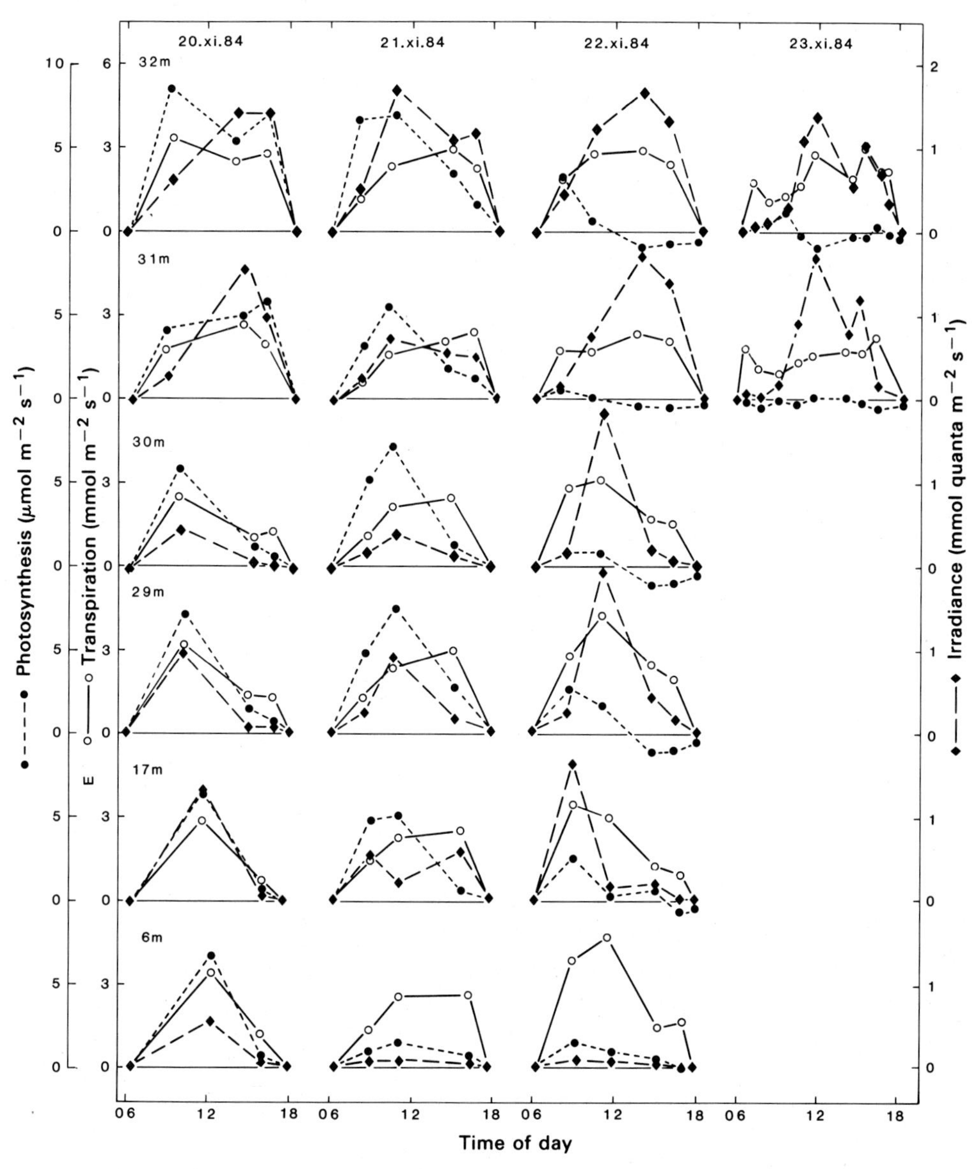

Fig. 3. Daily variation in mean irradiance (♦) transpiration (○) and photosynthesis rates (●) for leaves on five branches at different levels in the crown of an *Argyrodendron peralatum* tree (17–32 m) and an understory sapling (6 m) during four successive days after a rain event.

(17 m) and understory (6 m) intercepted higher proportions of the irradiances that arrived at nearby horizontal sensors (22 and 10 m) than was the case in the upper crown. However, the most frequent irradiances at leaf surfaces in the lower crown and understory were less than 100 μmol m^{-2} s^{-1}.

During the first 2 days after the first rain event, gas exchange rates in the upper crown positions (29–32 m) closely followed the daily course of irradiance (Figs 3*a*, 3*b*). In the lower crown (17 m) and understory (6 m), photosynthesis and transpiration rates were closely related to the occurrence of sunflecks, but the average daily courses of gas exchange at each level were again similar to that for the irradiance at a horizontal sensor above the forest canopy. The pattern was altered during the third and fourth days of a natural drying cycle, with a marked reduction in rates of photosynthesis during the middle of the day (Figs 3*c*, 3*d*).

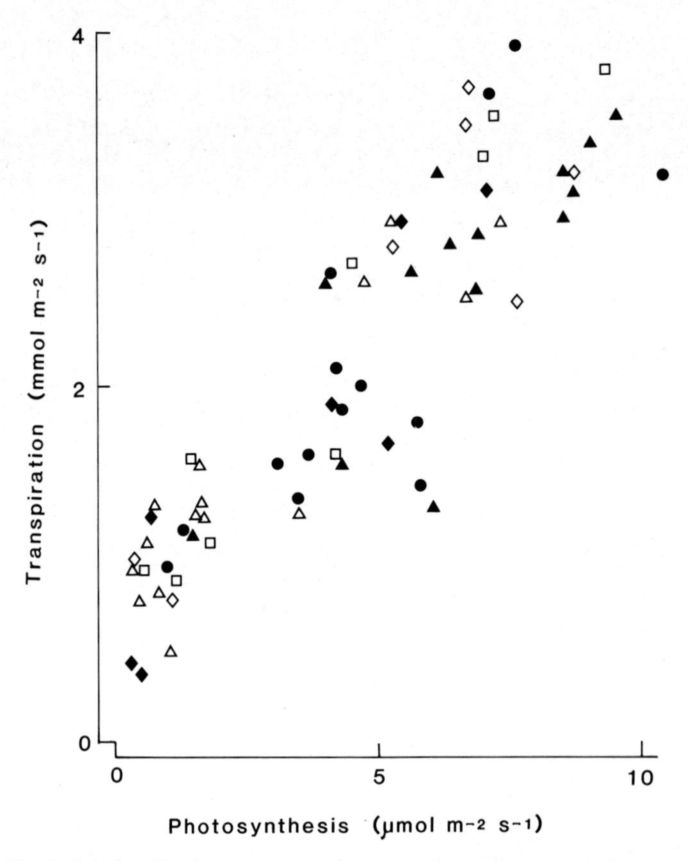

Fig. 4. Relationships between rates of transpiration and apparent photosynthesis in leaves of *Argyrodendron peralatum* from branches in the upper canopy at 32 m (●), 31 m (▲), 30 m (△), 29 m (□) and in the lower crown at 17 m (♦) and understory at 6 m (◇) on the first day after a rain event.

During the first day of the drying cycle, there was no distinction between the relationships between assimilation (A) and transpiration (E) for leaves at any of the sampling positions, either in the upper crown or understory leaves (Fig. 4), and there was no difference between morning and afternoon readings (not shown). On the following day, the relationship between E and A in the morning was similar to that for day 1, but during the afternoon the rates of transpiration were maintained or increased, while rates of assimilation decreased. Again, there was no clear distinction in the relationships between E and A for upper crown and understory leaves. After a second rain event and several days of cloudy or wet weather, the relationship between E and A was very similar to that at the beginning of the first drying cycle.

Apparent photosynthesis showed linear relationships with stomatal conductance under all conditions when the irradiance at a horizontal sensor above the canopy exceeded 200 μmol $m^{-2} s^{-1}$ (Fig. 5). The increments in photosynthesis per unit

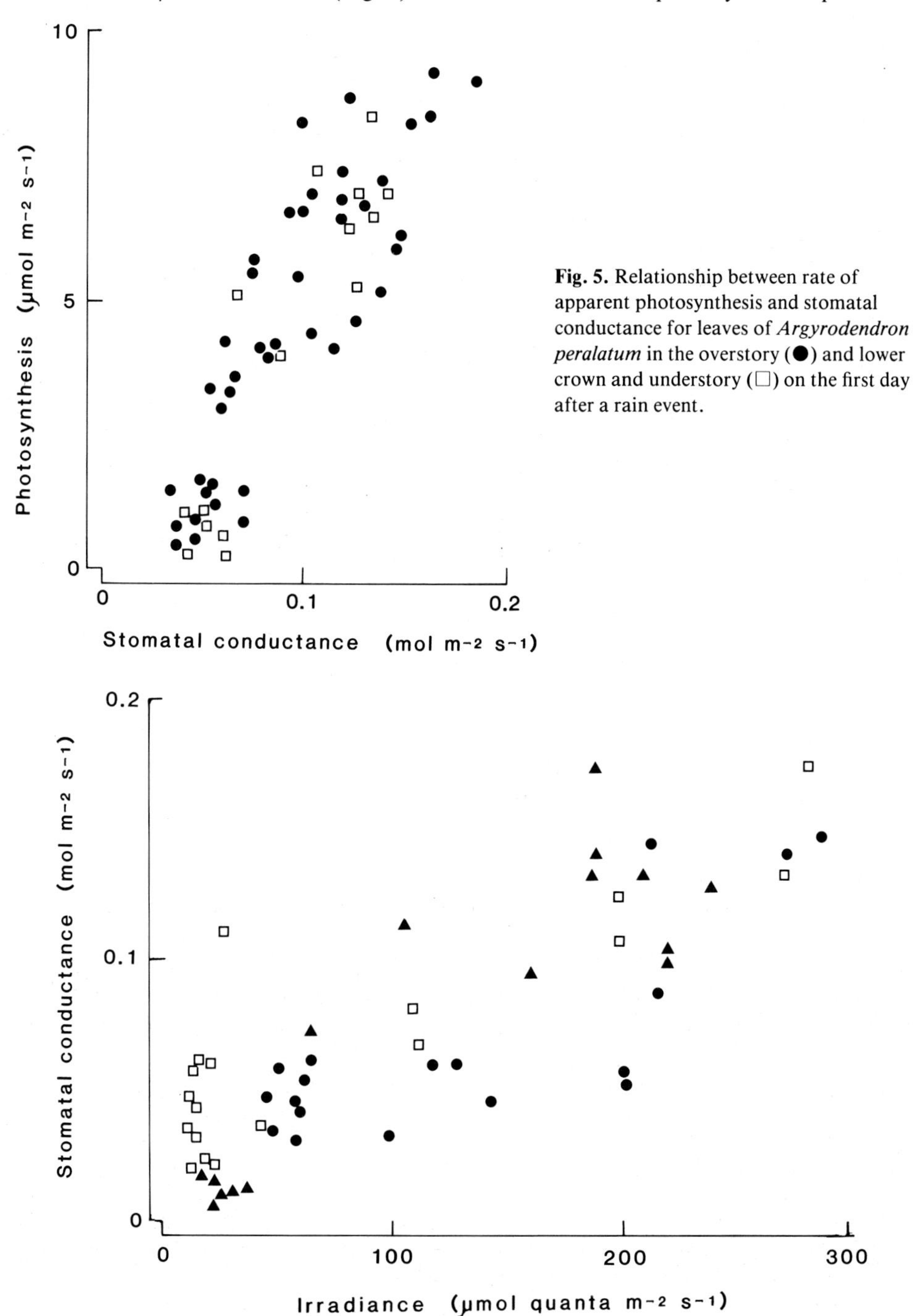

Fig. 5. Relationship between rate of apparent photosynthesis and stomatal conductance for leaves of *Argyrodendron peralatum* in the overstory (●) and lower crown and understory (□) on the first day after a rain event.

Fig. 6. Relationship between stomatal conductance to water vapour and irradiance for leaves of *Argyrodendron peralatum* from branches in the upper crown at 31–32 m (●), 29–30 m (▲) and lower crown and understory at 6 and 17 m (□) on the first day after a rain event.

increment in conductance (dA/dg) were similar for upper crown (overstory) and lower crown and understory leaves. There was no change in the relationship after the second rain event, and dA/dg appeared to be constant over the range of conductances observed.

The relationships between stomatal conductance and irradiance at irradiances up to 300 μmol $m^{-2} s^{-1}$ were not close, either for upper canopy or understory leaves after the first rain event (Fig. 6). There was no significant difference between the relationships derived for any of the leaf categories.

Photosynthesis responded to irradiance in a similar manner for leaves from the upper crown and understory during the first day after the first rain event. There was no evidence that leaves could be segregated into sun and shade types on the basis of photosynthetic rates (Fig. 7).

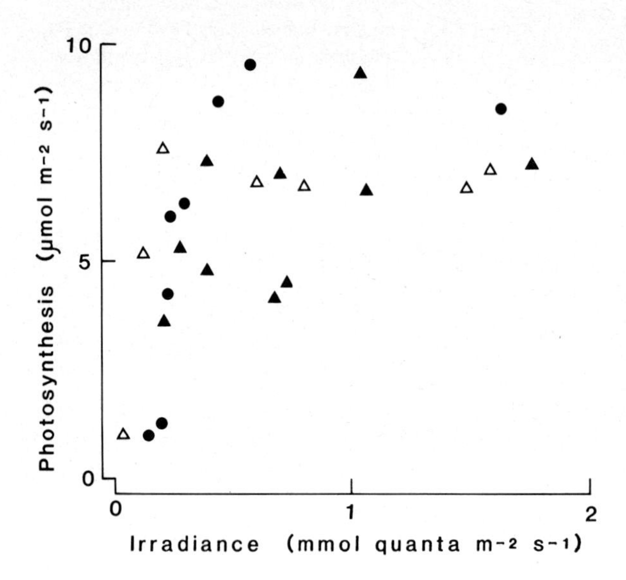

Fig. 7. Relationship between apparent photosynthesis and irradiance in leaves of *Argyrodendron peralatum* from branches in the upper crown at 31–32 m (●) and 29–30 m (▲) and in the lower crown and understory at 6 and 17 m (△) after a rain event.

Discussion

Light reaching the upper crown of a mature *A. peralatum* tree was attenuated so that, at the base of the crown the irradiance at a horizontal sensor was about 11–12% of that above the canopy (Table 1). However, below the level of the main crown, there was an increase in irradiance to about 15% of that about the canopy. A similar response was observed in a layered tropical forest by Torquebiau (1988). Although the forest studied in the present work was not obviously layered, the space between about 20 m and 3 m above the ground was almost devoid of foliage, and the occurrence of gaps in the canopy allowed the penetration of a substantial amount of light to the understory. This light did not appear to be sufficient to sustain rapid height growth in understory saplings of *A. peralatum* or of other species in the area.

Leaves in the upper crown intercepted less than the possible irradiance because of their orientations (Fig. 2). During the course of a day, the irradiance incident on leaf surfaces exceeded 1 mmol $m^{-2} s^{-1}$ on 54% of the occasions on which photosynthesis was estimated. On 23% of occasions, the irradiance was less than 200 μmol $m^{-2} s^{-1}$, and on 35% of occasions it was less than 500 μmol $m^{-2} s^{-1}$. Therefore, on a cloudless day, most of the foliage in the upper portion of the crown of this tree functioned for most of the day at irradiances that were above the saturating irradiance for photosynthesis of about 500 μmol $m^{-2} s^{-1}$ (Fig. 7). There was considerable variation in irradiance at different leaves within a single cluster on any occasion, indicating that leaves were functioning in continually fluctuating environments and the majority were not exposed to very high irradiances for long periods of the day.

The light responses of photosynthesis did not appear to vary with position of leaves in the canopy (Figs 5, 7). This conclusion may be affected by the method of measurement of photosynthesis, which was by short-term estimation on a large number of leaves under variable conditions. Pearcy (1988) observed different light responses of photosynthesis in upper canopy and understory leaves from this species, but in his study single leaves were used throughout the day under closely controlled conditions, and the understory plant received about 3·5% as much radiation during the course of the day as the upper canopy foliage. The differences between irradiances reaching the upper crown and understory leaves in the present study were considerably smaller, so that the differences in physiological characteristics may be expected to be smaller.

The distribution of photosynthetic rates in the crown of *Argyrodendron peralatum* at the beginning of a natural drying cycle was similar to that described for *Picea abies*, a temperate forest conifer (Schulze *et al.* 1977). On the first day of the drying cycle, rates of photosynthesis in the predominantly sunlit and upper shaded portions of the crown did not differ markedly, and they were substantially greater than the rates in the more extensively shaded lower crown and understory (Fig. 3). Leaves of the upper crown, whether mostly exposed to sunlight or partly shaded, all appeared to function in a similar manner, despite differences in their orientations and appearances.

It may be concluded that, if *A. peralatum* is grown in deep shade, the physiological characteristics of the leaves are different from those of mature trees but, where foliage is exposed frequently to extended sunflecks or sunpatches, these differences do not occur. Therefore, all leaves in the upper crown of a mature *A. peralatum* tree in the environment of the present study can be assumed to have similar photosynthetic characteristics. Estimates of daily photosynthetic integrals for the purpose of calculating total dry matter production by the whole crown under conditions of adequate water supply could then be derived from measurements of light interception by the crown and estimates of leaf area index (Monteith 1977). Such a procedure would allow a much smaller sample of leaves to be used for the determination of canopy photosynthesis than if there were several distinct categories of leaf function within the crown of a single tree. In fact, it would be necessary only to establish the photosynthetic characteristics of leaves from the upper crown and from the more or less continuously shaded understory. This has important consequences for the logistics of ecophysiological studies in these forests, but it still entails the non-trivial matter of access to the upper canopy.

Acknowledgments

The work was made possible by support from the Australian Research Grants Scheme, CSIRO Division of Forest Research, Dr G. C. Stocker, and Messrs K. Sanderson and L. Hutley.

References

Bazzaz F. A., and Carlson R. W. (1982). Photosynthetic acclimation to variability in the light environment of early and late successional plants. *Oecologia,* **54**, 313–16.

Kwesiga, F. R., Grace, J., and Sandford, A. P. (1986). Some photosynthetic characteristics of tropical timber trees as affected by the light regime during growth. *Ann. Bot.* **58**, 23–32.

Monteith, J. L. (1977). Climate and efficiency of crop production in Britain. *Phil. Trans. R. Soc., Lond. B* **281**, 277–94.

Mooney, H. A., Field, C. D., and Vasquez-Yanes, C. (1984). Photosynthetic characteristics of wet tropical forest plants. In 'Physiological Ecology of Plants of the Wet Tropics'. (Eds E. Medina, H. A. Mooney and C. Vasquez-Yanes.) pp. 129–38. (Dr W. Junk: The Hague.)

Oberbauer, S. F., and Strain, B. R. (1985). Effects of light regime on the growth and physiology of *Pentaclethra macroloba* (Mimosaceae) in Costa Rica. *J. Trop. Ecol.* **1**, 303–20.
Pearcy, R. W. (1988). Photosynthetic gas exchange responses of Australian tropical forest trees in canopy, gap and understory microenvironments. *Funct. Ecol.,* in press.
Schulze E.-D. (1970). Der CO_2-Gaswechsel der Buche (*Fagus silvatica* L.) in Abhangigkeit von den Klimafaktoren im Freiland. *Flora* **159**, 177–232.
Schulze, E.-D., Fuchs, M., and Fuchs, M.I. (1977). Spatial distribution of photosynthetic capacity and performance in a mountain spruce forest of Northern Germany. I. Biomass distribution and daily CO_2 uptake in different crown layers. *Oecologia* **29**, 43–61.
Torquebiau, E. (1988). Photosynthetically active radiation environment, patch dynamics and architecture in a tropical rainforest in Sumatra. *Aust. J. Plant Physiol.* **15**, 327–42.
Watts, W. R., Neilson, R. E., and Jarvis, P. G. (1976). Photosynthesis in sitka spruce (*Picea sitchensis* (Bong.) Carr.). VII. Measurements of stomatal conductance and $^{14}CO_2$ uptake in a forest canopy. *J. Appl. Ecol.* **13**, 623–38.
Whitmore, T. C. (1975). 'Tropical Rainforests of the Far East.' (Clarendon Press: Oxford.)

Photosynthetically Active Radiation Environment, Patch Dynamics and Architecture in a Tropical Rainforest in Sumatra

Emmanuel F. Torquebiau

SEAMEO-BIOTROP, P.O. Box 17, Bogor 16001, Indonesia;
present address: International Council for Research in Agroforestry,
P.O. Box 30677, Nairobi, Kenya.

Abstract

The photosynthetically active radiation (PAR) environments of three sites within a tropical rainforest in Sumatra were characterised. The sites and sensor locations were chosen according to a previous study of forest dynamics based on forest mosaic maps and an analysis of forest architecture by means of detailed profile diagrams. Two sites were located in mature forest patches: one showed a layered architecture while the other had a continuous vertical distribution of foliage. A third site represented a 320 m^2 treefall gap.

Vertical gradients of PAR from above canopy to ground level were studied in the mature forest zones, while a horizontal gradient of PAR at ground level was studied across a gap–mature forest boundary.

Daily total PAR at ground level in the mature forest sites was 0·78% and 0·57% of incident radiation above the forest. The decrease of radiation from above canopy downwards was strikingly different between the sites: the layered site showed a 'PAR inversion' between successive layers, the amount of radiation under a given layer being less than at the top of the next lower layer. The non-layered site did not show such an inversion. The treefall gap radiation received was approximately 20% of the radiation recorded above the canopy.

The spatial distribution of PAR in the forest is highly variable but can be usefully analysed against a framework of forest dynamics and architecture. Temporal distribution of PAR is also highly variable, and both distributions need to be analysed to understand photosynthetic responses and growth rates of rainforest species.

Introduction

Tropical rainforests are mosaics of patches at different stages of development (Whitmore 1975; Oldeman 1983; Torquebiau 1986). The mosaic-dependent canopy heterogeneity results in highly variable patterns of light availability in the underlayers. Within similar patches, there are differences in the proportion of radiation transmitted which are likely to result from differences in the architecture of both forest canopy and underlayers (Oldeman 1974; Yoda 1974; Chazdon and Fetcher 1984). Very little is known, however, about how radiation regimes vary in relation to forest architecture and the nature of the radiation environment experienced by species in relation to forest dynamics. Such data are necessary for further studies on photosynthesis or ecophysiology of rain forest species and for forest management practice. Information regarding the radiation environment above and within forests is also of great value if we are to understand energy exchanges between the forest and the atmosphere (Baldocchi *et al.* 1984). This paper describes such observations in lowland dipterocarp forest in Sumatra.

0310-7841/88/010327$03.00

Forest Dynamics, Forest Mosaics and Radiation Measurements

Forest mosaics, which are the visible outcome of the patchy functioning of forests (i.e. of forest dynamics) are an important framework within which a number of forest ecological studies can usefully be linked (e.g. see Charles-Dominique *et al.* 1981; Bourlière 1983; Salick *et al.* 1983 for animal distribution in relation to patchiness; Granville 1978; Denslow 1980; Martinez 1980; Kahn 1982, 1986; Brokaw 1985, 1987 for plant diversity in relation to patchiness).

The patch (a spatial concept, the temporal equivalent of which is 'phase') is an ecologically uniform unit (or 'eco-unit', Oldeman 1983) which usually originates as a treefall gap (natural disturbance, see Pickett and White 1985) and subsequently develops through successional stages as an independent forest entity with a limited number of interactions with bordering patches. Eco-units differ among themselves at least from floristic and structural points of view, but only their young stages (i.e. the gaps and early successional phases) are well documented (e.g. see Brokaw 1983, 1985). Mature forest zones are seldom described in terms of patches.

Dynamic communities with a high frequency of natural disturbance display heterogeneous patterns. It is important to recognise the scale of this heterogeneity in order to adjust observations to it. Statistical samplings over large surfaces may include several different eco-units and hide ecologically significant heterogeneity behind meaningless averages.

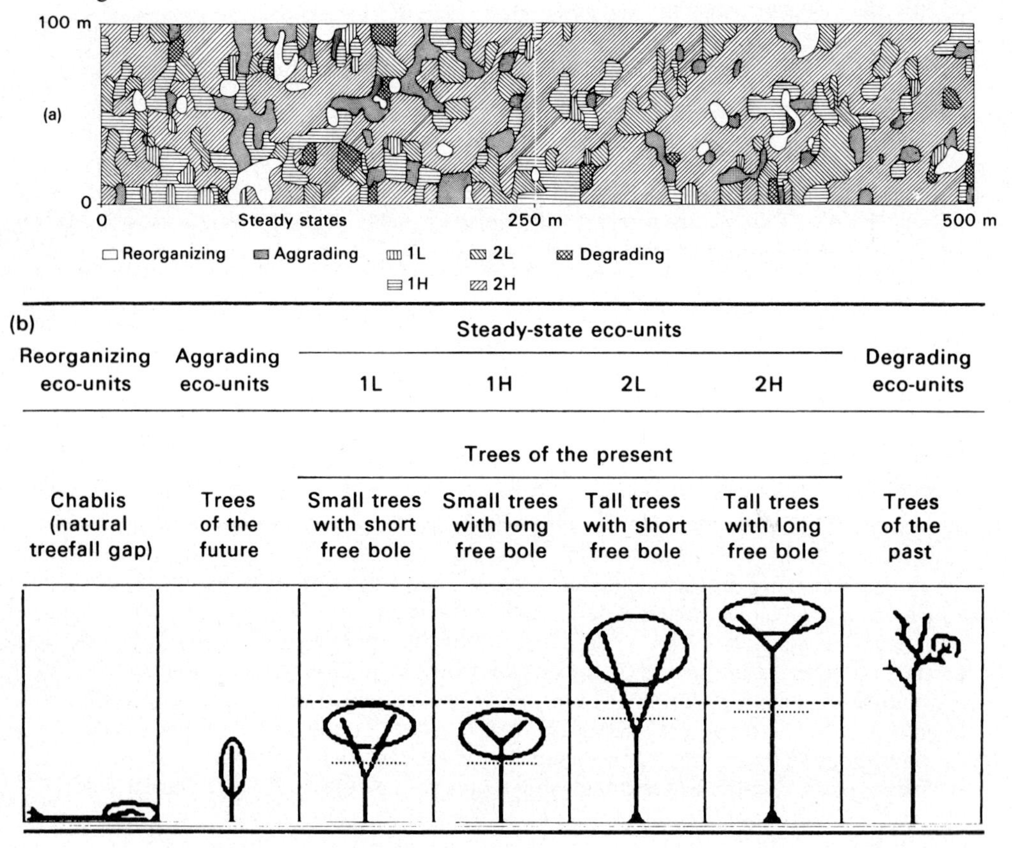

Fig. 1. (*a*) Forest mosaic. The eco-units are drawn according to the categories shown in (*b*). For steady-state eco-units, 1 and 2 indicate first and second canopy level, L and H indicate low and high inversion level. From Torquebiau (1986).

Size and architecture of canopy trees can be used as easily observable criteria to identify eco-units in the tropical rain forest. Seven kinds of eco-units were recognised in tropical rain forests in Indonesia by Torquebiau (1986) as shown in Fig. 1. The size range of the eco-units starts from very small values of the order of some square metres (partial treefall gaps, small trees of the future; see Hallé *et al.* 1978), to large eco-units of several hundred square metres (tall canopy trees) which often result from the connection of smaller similar eco-units and appear as 'networks' rather than units (see Fig. 1).

Characteristically light, both irradiance and spectral distribution, and therefore photosynthetically active radiation (PAR), is an ecological factor which is linked with forest mosaic patterns. For example, radiation regimes in a gap, under canopies of pioneer trees or under primary species will be strikingly different. Differences in the amount, and in the spatio-temporal and spectral distribution of radiation reaching different levels of the forest depend not only on the quantity of foliage but on the architecture and arrangement of tree crowns.

The present study was undertaken to compare patterns of PAR in and between forest eco-units. First, two vertical gradients of PAR, from above the forest to ground level, were studied in two sites both belonging to mature, dense patches (2H steady-state eco-units; see Fig. 1). These eco-units have been found consistently to be the most widespread type in different tropical rain forest localities in Indonesia (57–69% of the sampled area; Torquebiau 1986). They are homogeneous with respect to the criteria of size and architecture of canopy trees but may differ among themselves in terms of floristics, underlayer architecture, plant density and distribution pattern, etc. The two sites chosen for this study differed regarding vertical foliage layering: one had obvious tree layers whereas the other showed continuous foliage distribution from the top to the ground. It was thought that these differences in layering would induce differences in the penetration of incident PAR to ground level. These properties were also examined in another site selected to record a horizontal gradient of PAR at ground level across a gap–mature forest interface.

Materials and Methods

Study Area

The study was conducted in the lowland dipterocarp forest of Sumatra (Indonesia) at Pasir Mayang, near Muarabungo, Jambi Province (long. 102°10′ E., lat. 1°5′ S., alt. 100 m a.s.l.). The climatic zone corresponds to the very humid bioclimate of Fontanel and Chantefort (1978), i.e. the mean temperature of the coldest month is above 20°C, annual rainfall between 2000 and 2500 mm, and there is no month with less than 100 mm of rain. Soils in the area are poor kaolinitic ferralitic soils developed on sedimentary bed rocks. The topography is regularly undulating and the drainage is good. Both study sites were situated on the same small ridge.

The forest type is the tropical lowland evergreen formation of Whitmore (1984) with an upper canopy height of about 50 m (Torquebiau 1986) and with the following conspicuous tree species: *Dipterocarpus baudii, D. lowii, Shorea acuminata, S. macroptera, Scorodocarpus borneensis, Dialium platysepalum, Koompassia malaccensis* and *Dyera costulata*. Voucher specimens for the species mentioned in Figs 2 and 3 are located at BIOTROP, Bogor, some of them still in the process of identification; they will later be deposited at the Herbarium Bogoriense (BO).

Equipment and Sampling

Measurements of PAR were made using NEC PH 201 gallium arsenide photosensors connected to a Campbell CR-21 data-logger. The spectral response of these sensors is somewhat similar to a 400-700 nm quantum sensor except that they are less sensitive below 450 and above 650 nm and the cosine error is large above 70° angle of incidence (R. Pearcy, personal communication). They are cheap and adequate substitutes for quantum sensors and remain at acceptable levels of error for field measurements. Sensors were calibrated against a Li-Cor Ll 190 SB quantum sensor.

PASIR MAYANG, JAMBI, SUMATRA.
Lat 1°5'S. Long 102° 10'E. Alt. 100 m.
December 1984.
45m
40
30
20
10
0

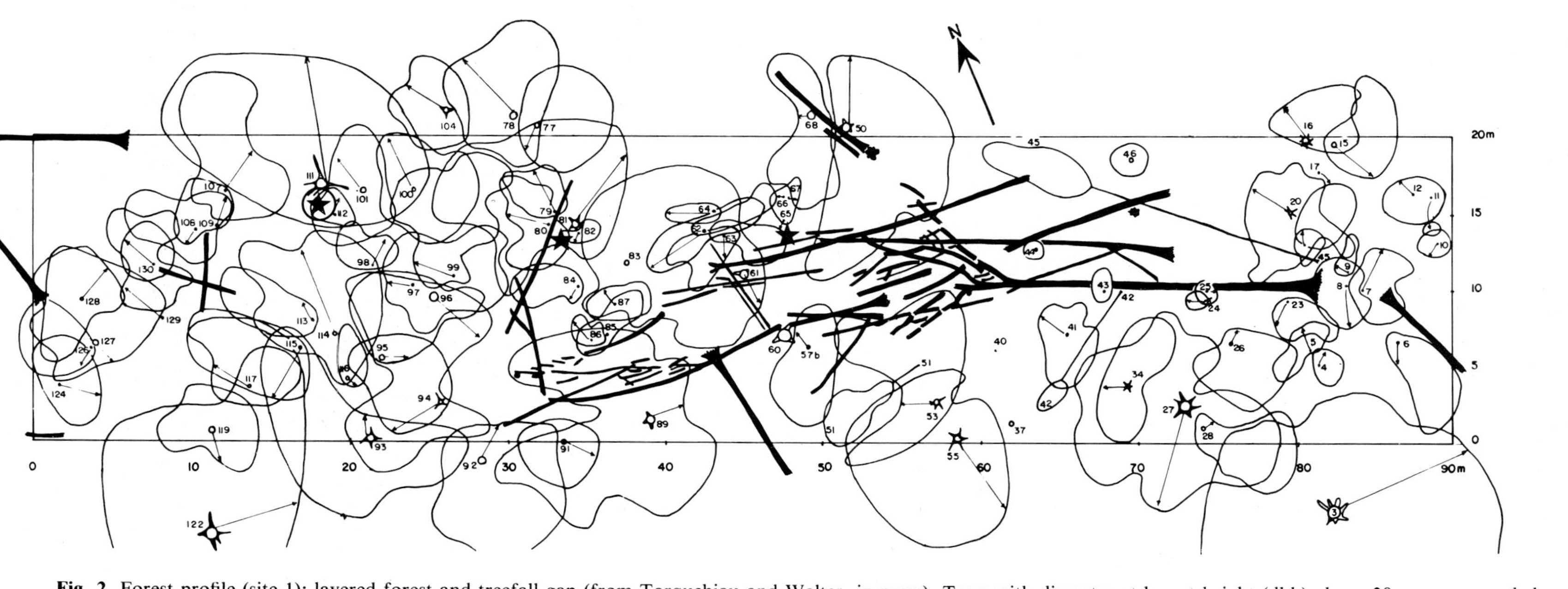

Fig. 2. Forest profile (site 1); layered forest and treefall gap (from Torquebiau and Walter, in press). Trees with diameter at breast height (dbh) above 30 cm are recorded on the whole 20 m width of the plot, the criterion for inclusion in the drawing being the presence of the tree's crown above the plot, i.e. the trunk may be outside the plot. Trees with dbh between 10 and 30 cm are recorded on a central strip 10 m in width. Trees taller than 2 m and with dbh below 10 cm are recorded on a width of 5 m. These small trees do not appear on the map but only on the profile, without numbers.

3 *Shorea acuminata* (Dipterocarpaceae)
5 *Dacryodes* (Burseraceae)
6 cf. *Baccaurea* (Euphorbiaceae)
7 *Palaquium gutta* (Sapotaceae)
8 *Durio* sp. (Bombacaceae)
9 *Diospyros* sp. (Ebenaceae)
10 *Calophyllum* sp. (Guttiferae)
11 *Santiria* sp. (Burseraceae)
12 *Polyalthia sumatrana* (Annonaceae)
14 cf. *Litsea* (Lauraceae)
15 *Aromadendron* (Magnoliaceae)
16 Myrtaceae
17 *Gardenia anisophylea* (Rubiaceae)
19 *Horsfieldia* cf. *tomentosa* (Myristicaceae)
20 *Eugenia* sp. (Myrtaceae)
23 *Eugenia* sp. (Myrtaceae)
24 *Santiria* sp. (Burseraceae)
25 *Palaquium gutta* (Sapotaceae)
26 *Canarium* sp. (Burseraceae)
27 *Shorea macroptera* (Dipterocarpaceae)
28 cf. *Eugenia* (Myrtaceae)
34 *Xerospermum* sp. (Sapindaceae)
37 Fagaceae
40 ?
41 cf. *Litsea* (Lauraceae)
42 *Dipterocarpus baudii* (Dipterocarpaceae)
43 *Palaquium obovatum* (Sapotaceae)
44 *Santiria* sp. (Burseraceae)
45. *Palaquium gutta* (Sapotaceae)
46 *Ochanostachys amentacea* (Olacaceae)
50 ?
51 cf. *Vatica* (Dipterocarpaceae)
53 Sapindaceae
55 *Dipterocarpus lowii* (Dipterocarpaceae)
57b cf. *Trigoniastrum* (Trigoniaceae)
60 ?
61 *Sarcotheca* sp. (Oxalidaceae)
62 *Shorea acuminata* (Dipterocarpaceae)
63 Sterculiaceae
64 *Santiria* sp. (Burseraceae)
66 *Ptychopyxis* cf. *kingii* (Euphorbiaceae)
67 *Dysoxylum* sp. (Meliaceae)
68 *Shorea macroptera* (Dipterocarpaceae)
77 *Xanthophyllum rufum* (Polygalaceae)
78 *Dipterocarpus lowii* (Dipterocarpaceae)
79 *Shorea macroptera* (Dipterocarpaceae)
80 *Castanopsis* cf. *inermis* (Fagaceae)
81 *Vatica* (Dipterocarpaceae)
82 *Ochanostachys amentacea* (Olacaceae)
83 ?
84 Euphorbiaceae
85 cf. *Sindora* (Leguminosae)
86 *Fordia* cf. *gibbsiae* (Leguminosae)
87 *Sandoricum emarginatum* (Meliaceae)
89 *Shorea* sp. (Dipterocarpaceae)
91 *Lophopetalum beccarianum* (Celastraceae)
92 *Calophyllum* (Guttiferae)
93 cf. Rhizophoraceae
94 *Trigoniastrum* (Trigoniaceae)
95 *Shorea acuminata* (Dipterocarpaceae)
96 *Shorea* (Dipterocarpaceae)
97 *Polyalthia hypoleuca* (Annonaceae)
98 *Milletia* (Leguminosae)
99 cf. *Eugenia* (Myrtaceae)
100 *Parashorea* (Dipterocarpaceae)
101 *Santiria griffithii* (Burseraceae)
104 Burseraceae
107 cf. *Eugenia* (Myrtaceae)
108 *Knema* cf. *malayana* (Myristicaceae)
109 *Palaquium* sp. (Sapotaceae)
111 *Shorea acuminata* (Dipterocarpaceae)
112 cf. *Hydnocarpus* (Flacourtiaceae)
113 *Shorea macroptera* (Dipterocarpaceae)
114 *Ixonanthes icosandra* (Linaceae)
115 *Ixonanthes icosandra* (Linaceae)
116 *Scaphium* sp. (Sterculiaceae)
117 *Parashorea* (Dipterocarpaceae)
119 *Lophopetalum beccarianum* (Celastraceae)
122 *Dipterocarpus baudii* (Dipterocarpaceae)
124 *Palaquium gutta* (Sapotaceae)
126 *Polyalthia hypoleuca* (Annonaceae)
127 *Shorea macroptera* (Dipterocarpaceae)
128 *Pellacalyx lobbii* (Rhizophoraceae)
129 *Dacryodes rugosa* (Burseraceae)

Sensors were monitored at 1-min intervals and the data logger computed 5- or 10-min averages, maxima and minima of PAR from 6 a.m. to 6 p.m. The vertical gradient data discussed in the present paper are limited to the averages during recording periods of 5–6 days for each site. Data of the horizontal gradient are 15-min interval instantaneous readings over 2 days.

Each sensor was tied at the top of a bamboo stick which was itself tied to a branch by a tree climber. The sensor above the forest canopy was also at the extremity of a long bamboo stick emerging above the tree; one technical limitation was that the horizontal levelling of this last sensor could not be guaranteed.

Sampling days spanned different weather conditions which varied irregularly throughout each day, with alternately clear and cloudy (or rainy) periods. Since one sensor was permanently monitored above the forest canopy (macroclimate), data from sensors inside the forest (microclimate) can be discussed in terms of relative values, whatever the weather conditions.

Data Analysis

For the vertical gradients, daily averages were calculated for each sensor from the 5- or 10-min average data and then multiplied by total day length to obtain daily total PAR, following Chazdon and Fetcher (1984). The percentage transmission of PAR was then calculated from daily totals for each sensor inside the forest (microclimate), using data from the sensor above the forest as a macroclimatic reference. These percentages were then averaged over 5–6 days to produce charts of PAR transmission in the forest sites.

For the horizontal gradient, daily averages, daily totals and comparative daily regimes of PAR were calculated and plotted from the 15-min interval data over 2 days.

Student's *t*-test was used to compare PAR transmissions or daily total PAR at different heights of the forest and between forest sites.

Sites

Two sites selected for the comparison of vertical gradients of PAR were 300 m apart and both belonged to the most mature zone of the forest mosaic, as discussed in the previous section. At first glance, the sites appeared similar, both dim and with relatively few small plants/seedlings at ground level. However, they showed contrasting architectures at upper levels, as seen in Figs 2 and 3.

Site 1 (left part of Fig. 2) was obviously layered: the upper level consisted of a 45 m high *Shorea acuminata* (tree No. 111). Although not very tall, this tree can be considered as an emergent since its crown is not adjacent to other trees at the same height (Nos 89, 119, 122). An empty space of about 6–8 m appears below the crown of this *Shorea*, before reaching the crown of the *Santiria griffithii* No. 101; this last had a 15 m high, narrow crown which made, together with trees Nos 98 and 112, a dense median layer. Another empty zone was visible downwards, from 12 to 3 m, before reaching an understory layer of palms (*Licuala* spp. and some rattan-like palms) and saplings.

Site 2 (Fig. 3) did not show any layering but a regular distribution of foliage from the upper canopy *Dipterocarpus lowii* No. 25 to the ground. Several crowns of mature trees were intermixed from the top of the forest to the lower branches of tree No. 30, at 13 m. A rather dense zone of saplings and a few rattan-like palms followed immediately below and continued down to some centimetres above ground. The *Licuala* palm layer found in site 1 was not observed here.

Since the primary question was 'how does PAR gradient vary according to forest structure?', the sensors were not arranged haphazardly nor regularly spaced but arranged according to the structure recognised on the forest profiles. In the layered forest (site 1), the sensors were placed in order to 'border' the three recognised layers and record PAR variations in and between the layers: one sensor was placed above and one sensor below each layer, as shown by the stars in Fig. 2. The hypothesis made was that the vertical gradient of PAR would be modulated according to the forest structure, i.e. that there would be a rapid decrease of radiation in the layers and a slower decrease between the layers. Six sensors were necessary for this, numbered 1·1 to 1·6 downwards.

In the non-layered forest (site 2), sensors were placed at heights similar with those of site 1 (stars in Fig. 3), so as to make data comparable. However, locations were chosen to avoid having the sensors completely inside a crown, which would have yielded very low radiation readings without meaning for the whole gradient. Instead, sensors were placed at the periphery of the crowns or close to main branches, in zones with rather low foliage density, so as to record as high a level of radiation as possible for a given height. The hypothesis made was that the gradient of PAR at site 2 would be less modulated (if at all) than in site 1. The six sensors of site 2 were numbered 2·1 to 2·6 downwards.

The horizontal gradient of PAR was monitored by three sensors at 0·3 m above the ground in site 1 (Fig. 2). One sensor was in the dense forest (base of tree No. 111, same place as sensor 1·6), one in a 320 m^2 treefall gap and one in the transition zone between the gap and the dense forest. The treefall gap is shown in the right part of Fig. 2 and the stars on the map of the site indicate the location of the sensors, numbered 3·1 to 3·3 from the gap to the dense forest, respectively.

Due to the limited number of sensors and of entries in the data-logger, it was not possible to duplicate the measurements at each sensor location, although this would have better accounted for spatial variation within each location (Pearcy 1983; Chazdon and Fetcher 1984). However, since the sensors were not placed at random but according to the previously studied forest structure, the duplication (Reifsnyder *et al.* 1971 / 1972) of measurements is not as critical as it would be for a purely stochastic sampling.

Results

Site 1

Measurements of PAR at site 1 are summarised in Table 1; the relative values of PAR at each forest level, calculated from daily totals averaged over 6 days are shown in Fig. 4. The vertical gradient of PAR in this site was strongly modulated by layering: values of PAR at the bottom of the superior layer were lower than values at the top of the median layer. The term 'PAR inversion' is proposed for this peculiarity. This

Table 1. Values of daily average (μmol m^{-2} s^{-1}), daily total (mol m^{-2}) and percentage of transmission of PAR over 6 days

Layered forest (site 1), vertical gradient

	Sensor 1·1	Sensor 1·2	Sensor 1·3	Sensor 1·4	Sensor 1·5	Sensor 1·6
Daily average	988·47	401·06	463·83	14·84	14·32	3·46
Daily total	42·7	17·33	20·04	0·64	0·62	0·15
%	100	40·59	46·93	1·5	1·45	0·35
Daily average	1144·27	435·36	530·08	13·1	15·54	
Daily total	49·43	18·81	22·9	0·57	0·67	
%	100	37·75	45·96	1·14	1·34	
Daily average	881·23	374·49	411·23	12·78	13·27	5
Daily total	38·07	16·18	17·77	0·55	0·57	0·22
%	100	42·5	46·68	1·44	1·5	0·58
Daily average	827·13	349·75	399·16	16·36	16·3	8·51
Daily total	35·73	15·11	17·24	0·71	0·7	0·37
%	100	42·29	48·25	1·99	1·96	1·04
Daily average	679·77	276·61	313·59	16·82	16·96	6·96
Daily total	29·37	11·95	13·55	0·73	0·73	0·3
%	100	40·69	46·14	2·49	2·49	1·02
Daily average	831·05	333·59	386·11	15·35	17·54	7·3
Daily total	35·9	14·41	16·68	0·66	0·76	0·32
%	100	40·14	46·46	1·84	2·12	0·89

inversion is highly significant (sensors 1·2 and 1·3, 40·66–46·74%, $t=7{\cdot}8$, d.f. = 10, $P=0{\cdot}001$). There was a tendency for the same phenomenon to occur between the second and the third layers, but it was not statistically significant (sensors 1·4 and 1·5, 1·73–1·81%, $t=0{\cdot}29$, d.f. = 10, $P=0{\cdot}05$).

The source of the increase in PAR with depth which produces the inversion is likely to be scattering of radiation by the different surfaces bordering the empty layer (trunks, twigs, leaves) particularly between the two upper layers where there were several big branches receiving large amounts of radiation.

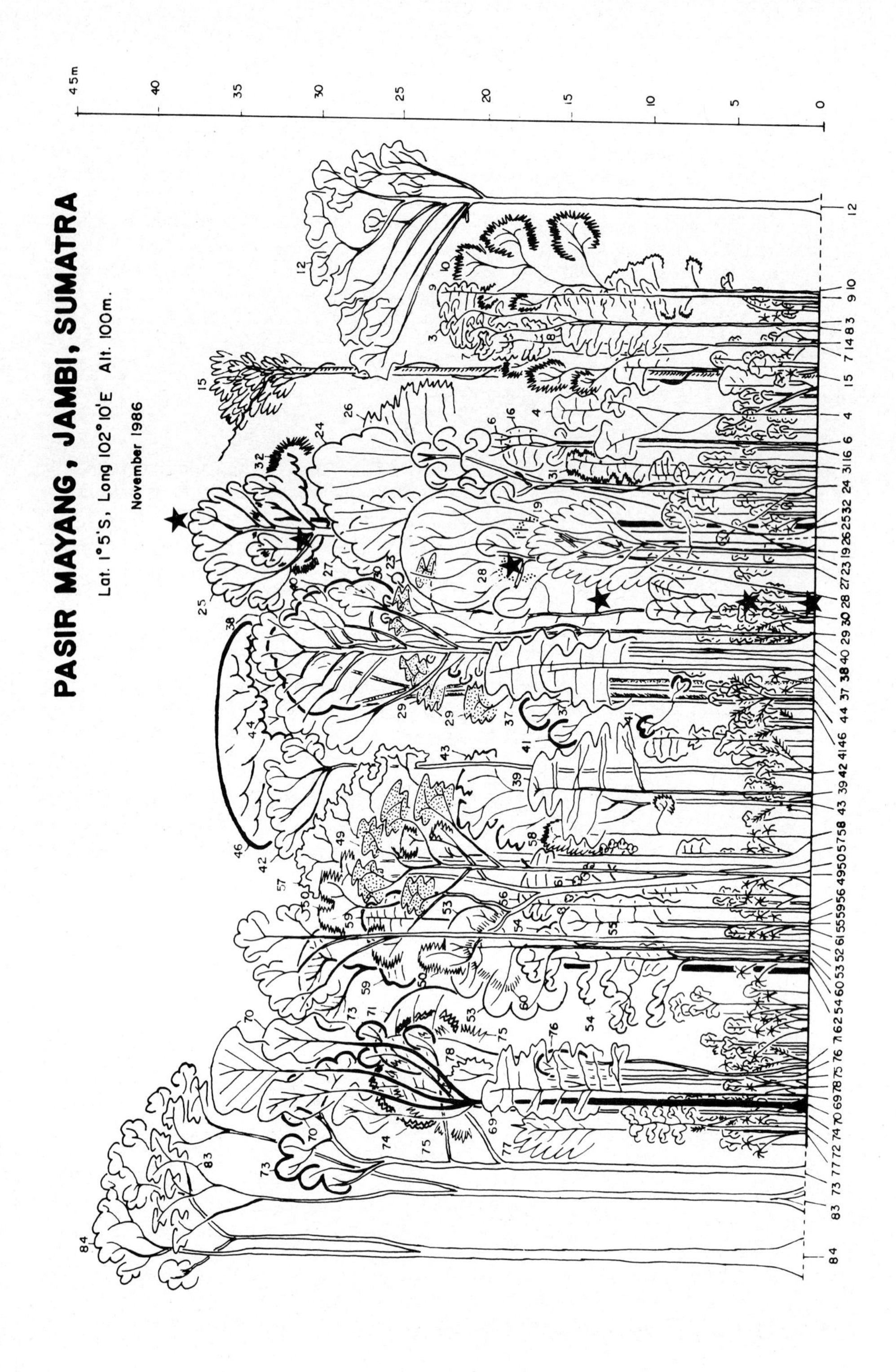
PASIR MAYANG, JAMBI, SUMATRA
Lat. 1°5'S, Long 102°10'E Alt. 100m.
November 1986
45m
40
35
30
25
20
15
10
5
0

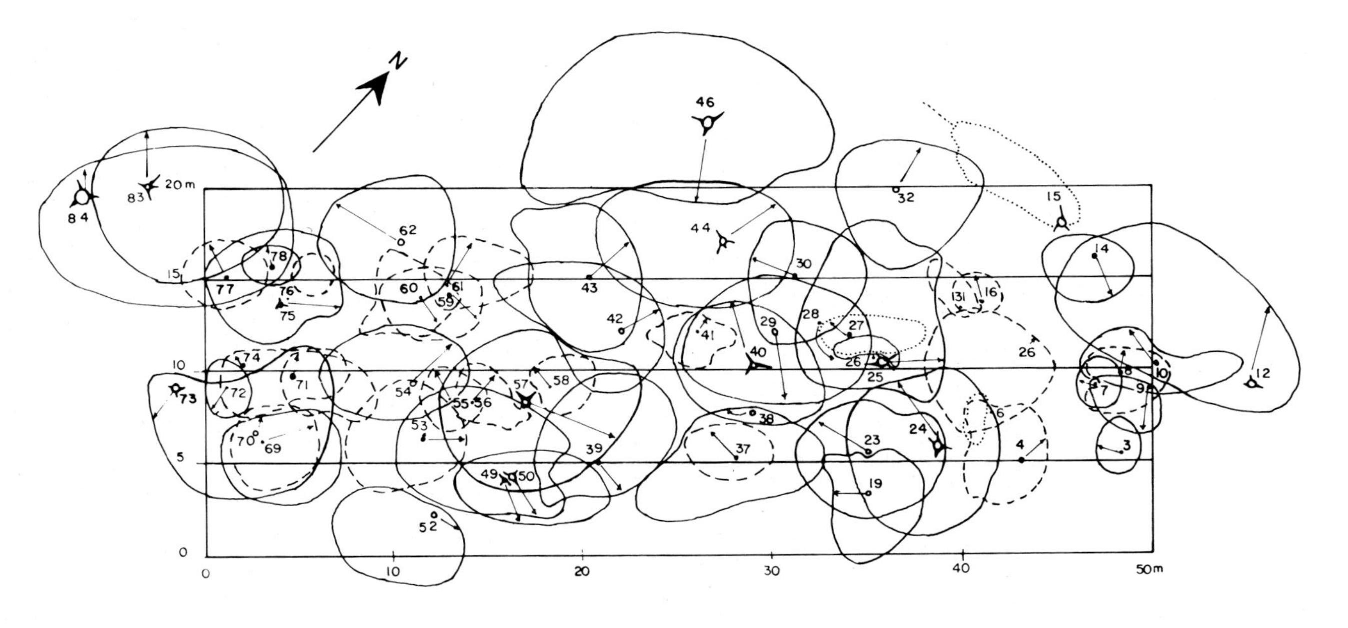

Fig. 3. Forest profile (site 2); non-layered forest. See technical data in legend of Fig. 2.

3 Leguminosae
7 *Diospyros*
8 *Hopea*
9 *Cyathocalyx pruniferus*
10 *Dacryodes*
12 *Lithocarpus*
14 *Dacryodes*
15 *Shorea*
16 Euphorbiaceae
19 *Bouea*
23 *Ixonanthes icosandra*
24 *Irvingia malayana*
25 *Dipterocarpus lowii*
26 *Lithocarpus encleisacarpus*
27 *Shorea macroptera*
28 ?
29 *Sindora*
30 *Sterculia*
31 Rhizophoraceae
32 *Shorea macroptera*
37 Rhizophoraceae
38 *Shorea macroptera*
39 *Knema conferta*
40 *Shorea*
41 *Ixonanthes icosandra*
42 *Parashorea*
43 *Durio*
44 *Santiria*
46 *Shorea acuminata*
49 Leguminosae
50 *Eugenia*
52 *Shorea macroptera*
53 *Xanthophyllum*
54 *Barringtonia*
55 ?
56 *Cyathocalyx*
57 *Palaquium*
58 *Hydnocarpus*
59 *Polyalthia hypoleuca*
60 Bombacaceae
61 *Quercus*
62 *Ochanostachys amentacea*
69 *Xanthophyllum*
70 *Ochanostachys amentacea*
71 *Xylopia ferruginea*
72 *Barringtonia*
73 *Shorea*
74 *Calophyllum*
75 Myrtaceae
76 *Parashorea*
77 Burseraceae
78 Euphorbiaceae
83 Leguminosae
84 *Shorea acuminata*

Measurements of PAR above the forest (macroclimate, sensor 1·1) differed by a factor of 1·7 between the highest and the lowest day. The highest daily total macroclimatic PAR was 49·43 mol m^{-2} during a day cloudy at dawn and until 9.30 a.m. but sunny afterwards. The lowest daily total PAR (29·37 mol m^{-2}) was recorded during a cloudy–rainy day, with only some brief periods of clear sky in the morning and a heavy storm at around 2 p.m.

Data in Table 1 show that the percentage transmission of PAR was in inverse ratio to the amount of radiation above the forest: the day with maximum total PAR had the lowest percentages of transmission on all sensors. The three lowest sensors (1·4, 1·5 and 1·6) had the highest values of transmission on the 2 days of lowest total PAR. The amount of radiation reaching the ground (sensor 1·6) varied between 0·3 and 1%.

At 35 m (i.e. after approximately one-third of the height of the forest), 60% of the incident radiation had been absorbed. The first plus second layers impose strong attenuation and very little radiation (1·7%) is left at 12 m. The difference from this level to the top of the lower layer is very small with an apparent PAR inversion which is not significant (see above), so that the similar values at 12 and 3 m indicate a constant radiation regime over these 9 m.

Site 2

PAR measurements for this site are summarised in Table 2 and the percentages of transmission are shown in Fig. 5. The PAR inversions observed in site 1 did not appear in this site. However, the vertical gradient of PAR still is slightly modulated, with 66% of the incident radiation absorbed at 31 m, i.e. in the first 9 m, a further 22% absorbed in the next 13 m, and 10% in the next 5 m. This suggests, although no layering was visible in this site, that there are levels of dense foliage at heights comparable to the layers of site 1.

Table 2. Values of daily average (μmol m^{-2} s^{-1}), daily total (mol m^{-2}) and percentage of transmission of PAR over 5 days

Non-layered forest (site 2), vertical gradient

	Sensor 2·1	Sensor 2·2	Sensor 2·3	Sensor 2·4	Sensor 2·5	Sensor 2·6
Daily average	820·93	252·06	59·09	15·39	5·95	
Daily total	35·46	10·89	2·55	0·66	0·26	
%	100	30·71	7·19	1·86	0·73	
Daily average	905·9	343·92	106·92	12·7	24·37	5·76
Daily total	39·13	14·86	4·62	0·55	1·05	0·25
%	100	37·98	11·81	1·41	2·68	0·64
Daily average	783·28	304·76	136·12	13·06	14·77	4·77
Daily total	33·84	13·17	5·88	0·56	0·64	0·21
%	100	38·92	17·38	1·65	1·89	0·62
Daily average	1098·76	416·1	165·65	16·85	13·97	8·51
Daily total	47·47	17·98	7·16	0·73	0·6	0·37
%	100	37·88	15·08	1·54	1·26	0·78
Daily average	458·64	116·05	45·29	9·73	6·1	1·1
Daily total	19·81	5·01	1·96	0·42	0·26	0·05
%	100	25·29	9·89	2·12	1·31	0·25

At 13 m, 98·3% of total PAR has been absorbed, a value similar to that of site 1 at the same height. Values below this level are similar to those of site 1. No PAR inversion appears between the middle and lower layers of this site when average values of daily totals PAR over 5 days are considered (Fig. 5), but an inversion does occur on some days (Table 2, sensors 2·4 and 2·5). The difference is not significant between these sensors (1·72–1·57%, $t=0{\cdot}4$, d.f.$=8$, $P=0{\cdot}05$) which indicates, like in site 1, a constant radiation regime over a height of 10 m. The amount of radiation reaching the ground varied from 0·2 to 0·8% and was not significantly different from the ground-level radiation at site 1 (sensors 1·6 and 2·6, 0·78–0·57%, $t=1{\cdot}12$, d.f.$=7$, $P=0{\cdot}05$).

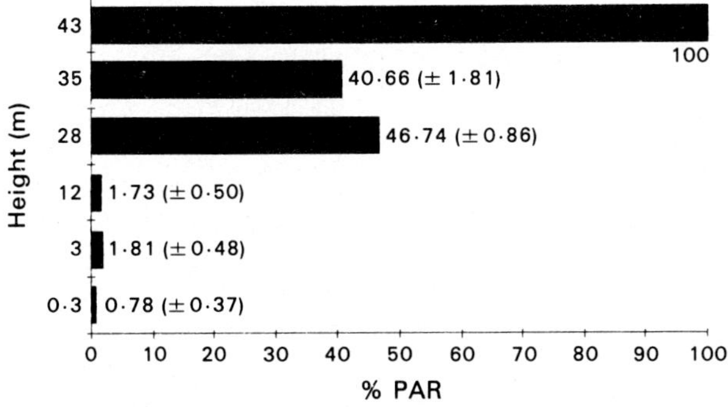

Fig. 4. Transmission of PAR: vertical gradient in layered forest; values calculated from average of daily totals over 6 days (see Table 1). Values in parentheses are confidence interval for t_5 (0·025), except for sensor at 0·3 m: t_4 (0·025).

In this site, measurements of total PAR above the forest differed by a factor of 2·4. A totally overcast–rainy day was experienced on which the daily total PAR above the forest was only 19·8 mol m^{-2}, compared with 47·47 mol m^{-2} on a day with irregular weather but frequent occurrence of highly reflective big white clouds. An inverse ratio between transmitted and total incident radiation was not observed in this site; values of transmission appeared more irregular. Any linkage with the absence of layering in this site remains to be investigated.

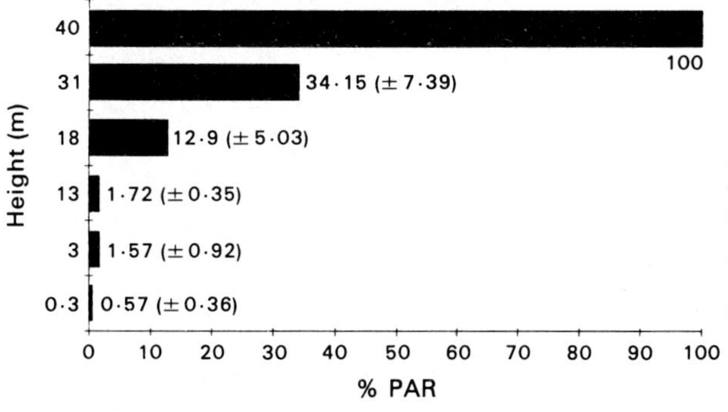

Fig. 5. Transmission of PAR: vertical gradient in non-layered forest; values calculated from average of daily totals over 5 days (see Table 2). Values in parentheses are confidence interval for t_4 (0·025), except for sensor at 0·3 m: t_3 (0·025).

Horizontal Gradient of PAR

Measurements of PAR for this gradient are summarised in Table 3 and the comparative values for daily regimes of PAR are shown for one day in Fig. 6. Sensor 3·3 (dense forest) was located exactly at the same place as sensor 1·6, and gave an average daily total PAR over 2 days not significantly different from the average value over 5 days for the vertical gradient study (0·3 and 0·27 mol m^{-2} respectively, $t=0{\cdot}32$, d.f.$=5$, $P=0{\cdot}05$). It was therefore possible to use the value of incident PAR above the forest over these 5 days to estimate transmission percentages of PAR at ground level for the horizontal gradient study.

Daily cumulative radiation penetrating this gap, which spreads over approximately 320 m^2, is about 20% of the incident PAR above the forest. A former study of this site by means of hemispherical photographs and pyranometer measurements (Torquebiau and Walter, in press) has shown that the location of sensor 3·1 was in the brightest part of the gap.

The radiation reaching the transition zone, 13 m away from the edge of the gap, is only 3% of the incident PAR. This sharp decrease shows that the macroclimatic perturbation caused by the treefall has limited influence outside the gap patch itself, a factor strengthening the concept of forest patchiness.

The daily regimes of PAR shown in Fig. 6 illustrate a day with irregular, mostly overcast weather. The general trend of variation is similar for all three sensors, but high values for the transition zone during the morning (sometimes higher than simultaneous readings in the gap) indicate that the influence of the gap on bordering zones depends on gap orientation. The length of this gap was oriented east–west and the radiation on its west side is consequently higher during the morning. This influence does not reach the dense forest zone, 25 m to the west.

Table 3. Values of daily average (μmol m^{-2} s^{-1}) and daily total (mol m^{-2}) PAR at ground level over 2 days (Walter and Torquebiau, unpublished data)

Average % of transmission calculated from data of average daily total of sensor 1·1 (Table 1, see text)

	Gap Sensor 3·1	Transition zone Sensor 3·2	Dense forest Sensor 3·3
Daily average	181·3	30·1	5·6
Daily total	7·8	1·3	0·24
Daily average	154	20·4	8·2
Daily total	6·6	0·9	0·35
Average daily total	7·2	1·1	0·3
Average % transmission	19·8	3	0·83

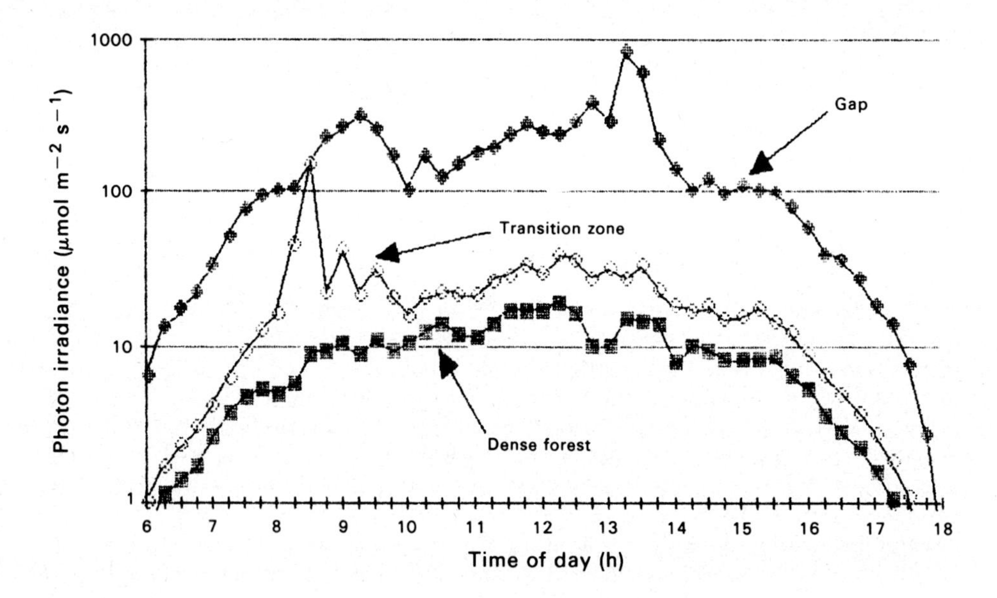

Fig. 6. Daily regimes of PAR at ground level; horizontal gradient. Log scale on ordinate (Walter and Torquebiau, unpublished data).

Discussion

Layering is a favourite theme among tropical rain forest ecologists (see Kira *et al.* 1969; Oldeman 1983) and discussion continues as to whether layering is an invariate feature of these forests or not. Oldeman (1983) has demonstrated that layering is a feature to be observed against a framework of forest dynamics: a layered architecture can be observed only in steady-state eco-units and this layering is defined exclusively by sets of trees of the present. Following this definition, the median layer described in site 1 is not, strictly speaking, a layer since it includes two small trees of the future (Nos 98 and 112), together with a tree of the present, No. 101. All three were, however, considered as a single layer unit for the purpose of the radiation measurement.

Torquebiau (1986) has also shown that layering is linked with the ability of trees to reiterate (see Hallé *et al.* 1978) at certain thresholds. These thresholds are likely (although not exclusively) to be radiation thresholds and are genetically determined, so that a tree is characteristic, when mature, of a particular forest layer. The light climate under the main forest canopy is in turn determined by the angle of incidence of sun radiation and the geometry of tree crowns and hence layering can be assessed through mathematical modelling (e.g. see Terborgh 1985).

The profile diagrams (Figs 2 and 3) show that subunits can be observed in a given steady-state eco-unit, based on the absence versus presence of layering. One aim of the present comparative study of PAR in layered and non-layered forest was to try to describe these subunits in terms of radiation. The strikingly different gradients of PAR between sites 1 and 2 confirm the existence of such subunits and imply that forest dynamics processes are acting at a fine grain level, i.e. inside eco-units recognised from upper canopy trees. This agrees with Oldeman's view (1983) that eco-units become smaller from pioneer to mature forest, with a greater number of species and providing a greater number of niches.

The profile diagrams (Figs 2 and 3) show that subunits studied are indeed very small, more or less limited to the vertical space under an upper canopy tree (e.g. see Oldeman's one tree 'climax' or 'primary' eco-units, 1983). Moving more than 5 or 10 m in any direction from the sensor locations, one would find very different architectures and, therefore, radiation regimes.

The PAR inversion observed in site 1 is an ecological factor which favours eco-unit development, allowing plants of the lower layers to grow faster: owing to this PAR inversion, plants at the bottom of an empty layer receive more radiation than plants at the base of the upper layer. Plants below the empty zone will consequently grow faster and tend to decrease the size of that empty layer. Layering thus appears to induce PAR inversion but not vice versa. Terborgh (1985) has also shown that midstory forest species regulate their heights in such a way as to maximise the interception of light energy contained in sunflecks, which suggests that stratification of forests has some adaptive significance.

Data on vertical illuminance modulation (selenium photoelectric cells) in a tropical forest in Malaysia by Yoda (1974) agree with the present work that the vertical light gradient is modulated by layering. No mention of any vertical light inversion was found in the literature, except data by Miller (1969) in openings in canopies of aspen and oak forests in North America which show increasing solar radiation (cal cm^{-2} min^{-1}) with depth. The increase occurs only with high direct and low sky radiation and is thought to be linked with the reflected and transmitted component of solar radiation from the leaves immediately above the opening. Early data by Allee (1926) in a tropical rainforest in Panama show an inversion of light intensity (foot-candles) linked with a greater importance of sunflecks at certain heights of the forest, suggesting layering.

Differences between vertical gradients do not necessarily imply that total amount of PAR reaching the ground will be different. As shown by the values at ground level in sites 1 and 2, these amounts are not significantly different despite dissimilar forest architecture. In this environment, temporal distribution of PAR in the form of sunflecks is highly heterogeneous and has greater consequences for leaf carbon gain than the total integrated PAR (Pearcy *et al.* 1985; Chazdon and Pearcy 1986). This same principle applies not only at ground level but in the whole lower third of the steady-state forest, where daily total PAR seldom exceeds 2% of incident radiation. Since the steady-state zones account for approximately 60% of the forest area (Torquebiau 1986), it is 20% of the above-ground forest volume which is under this dynamic photosynthetic environment.

Different studies (e.g. Cachan and Duval 1963; Björkman and Ludlow 1972; Chazdon and Fetcher 1984) have shown that radiation in the understory of tropical forests is not a simple function of the incident radiation but that the proportion of radiation in the understory is higher on cloudy days. A better transmission of diffuse radiation, relatively more important on overcast days, is generally thought to be the mechanism responsible (e.g. Cachan and Duval 1963).

The values for daily integrated PAR at ground level found in Sumatra are slightly lower than those published by Chazdon and Fetcher (1984) for La Selva tropical rainforest in Costa Rica: average 0·34 mol m^{-2} over 13 days in the wet and dry seasons. Björkman and Ludlow (1972) found daily total values of PAR of 0·21 mol m^{-2} (average over 12 days) in a subtropical rainforest in Queensland, Australia, where daily total in an open environment was 44·2 mol m^{-2}, giving a rather low transmission value of 0·48%. Data by Pearcy (1983) in a subtropical evergreen forest of Hawaii show higher transmission values, ranging from 1·5 to 3·8%. Data of Torquebiau and Walter (in press) on the transmission of solar radiation (pyranometers: 0·3–3 μm) in the Sumatra plot described in this paper show values which are always higher than transmission of PAR, e.g. 4% transmitted at ground level in site 1 (location of sensors 1·6 and 3·3), 8% in the transition zone (location of sensor 3·2) and 35% in the gap (location of sensor 3·1). Such high values confirm that leaves preferentially absorb PAR compared with near infrared radiation (see e.g. Baldocchi *et al.* 1984).

Treefall gaps in this Sumatra forest type have been found to range from 17 to 419 m^2, and mean gap size is about 100 m^2 (Torquebiau 1986 and unpublished data). This range extends from the very small, partial gaps, originating from the fall of heavy branches, to large gaps originating from the fall of several trees. Microclimatic perturbations following treefall are variable among these different gaps and doubtless generate varied mosaic patches. The 320 m^2 gap studied here is a rather large one for the area and was evidently recent (some months old) at the time of the field inventory since very little regrowth was observed in it.

In a tropical rainforest in Costa Rica, Chazdon and Fetcher (1984) found daily total PAR (average over 3 days) in a 400 m^2 gap varying from 5·85 mol m^{-2} (dry season) to 7·99 mol m^{-2} (wet season), values which agree with the 7·2 mol m^{-2} found in the 320 m^2 Sumatra gap. This last gap, which has been monitored for $2\frac{1}{2}$ years, has shown no subsequent development of pioneer trees. A review by Brokaw (1985) indicates that the critical gap size for invasion by pioneer trees in neotropical rainforests varies from 150 to 1000 m^2.

Using the profile diagrams as guides to select relevant examples of forest structure, field measurements of PAR show that the distribution of radiation is spatially heterogeneous. The values of PAR discussed in this paper are valid only for the immediate surroundings of the sensors, probably in a radius of the order of some metres at most. The profile diagrams show that forest structure changes rapidly when moving in any direction from the sensors (see also Pearcy 1983). In Fig. 4, for example, it is evident that the set of 10 m high trees is developed in the surroundings only of the sensors but not in the rest of the plot. Underlayer structure in other parts of site 2 is similar to structure observed in site 1 and probably accounts for the PAR inversion observed on some days between sensors 2·4 and 2·5 (Table 2), although the average values did not differ significantly. The aim of this work was to characterise PAR environments in relation to forest structure at this fine grain level, and no statistical sampling by several sensors at similar heights was planned. Rather it is suggested that, by combining the study of spatial and temporal variations of PAR in microenvironments and taking into account the dynamic nature of the photosynthetic apparatus (Pearcy, in press), it will be possible to obtain better understanding of photosynthetic processes of rainforest species.

Acknowledgments

The work described was supported by the French Cooperation Programme with the Regional Center for Tropical Biology (BIOTROP) of the South East Asia Ministries of Education Organization (SEAMEO). The Research School of Biological Sciences, Australian National University, Canberra, provided a travel grant to attend the Robertson Symposium. The author wishes to thank J. M. N. Walter (Strasbourg Botanical Institute, France) for his assistance on data analysis and for a helpful collaboration in forest microclimatology which goes well beyond the scope of the present paper. Thanks are also due to C. B. Osmond for useful comments on the manuscript.

References

Allee, W. C. (1926). Measurement of environmental factors in the tropical rainforest of Panama. *Ecology* **7**, 273–302.

Baldocchi, D. D., Matt, D. R., Hutchison, B. A., and McMillen, R. T. (1984). Solar radiation within an oak–hickory forest: an evaluation of the extinction coefficients for several radiation components during fully-leafed and leafless periods. *Agric. For. Meteorol.* **32**, 307–22.

Björkman, O., and Ludlow, M. M. (1972). Characterization of the light climate on the floor of a Queensland rainforest. *Carnegie Inst. Wash. Year Book* **71**, 85–94.

Bourlière, F. (1983). Animal species diversity in tropical forests. In 'Tropical Rainforest Ecosystems'. (Ed. F. B. Golley.) pp. 77–92. (Elsevier: Amsterdam.)

Brokaw, N. V. L. (1983). Treefalls: frequency, timing and consequences. In 'The Ecology of a Tropical Forest. Seasonal Rhythms and Longterm Changes'. (Eds E. G. Leigh Jr, A. S. Rand and D. M. Windsor.) pp. 101–8. (Oxford University Press: Oxford.)

Brokaw, N. V. L. (1985). Treefalls, regrowth and community structure in tropical forests. In 'The Ecology of Natural Disturbance and Patch Dynamics.' (Eds S. T. A. Pickett and P. S. White.) pp. 53–69. (Academic Press: New York.)

Brokaw, N. V. L. (1987). Gap-phase regeneration of three pioneer tree species in a tropical forest. *J. Ecol.* **75**, 9–19.

Cachan, P., and Duval, J. (1963). Variations microclimatiques verticales et saisonnières dans la forêt sempervirente de basse Côte d'lvoire. *Ann. Fac. Sci. Univ. Dakar* **8**, 5–87.

Charles-Dominique, P., Atramentowicz, M., Charles-Dominique, M., Gerard, H., Hladik, A., Hladik, C. M., and Prevost, M. F. (1981). Les mammifères frugivores arboricoles nocturnes d'une forêt guyanaise: inter-relations plantes–animaux. *Rev. Ecol. Terre Vie* **35**, 341–435.

Chazdon, R. L., and Fetcher, N. (1984). Photosynthetic light environments in a lowland tropical rainforest in Costa Rica. *J. Ecol.* **72**, 553–64.

Chazdon, R. L., and Pearcy, R. W. (1986). Photosynthetic responses to light variation in rainforest species. II. Carbon gain and light utilization during lightflecks. *Oecologia* **69**, 524–31.

Denslow, J. S. (1980). Gap partitioning among tropical rainforest. In 'Tropical Succession'. (Ed. J. Ewel.) *Biotropica*, Vol. 12, No. 2, Supplement, pp. 47–55.

Fontanel, J., and Chantefort, A. (1978). Bioclimates of the Indonesian Archipelago. [4 maps + book.] (Institut Français de Pondicherry: Pondicherry, India.)

Granville, J. J. de (1978). Recherches sur la flore et la végétation guyanaises. Thèse Doctorat d'Etat, Montpellier University.

Hallè, F., Oldeman, R. A. A., and Tomlinson, P. B. (1978). 'Tropical Trees and Forest, an Architectural Analysis.' (Springer-Verlag: Berlin.)

Kahn, F. (1982). La reconstitution de la forêt tropicale humide. Sud-Ouest de la Côte d'lvoire. ORSTOM, Report No. 97, Paris.

Kahn, F. (1986). Life forms of Amazonian palms in relation to forest structure and dynamics. *Biotropica* **18**, 214–18.

Kira, T., Shinozaki, K., and Hozumi, K. (1969). Structure of forest canopies as related to their primary productivity. *Plant Cell Physiol.* **10**, 129–42.

Martinez, M. (1980). Aspectos sinecológicos del proceso de renovación natural de una selva alta perennifolia. Tesis Profesional, UNAM, Mexico.

Miller, P. C. (1969). Solar radiation profiles in openings in canopies of aspen and oak. *Science* **164**, 368–9.

Oldeman, R. A. A. (1974). L'architecture de la forêt guyanaise. ORSTOM, Report No. 73, Paris.

Oldeman, R. A. A. (1983). Tropical rainforest architecture, silvigenesis and diversity. In 'Tropical Rainforest: Ecology and Management'. (Eds S. L. Sutton, T. C. Whitmore and A. C. Chadwick.) pp. 139–50. British Ecol. Soc. Spec. Publ. No. 2. (Blackwell: Oxford.)

Pearcy, R. W. (1983). The light environment and growth of C_3 and C_4 tree species in the understory of a Hawaiian forest. *Oecologia* **58**, 19–25.

Pearcy, R. W. (in press). Photosynthetic responses of tropical forest trees. Biotrop Symposium on Tropical Plant Ecophysiology, Bogor, Indonesia, December 1985.

Pearcy, R. W., Osteryoung, K., and Calkin, H. W. (1985). Photosynthetic responses to dynamic light environments by Hawaiian trees. *Plant Physiol.* **79**, 896–902.

Pickett, S. T., and White, P. S. (Eds) (1985). 'The Ecology of Natural Disturbance and Patch Dynamics.' (Academic Press: New York.)

Reifsnyder, W. E., Furnival, G. M., and Horowitz, J. L. (1971 / 1972). Spatial and temporal distribution beneath forest canopies. *Agric. Meteorol.* **9**, 21–37.

Salick, J., Herrera, R., and Jordan, C. F. (1983). Termitaria: nutrient patchiness in nutrient-deficient rainforests. *Biotropica* **15**, 1–7.

Terborgh, J. (1985). The vertical component of plant species diversity in temperate and tropical forests. *Am. Nat.* **126**, 760–76.

Torquebiau, E. F. (1986). Mosaic patterns in dipterocarp rainforest in Indonesia, and their implications for practical forestry. *J. Trop. Ecol.* **2**, 301–25.

Torquebiau, E. F., and Walter, J. M. N. (in press). Forest architecture and the transmission of solar radiation in a rainforest in Sumatra, first account. Biotrop Symposium on Tropical Plant Ecophysiology, Bogor, Indonesia, December 1985.

Whitmore, T. C. (1975). 'Tropical Rainforests of the Far East.' (Clarendon Press: Oxford.)

Whitmore, T. C. (1984). 'Tropical Rainforests of the Far East.' 2nd edn. (Clarendon Press: Oxford.)

Yoda, K. (1974). Three-dimensional distribution of light intensity in a tropical rainforest of West Malaysia. *Jpn J. Ecol.* **24**, 247–54.

On the Role of Photosynthetic Responses in Constraining the Habitat Distribution of Rainforest Plants*

Christopher B. Field

Department of Plant Biology, Carnegie Institution of Washington,
290 Panama Street, Stanford, CA 94305, U.S.A.

Abstract

Two rainforest species that differ in the range of light environments encountered by an individual provide a model system for illustrating implications of diurnal variation and spatial heterogeneity of the light environment within single canopies. The two species are similar in the acclimation of leaf-level photosynthetic characteristics to the leaf's past light environment, but are distinguished by the limited ability of the high-light species to decrease dark respiration in response to growth under shade. Over 24 h, total carbon gain from a leaf increases with increasing light, but may increase or decrease with increasing nitrogen. Intermediate nitrogen levels and photosynthetic capacities yield maximum daily carbon gain in only a narrow range of light environments. The ratio of daily photosynthesis to leaf nitrogen is a complex function of nitrogen, photosynthetic characteristics, and light availability in a microsite. Nitrogen levels in real leaves may be close to the optima, but leaf nitrogen levels are clumped relative to the optima. Daily carbon gain from a canopy of fixed total nitrogen tends to increase as the leaf area of the canopy decreases, a trend caused by two factors. First, photosynthetic capacity is a nearly linear function of leaf nitrogen up to the highest nitrogen levels observed in nature. Second, only leaves with nitrogen levels above a substantial threshold are competent to generate positive rates of net photosynthesis. Differences between the species in canopy architecture and leaf duration have counteracting effects on long-term nitrogen-use efficiency.

Introduction

Tropical rainforest light environments are spatially diverse and temporally complex. Light availability varies vertically, from the forest floor to the top of the canopy (Chiariello 1984); horizontally, from disturbed to undisturbed segments of forest (Chazdon and Fetcher 1984); and temporally, on several scales. Sunflecks, clouds, the diurnal cycle, and the seasonal cycle can each impose a fluctuation of one to two orders of magnitude in light availability. On a time scale of years to decades, whole forests are destroyed and rebuilt. The march of disturbance and revegetation continuously reconfigures the distribution of light-absorbing elements. Superimposed on all these patterns in the habitat, individual plants generate additional heterogeneity through their own growth and canopy structure.

As a result of this heterogeneity, information on the steady-state responses of single-leaf photosynthesis to environmental variables may be of limited utility in elucidating physiological determinants of ecological success. Studies that focus directly on the environmental heterogeneity and on plant responses to it offer opportunities for progress in identifying novel adaptations and extending mechanistic studies to ecologically relevant contexts. It is very difficult to identify, *a priori*, the sources of heterogeneity likely to impact most heavily on ecological success. Pearcy (1988) demonstrates that differences in responses to short light flecks can have large implications for the carbon

*C.I.W.-D.P.B. Publication No. 991

0310-7841/88/010343$03.00

balance. Chazdon (1986) found that the vertical gradient of light near the rainforest floor places some plants above and others below the level of whole-plant light compensation. Even gradients of light availability within leaves can have profound implications for photosynthesis (Terashima and Saeki 1985).

Here, I focus on two types of variation in the light environment; spatial heterogeneity within the canopies of individual plants, and the temporal heterogeneity of a diurnal cycle. Using as examples two rainforest species that differ in light requirements, I summarise the contribution of steady-state photosynthetic responses to explaining the differences in habitat distributions and then illustrate new considerations that emerge as the scale for temporal and spatial integration broadens. My emphasis on interactions between carbon and nitrogen economies emerges partly from the quantitative importance of those interactions, and partly because the basis of the interactions is well enough understood that nitrogen can be viewed a model system for resources affecting photosynthesis.

A Model System

Piper (Piperaceae) is a large (500–2000 species) pantropical genus of herbs, shrubs, vines and trees. Species are often locally diverse, with as many as 50 in a single forest. Some of the habitat partitioning among species may reflect differences in moisture or nutrient requirements, but most of the sorting appears to be along light gradients. *Piper* seeds are generally animal dispersed. Bats are the most important dispersing agents (Fleming 1986). However, *Piper* species that differ in habitat distribution typically share species of bat frugivores (Fleming 1985), and most of the seeds are dispersed to semi-permanent night-feeding roosts quite close to the site of harvest (Fleming 1981), indicating that habitat-selective dispersal is probably not the primary determinant of species' distributions. Germination requirements, in contrast, are very important. Large-seeded species, of which *P. nigrum*, common black pepper, is representative, usually have no light requirement for germination and little potential for seed dormancy (Vazquez-Yanes 1976). Many of these species are competent to grow to maturity in the heavily shaded rainforest understory. As seed size decreases, both the light requirements for germination and the potential for dormancy tend to be better developed. Of the many species that germinate in recent disturbances or gaps, some persist only until shaded (Gomez-Pompa 1971). Individuals of these species encounter only a limited range of light environments. Other species that germinate in gaps persist until the forest re-establishes, and individuals of these species experience both high-light and low-light environments.

Piper hispidum Swartz is a shrub that germinates shortly after natural or man-made disturbance but is also common in the understory of secondary forest and may persist into heavily shaded primary forest. *Piper auritum* Kunth is a pioneer tree that rarely occupies shaded sites (Gomez-Pompa 1971) but wilts rapidly when exposed to full sun (Chiariello *et al.* 1987). Germination in both species is phytochrome controlled (Vazquez-Yanes and Orozco-Segovia 1982; Vazquez-Yanes and Smith 1982), and both species germinate in nature only after repeated exposure to daylight unfiltered by vegetation or, in *P. hispidum*, after several months of storage in the soil (Orozco-Segovia 1986). Both species are abundant at the 'Los Tuxtlas' Biological Preserve, in Veracruz, Mexico, the site of the studies described here.

Photon flux in the photosynthetically active wavelengths (photon irradiance in the terminology of Bell and Rose (1981) — abbreviated PFD here) in the natural habitats of *P. hispidum* (the persistent or generalist species) and *P. auritum* (the ruderal or specialist species) was characterised from the computerised analysis of hemispherical photographs, as described by Chazdon and Field (1987*a*). This technique, based on the approach of Anderson (1964), combines several kinds of calculations. It compares the spatial distribution of PFD absorbing elements and position of the sun to estimate the time-dependent penetration of direct and diffuse PFD. The technique then combines the estimates of radiation penetration with estimates of direct and diffuse PFD above the vegetation to predict potential PFD in the microsite of the camera location. This estimate, corrected for the angular relationship between sun and leaf, yields a prediction, updated at 2-min intervals, of the direct plus diffuse PFD either directly above a plant or impinging on both sides of a particular leaf. As used here, the technique includes several simplifying assumptions: (1) that the sky is cloudless and that direct PFD always penetrates to the top of the canopy; (2) that all absorbing elements (mostly leaves and stems) have no reflectance or transmittance; and (3) that every pixel analysed is either completely open or completely obscured. These assumptions certainly result in some errors. However, estimates of daily PFD from the computerised analysis of hemispherical photographs are highly correlated with measurements from photovoltaic PFD sensors (Chazdon and Field 1987*a*). The technique also indicates the timing and intensity of large sunflecks with good accuracy, though it is less effective with short sunflecks (Chazdon and Field 1987*a*).

Table 1. Potential PFD in the microsites of randomly selected, naturally growing individuals of *P. hispidum* and *P. auritum*

Daily PFD is estimated for the summer solstice from the computerised analysis of hemispherical photographs. The estimates are specific for a horizontal surface at the top of each canopy. The means are different at the 0·1% confidence level

Species	Daily PFD (mol m^{-2} day^{-1}) Mean	Maximum	Minimum	Coefficient of Variation
P. hispidum (persistent)	8·9	19·3	0·5	79%
P. auritum (high-light)	26·5	42·9	11·9	40%

I analysed hemispherical photographs taken at the top of the canopy of 10 naturally growing, randomly selected individuals of *P. hispidum* and *P. auritum*. All individuals of the specialist, *P. auritum*, were in gaps and clearings while the persistent species, *P. hispidum*, was in gaps, clearings and understory. Daily potential PFD, on the summer solstice, was approximately three times higher for the open site specialist, *P. auritum*, than for the persistent species, *P. hispidum* (Table 1). The highest estimated PFD for any *P. hispidum* was somewhat below the mean for *P. auritum*. While individuals of both species occurred in recent gaps with very open canopies, maximum PFD for *P. auritum* was higher than that for *P. hispidum* as a result of *P. auritum*'s taller stature. Although the two species coexist in the most open rainforest sites, the persistent species, *P. hispidum*, rarely encounters the maximum daily PFD experienced by the open site specialist, *P. auritum*. The absolute range of daily PFD among *P. auritum* individuals was over 60% larger than that among *P. hispidum* individuals, but the proportional range was much smaller. Potential daily PFD ranged approximately 4-fold among individuals of the open site specialist but over 20-fold among individuals of the persistent species.

The analysis of the hemispherical photographs supports the hypothesis that, in comparison to the open site specialist, the persistent species *P. hispidum* must acclimate to a broader range of light environments. However, individuals of both species are successful over a range of light conditions, and that range would be even broader if it reflected seasonal as well as site to site variation. Thus differences in acclimation potential, if they exist, should be quantitative rather than qualitative.

Controlled-environment Photosynthetic Responses

Walters and Field (1987) compared the photosynthetic responses of naturally growing individuals of *P. hispidum* and *P. auritum* selected from each species' entire range of light environments. Within each species, the relationship between photosynthetic characteristics (assessed from response curves using the steady-state gas exchange system of Field *et al.* (1982)) and potential daily PFD (assessed from hemispherical photographs as described above) was similar to that described for many species grown under controlled conditions (Björkman 1981). A_{max}, the light-saturated photosynthesis rate at normal CO_2, temperature, and humidity increased with increasing daily PFD, as did the PFD for light saturation. The initial slopes of the PFD responses of photosynthesis were independent of PFD during growth. The initial slopes of the A–c_i responses

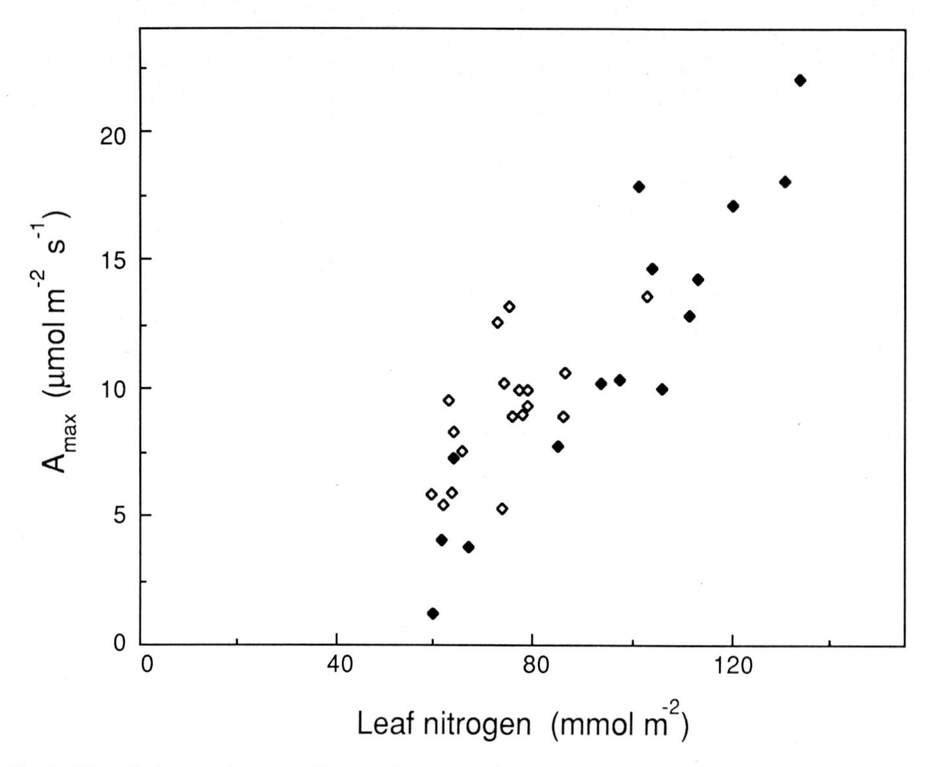

Fig. 1. The relationship between photosynthetic capacity (A_{max}) and leaf nitrogen for individual leaves of naturally growing *P. hispidum* (closed symbols) and *P. auritum* (open symbols). For *P. hispidum*, $A_{max} = -10{\cdot}51 + 0{\cdot}228N$, $r = 0{\cdot}92$, $P<0{\cdot}001$. For *P. auritum*, $A_{max} = -2{\cdot}16 + 0{\cdot}152N$, $r = 0{\cdot}65$, $P<0{\cdot}003$. Data from Walters and Field (1987) and Chazdon and Field (1987*b*).

(photosynthesis–intercellular CO_2) were also similar between species and greater in plants growing in sites of high than low PFD, indicating decreased investment in Rubisco (ribulose-1,5-bisphosphate carboxylase/oxygenase) in plants from low-PFD sites (Caemmerer and Farquhar 1981). In plants of high-PFD sites, the initial slope of the A–c_i response was sensitive to the PFD during the measurement, even at PFD levels above 300 μmol m^{-2} s^{-1}, suggesting the light-dependent inactivation of Rubisco described by Mott *et al.* (1984). The absence of this response in low-PFD grown plants may be adaptive if fully activated Rubisco increases the efficiency of sunfleck utilisation (Chazdon and Pearcy 1986).

In both species, A_{max} was highly correlated with leaf nitrogen content (Fig. 1) and with daily PFD during growth (Fig. 2). The responses for the two species were very similar. While there are no statistically significant differences between the linear regressions of A_{max} on leaf nitrogen for the two species, the regression lines are not coincident, and the difference, especially if future data demonstrate its statistical significance, may have important consequences. The regression analysis suggests that, at least at low and intermediate nitrogen levels, the high-light specialist, *P. auritum*, may realise a greater A_{max} for a given nitrogen investment than the persistent species. This pattern is consistent with the observation that high-light plants allocate proportionately more nitrogen to Rubisco and less to components of light harvesting than shade plants (Seemann *et al.* 1987). If Rubisco is the primary determinant of A_{max} (Woodrow and Berry 1988), then decreasing nitrogen investments in all other functions should tend to increase A_{max}

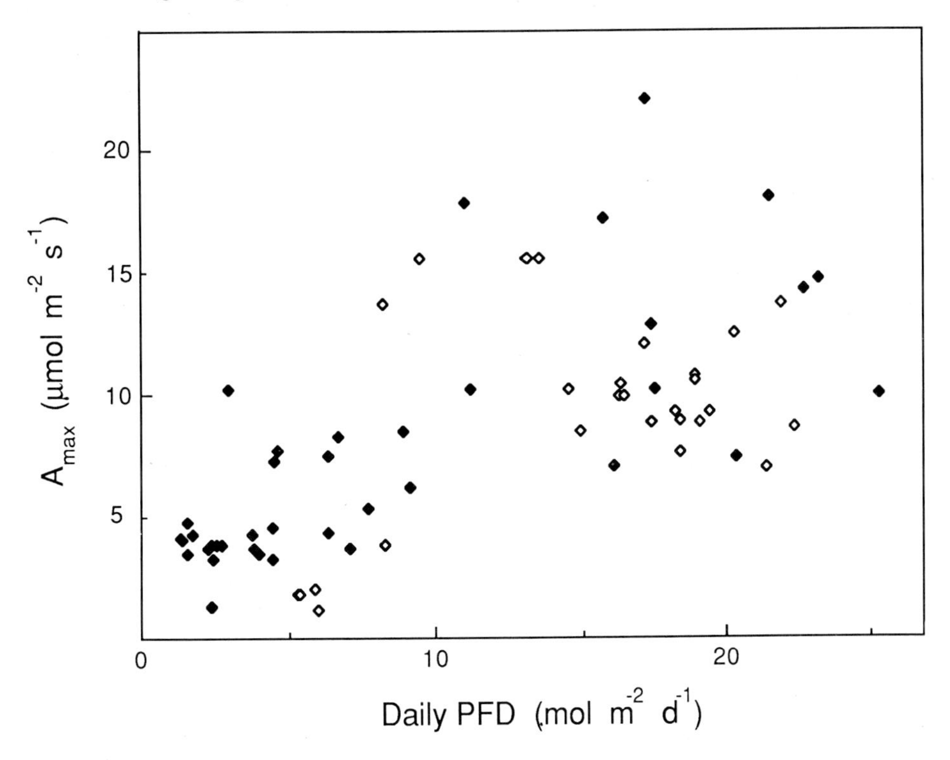

Fig. 2. The relationship between photosynthetic capacity (A_{max}) and daily PFD for individual leaves of naturally growing *P. hispidum* (closed symbols) and *P. auritum* (open symbols). Daily PFD, estimated from hemispherical photographs, is an average over 2 or 3 months preceding the photosynthesis measurement. For *P. hispidum*, $A_{max} = 3{\cdot}24 + 0{\cdot}514\text{PFD}$, $r = 0{\cdot}75$, $P<0{\cdot}001$. For *P. auritum*, $A_{max} = 5{\cdot}18 + 0{\cdot}267\text{PFD}$, $r = 0{\cdot}65$, $P<0{\cdot}001$. Data from Walters and Field (1987) and Chazdon and Field (1987*b*).

at any nitrogen level. However, a test of this possibility depends on future biochemical data, and the indication from the gas exchange is that the two species, despite the large differences in natural distributions, share very similar sensitivities of photosynthetic characteristics to leaf nitrogen and the light environment. The species did, however, differ in one gas-exchange characteristic. Dark respiration was correlated with A_{max} in the persistent species, *P. hispidum*, but not in the open site specialist (Fig. 3; also Walters and Field 1987). Decreased CO_2 loss during darkness potentially yields a carbon balance benefit for the species that generally succeeds in low-PFD sites. The overall picture from the gas exchange under controlled conditions was that the species are generally very similar, but with subtle and potentially important differences. The similarity of the responses argues against the hypothesis that the persistent species is more competent to adjust photosynthetic characteristics in response to the light environment during growth. It also tends to argue against the hypothesis that photosynthetic characteristics play a role in determining the persistence of these two species through secondary succession. However, photosynthetic characteristics are several steps removed from ecological success. Spanning those steps is very difficult in the absence of information about the magnitude of a difference in photosynthetic characteristics sufficient to account for a particular difference in habitat breadth. A more accurate picture of the role of photosynthesis in success requires integrating the carbon income over biologically relevant time periods. Ultimately, it will be useful to extend the integration over the entire life of the plant. As a first step, I begin with annual integrations, comprised of simulations for one day from each month.

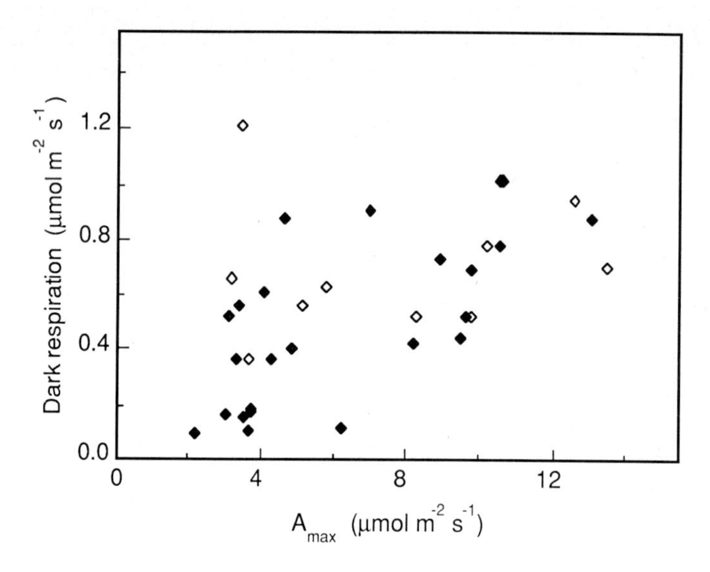

Fig. 3. The relationship between dark respiration and photosynthetic capacity (A_{max}) for individual leaves of naturally growing *P. hispidum* (closed symbols) and *P. auritum* (open symbols). For *P. hispidum*, dark respiration = $0{\cdot}114 + 0{\cdot}059A_{max}$, $r = 0{\cdot}65$, $P<0{\cdot}001$. For *P. auritum*, dark respiration = $0{\cdot}627 + 0{\cdot}013A_{max}$, $r = 0{\cdot}20$, $P>0{\cdot}05$.

Daily Photosynthesis

Simulation Methods

To estimate daily photosynthesis, I combine two types of simulation models. The daily course of PFD, at 2-min intervals, is estimated from the computerised analysis of hemispherical photographs, as described above and by Chazdon and Field (1987*a*). Each PFD estimate, specific to the location, angle, and orientation of a particular leaf, is one input for a photosynthesis model. By changing the parameters of the photosynthesis model, it is possible to assess the implications for daily photosynthesis of changing the photosynthetic characteristics of a leaf in a particular microsite.

The photosynthesis model used for these simulations combines elements of the biochemical model of Farquhar *et al.* (1980) and the empirical model of Johnson and Thornley (1984). It is a steady-state model and does not account for the evidence that carbon gain may be sensitive to fluctuating PFD (Chazdon and Pearcy 1986; Pearcy 1988). The basic equation for the response of photosynthesis to PFD is a nonrectangular hyperbola (equation 1 in Johnson and Thornley (1984)), but the parameters of that equation, the light-saturated photosynthesis rate and the quantum yield of photosynthesis, are derived from the model of Farquhar *et al.* (1980). This equation yields net CO_2 fixation in the absence of 'day' or mitochondrial respiration. 'Day' respiration is calculated from the empirically determined relationship between A_{max} and dark respiration (Fig. 3), assuming that the light dependence of 'day' respiration is the same as that reported by Brooks and Farquhar (1985) for spinach. I also assume that, over an entire night, the mean rate of dark respiration is 75% of the rate measured shortly after placing the leaf in darkness. This assumption, quantitatively supported by unpublished data on *P. hispidum* (T. Hennessey, personal communication), is intended to account for the dependence of dark respiration on carbohydrate reserves (Azcón-Bieto *et al.* 1983). None of the conclusions from the simulations depends on either of the assumptions concerning respiration. The combination of the two modelling approaches allows great flexibility in fitting light response curves, but fails to account for the temperature dependence of electron transport. Since daily and seasonal temperature variation in tropical rainforests is small, the simulations here assume constant temperature. To further simplify the model, I also assume that c_i, intercellular CO_2 concentration, is constant at 270 μmol mol^{-1}, the mean value observed in the field. The assumption of constant c_i is, for many species, quite robust across changes in PFD when temperature and humidity are constant, and when PFD is moderate to high (Wong *et al.* 1979). Temperature and humidity are reasonably constant in many tropical forest microsites, but stomatal conductance in both species is sensitive to humidity (Mooney *et al.* 1983; Chiariello *et al.* 1987)).

I simulated daily photosynthesis as a function of leaf nitrogen content for both species in the microsites of six leaves distributed throughout a single canopy of *P. hispidum* and six more in a single canopy of *P. auritum*. Both plants were growing naturally in a clearing, approximately 10 m apart. Photosynthetic capacity, leaf nitrogen, and the light environment of these leaves is presented by Chazdon and Field (1987*b*). Carbon balance is presented as an annual mean derived from monthly simulations. All simulations assume cloudless skies and do not allow for wilting in *P. auritum*. The simulations assume that leaf chlorophyll content is constant at 0·4 g m^{-2} and vary the ratio of Rubisco to chlorophyll. The nitrogen and PFD dependence of A_{max} for both species are given in Figs 1 and 2. The PFD dependence of leaf nitrogen was derived by combining the data in Figs 1 and 2. Fig. 3 gives the A_{max} dependence of dark respiration.

Simulation Results

Daily PFD on a horizontal surface above the vegetation varied from 47 mol m^{-2} day^{-1} at the summer solstice to 25 mol m^{-2} day^{-1} at the winter solstice. Daily PFD on individual leaves of *P. auritum*, the open site specialist, ranged from 8·1 to 33·3 mol m^{-2} day^{-1}. Annual mean PFD was more conservative, ranging from 13·5 to 19·9 mol m^{-2} day^{-1} with a mean of 17·1 and a coefficient of variation of 15%. Wilting in leaves of *P. auritum* exposed to full sun probably reduces the highest values to yield an even lower coefficient of variation under natural conditions (Chiariello *et al.* 1987). For the persistent species, *P. hispidum*, the range was 0·2–17·1 mol m^{-2} day^{-1}. Annual mean PFD ranged from 0·7 to 10·8 mol m^{-2} day^{-1} with a mean of 7·1 and a coefficient of variation of 55%. Even though the two plants were in essentially the same habitat, differences in leaf inclination and canopy architecture made daily PFD lower and more variable in the canopy of the persistent species.

Daily net photosynthesis (A_{day}) generally increased with increasing PFD in both species, but the pattern was not monotonic (Fig. 4). The deviation from a monotonic relationship is a consequence of the fact that, from the perspective of the carbon balance, PFD above the level sufficient to saturate photosynthesis is wasted. Thus, the ratio of A_{day} to daily PFD increases as the proportion of PFD above saturation decreases. The pattern for the two species was similar. However, when photosynthetic capacity was low, A_{day} was greater in the persistent species, *P. hispidum*, but the situation was reversed when photosynthetic capacity was high. This pattern is a consequence of the difference, between the species, in the relationship between respiration and A_{max}. Respiration in the high-light specialist, *P. auritum*, was constant, but it varied with A_{max} in the persistent species (Fig. 3).

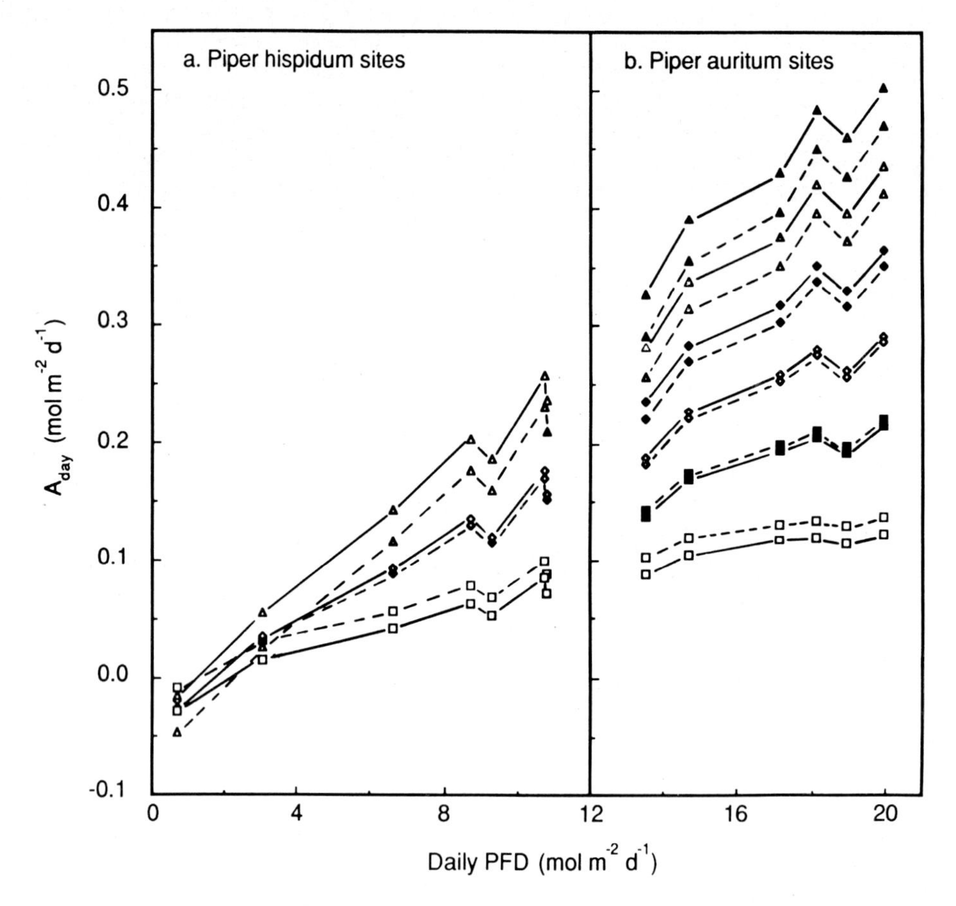

Fig. 4. Simulated relationship between daily carbon gain (A_{day}) and daily PFD (both are presented as annual means) for leaves with the photosynthetic characteristics of *P. hispidum* (dashed lines) and *P. auritum* (solid lines). Daily PFD was estimated from hemispherical photographs taken in the canopy of a single individual of *P. hispidum* (*a*) and *P. auritum* (*b*) and analysed with calculations specific to the angle and azimuth of particular leaves. The daily course of PFD is an annual mean calculated from monthly estimates. The lines with different symbols correspond to photosynthetic capacities of 4 (□), 8 (■), 12 (◇), 16 (♦), 20 (△), and 24 (▲) μmol $m^{-2} s^{-1}$.

For the high-light specialist, *P. auritum*, A_{day} increases with increasing A_{max} or leaf nitrogen (Fig. 4). This pattern holds for the persistent species in high PFD microsites but, at low PFDs, the carbon-gain benefits of increasing photosynthetic capacity become insufficient to offset the carbon losses caused by the higher dark and 'day' respiration rates that accompany higher A_{max} (Fig. 4). The range of PFDs where intermediate A_{max}

values yield maximum daily photosynthesis in *P. hispidum* is narrow, suggesting that whole canopy photosynthesis is maximised when all leaves have either high or low photosynthetic capacities, depending on their microsite. In the high-light specialist, *P. auritum*, A_{day} always increases with A_{max}, suggesting that canopy photosynthesis is maximised when all leaves have the highest possible photosynthetic capacities. Of course, neither pattern actually occurs. A_{max} increases gradually with PFD (Fig. 2).

Does the difference between the observed distribution of A_{max} and the distribution that maximises daily carbon gain indicate that real canopies are not configured to maximise carbon gain? Several considerations are relevant. First, the estimates of PFD from canopy photographs tend to underrepresent intermediate light levels (Chazdon and Field

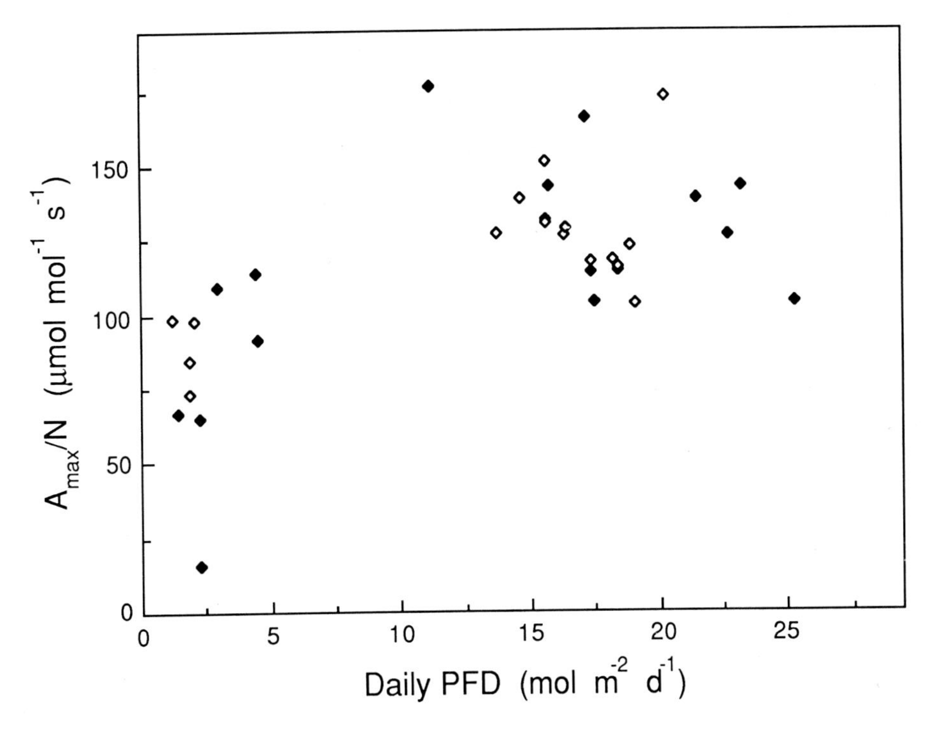

Fig. 5. The relationship between instantaneous nitrogen use efficiency (A_{max}/N) and daily PFD for individual leaves of naturally growing *P. hispidum* (closed symbols) and *P. auritum* (open symbols). Daily PFD, estimated from hemispherical photographs, is an average over 2 or 3 months preceding the photosynthesis measurement. For *P. hispidum*, $A_{max}/N = 77{\cdot}5 + 2{\cdot}72\text{PFD}$, $r = 0{\cdot}58$, $P<0{\cdot}022$. For *P. auritum*, $A_{max}/N = 86{\cdot}9 + 2{\cdot}39\text{PFD}$, $r = 0{\cdot}69$, $P<0{\cdot}001$. Data from Walters and Field (1987) and Chazdon and Field (1987*b*).

1987*a*), the light levels most likely to favour intermediate A_{max}. Second, these simulations including a cloudless sky assumption cannot illustrate the frequency with which individual leaves encounter PFD levels above and below the boundary for the carbon gain superiority of high A_{max}. Third, I have to this point ignored the costs of increasing A_{max}. If the costs of a high A_{max} offset the increased carbon gain a plant may, in fact, maximise the availability of photosynthate for growth and reproduction with photosynthetic characteristics that do not maximise daily photosynthesis. Among the costs of achieving a high A_{max} are the costs of transpiration (Cowan 1986; Givnish 1986), leaf nutrients (Field and Mooney 1986), and leaf construction and maintenance

(Penning de Vries *et al.* 1974; Penning de Vries 1975), as well as increased risk of damage by herbivores (Mattson 1980). Bloom *et al.* (1985) develop some of the rules necessary for accounting costs in different resources.

Nitrogen-use Efficiency

Leaf nitrogen is highly correlated with A_{max} in many species (Field and Mooney 1986), including naturally growing *Piper* species (Fig. 1). Nitrogen availability is frequently high in tropical rainforest soils (Vitousek and Denslow 1986; Vitousek and Sanford 1986). However, even when availability is high, nitrogen nutrition can consume 20–50% of a plant's carbon budget (Chapin *et al.* 1987). With costs this high, returns in photosynthesis per unit of nitrogen investment may be a major component of acclimation to contrasting light environments.

Instantaneous Nitrogen-Use Efficiency

The ratio of A_{max} to leaf nitrogen is an instantaneous index of the efficiency with which nitrogen is partitioned among components of photosynthesis, and between photosynthesis and other processes in the leaf. However, this index is specifically relevant only to the conditions under which A_{max} is measured. It cannot be used to evaluate components of nitrogen-use efficiency (NUE) related to leaf longevity or nutrient retranslocation (Rundel 1982) or to the ratio of photosynthesis to leaf nitrogen realised under natural conditions.

In both the persistent species, *P. hispidum*, and in the open-site specialist, *P. auritum*, the ratio of A_{max} to N is low in plants or leaves grown in low-PFD environments and higher in leaves or plants grown in high-PFD environments (Fig. 5). The low values for low-PFD grown leaves, an algebraic consequence of the negative *y*-intercept in the A_{max}–N relationship, may indicate a lower relative investment of nitrogen in A_{max}. It is possible that decreased investment in A_{max}, of which Rubisco is a major determinant (Woodrow and Berry 1988), reflects increased investment in light harvesting and electron transport. Seemann *et al.* (1987) observed exactly this phenomenon, but with different relative sensitivities of Rubisco and chlorophyll to light availability in sun and shade species. An alternative possibility is that decreased investment in A_{max} reflects increased nitrogen investment in non-photosynthetic reactions.

Daily Nitrogen-use Efficiency

A more ecologically relevant, though still incomplete, index of nitrogen-use efficiency is the ratio of A_{day} to N. Combining the simulations for A_{day} as a function of A_{max} with the empirically determined A_{max}–N and A_{max}–daily PFD relationships yields daily NUE for leaves with the photosynthetic characteristics of *P. hispidum* and *P. auritum* through the canopies of both species (Fig. 6). For leaves with the photosynthetic characteristics of the high-light specialist, *P. auritum*, daily NUE decreases with increasing leaf nitrogen in all microsites with daily PFD between 6 and 18 mol m^{-2} day^{-1}. In microsites with higher or lower daily PFD, daily NUE first increases, then decreases, with increasing leaf nitrogen (Fig. 6). Thus, the leaf nitrogen levels that maximise daily NUE in leaves of *P. auritum* are high in microsites of low or high daily PFD and low in sites of intermediate PFD. Leaf nitrogen levels predicted from daily PFD, using the relationships in Figs 1 and 2, are close to the optima for several leaves, especially those in the microsites of highest PFD (Fig. 6). For the low PFD microsites, microsites occupied by *P. hispidum* leaves but darker than those occupied by *P. auritum*, the nitrogen levels that maximise daily NUE in *P. auritum* may be higher or lower than the nitrogen levels predicted from regressions through real data.

For leaves with the photosynthetic characteristics of the persistent species, *P. hispidum*, daily NUE increases with leaf nitrogen in all microsites with daily PFD greater than 7 mol m^{-2} day^{-1} (Fig. 6). In these brighter microsites, the second derivatives of the daily NUE–leaf nitrogen relationships are negative, but daily NUE does not reach a maximum in the range of nitrogen concentrations simulated, which extends about 25% higher than the highest nitrogen levels measured in real leaves (Fig. 1). Thus, daily NUE in this species is maximised when leaf nitrogen levels in all but the shadiest microsites are very high.

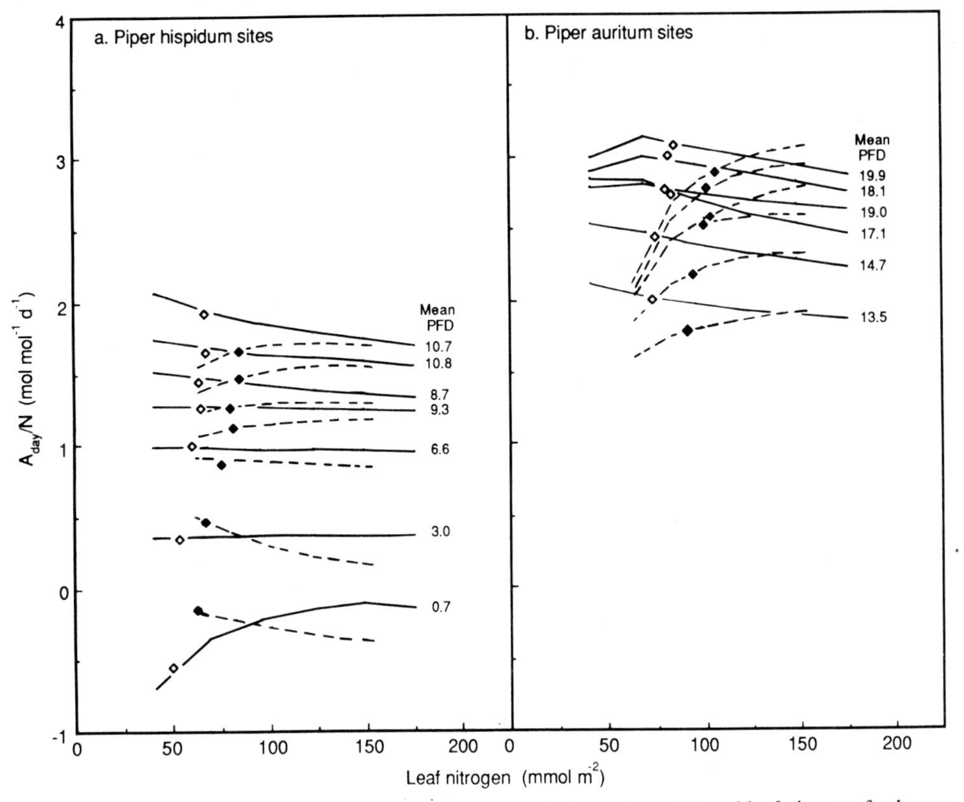

Fig. 6. Simulated relationship between daily nitrogen use efficiency (A_{day}/N) and leaf nitrogen for leaves with the photosynthetic characteristics of *P. hispidum* (♦, dashed lines) and *P. auritum* (◇, solid lines). Simulations are for the same sites and conditions as in Fig. 4. Daily PFD (annual mean) in each of the microsites is indicated to the right of the simulation for leaves with the photosynthetic characteristics of *P. auritum*. Daily PFD and the ranking among A_{day} was the same for the simulations using the photosynthetic characteristics of *P. hispidum*. Diamonds indicate the pattern expected for leaves with the N–daily PFD relationship derived from Figs 1 and 2. Solid diamonds indicate the pattern for *P. hispidum* and open diamonds indicate the pattern for *P. auritum*.

In the microsites of highest PFD, leaves with the photosynthetic characteristics of the persistent species, *P. hispidum*, tend to realise lower daily NUE at low leaf nitrogen and greater daily NUE at high leaf nitrogen than leaves of the high-light specialist. In the shadiest sites, the situation is reversed. And in microsites of intermediate PFD, leaves with the photosynthetic characteristics of the high-light specialist, *P. auritum*, tend to realise greater daily NUE at all nitrogen levels (Fig. 6). This pattern suggests that the restricted range of photosynthetic capacities and nitrogen levels in *P. auritum*, relative to *P. hispidum* (Fig. 1), may have functional significance.

The predicted response of leaf nitrogen to daily PFD for *P. hispidum* (from the data in Figs 1 and 2) is similar to that for *P. auritum*, but predicted nitrogen levels are somewhat greater in the persistent species, *P. hispidum*. For both species, the trends in leaf nitrogen as a function of daily PFD predicted from real data place some but not all leaves close to the optimum nitrogen levels for maximising daily NUE. However, nitrogen levels predicted from the empirical N–daily PFD relationship are clumped relative to the levels that maximise daily NUE (Fig. 6), a result similar to that Hirose and Werger (1987) obtained in studies on *Solidago altissima*. To maximise daily NUE, leaf nitrogen should be more sensitive to daily PFD, both in *P. auritum* and in *P. hispidum*.

Canopy-level NUE

The analysis of daily NUE identifies the nitrogen level that maximises A_{day} for each leaf in a canopy, but it does not address the possible effects on whole-canopy photosynthesis of adjusting the distribution of nitrogen among leaves. Mooney and Gulmon (1979) first suggested that returns on nitrogen investments should be maximised when those investments are tuned to light availability. Field (1983) formalised this notion, arguing that canopy-level daily NUE is maximised when A_{day} in every leaf is equally sensitive to a small change in N. This is equivalent to the condition that

$$A_{day}/N = \lambda, \qquad (1)$$

where λ is an unspecified Lagrange multiplier that changes with canopy structure and total nitrogen investment.

For the canopies of *P. hispidum* and *P. auritum* described above, A_{day} increases with daily PFD (Fig. 4) and, for all but the leaves in the shadiest microsites, increases with N (Fig. 7). The slope of the A_{day}–N relationship is greatest for leaves in high-PFD microsites and decreases with increasing N. Canopy-level A_{day} is maximised when the N investment in each leaf is at a point of constant slope on the A_{day}–N relationship. For the simulations in Fig. 7, the nearly linear responses of A_{day} to N insure that an optimum canopy for the maximisation of daily photosynthesis does not simultaneously include leaves in microsites of widely varying PFD, for either of the two species, within the constraints of the modelling approach used here.

Using the criterion in equation (1), the optimal N distributions omit leaves in microsites receiving low PFD. With increasing canopy N, optimal distributions begin calling for leaves in the microsites of highest daily PFD and only gradually add leaves in darker microsites deeper in the canopy. The criterion in equation (1) yields an N distribution specific not only to the total nitrogen in the leaf array but also to the total leaf area in the array. Because the A_{max}–N relationship for both species has a negative *y*-intercept, leaves require a substantial N investment to reach positive carbon balance in the light and an even greater investment to realise a positive carbon balance over 24 h. On the other hand, the empirical A_{max}–N relationship for the *Piper* species shows no sign of saturating at the highest N levels observed in nature. Thus, diminishing returns on increasing N investment in leaves already produced result solely from light limitation and not from constraints on photosynthesis. In some plants, for example wheat (Evans 1983) and soybean (Lugg and Sinclair 1981), the A_{max}–N relationship does begin to saturate at high N, and it is conceptually reasonable that it should in all plants. However, many species fail to increase N levels beyond the level of diminishing returns in A_{max} (Field and Mooney 1986), thus making it very difficult to predict the N level above which they should occur. The simulations here allow N to range up to 25% higher than the highest levels observed in the field in *P. hispidum* and 50% above the levels observed in *P. auritum*. If the A_{max}–N relationship does begin to saturate at these levels, the optimal distributions will be more constrained, and could approach the actual distributions.

The high nitrogen costs of adding leaves deep in the canopy mean that, for a given total canopy N, total photosynthesis is maximised when the leaf area is minimised, and leaves in the darkest microsites are eliminated. However, if two canopies add leaves at the top at the same rate but differ in leaf longevity, the canopy with the greater leaf longevity will be deeper, at least in terms of PFD interception. Since long-term NUE depends on leaf longevity and N recovery as well as on daily NUE, a species with increased leaf longevity has an intrinsic NUE advantage (Loveless 1961; Chabot and Hicks 1982). Thus a trade-off exists. A shallow canopy confers the possibility of a high daily NUE at the canopy level but limits long-term NUE as leaf longevity decreases. A deep canopy decreases canopy-level daily NUE but allows longer leaf longevities.

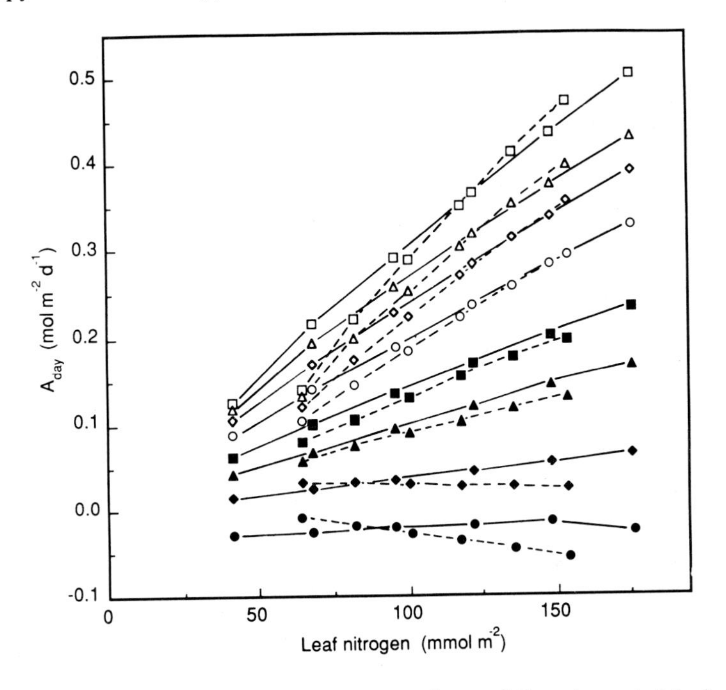

Fig. 7. Simulated relationship between annual mean daily carbon gain (A_{day}) and leaf nitrogen for leaves with the photosynthetic characteristics of *P. hispidum* (dashed lines) and *P. auritum* (solid lines). Daily PFD (annual mean) in the microsites was 19·9 (□), 17·1 (△), 14·7 (◇), 13·5 (○), 8·7 (■), 6·6 (▲), 3·0 (◆), and 0·7 (●) mol m^{-2} day^{-1}. Solid symbols indicate microsites in *P. hispidum* canopies and open symbols indicate microsites in *P. auritum* canopies. Simulations are for the same sites and conditions as in Fig. 4.

The two *Piper* species solve the trade-off in different ways. The open-site specialist, *P. auritum*, maintains an umbrella shaped crown with little leaf overlap. Leaves last less than 6 months. The persistent species, *P. hispidum*, has a much deeper canopy and much more internal shading. Leaf durations average more than 1 year (Williams *et al.* 1988).

The two species show similar acclimation of photosynthesis to PFD during growth, and both possess traits that insure high NUE. However, the pattern in *P. auritum*, which relies on high PFD to generate high NUE, is effective only in sunny sites. *P. hispidum*'s combination of responsiveness of leaf nitrogen to PFD and long leaf duration allows high NUE in sunny and shady sites.

Acknowledgments

Thanks to Joe Berry and members of the Carnegie photosynthesis modelling seminar for collaborating in the development of the photosynthesis model and to the staff of the Los Tuxtlas Biological Preserve for assistance in many phases of the field research. The *Piper* research has been supported by the National Science Foundation.

References

Anderson, M. C. (1964). Studies of the woodland light climate. 1. The photographic computation of light conditions. *J. Ecol.* **52**, 27–41.

Azcón-Bieto, J., Lambers, H., and Day, D. A. (1983). Effect of photosynthesis and carbohydrate status on respiratory rates and the involvement of the alternative pathway in leaf respiration. *Plant Physiol* **72**, 598–603.

Bell, C. J., and Rose, D. A. (1981). Light measurement and the terminology of flow. *Plant Cell Environ.* **4**, 89–96.

Björkman, O. (1981). Responses to different quantum flux densities. In 'Physiological Plant Ecology I. Responses to the Physical Environment'. (Eds O. L. Lange, P. S. Nobel, C. B. Osmond and H. Ziegler.) Encycl. Plant Physiol. New Ser., Vol. 12A, pp. 57–107. (Springer-Verlag: Berlin.)

Bloom, A. J., Chapin, F. S. III, and Mooney, H. A. (1985). Resource limitation in plants — an economic analogy. *Annu. Rev. Ecol. Syst.* **16**, 363–92.

Brooks, A., and Farquhar, G. D. (1985). Effect of temperature on the CO_2/O_2 specificity of ribulose-1,5-bisphosphate carboxylase/oxygenase and the rate of respiration in the light: estimates from gas exchange measurements on spinach. *Planta* **165**, 397–406.

Caemmerer, S. von, and Farquhar, G. D. (1981). Some relationships between the biochemistry of photosynthesis and the gas exchange of leaves. *Planta* **153**, 376–87.

Chabot, B. F., and Hicks, D. J. (1982). The ecology of leaf life spans. *Annu. Rev. Ecol. Syst.* **13**, 229–59.

Chapin, F. S. III, Bloom, A. J., Field, C. B., and Waring, R. H. (1987). Plant responses to multiple environmental factors. *Bioscience* **37**, 49–57.

Chazdon, R. L. (1986). Light variation and carbon gain in rain forest understory palms. *J. Ecol.* **74**, 995–1012.

Chazdon, R. L., and Fetcher, N. (1984). Photosynthetic light environments in a lowland tropical rain forest in Costa Rica. *J. Ecol.* **72**, 553–64.

Chazdon, R. L., and Field, C. B. (1987*a*). Photographic estimation of photosynthetic light conditions: evaluation of a computerised technique. *Oecologia* **73**, 525–32.

Chazdon, R. L., and Field, C. B. (1987*b*). Determinants of photosynthetic capacity in six rainforest *Piper* species. *Oecologia* **73**, 222–30.

Chazdon, R. L., and Pearcy, R. W. (1986). Photosynthetic responses to light variation in rainforest species. II. Carbon gain and photosynthetic efficiency during sunflecks. *Oecologia* **69**, 524–31.

Chiariello, N. (1984). Leaf energy balance in the wet lowland tropics. In 'Physiological Ecology of Plants of the Wet Tropics'. (Eds E. Medina, H. A. Mooney and C. Vazquez-Yanes.) pp. 85–98. (Dr W. Junk: The Hague.)

Chiariello, N. R., Field, C. B., and Mooney, H. A. (1987). Midday wilting in a tropical pioneer tree. *Funct. Ecol.* **1**, 3–11.

Cowan, I. R. (1986). Economics of carbon fixation in higher plants. In 'On the Economy of Plant Form and Function'. (Ed. T. J. Givnish.) pp. 133–70. (Cambridge University Press: Cambridge.)

Evans, J. R. (1983). Nitrogen and photosynthesis in the flag leaf of wheat (*Triticum aestivum* L.). *Plant Physiol.* **72**, 297–302.

Farquhar, G. D., Caemmerer, S. von, and Berry, J. A. (1980). A biochemical model of photosynthetic CO_2 assimilation in leaves of C_3 species. *Planta* **149**, 78–90.

Field, C. (1983). Allocating leaf nitrogen for the maximization of carbon gain: leaf age as a control on the allocation program. *Oecologia* **56**, 341–7.

Field, C., Berry, J. A., and Mooney, H. A. (1982). A portable system for measuring carbon dioxide and water vapour exchange of leaves. *Plant Cell Environ.* **5**, 179–86.

Field, C., and Mooney, H. A. (1986). The photosynthesis–nitrogen relationship in wild plants. In 'On the Economy of Plant Form and Function'. (Ed. T. J. Givnish.) pp. 25–55. (Cambridge University Press: Cambridge.)

Fleming, T. H. (1981). Fecundity, fruiting pattern, and seed dispersal in *Piper amalago* (Piperaceae), a bat-dispersed tropical shrub. *Oecologia* **51**, 42–6.

Fleming, T. H. (1985). Coexistence of five sympatric *Piper* (Piperaceae) species in a dry tropical forest. *Ecology* **66**, 688–700.

Fleming, T. H. (1986). Opportunism versus specialization: the evolution of feeding strategies in frugivorous bats. In 'Frugivores and Seed Dispersal'. (Eds A. Estrada and T. H. Fleming.) pp. 105–18. (Dr W. Junk / Nijhoff: Dordrecht.)

Givnish, T. J. (1986). Optimal stomatal conductance, allocation of energy between leaves and roots, and the marginal cost of transpiration. In 'On the Economy of Plant Form and Function'. (Ed. T. J. Givnish.) pp. 171–214. (Cambridge University Press: Cambridge.)

Gomez-Pompa, A. (1971). Posible papel de la vegetacion secundaria en la evolucion de la flora tropical. *Biotropica* **3**, 125–35.

Hirose, T., and Werger, M. J. A. (1987). Nitrogen use efficiency in photosynthesis of a single leaf in a canopy of a *Solidago altissima* population. *Physiol. Plant.* **70**, 215–22.

Johnson, I. R., and Thornley, J. H. M. (1984). A model of instantaneous and daily canopy photosynthesis. *J. Theor. Biol.* **107**, 531–45.

Loveless, A. R. (1961). A nutritional interpretation of sclerophylly based on differences in the chemical composition of sclerophyllous and mesophytic leaves. *Ann. Bot.* **25**, 168–84.

Lugg, D. G., and Sinclair, T. R. (1981). Seasonal changes in photosynthesis of field-grown soybean leaflets. 2. Relation to nitrogen content. *Photosynthetica* **15**, 138–41.

Mattson, W. J. Jr (1980). Herbivory in relation to plant nitrogen content. *Annu. Rev. Ecol. Syst.* **11**, 119–61.

Mooney, H. A., Field, C. B., Vazquez-Yanes, C., and Chu, C. C. (1983). Environmental controls on stomatal conductance in a shrub of the humid tropics. *Proc. Natl Acad. Sci. U.S.A.* **80**, 1295–7.

Mooney, H. A., and Gulmon, S. L. (1979). Environmental and evolutionary constraints on the photosynthetic characteristics of higher plants. In 'Topics in Plant Population Biology'. (Eds O. T. Solbrig, S. Jain, G. B. Johnson and P. H. Raven.) pp. 316–37. (Columbia University Press: New York.)

Mott, K., Jensen, R., and Berry, J. A. (1984). Photosynthesis and ribulose-1,5-bisphosphate concentrations in intact leaves of *Xanthium strumarium* L. *Plant Physiol.* **76**, 968–71.

Orozco-Segovia, A. D. L. (1986). Fisiología ecología del photoblastismo en semillas de cuatro especies del género *Piper* L. Ph.D. Thesis, Universidad Nacional Autonoma de Mexico.

Pearcy, R. W. (1988). Photosynthetic utilisation of lightflecks by understory plants. *Aust. J. Plant Physiol.* **15**, 223–38.

Penning de Vries, F. W. T. (1975). The cost of maintenance processes in plant cells. *Ann. Bot.* **39**, 77–92.

Penning de Vries, F. W. T., Brunsting, A. H. M., and van Laar, H. H. (1974). Products, requirements and efficiency of biosynthesis: a quantitative approach. *J. Theor. Biol.* **45**, 339–77.

Rundel, P. W. (1982). Nitrogen utilization efficiencies in mediterranean-climate shrubs of California and Chile. *Oecologia* **55**, 409–13.

Seemann, J. R., Sharkey, T. D., Wang, J. L., and Osmond, C. B. (1987). Environmental effects on photosynthesis, nitrogen-use efficiency, and metabolite pools in leaves of sun and shade plants. *Plant Physiol.* **84**, 796–802.

Terashima, I., and Saeki, Y. (1985). A new model for leaf photosynthesis incorporating the gradients of light environment and of photosynthetic properties of chloroplasts within a leaf. *Ann. Bot.* **56**, 489–99.

Vasquez-Yanes, C. (1976). Estudios sobre ecofisiologia de la germinacion en una zona calido-humeda de Mexico. In 'Regeneracion de Selvas'. (Eds A. Gomez-Pompa, S. del Amo S., C. Vazquez-Yanes and A. Butando.) pp. 279–381. (Editorial Continental: Mexico.)

Vazquez-Yanes, C., and Orozco-Segovia, A. (1982). Germination of the seeds of a tropical rain forest shrub, *Piper hispidum* Sw. (Piperaceae) under different light qualities. *Phyton* **42**, 143–9.

Vazquez-Yanes, C., and Smith, H. (1982). Phytochrome control of seed germination in the tropical rain forest pioneer trees *Cecropia obtusifolia* and *Piper auritum* and its ecological significance. *New Phytol.* **92**, 477–85.

Vitousek, P. M., and Denslow, J. S. (1986). Nutrient dynamics in treefall gaps in a lowland tropical rainforest. *J. Ecol.* **74**, 1167–78.

Vitousek, P. M., and Sanford, R. L. Jr (1986). Nutrient cycling in moist tropical forest. *Annu. Rev. Ecol. Syst.* **17**, 137–67.

Walters, M. B., and Field, C. B. (1987). Photosynthetic light acclimation in two rainforest *Piper* species with different ecological amplitudes. *Oecologia* **72**, 449–56.
Williams, K., Field, C. B., and Mooney, H. A. (1988). Relationships among leaf construction cost, leaf longevity, and light environment in rainforest plants of the genus *Piper*. *Am. Nat.*, in press.
Wong, S. C., Cowan, I. R., and Farquhar, G. D. (1979). Stomatal conductance correlates with photosynthetic capacity. *Nature* **282**, 424–6.
Woodrow, I. E., and Berry, J. A. (1988). Enzymatic regulation of photosynthetic CO_2 fixation in C_3 plants. *Annu. Rev. Plant Physiol.* **39**, 533–94.